SEMICONDUCTOR DEVICES

A SIMULATION APPROACH

Kevin M. Kramer • W. Nicholas G. Hitchon

To join a Prentice Hall Internet mailing list,
point to http://www.prenhall.com/mail_lists

Prentice Hall PTR
Upper Saddle River, New Jersey 07458
http://www.phptr.com

ISBN 0-13-614330-X

90000

9 780136 143307

Editorial/Production Supervision: Joe Czerwinski
Acquisitions Editor: Russ Hall
Manufacuring Manager: Alexis R. Heydt
Cover Design Director: Jerry Votta
Cover Design: Wee Design Group
Marketing Manager: Betsy Carey

© 1997 by Prentice Hall PTR
Prentice-Hall, Inc.
A Division of Simon and Schuster
Upper Saddle River, NJ 07458

Prentice Hall books are widely used by corporations and government agencies for training, marketing, and resale. The publisher offers discounts on this book when ordered in bulk quantities. For more information, contact:

Corporate Sales Department
Phone: 800-382-3419
Fax: 201-236-7141;
e-mail: corpsales@prenhall.com

Or write

Prentice Hall PTR
Corp. Sales Dept.
One Lake Street
Upper Saddle River, NJ 07458

Printed in the United States of America

10 9 8 7 6 5 4 3 2 1

ISBN: 0-13-614330-X

Prentice-Hall International (UK) Limited, London
Prentice-Hall of Australia Pty. Limited, Sydney
Prentice-Hall of Canada Inc., Toronto
Prentice-Hall Hispanoamericana, S.A., Mexico
Prentice-Hall of India Pte. Ltd., New Delhi
Prentice-Hall of Japan, Inc., Tokyo
Simon & Schuster Asia Pte. Ltd., Singapore
Editora Prentice-Hall do Brasil, Ltda., Rio de Janeiro

Contents

III   Semiconductor Devices

7   PN Junction Diodes

Acknowledgements

This book is dedicated to our families;
and ad majorem Dei gloriam.

We should also like to thank the people who helped with the creation of the software, including Eric Keiter, Andrew Christlieb and Josh Feng.

Kevin would like to especially thank his wife, Jenifer, for her patience and support. He would also like to give thanks to his Lord, Savior, and Life – Jesus Christ.

Nick thanks Adam and Jackie for all their encouragement.

Chapter 1

Introduction

This book describes the physics and mathematics needed for semiconductor device simulation and provides tools which make the simulation straightforward. This chapter describes the book, as well as the advantages and the problems of simulation of physical systems. There are several ways the book could be used, depending on the needs of the user and how much the user knows about devices and device simulation.

1. A person who is not familiar with the physics of semiconductors or the simulation of physical systems will find Chapters 2 and 3 review electromagnetism and transport processes and the simulation of both of them. Sections 2.1, 2.8 and 3.2 in particular and other related sections, discuss a number of important issues in simulation - and these sections provide simple, clear examples of how to do simulations. They could then proceed to Part II, the middle block of chapters which describe simulation and our software, in more detail. Finally they could go on to the chapters on detailed device modeling.

2. Someone familiar with basic ideas of numerical modeling would find the early chapters useful as a quick introduction to how the software works and as a review of how physical principles are built into simulations. (Sections 2.1, 2.8 and 3.2 among others would again be helpful.) They could read parts of this material and then move on to Part II. The details of setting up device models which are given in Part III should be read next.

3. A person who is familiar with device modeling and simulation might prefer to start looking at Part III, to get a flavor of how the software works. (Even an expert might prefer to start with the simpler examples, in the sections mentioned above, however.) They could then go to Part II for more details as needed.

   Section 1.5 gives an example simulation. Readers who want to learn how to set up simulations first and who will worry afterwards about the role of simulations and the things that can go wrong with them, could start with Section 1.5 (with its illustrative example simulation), then go to Section 1.6 (which outlines the book) and then move on to Chapter 2.

1

1.1 Why Simulation?

'Now the elements of the art of war are first, measurement of space; second, estimation of quantities; third, calculations; fourth, comparisons; and fifth, chances of victory. ' [1]

This book is concerned with taking physical data (including data on dimensions and quantities) and performing analytic estimates and numerical calculations based on them. By comparing them we should be able to understand what the data mean and design a workable product. This procedure is valuable as an educational tool, as well as during the design process. A deeper understanding of how a device works will certainly help during design, as will access to numerical tools which predict how the design will function.

'Who excels at resolving difficulties does so before they arise.'

'In planning, never a useless move; in strategy, no step taken in vain. ' [1]

Analytic descriptions of the physical universe have been the rule in education. This has drastically limited the type of problems that can be addressed in the classroom.

'From materials so slight and so scanty it is impossible to extract a solution of the problem. It remains to try whether the survey of a broader field may not yield us the clue we seek.' [2]

Numerical methods offer the possibility of tackling much more realistic physical problems, but their success has been limited by a number of factors which have prevented them being as widely accessible as they might be.

What Is Wrong with Simulation

One problem is that computer programs are often extremely complex to write, to use (including to modify them) and simply to understand what they are doing. We believe that we have a solution to most of these problems in the software package we provide here.

Another major drawback with computer models is the way they are used. It is not difficult to get wrong results from simulations. Using the wrong input parameters, a mesh which is too coarse, or using a model outside its range of validity can each lead to a failure to converge, or to incorrect answers. If the convergence is not complete, the code may stop before it reaches an accurate result.

The problems mentioned so far are the most obvious things which can go wrong. There are other more subtle difficulties with simulations. The models of the physical system which we choose to use can have unexpected weaknesses. They cannot be trusted without extensive validation. The way one goes about testing a computational model such as those used here is usually in comparison with experiments and simpler models. In the later chapters there will be some opportunity to illustrate this by example, but for now we should make it clear that it is essential to subject the results of simulations to very rigorous scrutiny. In this way unexpected problems will often be turned up, of various kinds. These may be with the implementation of the simulation, or they may be due to a shortcoming in the physical picture which is being used to describe the real situation. This testing procedure is discussed in some detail in Chapter 5.

> 'Not that the incredulous person doesn't believe in anything. It's just that he doesn't believe in everything. Or he believes in one thing at a time. He believes a second thing only if it somehow follows from the first thing. He is nearsighted and methodical, avoiding wide horizons. If two things don't fit, but you believe both of them, thinking that somewhere, hidden, there must be a third thing that connects them, that's credulity. ' [3]

The Pointless Simulation

There is another common failing of many simulation projects, which is in a way rather subtle. Quite often published results of simulations include detailed and complicated plots of the variables which were calculated, without its being clear what has been learned from the simulation. It is essential that, in addition to the simulation being correct both physically and mathematically, and appropriate to the problem it is intended to describe, it provides useful information about it.

One way in which the simulation might provide useful information is if we can learn from it what the really crucial physical effects are. This can be helpful so we can develop simpler models which work well and which give us insight into what is really happening. This might be achieved by looking at the simulation results and comparing them to analytic models which are set up to include those processes which the simulation showed to be most important.

Simple models are not always available. Nevertheless, we might be able to use the simulation to show that if a particular process is included, the simulation agrees with reality, and otherwise it does not; this conclusion would need to be tested very carefully, however. A tabulation of data with no physical explanation or additional insight may be useful for specific purposes, if the data calculated has been chosen appropriately with a goal in mind. What is generally not useful is a calculation which is done without a clear objective for what is to be learned from it. An experiment which is done without much thought being given to which variables to measure, or which points to take data at, is much less useful than a carefully designed experiment [4]. The same is true of simulations.

This point might be made best by examples. Suppose a claim has been made that shockwaves travel around a mechanical structure like waves on a network of rubber bands. The basis for the claim in this hypothetical situation is the comparison of measurements of shockwaves on the actual structure to simulations of a network of rubber bands. It is an interesting exercise to try to decide how useful the rubber-band model has been shown to be.

The answer will depend on other information which has not been provided yet, including whether it is truly plausible that the structure behaves simply like the network of rubber bands. Before the experiment was done it was presumably unclear how the structure would behave. This may have been because its mechanical properties were simply unknown. The rubber-band model's success might then be plausibly interpreted as meaning that the structure has properties similar to rubber.

However, if the structure's mechanical properties were known, at least to some extent, and they are not similar to those of rubber bands, then the situation is quite different. In this case the agreement of experiment and the model is quite likely to be simply a coincidence. Using a more detailed model to give an intermediate point of comparison might help to explain what is really happening.

As another example, suppose we calculate the state of a piece of semiconductor which is being etched, using a model which is known to be at best only approximately correct and is often quite wrong. Suppose further that our results agree with experiment in some range of parameters but there is no theoretical basis for expecting agreement. It is difficult to see what has been learned from this, because the agreement is presumably a coincidence and so we should not expect agreement for other ranges of parameters. Yet much of the literature on modeling does this (not necessarily in the context of semiconductor etching). Or suppose that we show that a simulation of an idealized reactor, with a gas whose reaction rates were invented so that we can do a simulation, predicts that the reactor deposits film at a certain rate. It may be that we had a plan when we did this and that something helpful was learned, but often simulations seem to be done for the sake of doing something rather than because they are really likely to serve a purpose.

Optimal Simulation

Simulation offers a powerful tool, if we use it creatively and appropriately. In this text we describe many standard and nonstandard computational techniques. In attempting to understand the real world, there is no reason to think that these are the only ways to go about things.

> 'If wise, a commander is able to recognize changing circumstances and act expediently.' [1]

An important point to notice about simulations is that the optimum simulation is one which captures the right physical process, and does so using a method which reflects the reality. Our physical picture of reality should guide us and show how to set up a calculation which has built into it the same underlying behavior. Use of standard numerical techniques can often blind us to the reality we are trying to build in to our simulation. We should be open to inventing new simulations which mimic the real situation as closely as possible. A powerful tool such as the one we use here is of real benefit in helping to set up such new types of simulation.

1.2   Purpose of the Book

This book is intended, among other things,

1. to provide a set of software which can be used easily to solve a very wide range of physical problems (including those arising from semiconductor devices),

2. to show how the physical processes in these systems can be understood,

3. and to explain subject matter ranging from introductory electromagnetism, transport theory and semiconductor device behavior through to advanced topics in device design and in kinetic theory simulation, in a new and powerful way which makes these ideas more accessible to readers at all levels.

Simulation

Computer models can provide an enormous amount of insight into how physical systems work, if the models are used properly. The software provided here allows us to set up solution schemes for partial differential equations. The power of the approach is due to the fact that it permits us to set up a method of solution simply and in a form that we can easily understand, even in situations which are physically very complex and difficult. Setting up the solution scheme gives us considerable insight into the mathematics which describes the problem. For instance, much of the effort usually goes into getting the boundary conditions right, and much of the space in the input files we create is devoted to boundary conditions. This comes as a surprise in the light of the training we are given in analytic methods where the boundary conditions are not always stressed. The emphasis on the boundary conditions does reflect the physical reality that the boundary conditions control the behavior of a system, and it shows us graphically how the mathematics captures the physical situation. Once the equations have been solved, we begin the process of looking at the graphical output in detail and using simple intuitive models to try to understand it. This reality check is essential if we are to be confident the results are right and it gives us a very effective training in how the physical system behaves.

In summary, setting up a simulation of a physical system and studying the results of that simulation is a powerful way of gaining insight into many aspects of the system. Setting it up involves being able to specify mathematically the equations and boundary conditions which describe the physical behavior. Studying the results shows how those inputs control the response of the system being studied. The ability of this software to handle complex situations means that a straightforward and highly flexible method is available for use in even very challenging model applications. The input file used to specify a problem is extremely compact and easy to read and understand. The compactness of the input file also allows entire input files to be given in the text, when this is the clearest way to present them. Most of the files have large sections in common with each other, and it is valuable to see the repetition of (for instance) the physical models employed. On occasion, however, if a file has recently been shown and then modified slightly for a different calculation, we shall only give the parts of the input file which were changed.

Layout of the Book

This book is in three main parts.

Part I

In the first part, basic concepts from physics and computational modeling are described at a relatively simple level which could be read by somebody who is unfamiliar with the material covered. The material is organized according to the physics topics. There is no separate chapter on numerical methods in this part of the book. The numerical methods are introduced in a natural way as needed throughout the discussion of the physics.

Part I consists of the first two chapters of this book. As mentioned above, these chapters are used to introduce several topics at the same time. Basic ideas from electromagnetism and from transport theory are covered, but these topics also serve as a vehicle to explain first the mathematical basis for the theory, and second the numerical implementation of the equations. In fact many of the mathematical ideas are developed with the main emphasis being on their discrete implementation; that is to say, their implementation in a form suitable for use in numerical solution.

Electromagnetism is important for the modeling task ahead. This material is intended to explain concepts such as div and curl and conservation equations. At the same time, an aspect of the concepts which is given particular emphasis is their numerical implementation. The operators grad, div and curl; conservation equations; and various important classes of boundary condition will be described in physical terms and in terms of their numerical implementation on a discrete mesh. The list of topics covered includes all the essentials that are required for many computer engineering and electrical engineering undergraduates. That is, an introduction to electrostatics and transmission lines is given.

The next chapter provides an overview of transport processes. The role of the divergence in the conservation equation and the numerical implementation of both of these is again stressed. In later chapters a more sophisticated implementation on an irregular two-dimensional mesh is introduced. After the basic concepts of the divergence and conservation equations are presented, we show how different types of transport processes are described mathematically and how the equations describing transport processes can be solved using a computer. In this context 'transport' means the types of processes such as diffusion and drift which cause carriers to move around inside semiconductor materials. Unfortunately the equations describing transport in semiconductors are genuinely very difficult to solve accurately. For this reason some very powerful tools are necessary to make their solution accessible.

Part II

Software is provided with this text which we believe makes the solution of the semiconductor transport equations not only possible but straightforward and readily understandable. Part II describes the more advanced numerical and physical models and the software tools we need to use to employ those models. The level of treatment in this part of the book is more advanced.

Part III

After transport equations and their solution have been reviewed, we turn in Part III to semiconductor devices. The strategy here is to review the analytic theory of each device and to compare the theory to simulations. This approach is beneficial since it validates and motivates the analytic theory and it provides insight into the modeling results.

Interpretation of Simulation Results

In the introduction we have stressed the importance of using all the tools at our disposal to check whether our simulations make sense and (if they do seem to be correct) to try to understand the results they are giving. A list of some of the places in the text where we make these kinds of comparisons of simple analytic models with simulations is given next. In each case, the simulation results are compared to analytic estimates, to check the simulations and to gain understanding as to what physical effects are most important in determining the outcome.

1. In Section 7.1.6 the file pn01.sg uses the nonlinear Poisson's equation to describe a pn junction.

2. In Section 7.2.1 pniv2.sg uses a 'simplified' version of the 'Scharfetter-Gummel' method to find (i, v) characteristics of a pn junction.

3. In Section 7.2.2 an accurate, general model of the 1D pn junction (pn03a.sg) is used to find (i, v) characteristics.

4. In Section 7.4.3 pn04.sg sets up a small signal model of a pn junction.

5. In Section 7.6.3 pniv3.sg examines the pn junction in the presence of photo-generated carriers.

6. In Section 10.1.1 moscap1.sg uses the nonlinear Poisson's equation to describe the electrostatics of a MOS capacitor.

7. In Section 10.6 mos.sg provides a full 2D model of a MOSFET which was used in this section to find the MOSFET currents and their dependence on the applied voltages.

We then turn to simulation of more detailed examples which go beyond what can be done analytically. The level of treatment is again somewhat beyond introductory, in that some familiarity with the purpose and basic function of the various devices is assumed.

Part IV

The last part of the book is devoted to more advanced topics in device simulation.

In the remainder of this introduction, we describe the motivation for and the philosophy behind the software used in this book (Section 1.3) and its relevance to semiconductors (Section 1.4) and we give a simple example of its use (Section 1.5). Section 1.6 gives a more detailed explanation of the layout of the book.

1.3   Physical Processes in Semiconductors and Their Numerical Description

In the first part of this book we describe the basic ideas which we need to draw from areas such as electromagnetism and particle transport to be able to understand the functioning of semiconductors. We explain Maxwell's equations and the equations which describe particle transport. Numerical solution of these and of general classes of partial differential equations is discussed. Simple electrostatic problems are solved, with emphasis on the numerical form of the equations and of the boundary conditions. Next, the methods are applied to the relatively simple problem of dopant diffusion in a semiconductor. This is followed by a more detailed discussion of numerical solution of more challenging problems and in particular of carrier transport in semiconductors.

This leads naturally to the rationale for the development and implementation of a highly flexible simulation tool, the *simulation generation framework (SGFramework)*. SGFramework was designed for the numerical solution of large-scale computational problems in applied physics. It has been used in the simulation of complex problems in physics and engineering. The features of this tool were designed in order to respond in a general way to the problems arising in such simulations. Performance of semiconductor devices is the principal area which motivated its development and which has been investigated using SGFramework. The SGFramework does not change the methodology of solving complex numerical problems in principle. In practice, SGFramework has had a significant impact.

We will now outline the nature of this software and why it is especially useful for semiconductor device simulation and as a tool for learning about semiconductor behavior. This discussion beings with an outline of why there is a need for such a software tool. Some features of the internal workings of the software will be indicated. In the next section we discuss the semiconductor applications which will be treated in the text.

The development of the software used here was prompted by the recognition of the need for a general purpose simulation tool which seamlessly integrates all the relevant components required to construct a complex simulation based on the solution of a system of discretized PDEs.

Once the tool had been implemented, it was applied to the modeling and simulation of complex physical systems. This process demonstrated a large number of features which are necessary in such a simulation. These features have been recognized and incorporated into SGFramework in the most general fashion possible. Finally it was possible to move on to the detailed modeling of semiconductor devices and to use the results to optimize devices to operate more efficiently under certain operating conditions.

The Solution of PDEs

Many interesting problems in science and engineering can be mathematically modeled by a system of partial differential equations (PDEs) with appropriate boundary and/or initial conditions. Unfortunately, analytical solutions are not available to most systems of PDEs. Therefore mathematicians, scientists and/or engineers must solve these systems of PDEs numerically using powerful computational techniques to obtain accurate approximations to the exact solution. Although many researchers in these fields understand these computational techniques in theory, far fewer of them are able to efficiently implement these techniques in practice. Despite the seeming simplicity of the techniques, their computer implementation is a time-consuming and error-prone task. Consequently many workers who would benefit from the numerical solution of systems of PDEs by gaining physical insight into their problem do not do so because of the intensive coding and implementation requirements.

The SGFramework is a general-purpose simulation tool for the numerical solution of PDEs and can be used to construct simulations for a variety of scientific and engineering problems. It allows users to concentrate on the physical and computational issues connected with their problem rather than on laborious computer-implementation considerations. The SGFramework performs many of the time-consuming and error-prone tasks involved in setting up simulations based on the solution of PDEs. It allows the user to construct an entire simulation in one step (two steps, if an irregular mesh is needed). Despite SGFramework's generality, it is capable of solving complex systems of coupled, nonlinear PDEs such as the semiconductor equations. The elegance and power of SGFramework will become apparent in the following chapters.

SGFramework is novel in that it combines all the numerical tools required for finite-difference (or similar) solution of complex PDEs in a single package. The SGFramework integrates several 'standard' computer science and/or numerical algorithms and techniques into a single software package. For instance, the SGFramework translator brings together algorithms and techniques from compiler design, symbolic algebra and differentiation and sparse matrix techniques. Although many of the components of the SGFramework are 'standard', the combination of these components is novel. The integration of these 'standard' techniques raises new issues that do not occur when considering the elements of the package individually. For instance, several symbolic algebra and calculus programs exist. However, at present none of them are able to efficiently support large arrays of variables and complex user-defined functions, which are necessary for the finite-difference solution of complex systems of PDEs.

In order to create this framework, several issues had to be resolved. First, as mentioned above, algorithms and techniques had to be integrated from several computer science disciplines including compiler design, symbolic algebra and differentiation, numerical algorithms, mesh generation and computer graphics. Second, in order to solve an arbitrary system of PDEs and construct a simulation, the SGFramework specification language was developed as well as a program to translate simulations written in this language to C++ code. Finally, in addition to the standard compiler features such as a scanner, parser, symbol table, etc., this program combines symbolic algebra, symbolic differentiation and sparse matrix techniques.

1.4   Simulation of Semiconductor Devices

The major topic of this book is the application of SGFramework to the solution of the semiconductor equations. This application is important in part because it impelled development of several new features of very general utility in SGFramework; an example of this is the availability of a naming/labeling procedure for regions which is used in the equation specification. The labeling procedure is novel in that it is introduced in a context other than finite elements. A second example would be the mesh generation and refinement procedures implemented in SGFramework. Another example is the ability to easily specify complex parameter models as user-defined functions and the efficient evaluation of these models using the function-precompute feature. Semiconductors themselves are important because they are virtually ubiquitous.

Efficient exploitation of new technologies such as silicon carbide devices requires cost-effective design techniques such as computer simulation. The SGFramework is well suited for such applications, because the user can easily specify arbitrary models for device parameters such as carrier mobilities, impact ionization rates, band gap narrowing, etc. This flexibility provides advantages over the existing device simulators that hard-code such parameters. As researchers derive new models for new semiconductor materials, one can immediately and easily incorporate the models into their device description, using the SGFramework to more accurately simulate the semiconducting device.

By using SGFramework to simulate semiconductors, it has been shown that the framework is very flexible and general yet powerful enough to handle extremely computationally complex physical systems. These simulations demonstrate that SGFramework can be used as a design tool as well as to understand device performance.

1.5 An Example Simulation: The Game of Life

In order to give an example of the way the software is used, which requires no knowledge of physics or of partial differential equations, we can use the computer to run a simulation based on the Game of Life [5]. The rules for setting up such an input file are stated in the user's manual, which is Chapter 15. To learn how a file is set up, however it is easiest to start with examples such as this. The meaning of the elements of the file should not be difficult to understand.

The idea of the game is that colonies of living creatures (probably bacteria) exist at points on a mesh. The colony (or the single bacterium) at each point is alive (in which case that point is assigned a value of 1) or dead (so the point would be given a value of zero). Time is allowed to pass in discrete intervals. After each step forward in time, each mesh point is checked to see if there is life there. Whether there is life at a point on the mesh at a given time depends on the previous state of that point and the previous states of its neighbors. If too few neighbors are alive, it dies. If too many are alive, it also dies. The rule for the cell's being alive at some time is explained in the comments within the file.

Input File life.sg

```
// FILE:  sgdir\ch01\life\life.sg
//
// This example is intended to show how discrete equations, similar to
// finite-difference equations, can be stepped forward in time.
//
// The equation used here is for the 'Game of Life'.
// If a cell has A = 1 it is 'alive' . If A = 0 it is not.
// If the cell is dead and has 3 living neighbors, a colony is 'born' in it.
// If it is alive, it stays alive if it has 2 or 3 living neighbors.
//
// The initial state used here starts a colony called a 'stop light',
// and a moving colony called a 'glider', which hits the edge and gets stuck.

const NX = 19  ;               // number of x-mesh points
const LX = NX-1;               // index of the last x-mesh point
const NY = 19  ;               // number of y-mesh points
const LY = NY-1;               // index of the last y-mesh point
const DT   = 1;                // time step
const TMAX = 40;               // maximum time
const IS   = 6 ;               // index of stop light
const JS   = 8 ;               // index of stop light
const IG   = 8 ;               // initial index of glider
const JG   = 15;               // initial index of glider

var t,A[NX,NY],count[NX,NY];

begin main
```

```
assign t=0.0 ;
assign A[i=IS,j=JS+1]   = 1 ;        // launch stop light
assign A[i=IS-1..IS+1,j=JS] = 1 ;
assign A[i=IG,j=JG+2]   = 1 ;        // launch glider
assign A[i=IG+1,j=JG+1] = 1 ;
assign A[i=IG-1..IG+1,j=JG] = 1 ;

while (t < TMAX) begin

  // count IS the number of live neighbors.
  assign count[i=1..LX-1,j=1..LY-1] = A[i-1,j-1] + A[i-1,j] + A[i-1,j+1]
                                    + A[i,  j-1]            + A[i,  j+1]
                                    + A[i+1,j-1] + A[i+1,j] + A[i+1,j+1];

  // If count = 3 or count + A = 3 , the cell will be alive at the new time.
  assign A[i=1..LX-1,j=1..LY-1] = (count[i,j]==3) or (count[i,j]+A[i,j]==3);

  write;
  assign t = t+DT ;
 end
end
```

A DOS script to run this is given in a table. Windows95 and NT users must open an MS-DOS prompt to run the SGFramework executables.

DOS Script to Run life.sg from Command Line

(First, open MS-DOS prompt.)

```
sgxlat life.sg
sgbuild sim life
life -vs2
```

This file can be run and the different outputs for different times extracted on a UNIX machine using the commands listed in the next table. In UNIX, a shell must be opened before giving these commands. In DOS the outputs from each timestep can be extracted one at a time using the extract command, as in the UNIX script but with a specific integer following the word life.res:

```
extract life.res 10
```

would extract the tenth set of results.

The first two lines of the UNIX script are to tell the computer that this is a k-shell and put it in verbose mode. The next three lines are common to both scripts; they

1. call SGFramework

UNIX Script to run life.sg

```
#!/bin/ksh
set -v

sgxlat life.sg
sgbuild sim life
life -vs2
typeset -i i
i=0
while (( $i < 41 ));
do
extract life.res $i
((i=$i+1));
done
```

    2. build the simulation and

    3. execute the simulation.

These three lines could all be typed in by hand just as well. The reason for the UNIX shell script is only to extract the output in this case. To do this we need to loop over 40 different sets of data, one for each time step. This is done next in the UNIX script. The command

```
extract life.res $i
```

takes the data for the ith time step and writes it to a file.

    In general SGFramework can plot the results it obtains if it has a 'skeleton file', by means of commands such as 'triplot'. The skeleton file contains the description of the mesh in a suitable form. This is described in later chapters and in the user's manual, Chapter 15. However in this case we want to make a movie of the results.

    To plot the results using MATLAB, a file called life.m was created, which is given in a table.

    If you have matlab, you can give the matlab command followed by the word life (with no .m on the end) and it will show a movie of the evolution of this life form.

    Two features are present, a stop light and a glider. A sample of the output from four successive frames, plotted without use of a graphics package, is shown in Figure 1.1. After an initial transient, the stop light alternates between two states, as seen here. The glider moves across the frame until it hits the edge.

```
        life.021                      life.022
. . . . . . . . . . . . . . . . . . .     . . . . . . . . . . . . . . . . . . .
.                         .     .                                 .
.                         .     .                                 .
.         X               .     .                                 .
.         X               .     .           XXX                   .
.         X               .     .                                 .
.                         .     .         X       X               .
.     XXX     XXX         .     .         X       X               .
.                         .     .         X       X               .
.         X               .     .                                 .
.         X               .     .           XXX                   .
.         X               .     .                                 .
.                         .     .                                 .
.             X           .     .                                 .
.             X           .     .               X X               .
.             XXX         .     .               XX                .
.                         .     .                X                .
.                         .     .                                 .
.                         .     .                                 .
. . . . . . . . . . . . . . . . . . .     . . . . . . . . . . . . . . . . . . .

        life.023                      life.024
. . . . . . . . . . . . . . . . . . .     . . . . . . . . . . . . . . . . . . .
.                         .     .                                 .
.                         .     .                                 .
.         X               .     .                                 .
.         X               .     .           XXX                   .
.         X               .     .                                 .
.                         .     .         X       X               .
.     XXX     XXX         .     .         X       X               .
.                         .     .         X       X               .
.         X               .     .                                 .
.         X               .     .           XXX                   .
.         X               .     .                                 .
.                         .     .                                 .
.                         .     .                                 .
.             X           .     .                   X             .
.             X X         .     .                   XX            .
.             XX          .     .                   XX            .
.                         .     .                                 .
.                         .     .                                 .
. . . . . . . . . . . . . . . . . . .     . . . . . . . . . . . . . . . . . . .
```

Figure 1.1. Several successive frames of an example of the 'Game of Life'.

File life.m

```
for i=0:39
  if(i<10)
  s = sprintf('load A.00%d;',i);
  else
  s = sprintf('load A.0%d;',i);
  end
  eval(s);
  contour(A);
  pause(1);
end
```

1.6 Organization of The Text

This book is organized as follows. Chapters 2 and 3 form Part I of the book, which introduces in a straightforward fashion the basic ideas which are needed later. Chapter 2 describes basic electromagnetism, the mathematical concepts involved and their use in computer simulations. Chapter 3 is a similar review of the ideas of transport theory and uses them to show how the equations of particle transport are derived and again how simple partial differential equations may be solved numerically. Diffusion of dopants in semiconductors is an important example which illustrates many of the points raised, so it is discussed in some detail at this point.

Part II goes over similar material at a more advanced level. Chapter 4 reviews the semiconductor equations and the models of the physical parameters used throughout this book. Derivations of basic material are given but detailed explanations of the models have been omitted since such presentations are readily found in the literature. Instead, this chapter is an overview of the relevant equations and parameters used in the modeling and simulation of semiconductor devices. For those seeking more information, ample references will be given. The material covered in this chapter will be used in examples in the chapters that follow.

Chapter 5 discusses in detail the numerical solution of a system of PDEs. This chapter will discuss several techniques for solving PDEs such as finite-difference methods and finite-element methods. Special emphasis will be given to finite-difference methods, since the semiconductor equations are most easily discretized via finite-difference methods. As an example, the semiconductor equations will be finite-differenced via the box integration method. Discretizing a system of PDEs results in a massive system of algebraic equations. Techniques for solving these systems will be discussed. In addition, a review of the software packages capable of solving PDEs is presented.

Chapter 6 begins the presentation of the truly novel contribution of this book.

It describes the SGFramework. The SGFramework translator, mesh generation program, mesh refinement program, numerical algorithm module and supporting programs will be discussed. SGFramework's symbolic-mathematics capabilities and the solution of PDEs on irregular meshes will be emphasized. This chapter also describes the novel features of the SGFramework such as the labeling scheme which allows very general classes of PDEs and boundary conditions to be specified on irregular meshes, the precomputed user-defined functions which improve the efficiency of the generated code and the ability to efficiently implement complex parameter models and discretization schemes via user-defined functions. Throughout this chapter, a mesh and equation specification file of a metal-oxide-semiconducting field-effect transistor will serve as an example.

The next chapters are Part III of the book, in which we turn to the discussion of different classes of semiconductor device. In each case the analytic theory of the device is given. The steady-state, small-signal and transient behaviors are all derived. Simulations of each device are set up, which are suitable for comparison to these analytic models.

In Chapter 7 we start with the pn junction diode. Since many device concepts are best explained in the context of the pn junction, we devote a considerable amount of space to graphing numerical results for pn junctions under different conditions, and explaining them in terms of basic concepts. We begin with 1D junction electrostatics, then go on to steady-state (i, v) characteristics and a 2D device model. The time dependent case is considered next (small signal analysis and transients) followed by a discussion of photodiodes.

This is followed in Chapter 8 by the bipolar junction transistor. We discuss static characteristics, the Ebers-Moll model, and the time-dependent behavior of the BJT. We then give examples of modeling of the BJT in a variety of circuits, representing it using lumped parameters. Finally we model the BJT microscopically.

Chapter 9 is a very brief presentation of the junction field effect transistor.

Chapter 10 discusses the metal-oxide-semiconductor transistor. We begin with 1D electrostatics of the MOS structure, including the (C, V) characteristics. We then turn to the full MOSFET and discuss its (i, v) characteristics and small signal behavior. We discuss sub-micron devices and VLSI. We then turn to numerical simulation of the MOSFET, which is a major part of this book, describing the approach and the results in some detail. Finally we turn to modeling the MOSFET as a lumped circuit element in important examples of circuits.

Chapter 11 is concerned with the simulation of power semiconductor devices. We take the opportunity to discuss in considerable detail how we go about device design and optimization. We first present an overview of the steady-state characteristics of power diodes, which are usually pin diodes. This chapter then discusses the computer-aided design of power semiconductor devices by means of a detailed example. We present and compare methods of simulating the breakdown behavior of pin diodes. These methods illustrate the flexibility of the SGFramework. The first method developed turned out to require very extensive computer resources. SGFramework's flexibility allowed other methods to be implemented easily. This chapter continues with a discussion of the various methods of 'terminating' semi-conducting junctions. Junction termination techniques are essential to increase the breakdown voltage of semiconducting devices, since the breakdown voltage of devices without junction terminations is significantly smaller than that of devices with junction terminations. The second part of the chapter focuses on the optimization of a semiconductor device, using statistical design techniques in conjunction with the simulations. We consider the example of pin diodes with a particular type of junction termination, field-limiting rings. The optimization of junction terminations is an important problem in the design of power semiconducting devices. The statistical design methods applied to an optimization are very widely applicable. The specific techniques presented in this chapter which were developed for field-limiting rings may be applied to other types of junction terminations. After an overview of field limiting rings, the optimization objectives are stated and two optimization techniques are described. This section of the chapter concludes with a discussion of the results. After this detailed examination of power diodes, with its emphasis on optimization of their design, most of the rest of the chapter is devoted to the most common other power semiconductor devices: power MOSFETs, thyristors and IGBTs.

Chapter 12 begins to address the issue of mixed-mode simulations, where the possibility of a time-varying external circuit is included which provides boundary conditions on the device simulation which change in response to the behavior of the device.

Chapter 13 discusses other topics in device simulation, which are not described most easily using partial differential equations. These include problems which are most frequently tackled using Monte Carlo methods or integral equation methods. An introduction is given to the simulation methods involved as well as to the physical processes. The examples include diverse topics such as ion implantation and hot-carrier behavior in semiconductors.

Chapter 14 describes some other applications of the software which are only peripherally connected to semiconductors.

PROBLEMS

1. Most initial conditions for the Game of Life do not lead to persistent solutions like the stop light or the glider. Choose several other initial conditions, set them up in input files, run them and plot the results. Note that isolated 'live' cells with no 'live' neighbors will not survive. Cells must be started in 'colonies' if they are to be able to persist. Try different patterns, such as a single short row of cells; then two or more rows separated by various numbers of spaces. Other possible patterns to try include intersecting lines at various angles and hollow or full squares. Predict what will happen at the first generation in each case.

2. Change the rule for survival in the Game of Life. At present a cell will be alive at the next time if there are 3 live neighbors, or if there are 2 live neighbors and the cell itself is alive. Allow the cell to survive under these conditions, and in addition let it

survive with one extra live neighbor in each case. Run the simulation with the stop light and glider and plot the results.

3. Launch two gliders, aimed to collide in the center of the simulation region and plot the results.

4. Replace the rule for finding the new value of A in each cell with a function $A = \exp\left[-(COUNT[i, j] - 3.5)^2\right] / \exp[-0.25]$. What is the relation of this rule to the previous rule? Run the simulation and plot the results.

Part I

Basic Numerical Analysis
and Semiconductor Physics

Chapter 2

Fundamentals of Electromagnetism and its Numerical Analysis

The basic ideas of electromagnetism (and particularly electrostatics) are perhaps the most crucial topic that must be mastered before semiconductor devices can be understood. Electromagnetism also provides an excellent opportunity to learn how to solve the sort of equations we need to deal with, both numerically and analytically. The purpose of this chapter is (1) to describe the basic concepts of electromagnetism, and (2) to show how those concepts dictate how to set up the equations of electromagnetism in a numerical form. Basic numerical analysis is explained. Vector operators (*grad, div* and *curl*) are described, with emphasis on how to set them up numerically. The equations of electromagnetism can then be introduced and their boundary conditions found. Finally a number of numerical examples of solving those equations are given.

Electromagnetism will be reviewed here, because of the importance of the topic for semiconductor modeling. In addition, the software used here is itself illustrated vividly in electromagnetics examples, and new insights into the concepts of electromagnetism are provided by the use of the software. Because of this, a second motivation for including this material is to introduce the computational tool we shall be using, and to use the tool to show electromagnetism in a new light. The equations used to describe semiconductor devices are primarily from electrostatics (which is outlined here) and transport theory (which is considered in the next chapter.) Between these two chapters the basis will be provided for the physical models to be used later, as well as their computational implementation.

Some of the topics covered in this chapter are primarily mathematical. These include calculus, solution of differential equations and vector calculus. Others are more physics related. They can all be introduced and explained in a graphic way with the software used here. The operators *grad, div* and *curl*, their meaning and their numerical implementation are explained first. Next we introduce Maxwell's equations, followed by the boundary conditions they imply and the numerical form of the boundary conditions. Much of the material in later chapters involves solving partial differential equations. The boundary conditions on those equations determine the solution to a great extent, so it will be of great importance to be able to specify the boundary conditions properly in the numerical solution of the equations.

The way in which differentiation is done numerically is explained first, in Section 2.1. This section includes the solution of sets of equations of various kinds using SGFramework. We will see that we can easily solve differential equations. Section 2.2 considers vectors as a prelude to vector calculus. We begin this section by emphasizing different types of graphical presentation of vector fields. Then in Section 2.3 we can introduce the 'gradient' (*grad*) and in Section 2.4 the 'divergence' (*div*). We introduce the operators *grad, div* and *curl* and the Laplacian and describe their properties. Solutions of equations involving these operators are given. The *grad* and *div* operators are vital in writing conservation equations. The equations describing the carrier densities in semiconductors, which are obtained from the equation of conservation of particles, and Poisson's equation for the electrostatic potential are all of this type. Understanding the meaning of *div* makes the form of the conservation equations clear. On the other hand, understanding *grad* and *div* makes the numerical implementation of the conservation equations straightforward. The operator *curl* is described in Section 2.5. The vector operators in curvilinear coordinates are described in Section 2.6.

An overview of electrostatics (Section 2.7) is followed by an introduction to the use of irregular meshes to solve PDEs, since electrostatics provides excellent examples of complex geometries where such a mesh is needed. Then the full set of Maxwell's equations (Section 2.9) is given, again with emphasis on the connection to the boundary conditions and their numerical form. *Div* and *curl* lead naturally (in Section 2.10) to the use of Maxwell's equations to obtain the boundary conditions on the electric (and magnetic) fields. In Section 2.11 it is shown how the numerical implementation of those boundary conditions also follows from the definitions of the vector operators.

In Section 2.12 we describe 'propagators' and how they are used in electromagnetism. A 'propagator' in electromagnetism is essentially the field due to a point charge, so if we know the field due to one point charge we can often use that to find the field due to many charges.

We then turn in Section 2.13 to example problems, most of which are drawn from electrostatics. These examples make it clear that the great majority of the effort involved in setting up these models is devoted to the boundaries.

2.1   Introduction to Numerical Differentiation

This section gives a simple treatment of a complex topic, numerical solution of differential equations. We will readdress this topic many times throughout this book. Each successive time we describe numerical methods, the treatment will be more sophisticated, until we can handle the full semiconductor equation in complex geometries.

To introduce numerical differentiation we will make use of variables that are set up as arrays. One of the features of SGFramework that makes it useful is that it can solve simultaneous equations when the quantity that is being calculated is an element of an array. For example, if we want to find the voltage V inside a rectangular box, we would often have to settle for finding it at a set of points inside the box. The values of V at these points may be represented by an array. The points will form a grid or mesh of points.

If the coordinate employed is x, and each mesh point is at $x = i\Delta x$, where i is an integer and Δx is the change in x between successive mesh-points, then we can label each x-value we use by its i-value. Similarly, we can let $y = j\Delta y$ on the mesh, and so each point on the mesh has an index i and an index j. Then the voltage V at a point (i, j) can be written as $V[i, j]$.

Differentiation is straightforward, using arrays like V. If the problem is one-dimensional, so we only use one coordinate $x = i\Delta x$, the array is just $V[i]$. From the definition of the derivative,

$$\frac{dV}{dx} = \lim_{\Delta x \to 0} \frac{\Delta V}{\Delta x} \tag{2.1}$$

we can find an approximate expression for $\frac{dV}{dx}$ at the point i. The difference in V between two points $i + 1$ and $i - 1$ is $V[i + 1] - V[i - 1]$ and the distance is just $2\Delta x$ so

$$\frac{dV}{dx} = \frac{V[i + 1] - V[i - 1]}{2\Delta x}. \tag{2.2}$$

This is a central difference, since it uses values of V at points $i \pm 1$ on either side of the point i. We could also have used one-sided differences such as

$$\frac{dV}{dx} = \frac{V[i + 1] - V[i]}{\Delta x} \tag{2.3}$$

or

$$\frac{dV}{dx} = \frac{V[i] - V[i - 1]}{\Delta x}. \tag{2.4}$$

These equations are usually less accurate than the first form given, since they give weight to one side of the point i and ignore the other side, but sometimes they are useful. The spacing between mesh points has to be chosen so as to resolve the actual shape of the curve, which in this case is the shape of V. The replacement of the derivative with a ratio of finite differences leads to errors. If the mesh is not fine enough, the derivatives found by finite differencing will be inaccurate.

In the file ex1.sg, we give a SGFramework input file which will allow us to differentiate a one-dimensional function. To do this we use two arrays. We start with $V[i]$, the voltage (or electrostatic potential). The electric field is given by

$$E = -\frac{dV}{dx} \tag{2.5}$$

Input File ex1.sg

```
// FILE:  sgdir\ch02\ex1\ex1.sg

const PI  = 3.14159265358979323846;    // pi
const DIM = 101;                        // number of mesh points
const DX  = 2.0*PI/(DIM-1);             // mesh spacing (cm)

var x[DIM], V[DIM], E[DIM];

// these equations could be assignment statements
equ E[i=0]       -> E[i] = -(V[i+1]-V[i] ) /(1.0*DX);
equ E[i=1..DIM-2] -> E[i] = -(V[i+1]-V[i-1])/(2.0*DX);
equ E[i=DIM-1]    -> E[i] = -(V[i]  -V[i-1])/(1.0*DX);

begin main
  assign x[i=0..DIM-1] = i*DX;
  assign V[i=0..DIM-1] = cos(x[i]);
  solve;
  write;
end
```

so if we know the values in the array $V[i]$ we can find the values in the electric field array $E[i]$ from it.

The assign command was used to initialize $V[i]$. An assign statement could equally well have been used to find $E[i]$ from $V[i]$, in this case where all the quantities on the right-hand side of the equation are known, but for demonstration purposes this was not done here.

The equ statement implies that those values of E indicated are to be found from that equation. The solve statement tells the software to solve the equations appearing in the equ statement(s). The equ statement is a more powerful way of writing an equation than the assign statement. In an assign statement the right hand side of the equation must be an explicit expression in terms of known quantities, which will be evaluated to give the value of the left-hand side. The expression to the right of the arrow in an equ statement is evaluated to give the value of the quantity on the left of the arrow. The quantity to the left of the arrow must appear in the expression to the right. However, in an equ statement the expression to the right need not be an explicit expression for the quantity on the left. It could (for instance) be a nonlinear equation for the quantity on the left which could not be solved analytically. The left and right sides of the expression can both contain the quantity being solved for, or just one of them may contain it. That is, the expression to the right of the arrow can read $f(x) = g(x)$ where x is the unknown. Finally, the expression may contain other unknown quantities which will only become known when a matrix is inverted, as is the case in the next example we shall be looking at.

The range of values where E is found stops one short of the range of values

Input File ex2.sg

```
// FILE:  sgdir\ch02\ex2\ex2.sg

const PI  = 3.14159265358979323846;   // pi
const DIM = 101;                      // number of mesh points
const DX  = 2.0*PI/(DIM-1);           // mesh spacing (cm)

var x[DIM], V[DIM], E[DIM];

// these equations could be assignment statements
equ V[i=0]         -> V[i] = 1.0;
equ V[i=1..DIM-1] -> E[i] = -(V[i]-V[i-1])/DX;

begin main
  assign x[i=0..DIM-1] = i*DX;
  assign E[i=0..DIM-1] = sin(x[i]);
  solve;
  write;
end
```

where V is known, at both ends. This is because to find $E[i = 0]$ we would need to know $V[i = -1]$, and similarly to find $E[i = DIM - 1]$ we would need to know $V[i = DIM]$, and we know neither of those V values.

If E and V are plotted, it should be clear that E does look like $-\frac{dV}{dx}$ with some small errors due to the way V and $\frac{dV}{dx}$ are represented as arrays at points $x = i\Delta x$.

Solution of Differential Equations

We can immediately turn around the equations we just solved and use the same expressions for the derivatives in a different way. Up to now we used the equations to find the derivative, e.g. we found $\frac{dV}{dx}$ from V. Instead we can easily use the same approach to find the quantity being differentiated, e.g. we can find V from $\frac{dV}{dx}$. In file ex2.sg, which does this, we are solving a differential equation. (In many of the numerical simulations given here, small changes in the input file can cause a lack of convergence. This in turn can lead to overflows or to domain errors. These will be reported as such, in most cases; but sometimes they will cause the simulation to crash.)

In this file, instead of knowing V at the start, and finding E from V, we knew E and we found V from it. As a result we solved a differential equation for V. There is only one 'independent variable', x, in this problem.

One thing to notice in this example is that although we do not explicitly use a boundary condition, there is a boundary condition built in to this simulation. It is just that $V[0] = 0$, since all variables and array elements are initialized to zero automatically until we explicitly change them and we never set $V[0]$ to any other value, with an assignment statement or an equation statement. Then all the other $V[i]$ can be found from this. The equation in this input file finds $V[i]$ from $E[i]$ and from $V[i-1]$, and it starts at $i = 1$, so $V[0]$ is needed before we can begin using this equation.

The other point worth noticing here is that one of the less accurate one-sided differences was used to represent $\frac{dV}{dx}$. This type of difference can be easier to handle than the central difference used in the previous example since it allows us to start calculating values at one side of the range of x and work across to the other side. A central difference scheme would have complicated this process, since $V[i+1]$ would appear in the equation for $V[i]$, and $V[i+1]$ is not found until after $V[i]$ is found, if we are working from one side to the other. The one-sided difference allows an explicit expression for $V[i]$ to be written in terms of known quantities.

Since we used the 'equ' statement to find $V[i]$ the use of the one-sided difference was not actually necessary; the equ statement calls matrix routines which could have handled the central difference. On the other hand, a little algebra would have to have been done before we could use an assignment statement. The algebra would have allowed us to rewrite our equation so that the right hand side was explicitly known, and $V[i]$ was on the left hand side.

The Second Derivative

The equation solved in the last example was a first-order ordinary differential equation, because the 'highest' derivative was a first derivative. The electrostatic potential may also be found from the charge density ρ, using Poisson's equation. In one dimension Poisson's equation is

$$\frac{d^2V}{dx^2} = -\frac{\rho}{\epsilon}. \tag{2.6}$$

This is a second-order ODE. To solve this we need to be able to rewrite $\frac{d^2V}{dx^2}$ using the differences between neighboring values of V, in the same way that we rewrote $\frac{dV}{dx}$.

If we use the same approach as before, the derivative evaluated at the point $x = i\Delta x$ is

$$\left(\frac{dV}{dx}\right)_{x=i\Delta x} = \frac{V[i+1] - V[i-1]}{2\Delta x}. \tag{2.7}$$

But if we use the same reasoning to find the derivative of $\frac{dV}{dx}$, then

$$\frac{d}{dx}\left(\frac{dV}{dx}\right)_{x=i\Delta x} = \left(\left(\frac{dV}{dx}\right)_{i+1} - \left(\frac{dV}{dx}\right)_{i-1}\right)/2\Delta x. \tag{2.8}$$

Now to find the derivative at the 'point $i+1$' (meaning at the point in space where $x = (i+1)\Delta x$) we take the expression for the derivative at the 'point i' and add one to i everywhere it appears, so that

$$\left(\frac{dV}{dx}\right)_{i+1} = \frac{V[i+2] - V[i]}{2\Delta x}. \tag{2.9}$$

Similarly, at the point $i-1$ we subtract one from i everywhere we find it in the derivative, so that

$$\left(\frac{dV}{dx}\right)_{i-1} = \frac{V[i] - V[i-2]}{2\Delta x}. \tag{2.10}$$

Using these in the approximate expression for the second derivative gives

$$\left(\frac{d^2V}{dx^2}\right)_{x=i\Delta x} = \frac{V[i+2] - 2V[i] + V[i-2]}{(2\Delta x)^2}. \tag{2.11}$$

This second derivative is evaluated using points $2\Delta x$ away from the central point $x = i\Delta x$ and dividing by $(2\Delta x)^2$. We can imagine using another mesh with half the spacing of this mesh. If we evaluated $\frac{d^2V}{dx^2}$ on the smaller mesh, we would take points two small mesh spaces away from the center point, which means one large mesh space away, and we would divide by $(\Delta x)^2$ instead of $(2\Delta x)^2$. This means that we can also say

$$\left(\frac{d^2V}{dx^2}\right)_{x=i\Delta x} = \frac{V[i+1] - 2V[i] + V[i-1]}{(\Delta x)^2}. \tag{2.12}$$

An alternative way to obtain the second derivative involves finding the first derivative at each of two fictional points, $i+\frac{1}{2}$ and $i-\frac{1}{2}$. These are found from the differences $V[i+1] - V[i]$ and $V[i] - V[i-1]$ respectively, each divided by Δx. To find the second derivative one then takes the difference between these first derivatives, and divides again by Δx, which gives the above equation again.

This way of representing the derivatives using the differences between the V values is called 'finite differencing'.

The only other thing we have to do before we can set up an input file to solve this equation is to notice that we will need two boundary conditions, because the equation is second order. A straightforward way to see this is to look at the finite-difference equation. Whichever value of V we are solving for, $V[i-1]$, $V[i]$ or $V[i+1]$, we need to know the other two to find the value we want. If for general i values we find $V[i+1]$ from $V[i]$ and $V[i-1]$, for instance, then at the smallest values of i we have to know $V[0]$ and $V[1]$ before we can find $V[2]$. Once we know $V[2]$, having found it from $V[1]$ and $V[0]$, we can then find $V[3]$ and so on. $V[0]$ and $V[1]$ are the boundary conditions in this case.

In the file ex3.sg we write the equation to be solved as an equation for $V[i]$, in terms of $V[i-1]$ and $V[i+1]$. In this case one cannot write an explicit expression for $V[i]$ in terms of known quantities. If we solved for each i-value in turn it would not be possible to find $V[i]$ from such an equation, since $V[i-1]$ would be known but $V[i+1]$ would not be known. An equation statement must be used. SGFramework solves the system of equations using (usually sparse) matrix techniques when told

Input File ex3.sg

```
// FILE:  sgdir\ch02\ex3\ex3.sg

const DIM = 100;                    // number of mesh points
const W   = 1.0;                    // width of mesh (cm)
const DX  = W/(DIM-1);              // mesh spacing (cm)
const EPS = 8.854e-14;              // permittivity of free space (F/cm)

var x[DIM], V[DIM], rho[DIM];

// this equation cannot be an assignment statement
// the boundary conditions are implicit (V[0] = 0.0, V[DIM-1] = 0.0)
equ V[i=1..DIM-2] -> (V[i+1]-2.0*V[i]+V[i-1])/sq(DX) = -rho[i]/EPS;

begin main
  assign x[i=all] = i*DX;
  assign rho[i=1*DIM/4..3*DIM/4] = 1.0e-10*sign(W/2.0-x[i]);
  solve;
  write;
end
```

to 'solve' an equation given in an 'equ' statement. To make the solution possible we need to specify $V[0]$ and $V[DIM-1]$, the two end values. To set them to zero, we do not initialize them or give an equation for them. The software sets them to zero unless told otherwise.

2.2   Vector Calculus

In the previous section we dealt with scalar quantities. We now turn to vectors vector quantities. The purpose of the next few sections, which cover what may be considered elementary material, is to develop vector analysis simply by providing physical understanding of the divergence and curl operators. This will allow us to implement these concepts numerically, drawing directly on the definitions and on our intuitive understanding of what they mean. For a more detailed explanation of the basic ideas of vector calculus etc. the books by Morse and Feshbach [6] and by D'haeseleer et al. [7] are useful.

It is much easier to make progress with the concepts of vectors if we can illustrate them graphically. We thus begin this section with the ways we draw pictures of vectors.

2.2.1   Plotting Vectors

If we have a vector $\mathbf{A}$ that we want to plot, there are several ways to do it. None of them is entirely satisfactory. In a figure where a dependent variable is shown for different values of two independent variables, we can plot:

1. $A_x(x, y), A_y(x, y)$ and $A_z(x, y)$ separately. This gives us a complete description of $\mathbf{A}$, but it may be difficult to understand and combine all the information. Another set of components such as A_r, A_ϕ and A_z may be more useful.

2. $|\mathbf{A}|$, which reduces the vector to a single quantity we can plot, but without knowing the direction of $\mathbf{A}$ we cannot tell much about it.

3. An 'arrow plot' which can show the magnitude and direction of two components of $\mathbf{A}$. This gives a good idea of the overall behavior of $\mathbf{A}$, but details are difficult to see.

One or the other of these plots may be better for different vectors, for example if we want to see what the vector's divergence or curl might be.

2.2.2   Plotting Field Lines

Another useful way to represent a vector is by plotting field lines. In a way, a field-line plot is similar to the arrow plot. If we want to draw a field line for a vector, we choose a starting point A and draw a short 'arrow' in the direction of the field at A with its tail at A. We move along the arrow to its tip; call it point B. Then we draw another arrow in the direction of the field at B with its tail at B, the tip of the first arrow, and its own tip at C. We move to C, and so on.

In the next frame, Figure 2.1, we illustrate the process of plotting a field line. One problem with field line plots is that the strength of the field is not shown. Underneath the normal field line plot we show a plot of the same field line, but with vertical lines added to show the strength of the field at various points along the line.

For magnetic fields we can use a compass to show the direction of the field. If we read the compass then move a short distance in the direction it points, read it again, move again, and so on, we will be moving along a field line.

The electric field $\mathbf{E}$ is defined as the electric force on one coulomb, or as the force $\mathbf{F}$ on a charge q divided by q. Then to find an electric field line we should draw arrows in the direction of the force on a small positive charge.

Now suppose we know a vector $\mathbf{A}$ for a range of (x, y) values and we want to use a computer to draw its field lines. If we start at a point (x_0, y_0) we should move a short distance in the direction of $\mathbf{A}(x_0, y_0)$. This will take us to (x_1, y_1). Then we will move in the direction of $\mathbf{A}(x_1, y_1)$ and so on.

The vector $\mathbf{A}$ at the ith point we come to is

$$\mathbf{A}(x_i, y_i) = (A_x(x_i, y_i), A_y(x_i, y_i)). \tag{2.13}$$

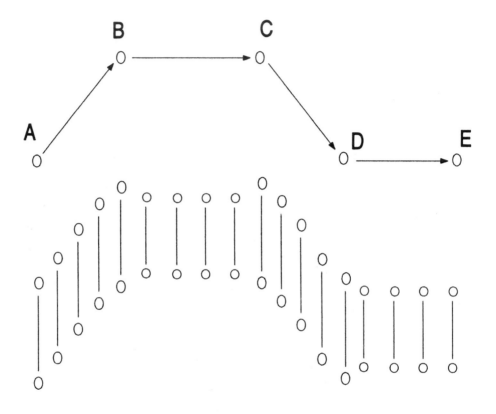

Figure 2.1. The process of plotting a field line. We 'move' along a series of short vectors, each of which points in the direction of the field, at the point where the vector starts. The top line shows the field line as a series of arrows in the direction of the field. The second picture shows the same line, but with vertical lines whose length indicates the strength of the field at different points along the field line.

We use the components of $\mathbf{A}(x_i, y_i)$ to decide the step $\mathbf{d}$ with components d_x in x and d_y in y. The step in x must be proportional to A_x and the step in y must be proportional to A_y if the overall step we take is to be along $\mathbf{A}$.

$$\mathbf{d} = \alpha \mathbf{A} \tag{2.14}$$

so

$$d_x = \alpha A_x, \; d_y = \alpha A_y. \tag{2.15}$$

These can be rewritten as

$$\frac{d_x}{d_y} = \frac{A_x}{A_y} \tag{2.16}$$

or if we use $d = |\mathbf{d}|$ and $A = |\mathbf{A}|$ we can say

$$\frac{d_x}{d} = \frac{A_x}{A} \tag{2.17}$$

and

$$\frac{d_y}{d} = \frac{A_y}{A}. \tag{2.18}$$

The last form is probably most useful, because it allows us to decide the size of the step d we want to use. It will also prevent us from dividing by zero if one (but not both) of the components of $\mathbf{A}$ happens to be zero at some point.

If we use this method to plot a field line, we record the values of (x_i, y_i) at all the points i and then represent them with dots on a plot. This is a very useful way of envisioning the field, but as explained above it does not tell us the magnitude of $\mathbf{A}$. Various better versions of a field-line plot might be invented. We have seen one where we draw a vertical line at each point (x_i, y_i) with height proportional to $A(x_i, y_i)$. Another possibility would involve using contours of the field strength, perhaps indicated by colors marked on the field lines, on the same plot as the field lines.

2.3   Gradient of Scalar Functions

To understand the equations of electricity and magnetism, we have to first understand the properties of three operators, *grad, div* and *curl*. *Grad* (which is short for gradient) operates on a scalar function. *Grad* is defined so that the small change (or differential) $d\Phi$ of a scalar Φ, which occurs when we move a distance $d\mathbf{r}$, is given by [7]

$$d\Phi = \nabla\Phi \cdot d\mathbf{r}. \tag{2.19}$$

Grad is consequently a vector which points in the direction in which the rate of increase of the scalar is largest. It is equal in magnitude to the rate of increase of the scalar in that direction. If we have a scalar Φ which depends on position, we calculate $\nabla\Phi$, for instance, in Cartesian coordinates to be

$$grad\Phi = \nabla\Phi = \mathbf{a}_x\frac{\partial\Phi}{\partial x} + \mathbf{a}_y\frac{\partial\Phi}{\partial y} + \mathbf{a}_z\frac{\partial\Phi}{\partial z}. \qquad (2.20)$$

For simplicity we restrict this discussion to two dimensions, x and y. Consequently, the gradient of $\Phi(x,y)$ is

$$\nabla\Phi = \mathbf{a}_x\frac{\partial\Phi}{\partial x} + \mathbf{a}_y\frac{\partial\Phi}{\partial y}. \qquad (2.21)$$

In the file exgrad.sg, the two components of the vector $\nabla\Phi$ are calculated, $\frac{\partial\Phi}{\partial x}$ and $\frac{\partial\Phi}{\partial y}$. These are calculated from Φ and then the magnitude of the gradient is found, $|\nabla\Phi|$. In this file, we are using a mesh whose grid points are defined by $x = i\Delta x$ and $y = j\Delta y$.

GRADPHIX in the specification file has been used to mean the x-component of $\nabla\Phi$, that is $\frac{\partial\Phi}{\partial x}$, and so on. MAGPHI is $|\nabla\Phi|$, which is

$$|\nabla\Phi| = \sqrt{\left(\frac{\partial\Phi}{\partial x}\right)^2 + \left(\frac{\partial\Phi}{\partial y}\right)^2}. \qquad (2.22)$$

As stated above, $\nabla\Phi$ is a vector which points in the direction in which Φ increases fastest. It is equal in magnitude to the rate of change of Φ with distance, in that direction.

A useful way to get a better idea of what $\nabla\Phi$ means is to choose a few points of the mesh and make a note of the components of $\nabla\Phi$ at each of them. Then the angle the contours of constant Φ make to the x-axis can be estimated from the plot of Φ. $\nabla\Phi$ will be perpendicular to the contours.

2.4   Divergence of Vectors

Divergence (*div* for short) and *curl* both are operations performed on vectors. $div\mathbf{A}$ (also written $\nabla\cdot\mathbf{A}$) is a scalar. If the vector $\mathbf{A}$ has Cartesian components (A_x, A_y, A_z), then

$$\nabla\cdot\mathbf{A} = \frac{\partial A_x}{\partial x} + \frac{\partial A_y}{\partial y} + \frac{\partial A_z}{\partial z}. \qquad (2.23)$$

This equation is not the definition of divergence, however [7]. It is only a formula for the divergence. In the material which follows it will be important to remember what is the meaning of the divergence. The meaning of $\nabla\cdot\mathbf{A}$ is as follows: $\nabla\cdot\mathbf{A}$ is the flux of $\mathbf{A}$ leaving a small volume ΔV, divided by ΔV. The small volume must then be let go to zero. In numerical approximations ΔV will be small but not zero. In the rest of this chapter we discuss the divergence without repeating the fact that the volume should be let go to zero, because this will not be done in our numerical approximations.

Input File exgrad.sg

```
// FILE:  sgdir\ch02\exgrad\exgrad.sg

const DIM = 30;                     // number of mesh points in x and y
const DX  = 1.0;                    // mesh spacing in x
const DY  = 1.0;                    // mesh spacing in y

var x[DIM], y[DIM], phi[DIM,DIM];
var gphix[DIM,DIM], gphiy[DIM,DIM], magphi[DIM,DIM];

const A  =  1.0, B  =  2.0;         // phi profile constants
const X0 = 50.0, Y0 = 50.0;         // phi profile constants
begin main
   // set up phi profile
   assign x[i=all] = i*DX;
   assign y[j=all] = j*DY;
   assign phi[i=all,j=all]=
   {A*[1.0-sq(x[i]/X0-1.0)]+B*[1.0-sq(y[j]/Y0-1.0)]} *
   {sq(x[i]/X0-1.0)+sq(y[j]/Y0-1.0)};

   // determine the x and y components of phi and its magnitude
   assign gphix[i=1..DIM-2,j=1..DIM-2] = (phi[i+1,j]-phi[i-1,j])/(2.0*DX);
   assign gphiy[i=1..DIM-2,j=1..DIM-2] = (phi[i,j+1]-phi[i,j-1])/(2.0*DY);
   assign magphi[i=all,j=all] = sqrt(sq(gphix[i,j])+sq(gphiy[i,j]));

   // write the results
   write;
end
```

The reason for introducing a concept such as divergence must be based on a belief that the flux of the vector is produced somewhat uniformly throughout a (small enough) volume. The amount of flux leaving the surface of a volume was presumed to be produced inside the volume. If we divide the flux leaving the volume by that volume, we are measuring how much flux is produced per unit volume. The rate of production of flux per unit volume is only a useful measure of how the vector behaves if the flux is being produced all throughout the volume (not just at a number of points, for instance) at a rate which varies somewhat slowly in space.

To make the idea of flux leaving a volume a little more graphic, imagine that the volume is a balloon full of gas. See Figure 2.2. (This does not correspond to a small volume suitable for defining divergence, and in the examples which follow the flow does not vary as smoothly or slowly as we need it to do when we define divergence.) The balloon may let gas leak out through its surface, in this example. Suppose that the gas inside (and outside) the balloon has a flow velocity **v** shown by the arrows in Figure 2.2. The density of the gas is (assumed to be) constant, so the flux is proportional to the velocity. The balloon does not change its shape or size in these examples.

a.) In the first example shown the gas circulates but it does not leave the volume. There is no flux out of the volume.

b.) In the next example which is on the top right the flow pattern does have a source in the center and the velocity has a component normal to the surface of the balloon, so gas must be leaving and there will be a net flux out.

c.) The case with a source in the center and purely radial flow, in the middle on the left, is the archetypal example of a net source in a volume. The balloon can not expand so it will have to let gas leak out. The outward leaking means there is a net flux out.

In the next few examples the balloon is inside a flowing gas. We will suppose first that the balloon is stationary. (Later we indicate what happens if it moves at the speed of the gas on the left hand side.) The simplest case is where the flow velocity is the same everywhere in both magnitude and direction. (Radially outward, as in example c.), is not the 'same direction'; the flow has to be parallel to itself everywhere.) This case is not shown in a figure. Suppose the flow velocity is constant and the balloon is stationary. Then the flux out balances the flux in. If it is moving with the flow and the flow velocity is constant everywhere, then the balloon moves at the same speed as the gas, so there is no flux in or out anywhere and no net flux.

d.) The next example (in the middle on the right) has a higher velocity on the right than on the left. If the balloon is stationary more leaves on the right than enters on the left.

e.) The example (on the bottom left) with a bigger flow velocity inside than out must have sources of gas at the bottom left and sinks which remove gas on the top right, inside the balloon, with no net source or sink.

f.) The final example has a higher flow velocity on the left than on the right. If the balloon is stationary, more enters on the left than leaves on the right.

(If the balloon moves at the speed of the flow on its left hand side, then in case d.) there will be a flux out on the right and so again there is a net flux out. In case f.) the moving balloon goes at the speed of the gas on the left which is faster than the gas on the right and so the gas flows in on the right. There is a next flux in.)

2.4.1   Calculation of the Divergence

Suppose we want to find the flux of **A** coming out of a rectangular volume with its sides parallel to the $x-$ or y-axis. We will assume there is no variation with z, so we can assume we are dealing with a unit length in the z-direction. The flux of **A** leaving the volume is the integral, over the sides of the volume, of the component of **A** along the outward normal to the surface.

If the rectangle is one of the cells of the two-dimensional mesh that we usually use, it might have corners at the points labeled by $(i, j), (i, j + 1), (i + 1, j)$ and

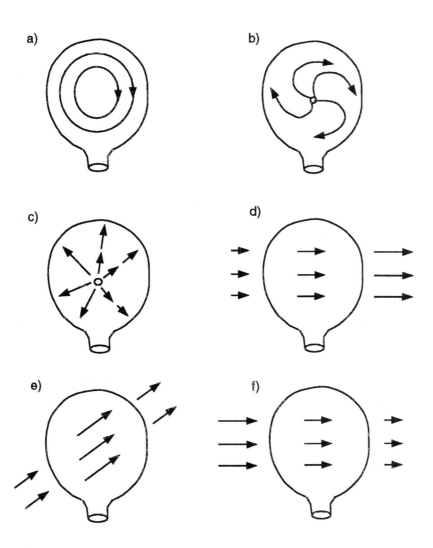

Figure 2.2. Illustration of the meaning of divergence, in terms of the flux (of gas) out of a volume (a balloon.)

$(i+1, j+1)$. The flux of $\mathbf{A}$ leaving the top is

$$F_A^{top} = \int_{x=i\Delta x}^{x=(i+1)\Delta x} A_y(x, y = (j+1)\Delta y)\, dx. \qquad (2.24)$$

As usual, the spacing in x is Δx and the spacing in y is Δy.

The flux leaving the bottom is

$$F_A^{bottom} = -\int_{x=i\Delta x}^{x=(i+1)\Delta x} A_y(x, y = j\Delta y)\, dx. \qquad (2.25)$$

This is negative, because if A_y is positive, the flux is not leaving but entering.

If these are combined, the net flux leaving in the y-direction is

$$F_A^y = \int_{x=i\Delta x}^{x=(i+1)\Delta x} [A_y(x, y = (j+1)\Delta y) - A_y(x, y = j\Delta y)]\, dx. \qquad (2.26)$$

If we assume Δx and Δy are very small, we can simplify this if we use two approximations. Before we approximate, we should notice that if the integrand is divided by Δy we obtain a 'new' integrand which is Δy multiplied by

$$\frac{[A_y(x, y = (j+1)\Delta y) - A_y(x, y = j\Delta y)]}{\Delta y}. \qquad (2.27)$$

This is exactly equal to one of our standard numerical approximations for $\frac{\partial A_y}{\partial y}$. This is an important point to note when setting up a numerical scheme. The approximation is replacing this numerical form of the derivative with the limiting form which is obtained when the intervals go to zero.

The integration over x can be done if Δx is small enough so that A_y hardly changes in that distance.

$$F_A^y \approx \frac{\partial A_y}{\partial y} \Delta x\, \Delta y. \qquad (2.28)$$

The net flux in the x-direction can be found in the same way.

$$F_A^x \approx \frac{\partial A_x}{\partial y} \Delta x\, \Delta y. \qquad (2.29)$$

This was all for a unit length in z, so $\Delta V = \Delta x \Delta y$, and if we divide by ΔV we find the flux leaving the cell, per unit volume, is

$$\nabla \cdot \mathbf{A} = \frac{F_A^x + F_A^y}{\Delta V} = \frac{\partial A_x}{\partial x} + \frac{\partial A_y}{\partial y}. \qquad (2.30)$$

We assumed there was no variation with z, so $\frac{\partial A_z}{\partial z}$ equals zero, and so this agrees with the original expression we gave for $\nabla \cdot \mathbf{A}$.

One important property of the fluxes between adjacent cells of this mesh is that, since the flux leaving one cell face enters the next cell through the same face, fluxes between adjacent cells cancel each other if we include a lot of adjacent cells. When we add all the net fluxes leaving cells, the total flux leaving is the total flux going out through the outer perimeter of all the cells combined. The total flux leaving the outer surface is the integral over the surface of the normal component of $\mathbf{A}$, which we find by taking the dot product of $\mathbf{A}$ with $\mathbf{n}$, the outward-pointing normal to the surface.

$$F^{Total} = \int_{outer\ surface} \mathbf{A} \cdot \mathbf{n}\, ds. \tag{2.31}$$

On the other hand, if we add all the fluxes leaving each individual cell, they should add up to the same answer

$$F^{Total} = \sum_{cells} \int_{cell\ surface} \mathbf{A} \cdot \mathbf{n}\, ds. \tag{2.32}$$

The fluxes leaving each single cell are summed, here. But the definition of $\nabla \cdot \mathbf{A}$ is that $\nabla \cdot \mathbf{A}$ equals the flux leaving the cell divided by the cell volume ΔV, so we can replace the flux leaving the cell by $(\nabla \cdot \mathbf{A})\Delta V$.

$$F^{Total} = \int_{outer\ surface} \mathbf{A} \cdot \mathbf{n}\, ds = \sum_{cells} (\nabla \cdot \mathbf{A})\Delta V. \tag{2.33}$$

Numerical Implementation of Divergence

If we specify an $\mathbf{A}$, we can evaluate the divergence of that $\mathbf{A}$ and check the last result, using an input file like exdiv.sg.

This file uses a small mesh, in the sense that it has few cells (or mesh points), so that the two ways of finding the total flux can be compared by hand. (Some error should be expected in comparing them, because we are not performing the integration over the cell faces exactly.) $\nabla \cdot \mathbf{A}$ is known in cells 0 to 3, in both i and j. The volume of a cell $\Delta V = \Delta x \Delta y = 0.02$, so we need to sum $\nabla \cdot \mathbf{A}$ over the cells and multiply by 0.02.

We defined $\nabla \cdot \mathbf{A}$ at the point (i, j) using the faces at $i, i+1$ and $j, j+1$, so the outer faces of the region bounding the cells at $i = 0$ to $i = 3$ are at $i = 0$ and $i = 4$ and the faces in j are at $j = 0$ and $j = 4$. Then we integrate A_x over the large face at $i = 4$ by summing $(A_x[0, 4] + A_x[1, 4] + A_x[2, 4] + A_x[3, 4])\Delta y$. The flux out of the face at $i = 0$ is, similarly, $-(A_x[0, 0] + A_x[1, 0] + A_x[2, 0] + A_x[3.0])\Delta y$. Fluxes at $j = 0$ and 4 can be found in the same way and added together, then compared to the sum of $(\nabla \cdot \mathbf{A})\Delta V$.

This exercise is worth repeating for some other vectors $\mathbf{A}$. For example,

1. $A_x = 2j - 0.2j^2, A_y = i + 0.4i^2.$

2. $A_x = 2\sin(\pi i/2) - 0.2j^2, A_y = \sin(\pi j/2) + 0.4i^2.$

Input File exdiv.sg

```
// FILE:  sgdir\ch02\exdiv\exdiv.sg

const DIM = 5;                          // number of mesh points in x and y
const DX = 0.2;                         // mesh spacing in x
const DY = 0.1;                         // mesh spacing in y

var x[DIM], y[DIM], Ax[DIM,DIM], Ay[DIM,DIM], divA[DIM-1,DIM-1];

begin main
  // initialize the x and y components of vector A
  assign    x[i=all] = i*DX;
  assign    y[j=all] = j*DY;
  assign    Ax[i=all,j=all] = 3.0*x[i]-0.2*sq(y[j]);
  assign    Ay[i=all,j=all] = y[j]+4.0*sq(y[j]);

  // compute the divergence and write the results
  assign divA[i=all,j=all] = (Ax[i+1,j]-Ax[i,j])/DX+(Ay[i,j+1]-Ay[i,j])/DY;
  write;
end
```

3. $A_x = 3x - 0.6xy$, $A_y = 0.3y^2 - 3y$.

The integral of $\nabla \cdot \mathbf{A}$ over this region in the first two cases is zero. The important point is to see why it is zero. In case 1, it is because A_x does not depend on x and A_y does not depend on y. If we think of the vector $\mathbf{a}_x A_x$, it gives the same flux entering at $x = 0$ as leaves at $x = 4\Delta x$, and similarly for $\mathbf{a}_y A_y$. In case 2, A_x is zero at $x = 0$ and $x = 4\Delta x$ and A_y is zero at $y = 0$ and $y = 4\Delta x$, so the overall flux leaving is zero. Adding a constant to any of these components does not change $\nabla \cdot \mathbf{A}$; it increases the flux out of one side by the same amount as it increases the flux entering the other so the net change in flux is zero.

Divergence and Conservation of Flux

We now compare two numerical implementations of the equation for conservation of flux: one done using an integral equation, and the other done using the divergence operator.

In static electrical problems, the flux of the electric field $\mathbf{E}$ leaving a volume is proportional to the charge Q inside the volume. If ρ is the density of charge per unit volume, then

$$Q = \int \rho \, dV. \tag{2.34}$$

In free space, the flux of $\mathbf{E}$ leaving a volume is equal to Q/ϵ_0 (ϵ_0 is the permittivity of vacuum); the flux is

$$\int \mathbf{E} \cdot \mathbf{n}\, ds = \frac{Q}{\epsilon_0} = \frac{1}{\epsilon_0} \int \rho\, dV \qquad (2.35)$$

with the same notation as before for $\mathbf{n}$, ds and V. This is the integral form of the equation for the conservation of flux.

Using the definition of divergence, we found that for any vector (such as $\mathbf{E}$) the flux of $\mathbf{E}$ leaving a volume obeys the equation

$$\int \mathbf{E} \cdot \mathbf{n}\, ds = \int \nabla \cdot \mathbf{E}\, dV \qquad (2.36)$$

so

$$\int \left(\nabla \cdot \mathbf{E} - \frac{\rho}{\epsilon_0} \right) dV = 0. \qquad (2.37)$$

Since this has to be true for any volume we choose to integrate over, it must be true that

$$\nabla \cdot \mathbf{E} = \frac{\rho}{\epsilon_0} \qquad (2.38)$$

at least in free space (because of the absence of material which can form electric dipoles). In general, we can use the 'electric displacement' $\mathbf{D} = \epsilon \mathbf{E}$, instead of $\mathbf{E}$. $\mathbf{D}$ obeys the equation

$$\nabla \cdot \mathbf{D} = \rho. \qquad (2.39)$$

If $\epsilon = \epsilon_0$, this is the same as the equation given earlier for $\mathbf{E}$.

This should be equivalent to the integral form of the equation for the conservation of flux, and this is what we shall check, by implementing them both, numerically.

We can find the electric field $\mathbf{E}$ (or $\mathbf{D}$) in two ways, using the two forms of the equation relating $\mathbf{E}$ to the charge, $\nabla \cdot \mathbf{E} = \frac{\rho}{\epsilon_0}$ or $\int \mathbf{E} \cdot \mathbf{n}\, ds = \frac{Q}{\epsilon_0}$, as is shown in the next example.

If the charge density ρ is independent of x and y but depends only on z, then we can have a situation where $\mathbf{E}$ is in the z-direction and we can write $\nabla \cdot \mathbf{E} = \frac{\rho}{\epsilon_0}$ as

$$\frac{dE}{dz} = \frac{\rho}{\epsilon_0}. \qquad (2.40)$$

To solve this we need a boundary condition. If $E(z = 0)$ is known, this will be a suitable boundary condition. Suppose it is zero, for example. If we define a mesh as usual, with $z = k\Delta z$, where k is an integer and Δz is the spacing of the mesh points, then this equation can be written in a suitable form for numerical implementation, for example as

$$\frac{E[k] - E[k-1]}{\Delta z} = \frac{\rho[k]}{\epsilon_0}. \tag{2.41}$$

E and ρ have been written as arrays, with their values defined at all the points k. However, $\rho[k]$ is defined inside the kth cell, whereas $E[k]$ is the electric field at the upper edge of the kth cell.

Another approach based on the other equation and on physical interpretation of the fluxes involved can also be employed. We are interested in the region $z > 0$, where there is a charge density, and we want to find the field **E** created by that charge. If $E(z = 0) = 0$, all the flux due to a charge Q to the right of $z = 0$ must leave to the right, since none is leaving through $z = 0$, because the field is zero there. That flux is equal to the charge divided by ϵ_0. The total charge (per unit area) in the cells from $z = 0$ to $z = k\Delta z$ can be evaluated numerically by adding up the charge in each cell from $z = 0$ to $z = k\Delta z$.

$$Q[k] = \sum_{m=0}^{k} \rho[m]\,\Delta z. \tag{2.42}$$

The flux of **E** leaving through $z = (k+1)\Delta z$ is just equal to this charge divided by ϵ_0, since the flux of **E** is given by $\int \mathbf{E} \cdot \mathbf{n}\,ds = \frac{Q}{\epsilon_0}$. **E** and **n** are in the same direction, and $\int E\,ds = E$, since E depends only on z and we used a unit area, so

$$E[k] = \frac{1}{\epsilon_0} \sum_{m=0}^{k} \rho[m]\Delta z \tag{2.43}$$

which gives us a second numerical way to find **E**.

We can show they are the same by starting at $k = 1$ and finding $E[k = 1]$ both ways. From the numerical form of the divergence equation above, the derivative is given by

$$\frac{E[1] - E[0]}{\Delta z} \stackrel{.}{=} \frac{\rho[0]}{\epsilon_0} \tag{2.44}$$

and $E[0] = 0$, so $E[1] = \frac{\rho[0]\Delta z}{\epsilon_0}$. Then

$$\frac{E[2] - E[1]}{\Delta z} = \frac{\rho[1]}{\epsilon_0} \tag{2.45}$$

so $E[2] = (\rho[1] + \rho[0])\frac{\Delta z}{\epsilon_0}$, and so on. This is in agreement with the other equation which is obtained by summing the charges and using the total charge to find the flux which is produced.

For a concrete example, suppose $\rho = \epsilon_0 \sin(2\pi z)$ for $0 \leq z < 1$ and $\rho = 0$ otherwise. We can find $E(z)$ from this, using the equations above. Since

$$\mathbf{E} = -\nabla V \tag{2.46}$$

Input File exemw.sg

```
// FILE:  sgdir\ch02\exemw\exemw.sg

const DIM = 101;                    // number of mesh points
const DZ  = 1.0/(DIM-1);            // mesh spacing

var z[DIM], phi[DIM], E[DIM], rho[DIM];

equ   E[kp=1..DIM-1] -> (  E[kp]- E[kp-1])/DZ = rho[kp-1];
equ phi[kp=1..DIM-2] -> (phi[kp]-phi[kp-1])/DZ =  -E[kp-1];

begin main
  assign z[k=all] = k*DZ;
  assign rho[kp=all]=sin(6.283*kp*z[kp]);
  solve;
  write;
end
```

in electrostatics, where V is the electrostatic potential, we could find V from **E** as well. The discrete version of this equation could be written as

$$V[k+1] - V[k] = -E[k]\Delta z. \tag{2.47}$$

The charge density in this example might describe an electrostatic wave in a plasma or in the electrons in a semiconductor. The wave has pushed electrons to the right, leaving a positive charge between $z = 0$ and $z = 1/2$ and creating a negative charge between $z = 1/2$ and $z = 1$. In the file shown, the index is one greater than in the previous equation, so $k + 1$ is replaced by k and k is replaced by $k - 1$. Both $E[KP = 0]$ and $V[KP = 0]$ are zero, since they are never set to any other value.

Divergence in Cylindrical and Spherical Geometries

This section uses the same ideas as the previous section, and applies them to finding the appropriate way to implement numerically (the radial part of) the divergence in cylindrical and spherical coordinates. This material could be skipped by those not wishing to solve conservation equations in cylindrical or spherical geometries.

The correct numerical expression for the divergence of a vector is that which follows from the definition of the divergence, as the flux of the vector leaving a small volume, per unit volume. In other words, we find the flux leaving a volume and we divide that flux by the volume (strictly as the volume goes to zero, but we cannot achieve this in our numerical scheme). We need to examine the surface that the flux passes through and the volume we shall be using, in the numerical scheme. Then we can set up a definition of the numerical divergence which obeys the numerical version of Gauss's law, i.e. the divergence multiplied by the volume of a cell must equal the net flux leaving the cell - exactly. Then we will have the ability to write conservation equations using the divergence and in a form where the flux is exactly conserved.

To do this we make use of the area $A(r)$ of a surface of radius r, and the volume $V(r)$ inside the surface. Since the numerical implementation of divergence is important in this work, we discretize the radius so $r = k\Delta r$. The area of the surface of radius $r = k\Delta r$ is then $A[k]$ and the volume inside the radius $r = k\Delta r$ is $V[k]$. The difference between two successive values of the volume is the cell volume, $\Delta V[m]$, given by

$$\Delta V[m] = V[m] - V[m-1]. \tag{2.48}$$

This is the small volume which will be critical in defining the divergence. The flux leaving this cell passes through two surfaces; the inner surface, at radius $(k-1)\Delta r$ with area $A[k-1]$ and the outer surface at radius $k\Delta r$ with area $A[k]$.

In cylindrical geometry, with symmetry in z and the azimuthal angle, or in spherical geometry with symmetry in the angles (θ, ϕ), the only independent variable we need to use is the radius r. If the charge per unit volume is $\rho(r)$, then at a radius $r = k\Delta r$ the total charge inside that radius per unit length in z (or the total charge inside that radius, in spherical geometry) is

$$Q[k] = \sum_{m=0}^{k} \rho[m]\,\Delta V[m]. \tag{2.49}$$

$\rho(r)$ has been replaced by $\rho[m]$.

[The cell volume $\Delta V[m]$ in the cylindrical case is

$$\Delta V[m] = \pi(r_m^2 - r_{m-1}^2) = \pi(\Delta r)^2(m^2 - (m-1)^2) = \pi(\Delta r)^2(2m-1) \tag{2.50}$$

so

$$Q[k] = \pi(\Delta r)^2 \sum_{m=0}^{k} \rho[m](2m-1). \tag{2.51}$$

The usual integral is obtained if we take the limit $\Delta r \to 0$, since $m\Delta r \to r$, $\Delta r \to dr$, and

$$Q(r) = \int \rho(r)2\pi r\,dr. \;] \tag{2.52}$$

The flux of E leaving through the radius $k\Delta r$ is equal to $\frac{Q[k]}{\epsilon_0}$. [In cylindrical coordinates the area the flux goes through is $A[k] = 2\pi k\Delta r$. This is essentially the same as saying that the area per unit length in z is $2\pi r$.] The flux per unit length in z (in cylindrical coordinates; in sphericals this quantity is the total flux) is the area multiplied by the field, so

$$A[k]\, E[k] = \frac{Q[k]}{\epsilon_0} = \frac{1}{\epsilon_0} \sum_{m=0}^{k} \rho[m]\Delta V[m]. \tag{2.53}$$

[In cylindrical coordinates this can be written as

$$2\pi k \Delta r E[k] = \frac{\pi}{\epsilon_0}(\Delta r)^2 \sum_{m=0}^{k} \rho[m](2m-1) \tag{2.54}$$

or

$$E[k] = \frac{\Delta r}{2\epsilon_0 k} \sum_{m=0}^{k} \rho[m](2m-1). \;] \tag{2.55}$$

Now to relate this approach to the differential equation for E, we note that since

$$A[k]E[k] = \frac{1}{\epsilon_0} \sum_{m=0}^{k} \rho[m]\,\Delta V[m] \tag{2.56}$$

then if k is decreased by one,

$$A[k-1]\, E[k-1] = \frac{1}{\epsilon_0} \sum_{m=0}^{k-1} \rho[m]\,\Delta V[m]. \tag{2.57}$$

In this expression the right-hand side is the same sum as before except it has one fewer term in it.

Using this in the equation above,

$$A[k]\, E[k] = \frac{1}{\epsilon_0}\rho[k]\,\Delta V[k] + A[k-1]\, E[k-1] \tag{2.58}$$

or

$$\frac{A[k]\, E[k] - A[k-1]\, E[k-1]}{\Delta V[k]} = \frac{\rho[k]}{\epsilon_0}. \tag{2.59}$$

We have derived on the left hand side of this equation, in a form which we could implement numerically, an exact expression for the one-dimensional divergence of the flux (in this case the flux of E). This expression shows the importance of the cell areas and volumes in calculating the divergence in non-Cartesian coordinates.

This result is valid for any one-dimensional problem, but for a cylinder the left-hand side is

$$\frac{kE[k] - (k-1)E[k-1]}{(k-\frac{1}{2})\Delta r}. \tag{2.60}$$

If we multiply the numerator by Δr then the numerator becomes $(rE)_k - (rE)_{k-1}$. At the same time we multiply the denominator by the same factor of Δr. The denominator is then $k - \frac{1}{2}(\Delta r)^2$. Since $k(\Delta r)^2$ equals $r\Delta r$, the denominator becomes $r\Delta r - \frac{1}{2}(\Delta r)^2$. In the limit as $\Delta r \to 0$, the second term in the denominator vanishes, yielding

$$\frac{1}{r}\frac{\partial}{\partial r}(rE(r)). \tag{2.61}$$

This shows why the divergence in cylindrical coordinates does not have a term $\frac{\partial E}{\partial r}$ but instead has $\frac{1}{r}\frac{\partial}{\partial r}(rE(r))$. It also shows that the numerical expression we get if we start by finding $E[k]$ from the flux and the one we obtain if we find $E[k]$ from the usual analytic form of $\nabla \cdot \mathbf{E} = \frac{\rho}{\epsilon}$ are the same, in the limit as $\Delta r \to 0$.

The full divergence of $\mathbf{A}$ in cylindrical coordinates (r, ϕ, z) is

$$\nabla \cdot \mathbf{A} = \frac{1}{r}\frac{\partial}{\partial r}(rA_r) + \frac{1}{r}\frac{\partial A_\phi}{\partial \phi} + \frac{\partial A_z}{\partial z}. \tag{2.62}$$

In spherical coordinates (r, θ, ϕ) then $\nabla \cdot \mathbf{A}$ is

$$\nabla \cdot \mathbf{A} = \frac{1}{r^2}\frac{\partial}{\partial r}(r^2 A_r) + \frac{1}{r\sin\theta}\frac{\partial A_\theta}{\partial \theta} + \frac{1}{r\sin\theta}\frac{\partial}{\partial \phi}(\sin\theta A_\theta). \tag{2.63}$$

We could again derive the first term using our one-dimensional result which was derived in this section. Instead of a volume element $\Delta V \simeq 2\pi r\Delta r$ we would have $\Delta V \simeq 4\pi r^2 \Delta r$, so the volume of the cell varies like r^2 instead of r. The flux is through an area equal to $4\pi r^2$ instead of $2\pi r$, so again r^2 replaces r. In terms of the indices used in the numerical expressions, the cell volume is

$$\Delta V[m] = V[m] - V[m-1] = \frac{4}{3}\pi(\Delta r)^3[m^3 - (m-1)^3] \tag{2.64}$$

and the area $A[k] = 4\pi(k\Delta r)^2$.

In the numerical implementation we should use the exact values for the area $A[k]$ and the volume $V[k]$ of the surface of radius $r = k\Delta r$. For illustration we shall approximate these here. Since $\Delta V[m] \simeq 4\pi(\Delta r)^3 m^2$, where we have kept the biggest term,

$$\frac{A[k]E[k] - A[k-1]E[k-1]}{\Delta V[k]} \simeq \frac{(k\Delta r)^2 E[k] - ((k-1)\Delta r)^2 E[k-1]}{(k\Delta r)^2 \Delta r} \tag{2.65}$$

which has been written in this form to show that it is approximately

$$\frac{1}{r^2}\frac{\partial}{\partial r}(r^2 E) \tag{2.66}$$

as expected.

In a problem where the densities of electrons and ions are known in cylindrical coordinates (for instance inside a glow discharge which is used to make a fluorescent light [8]) we can use the cylindrical form of the equations to find the radial electric field. The equation for $E[KP]$ in the last example must be changed, and ρ must be found from the charged particles' densities. $E = -\frac{\partial V}{\partial r}$ is still true, so the equation for V is still valid.

2.5   Curl of Vectors

In electromagnetic problems we often have to integrate a vector along a path. In this section we explain why these path integrals are important, and how they lead to the *'curl'* operator. To define *curl* of a vector $\mathbf{A}$ at a point p we have to integrate $\mathbf{A}$ around a path which forms a small, closed loop around p. (This integral measures the 'rotation' of the vector around that loop.) We then divide the integral by the area of the loop. (By dividing by the area we find the 'rotation per unit area'.) In the limit when the area of the loop goes to zero, this gives the component of *curl* $\mathbf{A}$ at point p (also written $\nabla \times \mathbf{A}$) normal to the plane that contains the loop. (See also [7].)

The integral of $\mathbf{A}$ around a loop is written as

$$I = \oint \mathbf{A} \cdot d\mathbf{r}. \tag{2.67}$$

Then the normal component of $\nabla \times \mathbf{A}$ is

$$(\nabla \times \mathbf{A})_n = \frac{\oint \mathbf{A} \cdot d\mathbf{r}}{\Delta s} \tag{2.68}$$

where Δs is the area enclosed by the integration loop.

The Role of Integrals around Loops

To see why $\nabla \times \mathbf{A}$ is a useful quantity, we need to first explain why line integrals around closed loops are important, and then why it is helpful to divide the integral around the loop by the area of the loop.

Figure 2.3 shows a loop of area ds. Arrows show the direction of integration. $\mathbf{n}$ is normal to the loop. The vector $d\mathbf{r}$ is a very short distance along the path of integration. A right-hand rule is used to define the direction in which the 'rotation vector' will point. (The rotation vector is normal to the loop, but we need to establish which normal we mean.) If the fingers of the right hand point in the direction of the integration, then the rotation vector is in the direction of the thumb. This means that when we divide the line integral by the area, which we do to find the *curl*, it is in the direction of the thumb if the field went around in the direction of the fingers of the right hand.

The two most common integrals around loops we will have to deal with in electromagnetism are the integrals of the electric field $\mathbf{E}$ and the magnetic field $\mathbf{H}$. The integral of $\mathbf{E}$ around a loop

$$I_E = \oint \mathbf{E} \cdot d\mathbf{r} \tag{2.69}$$

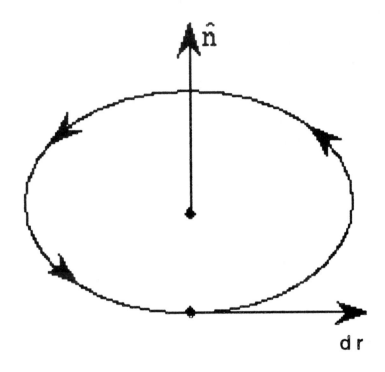

Figure 2.3. The line integral around a circular loop, showing the direction of integration. A right hand rule applies. When the fingers of the right hand point along the direction of integration, the thumb points along the normal.

is especially important because $q\mathbf{E}$ is the force the electric field exerts on a particle which has a charge q. The work done on the particle by the field when the particle moves a distance $d\mathbf{r}$ is $q\mathbf{E} \cdot d\mathbf{r}$. The work done by the field if the particle moves all the way around the loop is

$$W_{loop} = q\, I_E = \oint q\mathbf{E} \cdot d\mathbf{r}. \qquad (2.70)$$

If the $\mathbf{E}$ field is conservative, this will be zero. This is analogous to walking around on a hill and returning to the starting point — the change in potential energy will be zero. If $\oint \mathbf{E} \cdot d\mathbf{r}$ is zero, a charged particle has no change in electrostatic potential energy and so no change in kinetic energy if it moves around the loop, then returns to its starting point. Sometimes the field is not conservative, and a particle can go around a loop and return to the start with more (or less) kinetic energy than it started with. I_E is the energy a unit charge ($q = 1$) picks up if it goes around the loop.

The other integral we mentioned is

$$I_H = \oint \mathbf{H} \cdot d\mathbf{r}. \qquad (2.71)$$

This integral is important because according to 'Ampere's law' it is equal to the current through the loop, in a static field. This allows $\mathbf{H}$ and I to be found from each other in some cases.

The fact that I_H equals the current can be used to explain why we should take a very small loop to do the integration, and then divide I_H by the area of the loop. The current can be thought of as making $\mathbf{H}$ go around. The more current, the more $\mathbf{H}$ goes around and the bigger I_H is. If the current is distributed over the area of the loop, and J is the current per unit area going through the loop, then the total current through the loop I_{loop} is

$$I_{loop} = \int_s J\, ds \qquad (2.72)$$

where this integral is an area integral over the loop.

For a very small loop, of area Δs, J is roughly constant across the loop. If J is roughly constant, we can remove it from the integral; $I_{loop} = \int_s J\, ds \simeq J \int_s ds = J \Delta s$.

$$I_{loop} \simeq J \Delta s \qquad (2.73)$$

and this is equal to $I_H = \oint \mathbf{H} \cdot d\mathbf{r}$ according to Ampere's law.

$$J \Delta s = \oint \mathbf{H} \cdot d\mathbf{r} \qquad (2.74)$$

or if we divide by Δs

$$J = \frac{\oint \mathbf{H} \cdot d\mathbf{r}}{\Delta s} \qquad (2.75)$$

and since (the normal component of) *curl* is defined as the line integral divided by the area (in the limit when the area is very small),

$$J = (\nabla \times \mathbf{H})_n. \qquad (2.76)$$

In other words, the current I_{loop} can make $\mathbf{H}$ 'rotate' and is equal to $I_H = \oint \mathbf{H} \cdot d\mathbf{r}$, which measures the rotation rate. J is the current per unit area and J gives the rate of rotation of $\mathbf{H}$ per unit area, which is $(\nabla \times \mathbf{H})_n$.

This is analogous to the divergence, in a way. In each case, we take the amount of the quantity of interest which is produced in a region, and divide it by the 'size' of the region it was produced in. (In the case of *curl*, the quantity being produced is 'rotation' of a vector. In the case of divergence, it is flux of the vector.) This gives us the rate of production within that region. For instance, we know that the total flux of $\mathbf{E}$ leaving a volume is equal to Q/ϵ, where Q is the charge in the volume. So Q produces a flux of $\mathbf{E}$. The flux produced per unit volume is $\nabla \cdot \mathbf{E}$, from the definition of *div*, so this should equal the charge per unit volume, ρ, divided by ϵ.

Both *div* and *curl* are the result of taking an integral of a vector (flux through the surface of a volume for *div*, or line integral around a closed loop, for *curl*) and then saying that the integral measures something being produced at a certain rate, per unit volume for *div*, and per unit area for *curl*. So we divide the integral of the flux by the volume to get the production rate of flux per unit volume (which is *div*) and we divide the line integral by the area to get *curl*, which measures the rotation produced per unit area.

The trouble with saying that *curl* measures rotation is that there are some situations where the *curl* of a vector is not zero, even though the vector does not seem to be rotating in the usual sense.

We can express a vector $\mathbf{C}$ as the sum of two vectors $\mathbf{A}$ and $\mathbf{B}$ where the 'rotating' part $\mathbf{A}$ has a non-zero *curl* and the 'nonrotating part' $\mathbf{B}$ has $\nabla \times \mathbf{B} = 0$. (In some cases $\mathbf{A}$ and/or $\mathbf{B}$ will be zero.)

$$\nabla \times \mathbf{C} = \nabla \times \mathbf{A} \qquad (2.77)$$

and $\mathbf{C}$ is rotating because $\mathbf{A}$ is rotating while $\mathbf{B}$ contributes nothing to $\nabla \times \mathbf{C}$.

This is not a very satisfactory way of seeing that $\mathbf{C}$ is rotating, because we may not always want to have to split a vector up into simpler pieces like this. We need an intuitive way to do the test that always works. Before we can simply look at a picture of a vector $\mathbf{C}$ and say that $\mathbf{C}$ is rotating, we need a better definition of 'rotating'.

A simple example, which shows what we need to do to test for rotation is given next. Suppose you walk all the way around the edge of a field on a windy day. We are interested in the force the wind exerts on you as you walk. If the wind helps you on average then it has an overall rotation in the direction you are walking. If it hinders you on average, it has an overall rotation in the opposite direction. If it helps you part of the time and hinders you the rest and the work it does cancels out, it has no overall rotation. This is illustrated schematically in Figure 2.4, which shows a path around the field and several wind patterns which may help or hinder the walker or have no overall effect, despite helping at times and hindering at other times.

Rotation defined this way corresponds to the direction of the line integrals which are used to define *curl,* so this is the right definition to use.

Now suppose we had a broad pan full of water, and a large number of food mixers stirring the water. They will spin the water clockwise in the case shown in Figure 2.5. If we integrate the flow velocity $\mathbf{v}$ along a path around the outside of the pan (dashed line) the answer is proportional to the total number of mixers. $(\nabla \times \mathbf{v})_n$ is the integral of $\mathbf{v}$ round a small path divided by the area inside the path. This measures the number of mixers per unit area near the point where *curl* is evaluated.

2.5.1   Derivation of Curl

To find the usual expression for *curl* in a form which will make it clear how to implement it numerically, we can use our usual two-dimensional Cartesian mesh, where $x = i\Delta x$ and $y = j\Delta y$.

To obtain the component of *curl* in the z-direction, we integrate in the counterclockwise direction as it appears from above the page. The sign convention we introduced requires us to choose the component of *curl* we want, and then look along that axis. The integration must be clockwise, when we look along the axis corresponding to that component. The z-component must be found by looking along the z-axis, which is from below the paper. Our integration is counter-clockwise from above the paper, but when the integration is viewed along the z-direction the integration is clockwise.

The integral will be along four lines. We begin with line (1) from the corner at (i,j) to $(i+1,j)$, then along line (2) to $(i+1,j+1)$. Next we integrate along line (3) to $(i,j+1)$ and then along line (4) back to (i,j). The area enclosed is $\Delta s = \Delta x \Delta y$.

Along sections (1) and (3) $d\mathbf{r} = \mathbf{a}_x dx$ and along sections (2) and (4) $\mathbf{r} = \mathbf{a}_y dy$, although dx and dy may be negative.

$$I = \oint \mathbf{A} \cdot d\mathbf{r} = \int_1 A_x \, dx + \int_2 A_y \, dy + \int_3 A_x \, dx + \int_4 A_y \, dy. \qquad (2.78)$$

The integrals along sections (3) and (4) are in the negative directions, but this is taken care of by the limits. If (1) and (3) are given the same limits in x by reversing the limits in (3), we change the sign on the integral along (3) and combine (1) and (3) to get

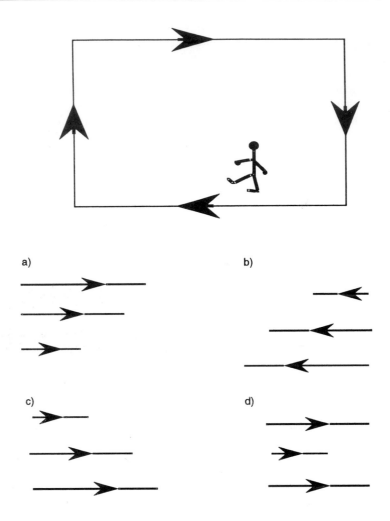

Figure 2.4. Illustration of the importance of the line integral around a closed loop, in the case of a walker going around a path which closes on itself. Cases a.)—d.) show possible wind patterns. The work the wind does on the walker is the line integral of the force it exerts, along the path taken. In this case the line integral is clockwise. The right hand rule says that the fingers of the right hand should be pointed along the line integral and the thumb will point in the direction of the rotation vector, or the axis of rotation. The rotation vector, and the *curl,* will be down into the paper in this case, if the line integral is positive.

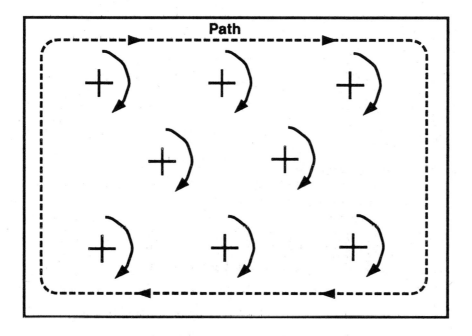

Figure 2.5. A source of rotation which gets bigger as the area inside a loop gets bigger; the bigger the area of the bath, the more food mixers can be squeezed into the area. The rotation per unit area is measured by the *curl,* which will measure the number of mixers per unit area, in this case.

$$\int_1 A_x \, dx + \int_3 A_x \, dx = \int_{x=i\Delta x}^{x=(i+1)\Delta x} dx \, [A_x(x, y = j\Delta y) - A_x(x, y = (j+1)\Delta y)].$$

(2.79)

The term in square brackets can be divided by Δy to give

$$-\frac{A_x(x, y = (j+1)\Delta y) - A_x(x, y = j\Delta y)}{\Delta y}$$

(2.80)

which is one of our numerical approximations which we can use to represent $-\frac{\partial A_x}{\partial y}$. So the integrals along (1) and (3) give

$$-\int_{x=i\Delta x}^{x=(i+1)\Delta x} \left(\frac{\partial A_x}{\partial y}\Delta y\right) dx.$$

(2.81)

If the derivative is roughly constant over this range in x, the integral just gives a factor of Δx times the integrand and so the result is

$$-\frac{\partial A_x}{\partial y}\Delta x\Delta y.$$

(2.82)

When we repeat this for the integrals along (2) and (4), we get

$$\frac{\partial A_y}{\partial x}\Delta x\Delta y.$$

(2.83)

Then the component of *curl* perpendicular to our loop is the sum of these divided by $\Delta s = \Delta x\Delta y$.

$$(\nabla \times \mathbf{A})_z = \frac{\oint \mathbf{A} \cdot d\mathbf{r}}{\Delta s} = \frac{\partial A_y}{\partial x} - \frac{\partial A_x}{\partial y}.$$

(2.84)

Numerical Implementation of *Curl*

In electrostatics the *curl* equations which are used are

$$\nabla \times \mathbf{E} = 0$$

(2.85)

and

$$\nabla \times \mathbf{H} = \mathbf{J}.$$

(2.86)

We will now do some example numerical problems using equations such as these and discuss their properties. The second equation, $\nabla \times \mathbf{H} = \mathbf{J}$, was our example which showed the physical significance of the *curl* operator. We will start with the integral version of this equation

$$\oint \mathbf{H} \cdot d\mathbf{r} = I,$$

(2.87)

Input File excurl.sg

```
// FILE:  sgdir\ch02\excurl\excurl.sg

const PI  = 3.14159265358979323846;    // pi
const DIM = 101;                       // number of mesh points
const DX  = 0.01;                      // horizontal mesh spacing
const DY  = 0.01;                      // vertical mesh spacing

var x[DIM], y[DIM], Hx[DIM,DIM], Hy[DIM,DIM], J[DIM,DIM];

equ J[i=1..DIM-2,k=1..DIM-2] ->
 J[i,k] = (Hy[i+1,k]-Hy[i-1,k])/(2.0*DX)-(Hx[i,k+1]-Hx[i,k-1])/(2.0*DY);

begin main
  assign x[i=all] = i*DX;
  assign y[k=all] = k*DY;
  assign Hx[i=all,k=all]=2.0*i*DX*cos(10*PI*k*DY);
  assign Hy[i=all,k=all]=10*PI*sq(i*DX)*sin(10*PI*k*DY);
  solve;
  write;
end
```

applied to a very small loop, and divide both sides of the equation by the area of the loop, to turn it into an equation per unit area which we can use at a point. As we said above, J is the current per unit area. $(\nabla \times \mathbf{H})_n$ is $\oint \mathbf{H} \cdot d\mathbf{r}/\Delta s$ for a small enough loop. When we evaluate $(\nabla \times \mathbf{H})_n$ in Cartesian coordinates for a loop in the (x, y) plane we find

$$(\nabla \times \mathbf{H})_n = (\nabla \times \mathbf{H})_z = \frac{\partial H_y}{\partial x} - \frac{\partial H_x}{\partial y}. \tag{2.88}$$

This is equal to J_z. It can be used at each point to find J_z from $\mathbf{H}$, whereas the integral form of the equation $\oint \mathbf{H} \cdot d\mathbf{r} = I$ cannot be used at a point.

The following example evaluates $(\nabla \times \mathbf{H})_z$ at each point of a rectangle in the (x, y) plane.

In this simulation, $\nabla \times \mathbf{H}$ is evaluated using a two-sided derivative. For example, we use values of H_y at $x = (i+1)\Delta x$ and at $x = (i-1)\Delta x$ to evaluate $\frac{\partial H_y}{\partial x}$ instead of the values at $x = (i+1)\Delta x$ and $x = i\Delta x$ which were called for in the derivation of $\nabla \times \mathbf{H}$. H_y is differentiated with respect to x, so the indices can be chosen to be $(i+1, j)$ and $(i-1, j)$ at the points where H_y is evaluated. Similarly, H_x is evaluated at $(i, j+1)$ and $(i, j-1)$.

The $\mathbf{H}$ used in the simulation has one special property; if

$$H_x = 2x \cos(31.415y) \tag{2.89}$$

and

$$H_y = -31.415x^2 \sin(31.415y) \tag{2.90}$$

then

$$\mathbf{H} = \nabla(x^2 \cos(31.415y)) \tag{2.91}$$

This example illustrates one important property of *curl*. Any vector which is the gradient of a scalar V has a *curl* of zero.

$$\nabla \times \nabla V = 0. \tag{2.92}$$

The numerical approximation to *curl* in our example introduces some error, so the answer will not be exactly zero, but it will be small.

If $\nabla \times \mathbf{H} = 0$ we know that $\mathbf{H}$ is the gradient of some scalar; all vectors that have zero *curl* can be written as gradients of scalars. This is useful to remember in electrostatics, because it means that saying $\nabla \times \mathbf{E} = 0$ is equivalent to saying $\mathbf{E} = \nabla$ (some scalar). In fact $\mathbf{E} = \nabla(-V) = -\nabla V$, where V is the electrostatic potential.

Another interesting example is obtained by using the same file, changing only H_x and H_y to

$$H_x = x \sin(62.83y) \tag{2.93}$$

$$H_y = 10y^2 \sin(62.83x). \tag{2.94}$$

These components are chosen so that the field tangent to the boundary of the region is (approximately) zero all along the boundary. Since

$$\oint \mathbf{H} \cdot d\mathbf{r} = \int_s (\nabla \times \mathbf{H})_n \, ds \tag{2.95}$$

then the average of $(\nabla \times \mathbf{H})_n$ should be zero.

2.5.2   The Importance of *Div* and *Curl*

We emphasize *div* and *curl* in part because the equations that describe the electric and magnetic fields (Maxwell's equations) have the *div* and *curl* operators in them. The divergence equations tell us how much flux of each field is produced per unit volume and what produces the flux. The *curl* equations tell us how much each field gets rotated per unit area and what rotates the field.

Later we will use the Maxwell equations to find boundary conditions on the fields. The divergence equations applied to a surface tell us how much extra flux is produced at the surface. Extra flux would mean that the component of the field normal to the surface was different on the two sides of the surface. The *curl* equations tell us how much extra rotation is produced at the surface. Extra rotation would lead to a difference in the fields tangential to the surface on either side of the surface.

2.6   Vector Operators in Curvilinear Coordinates

When a coordinate system is being used which is not simply the Cartesian set (x, y, z) the vector operators are a little more complicated. Expressions are given in this section for the vector operators, which are valid so long as the coordinates are orthogonal. These expressions can be used to derive numerical expressions for the operators in the usual way.

In orthogonal coordinates (u^1, u^2, u^3) the distance moved when the coordinates change by (du^1, du^2, du^3) is ds where

$$(ds)^2 = h_1^2(du^1)^2 + h_2^2(du^2)^2 + h_3^2(du^3)^2. \qquad (2.96)$$

The factors (h_1, h_2, h_3) are known as Lamé functions or scale factors [7]. To use the expressions below it is necessary to work out what the Lamé functions are for a given coordinate system. For example, cylindrical coordinates (r, ϕ, z), provide the simplest example we could choose. If we make a small change in the coordinate r by itself, the distance we move is $ds = dr$. If we change ϕ by itself, $ds = rd\phi$. If we change z by itself, $ds = dz$. If we change all of them at once by small amounts, then because they are orthogonal,

$$(ds)^2 = (dr)^2 + (rd\phi)^2 + (dz)^2. \qquad (2.97)$$

From this we can read off the values $h_1 = 1$, $h_2 = r$ and $h_3 = 1$, for this set of coordinates.

If vectors are written in the form

$$\mathbf{A} = \mathbf{a}_1 A_1 + \mathbf{a}_2 A_2 + \mathbf{a}_3 A_3 \qquad (2.98)$$

where the basis vectors $(\mathbf{a}_1, \mathbf{a}_2, \mathbf{a}_3)$ are unit vectors, then the expressions for *grad*, *div* and *curl* in orthogonal coordinates are:

$$\nabla \Psi = \frac{\mathbf{a}_1}{h_1} \frac{\partial \Psi}{\partial u^1} + \frac{\mathbf{a}_2}{h_2} \frac{\partial \Psi}{\partial u^2} + \frac{\mathbf{a}_3}{h_3} \frac{\partial \Psi}{\partial u^3}. \qquad (2.99)$$

$$\nabla \cdot \mathbf{C} = \frac{1}{J} \left[\frac{\partial}{\partial u^1} \left(J \frac{C_1}{h_1} \right) + \frac{\partial}{\partial u^2} \left(J \frac{C_2}{h_2} \right) + \frac{\partial}{\partial u^3} \left(J \frac{C_3}{h_3} \right) \right] \qquad (2.100)$$

where J is the 'Jacobian'. In general J is defined to obey $g = J^2$, where g is the determinant of the 'metric matrix', in any coordinate system. However, in orthogonal coordinates $J = h_1 h_2 h_3$ [7].

$$\nabla \times \mathbf{C} = \frac{1}{J} \left[h_1 \mathbf{a}_1 \left(\frac{\partial}{\partial u^2} (h_3 C_3) - \frac{\partial}{\partial u^3} (h_2 C_2) \right) + ... \right] \qquad (2.101)$$

The Laplacian $\nabla^2 \Psi$ can be derived from $\nabla^2 \Psi = \nabla \cdot \nabla \Psi$, and is given by

$$\nabla^2 \Psi = \frac{1}{J} \frac{\partial}{\partial u^1} \left[\frac{J}{h_1^2} \frac{\partial \Psi}{\partial u^1} \right] + ... \qquad (2.102)$$

2.7 Electrostatics

In electrostatic problems in two or three dimensions, the equations that describe the fields are

$$\nabla \cdot \mathbf{D} = \rho \qquad (2.103)$$

and

$$\nabla \times \mathbf{E} = 0 \qquad (2.104)$$

where $\mathbf{D} = \epsilon \mathbf{E}$. $\nabla \cdot \mathbf{D} = \rho$ means that flux of $\mathbf{D}$ is produced at a rate ρ per unit volume. $\nabla \times \mathbf{E} = 0$ means that $\mathbf{E}$ does not rotate, so $\oint \mathbf{E} \cdot d\mathbf{r} = 0$ round any loop, which also means that $\mathbf{E}$ is a conservative field and $\mathbf{E}$ can be written as $\mathbf{E} = -\nabla V$ with V being the electrostatic potential.

$\nabla \cdot \mathbf{D} = \rho$ is not enough, by itself, to determine $\mathbf{D}$ because (in two Cartesian coordinates) $\mathbf{D}$ has two components D_x and D_y, which can depend on x and y, and $\nabla \cdot \mathbf{D} = \rho$ is only one equation. $\nabla \times \mathbf{E} = 0$ provides the extra information that is needed. Either it can be used directly, or we can use it to say $\mathbf{E} = -\nabla V$, which then means

$$\nabla \cdot (-\epsilon \nabla V) = \rho \qquad (2.105)$$

and if ϵ is constant throughout the region

$$\nabla^2 V = -\frac{\rho}{\epsilon} \qquad (2.106)$$

since $\nabla \cdot (\nabla V) = \nabla^2 V$. This equation is Poisson's equation. ∇^2 is the 'Laplacian'. When there is no free charge density, $\rho = 0$ and

$$\nabla^2 V = 0 \,, \qquad (2.107)$$

which is Laplace's equation.

To understand the meaning of the parameter ϵ which appears in these equations we need to consider the polarization which is set up in a material by the electric field. This is the topic of Section 2.7.1.

2.7.1 The Electric Dipole

An electric dipole consists of two charges with equal magnitudes but opposite signs, some distance apart. The vector $\mathbf{r}$ is the position of the positive charge, measured from the negative charge. In other words $\mathbf{r}$ is equal in magnitude to the distance between the charges, and $\mathbf{r}$ points from the negative charge $-q$ to the positive charge q. The electric dipole moment $\mathbf{p}$ is defined to be

$$\mathbf{p} = q\mathbf{r}. \qquad (2.108)$$

Electric dipoles are particularly important for understanding dielectric materials. When an electric field is applied to many materials the molecules become polarized. This means that positive and negative charges are separated to some extent, as the charged particles in the molecule move away from their equilibrium positions. The charges separate because the electric field pushes positive charges in the direction of the electric field (since the electric field is defined to be the force on a unit positive charge). It pushes in the opposite direction on the negative charges. Molecules which were originally neutral everywhere are made to have 'electric dipole moments'.

If all the molecules in a slab of material have the same electric dipole moment, the overall effect is to set up a charge along the surface of the material. The inside stays neutral overall. If the electrons from one molecule move to the right (say), then electrons from all the molecules move to the right. So although a molecule may 'lose' some electrons which move towards its neighbor on its right, they are 'replaced' by electrons from its neighbor on its left. But if it is on a surface and it has no neighbor on the left to give it electrons, or no neighbor on its right to give them to, net charges are set up.

In this example, the electric field must have been pointing to the left, because it moved electrons to the right. By moving electrons to the right it set up a negative charge on the right of the dielectric and a positive charge on the left side of the dielectric. These charges partially shield the inside of the dielectric from the electric field outside. To see this, we need to know the direction of the electric field set up by this 'polarization charge' on the surface. The electric field is defined to point in the direction of the force on a positive charge. Inside the dielectric the extra charges on the surface will try to pull the positive charge to the right. The charges on the surface have created their own electric field, which is to the right. The original field was to the left, so the extra field opposes the original field. The net effect of the polarization is to decrease the field in the dielectric.

The decrease in the field is described by the relative dielectric constant ϵ_r, which appears when we write $\mathbf{D} = \epsilon\mathbf{E}$ and $\epsilon = \epsilon_r\epsilon_0$ and substitute into $\nabla \cdot \mathbf{D} = \rho$.

$$\nabla \cdot (\epsilon_r\epsilon_0 \mathbf{E}) = \rho \qquad (2.109)$$

so if ϵ_r is constant

$$\nabla \cdot \mathbf{E} = \frac{\rho}{\epsilon_r\epsilon_0}. \qquad (2.110)$$

$\mathbf{E}$ is smaller for the same ρ if $\epsilon_r > 1$ than if $\epsilon_r = 1$, which it is in a vacuum.

The dipole moment per unit volume is called the polarization $\mathbf{P}$. If we take a small volume Δv and sum the electric dipole moments inside Δv, then $\mathbf{P}$ is the sum of the dipole moments divided by Δv.

$$\mathbf{P} = \frac{\sum_{\Delta v} \mathbf{P}}{\Delta v} \qquad (2.111)$$

The electric displacement $\mathbf{D}$ is most correctly defined as

$$\mathbf{D} = \epsilon_0 \mathbf{E} + \mathbf{P}. \tag{2.112}$$

Now the polarization is usually proportional to $\mathbf{E}$ and the constant of proportionality is written so that

$$\mathbf{P} = \chi \epsilon_0 \mathbf{E} \tag{2.113}$$

which means that

$$\mathbf{D} = (1 + \chi)\epsilon_0 \mathbf{E} \tag{2.114}$$

so if we say $\epsilon_r = (1 + \chi)$ we get the usual definition of $\mathbf{D}$.

If we use $\mathbf{D} = \epsilon_0 \mathbf{E} + \mathbf{P}$ in $\nabla \cdot \mathbf{D} = \rho$ we find

$$\nabla \cdot (\epsilon_0 \mathbf{E}) = \rho - \nabla \cdot \mathbf{P} \tag{2.115}$$

so $-\nabla \cdot \mathbf{P}$ is a charge density created by polarization.

To find the electric field due to an electric dipole we can use two cells of our usual 2D mesh to contain the charges and impose equal and opposite potentials $\pm V_d$ in the cells. A long way from a dipole the electric field falls away rapidly with distance. This is because the distance between the charges is relatively small and so the dipole looks like a net charge of zero to a distant observer. For this reason we can set $V = 0$ on the outside boundary if it is far enough away from the dipole.

Inside a dielectric we can find the polarization from the electric field.

$$\mathbf{P} = \chi \epsilon_0 \mathbf{E} = (\epsilon_r - 1)\epsilon_0 \mathbf{E}. \tag{2.116}$$

In an example later in this chapter we find the electric field inside and outside a dielectric slab by finding the electrostatic potential V. The field $\mathbf{E}$ is given by $\mathbf{E} = -\nabla V$ everywhere in this example. In the vacuum outside the dielectric $\epsilon_r = 1$, so $\mathbf{P} = (\epsilon_r - 1)\epsilon_0 \mathbf{E}$ gives $\mathbf{P} = 0$, as it should. Inside the dielectric $\mathbf{P}$ is nonzero, and simply proportional to $\mathbf{E}$ by assumption.

2.7.2   Solution of Laplace's Equation

The simplest version of Poisson's equation is with $\rho = 0$, so that

$$\nabla^2 V = 0 \tag{2.117}$$

(called Laplace's equation). We use V as the potential, despite possible confusion with the volume inside a surface, because SGFramework does not support Greek letters. The solution of this equation in a rectangular box, with three of the sides held at $V = 0$ and the fourth side at a different voltage, is given in laplace.sg. See Figure 2.6. The electric-field components (E_x, E_y) can also be calculated. Solving this problem by solving

$$\nabla \cdot \mathbf{E} = 0 \tag{2.118}$$

Input File laplace.sg

```
// FILE:  sgdir\ch02\laplace\laplace.sg

const NX = 21;                    // number of x-mesh points
const NY = 21;                    // number of y-mesh points
const LX = NX - 1;                // index of last x-mesh point
const LY = NY - 1;                // index of last y-mesh point
const DX = 1.0;                   // mesh spacing

var V[NX,NY];

// Laplace's equation
equ V[i=1..LX-1,j=1..LY-1]  -> {V[i-1,j] + V[i,j-1] - 4*V[i,j] +
                                V[i+1,j] + V[i,j+1]} / sq(DX) = 0.0;

begin main
  assign V[i=all,j=all] = 0.0;
  assign V[i=0,  j=all] = 10.0;
  solve;
end
```

and

$$\nabla \times \mathbf{E} = 0 \qquad\qquad (2.119)$$

would not be convenient. This is because the boundary condition we need to apply is a condition on V, not $\mathbf{E}$.

2.8   Solution of PDEs on Irregular Meshes

SGFramework can solve PDEs on irregular meshes designed to handle physically complex geometries. Irregular means the mesh has no regular pattern, so it is not possible to lay the mesh out in rows and columns in which neighboring nodes may be given sequential indices. For such a mesh a 2D array is not appropriate and the mesh points are stored in a 1D array.

As an example problem we shall continue with Laplace's equation, but this time making use of a more general way of handling the mesh.

Irregular mesh geometries can be set up using a mesh specification file (or a mesh 'skeleton file'; hence the .sk suffix on the file name.) Using appropriate SGFramework executables the mesh is generated and refined. The other type of specification file we use is an equation specification file. If there is no need for a mesh specification file we omit the word 'equation' when we refer to the equation specification file. The equation specification file reads the mesh information provided by the skeleton

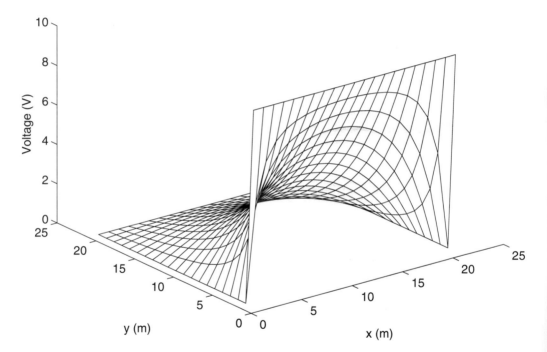

Figure 2.6. A classic example of the solution of Laplace's equation, in a rectangular box with one side held at 10 volts, all the other sides being held at zero. The input file which generated this is given in the text. Plots of results can be generated by SGFramework, provided a skeleton file is used to define the mesh.

Skeleton File sq.sk

```
// FILE:  sgdir\ch02\lapsk\sq.sk
// Mesh file for simple square mesh

const dx =  1.0;
const W  = 20.0;

point pA = (0.0,  W);
point pB = (W  ,  W);
point pC = (W  ,0.0);
point pD = (0.0,0.0);

edge eAB = METAL1 [pA,pB] (dx, 0.0);
edge eBC = METAL2 [pB,pC] (dx, 0.0);
edge eDC = METAL2 [pD,pC] (dx, 0.0);
edge eAD = METAL2 [pA,pD] (dx, 0.0);

region rABCD = AIR {eAD,eDC,eBC,eAB} RECTANGLES;

coordinates x, y;
```

file.

One of the advantages of using a mesh skeleton file is that SGFramework provides its own (rather limited) plotting program, triplot, which can be used if it is provided with a mesh file. In the example that follows, the mesh specification file is called 'sq.sk', meaning 'square skeleton'. SGFramework sets up the mesh, refines it (which is not really necessary in the examples given here - see the User's Manual and later chapters) and generates the mesh file 'sq.msh' for later use. This file is mostly self-explanatory. The points pA through pD are defined by their coordinates. The edges are defined by the points. The order of the edges (running counterclockwise) is important when defining regions, and if the edges are to include a region which is divided into rectangles, then opposite edges must be parallel (and point in the same direction) and have the same spacing. The refinement function is not used here. (The refinement function name is arbitrary. C is used in the semiconductor simulations since the refinement is based on the doping concentration.) See the User's Manual, Chapter 15, and Chapter 7, for more details.

The input file which uses this is lapsk.sg (which is short for 'Laplace solution using skeleton file'): This reads the mesh file 'sq.msh'. From sq.msh it obtains the coordinates of all the nodes on the mesh, (and where applicable the refinement function which was used to refine the mesh). In sq.sk the coordinates were identified as (x, y). The arrays $x[j]$ and $y[j]$ are 'automatically' passed to lapsk2.sg when it reads sq.msh. (This is important, because otherwise there would be no way of

Input File lapsk.sg

```
// FILE:  sgdir\ch02\lapsk\lapsk.sg

mesh "sq.msh";

var V[NODES];
unknown V[all];
known V[METAL1], V[METAL2];

equ V[i=AIR] -> nsum(i,j,all,{V[i] - V[node(i,j)]}) = 0.0;

begin main
  assign V[i=METAL1] = 10.0;
  assign V[i=METAL2] =  0.0;
  solve;
  write;
end
```

finding the positions of the nodes.) See Sections 15.2.3, and 15.3.1 for details.

In the equation statements, the equation for $V[i = METAL1]$ comes first, so it overrides later equations in cases where the regions overlap. This is important, because the edges are actually part of the region AIR, but the equation used for $i = AIR$ does not apply to the edges.

A function 'nsum' is used which performs a sum over neighbors (hence the name). The mesh file provides information on which nodes are neighbors of any node i. This is essential on an irregular mesh. The index j indicates the jth neighbor, so j runs from 0 to 3 on our regular mesh. The index i runs from 0 to one less than the total number of nodes, and the jth neighbor of node i is itself a node with an index k between 0 and one less than the number of nodes. That index is obtained using the function $NODE(i, j)$.

The neighbor summation function in this file evaluates and sums the expression $V[i] - V[NODE(i, j)]$ for the node i and all of its neighbors j (as specified by the statement $nsum(i, j, all, ...)$. $(V[i] - V[k])/\ell$, where ℓ is the distance between i and k, is equal to the component of the electric field $\mathbf{E}$ pointing from i to k, because $\mathbf{E} = -\nabla V$. From $\mathbf{E}$ we find the divergence of $\mathbf{E}$, as described below. There is no need to divide by the distance between neighbors, in this file, because all the distances are equal and the right hand side of the equation is zero. Dividing by an appropriate factor shows that this equation is equivalent to Laplace's equation, on this regular mesh.

These files are run using the script given in the table. Build scripts will generally be called RUN.bat in Win32 and RUN in UNIX.

The first line defines the type of shell script used. This line is not necessary if

Script File for PDE on Irregular Mesh

```
#!/bin/sh
mesh sq.sk
sggrid -qn10 -vr sq.xsk
sgbuild ref sq_ref
sq_ref -dadj1 -qn10 -vr
sgxlat -p16 lapsk.sg
sgbuild sim lapsk
order lapsk.top lapsk.prm
lapsk -vs2 -vl2 -Llapsk.log
extract lapsk.res
```

Note that each time a window is opened to show the mesh, that window must be closed before the run will proceed.

the other commands are typed in by hand. The next two lines handle generating the initial mesh. The next two lines refine the mesh. (Twice during the mesh generation, a window is opened to show the current mesh. These windows must be closed before the next command can be executed.) The rest of the commands are similar to those used previously.

The results can be plotted using the SGFramework plotter triplot, with a command such as

```
triplot lapsk V -ps
```

The '-ps' instructs the triplot program to make a postcript file as well as plotting the array V to the screen. If it is left off, the program only plots to the screen. The postscript file is shown in Figure 2.7.

When the word RECTANGLES is left out of the skeleton file, a triangular mesh is used. However, the equation implied by the nsum is no longer Laplace's equation, once the lengths between nodes become variable; see Figure 2.8.

2.8.1   The Functions *elen* and *ilen*

Once the distances between cells are not equal, we need to make use of two new functions, $elen(i,j)$ and $ilen(i,j)$ (and often a function $area(i)$). $elen(i,j)$ is the distance between node i and its jth neighbor. $ilen(i,j)$ is the length of the perpendicular bisector of the line from node i to its jth neighbor. These perpendicular bisectors end where they meet each other - see Figure 5.1.

The difference in the potential V between the two points, divided by the distance between the points, is the component of ∇V along the line between the points - in this case, the line pointing from i to its jth neighbor. In what follows we need $-\nabla V$, so we use $-\frac{\partial V}{\partial \ell} \simeq (V[i] - V[node(i,j)]) / elen(i,j)$.

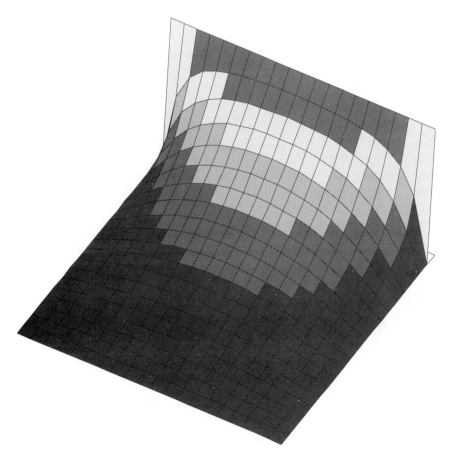

Figure 2.7. Laplace equation solution produced using a skeleton file.

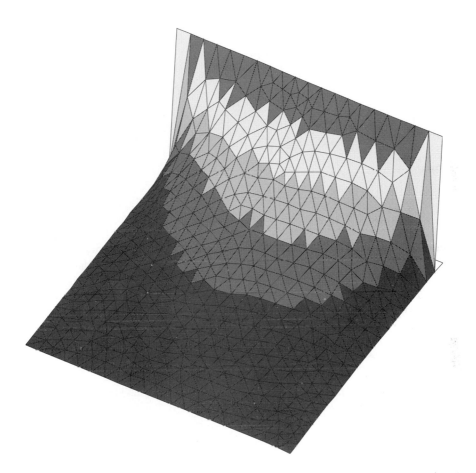

Figure 2.8. Solution of an equation (which is similar to Laplace's equation) using a triangular mesh - by leaving out the word RECTANGLES in the skeleton file.

Input File lapsk2.sg

```
// FILE:  sgdir\ch02\lapsk2\lapsk2.sg

mesh "sq.msh";

var V[NODES];
unknown V[all];
known V[METAL1], V[METAL2];

equ V[i=AIR] -> nsum(i,j,all,{(V[i]-V[node(i,j)])/elen(i,j)}*ilen(i,j)) = 0.0;

begin main
 assign V[i=METAL1] = 10.0;
 assign V[i=METAL2] =  0.0;
 solve;
 write;
end
```

Now the electric field is $\mathbf{E} = -\nabla V$ and the flux of $\mathbf{E}$ is $-\frac{\partial V}{\partial \ell}$ times the length of the line the flux passes through, which is $ilen(i,j)$. (This assumes a 2D problem, where the length $ilen(i,j)$ in the plane we are working with, 'the plane of the paper', has a unit length out of the plane associated with it. This unit length converts $ilen(i,j)$ into an area, equal in magnitude to $ilen(i,j)$.) Summing this over all the neighbors, using $nsum$, the total flux leaving node i is obtained. In the file which follows the flux is set to zero.

This is implemented in the file lapsk2.sg, which uses the same mesh file as before.

The output from this file, obtained using the skeleton file sq.sk with the word RECTANGLES left out, is shown in Figure 2.9. If RECTANGLES were left in, the figure would be the same as Figure 2.7.

A more ambitious skeleton file 'tee.sk' is given next. It describes a slightly more complicated shape, and the boundary is divided into three metal regions. (This shape is intended to be useful for waveguide simulations.)

The input file 'lapsk2.sg' must be modified slightly to deal with this mesh specification file, since there are now three portions to the boundary, instead of two. This is done in 'laptee.sg'.

The Cell Area

The equation which was implemented in the last input file relied on finding the flux leaving the cell around the point i, and setting that flux to zero. Often an equation is written in terms of the divergence of the flux. We can calculate the divergence by computing the flux leaving the cell and dividing by the volume of the cell. (This

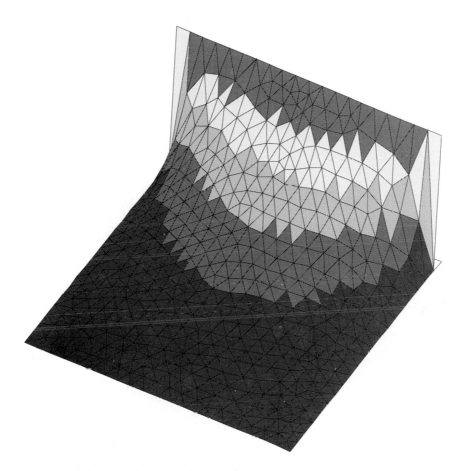

Figure 2.9. Solution of Laplace's equation on a triangular mesh, obtained using the functions $elen(i, j)$ and $ilen(i, j)$.

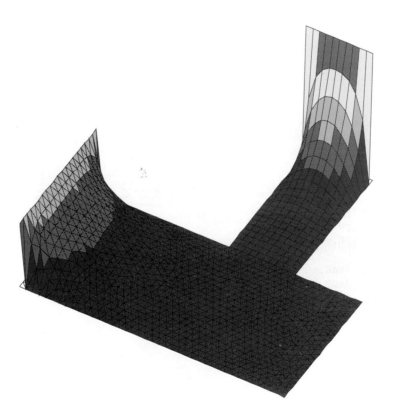

Figure 2.10. Solution of Laplace's equation in a more complex geometry, using the skeleton file 'tee.sk'.

Skeleton File tee.sk

```
// FILE:  sgdir\ch02\laptee\tee.sk
// Mesh file for tee-shaped mesh

const dx = 1.0e-2  ;
const W1 = 20.0*dx ;
const W2 = 30.0*dx ;
const W3 = 40.0*dx ;
const t1 = 20.0*dx ;
const t2 = 40.0*dx ;

// Layout of points in this file illustrates the actual geometry
point              pG=(W1,t2),  pH=(W2,t2) ;
point pA=(0.0,t1),  pB=(W1,t1),  pC=(W2,t1),  pD=(W3,t1);
point pE=(0.0,0.0),                           pF=(W3,0.0);

edge eAB = METAL3 [pA,pB] (dx, 0.0);
edge eBC =        [pB,pC] (dx, 0.0);
edge eCD = METAL3 [pC,pD] (dx, 0.0);
edge eEF = METAL3 [pE,pF] (dx, 0.0);
edge eGH = METAL1 [pG,pH] (dx, 0.0);
edge eAE = METAL2 [pA,pE] (dx, 0.0);
edge eDF = METAL3 [pD,pF] (dx, 0.0);
edge eGB = METAL3 [pG,pB] (dx, 0.0);
edge eHC = METAL3 [pH,pC] (dx, 0.0);

region rADEF = AIR {eAE,eEF,eDF,eCD,eBC,eAB} ;
region rGHBC = AIR {eGB,eBC,eHC,eGH} RECTANGLES;

coordinates x, y;
```

is approximate because the divergence of the flux is defined so that if we find the flux leaving a small volume, and we divide that flux by the volume, then the result is the divergence, provided we use an infinitesimal volume.) To find the divergence we need the area of the cell surrounding node i. Again, because we assume each cell has a unit height 'out of the plane', the area in the plane, which we call $area(i)$, is also equal to the volume of the cell. $area(i)$ is the area enclosed by the same lines (the perpendicular bisectors of the lines from node i to its neighbors j) of length $ilen(i,j)$, through which the flux leaves the cell.

Once the flux we found before is divided by $area(i)$, we have the divergence of the flux. It is important to notice that the flux we are finding in files such as lapsk2.sg and laptee.sg is the flux of $-\nabla V$, with emphasis on the minus sign.

The file wavtee.sg does this in a problem having time dependence. Strictly this

Input File laptee.sg

```
// FILE:  sgdir\ch02\laptee\laptee.sg

mesh "tee.msh";

var V[NODES];
unknown V[all];
known V[METAL1], V[METAL2], V[METAL3];


equ V[i=AIR] ->
  nsum(i,j,all,{(V[i]-V[node(i,j)])/elen(i,j)}*ilen(i,j)) = 0.0;

begin main
  assign V[i=METAL1] = 10.0 ;
  assign V[i=METAL2] =  5.0 ;
  assign V[i=METAL3] =  0.0 ;
  solve;
  write;
end
```

does not belong in a section on electromagnetism, since the file is set up to treat
a diffusion problem, with a diffusion coefficient of 1 (see the next chapter), as well
as a wave propagation problem, with a wave speed of 1. However, the solution of
Laplace's equation is often obtained by running a diffusion equation to steady state,
and this illustrates the power of the approach we are using to handle a wide variety
of problems.   The diffusion equation being solved is

$$\frac{\partial V}{\partial t} = -\nabla \cdot (-\nabla V) = \nabla^2 V \, . \tag{2.120}$$

We can do a simple check on the signs of the terms. If the right hand side is
negative then V will decrease. For the right hand side to be negative, the curvature
must be negative. For the initial condition of a spike in V, the curvature is certainly
negative. The peak value of V does decrease in reality, so the signs in the equation
are correct. If we had put the opposite sign, an initial spike would get bigger.

The wave equation which is solved is:

$$\frac{\partial^2 V}{\partial t^2} = -\nabla \cdot (-\nabla V) = \nabla^2 V \tag{2.121}$$

In the file wavtee.sg, the right hand side of the equation being solved is $-\nabla \cdot \mathbf{E}$.
The function nsum is used to find the flux of $\mathbf{E}$ leaving the cell. It is then divided
by the cell area to obtain the divergence. Since $\mathbf{E} = -\nabla V$, this is equal to $\nabla^2 V$, as

required. To check the signs used in the file, suppose that the variable V is a density which obeys a diffusion equation. The flux out of the cell would be proportional to $-\nabla V$. If $-\nabla V$ points out of the cell, then V should go down. In the file wavtee.sg, the right hand side evaluates V at the central node and subtracts V at the neighbor, to find the flux. This is positive if V is bigger at the center. It sums these fluxes and divides by the area. This yields the divergence of the flux. A negative sign is put in front of the divergence, so that a positive divergence leads to a decreasing density (as it should.)

The time step used in the diffusion problem is orders of magnitude too large for accuracy (see the next chapter) but the solution converges to the correct steady state. The steady-state solution is also a solution of Laplace's equation.

The time step in the wave problem is accurate enough to resolve the wave. Again, the proper choice of step is discussed in the next section.

(The variable $V1$ is the 'old' value of V, from one time step before. $V2$ is from two times steps before, and is only needed when we modify the file to solve the wave equation.)

Input File wavtee.sg

```
// File: sgdir\ch02\wavtee\wavtee.sg
//
// Diffusion (A) and Wave (B) equation solution using skeleton file
// Comment/uncomment the appropriate statements to solve either the
// diffusion or the wave equation.

mesh "tee.msh";

const om = 0.3/dx;
//const tmax = 0.05; // A
//const dt   = 0.01; // A
const tmax = 12.66/om; // B
const dt   = 0.5*dx; // B

var V[NODES], V1[NODES], V2[NODES], t, Vapp1, Vapp2;
unknown V[all];
known V[METAL1], V[METAL2], V[METAL3];

set NEWTON ACCURACY = 1.0e-12;

equ V[i=AIR] ->
// (V[i] - V1[i])/dt =                    // A
   (V[i] - 2.0*V1[i] + V2[i])/sq(dt) =     // B
  - nsum(i,j,all,{(V[i]-V[node(i,j)])/elen(i,j)}*ilen(i,j))/area(i);

begin main
 assign V[i=all]     = Vapp2;
 assign V1[i=all]    = Vapp2;
```

```
 while (t < tmax) begin
    assign t = t + dt ;
    assign V2[i=all] = V1[i] ;
    assign V1[i=all] = V[i]  ;
// assign Vapp1 = 0.0 ; // A
// assign Vapp2 = 1.0 ; // A
    assign Vapp1 = sin(om*t) ; // B
    assign Vapp2 = 0.0 ;        // B
    assign V[i=METAL1]  = Vapp1;
    assign V[i=METAL2]  = Vapp1;
    assign V[i=METAL3]  = 0.5*Vapp2;
    solve;
    write;
  end
end
```

The output after four time steps of the diffusion problem is shown in Figure 2.11
— with lines denoted A implemented. Figure 2.12 shows the output from time step
80 of the wave equation solution, with lines denoted B employed.

2.8.2   3D Meshes

Setting up a nonuniform 3D mesh is a topic beyond the scope of this book. Simple
3D meshes are straightforward, since a mesh which is basically cubic can be set
up using three indices in all the arrays instead of two as was done in the earlier
2D examples. It should be clear, for instance, how to extend the examples where
grad, div and *curl* are evaluated from 2D to 3D, as long as the three coordinates are
orthogonal. In this section we will show how to deal with a case where we have a
complex mesh in two dimensions but a third coordinate which is orthogonal to the
other two.

The input file 3d.sg which follows solves Laplace's equation in 3D, using the
skeleton file 3D.sk (which is almost the same as tee.sk) to define the mesh in the
(x,y) plane and a simple mesh in the z-direction. The equation being solved is not
written as the divergence of the flux of the electric field, but as the total flux of
the field leaving a 3D cell of the mesh. For this reason, the flux per unit area is
multiplied by the area of the cell face which the flux is passing through. The nsum
function handles the flux within the (x,y) plane, and is not used to handle the z
flux. The net flux in the z-direction should be recognisable as something which
reduces to the usual finite-difference second derivative in z. Finally, if there were a
source term on the right hand side, it would consist of the source per unit volume
multiplied by the cell volume.

It is also possible to use one equation for part of the range in z and another
elsewhere. This means that we could have used Laplace's equation up to z_1 (say)
then added a source term from z_1 to z_2 and reverted back to Laplace's equation
for the rest of the range of z. Similarly we could have divided the 2D plane into

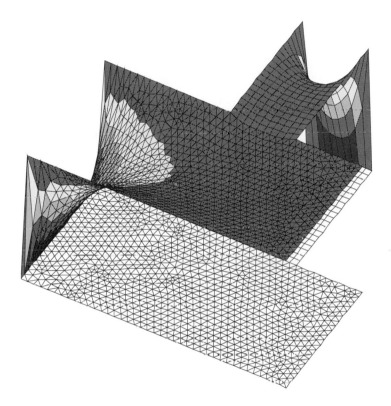

Figure 2.11. The solution of a diffusion equation in the same region as before, after a small number of time steps.

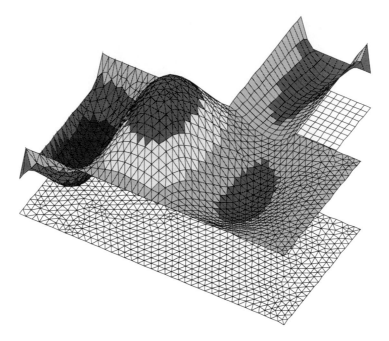

Figure 2.12. Output after 80 time steps of the wave equation solution.

Input File 3d.sg

```
// FILE:  sgdir\ch02\3d\3d.sg

mesh "3d.msh";

const ZDIM = 7 ;
const DZ   = 1.0e-2;

var V[NODES,ZDIM];
unknown V[all,1..ZDIM-2];
known V[METAL1,all], V[METAL2,all], V[METAL3,all];

equ V[i=AIR,k=1..ZDIM-2] ->
 nsum(i,j,all,{(V[i,k]-V[node(i,j),k])/elen(i,j)} * {ilen(i,j)*DZ}) +
            {(V[i,k]-V[i,k-1])/DZ} * {area(i)} +
            {(V[i,k]-V[i,k+1])/DZ} * {area(i)} = 0.0;

begin main
 assign V[i=METAL1,k=1..ZDIM-2] = 10.0;
 assign V[i=METAL2,k=1..ZDIM-2] =  5.0;
 assign V[i=METAL3,k=1..ZDIM-2] =  0.0;
 assign V[i=ALL,k=0]            =  0.0 ;
 assign V[i=ALL,k=ZDIM-1]       =  0.0 ;
 solve;
 write;
end
```

regions, and the z axis into ranges and used different equations in each combination we made.

In this example the boundary conditions are all put on using 'assign' statements. As is discussed in a later chapter, there are circumstances where the boundaries of the mesh defined by the skeleton file are treated by default as 'no-flux' boundaries even if no explicit boundary condition is imposed. If a boundary condition is imposed it supersedes this default. The boundaries in the third coordinate, which is not part of the mesh which is defined by the mesh file, must have boundary conditions put on them explicitly.

This input file uses a slightly modified version of the skeleton file tee.sk, called 3d.sk: Unfortunately, plotting more than a cross section of the results of this run is also beyond the scope of this book.

Skeleton File 3d.sk

```
// FILE:  sgdir\ch02\3d\3d.sk
// Mesh file for tee shaped mesh

const dx = 1.0e-2  ;
const W1 = 10.0*dx ;
const W2 = 15.0*dx ;
const W3 = 20.0*dx ;
const t1 = 10.0*dx ;
const t2 = 20.0*dx ;

// Layout of points in this file illustrates the actual geometry
point                 pG=(W1,t2),  pH=(W2,t2) ;
point pA=(0.0,t1),  pB=(W1,t1),  pC=(W2,t1),  pD=(W3,t1);
point pE=(0.0,0.0),                            pF=(W3,0.0);

edge eAB = METAL3 [pA,pB] (dx, 0.0);
edge eBC =        [pB,pC] (dx, 0.0);
edge eCD = METAL3 [pC,pD] (dx, 0.0);
edge eEF = METAL3 [pE,pF] (dx, 0.0);
edge eGH = METAL1 [pG,pH] (dx, 0.0);
edge eAE = METAL2 [pA,pE] (dx, 0.0);
edge eDF = METAL3 [pD,pF] (dx, 0.0);
edge eGB = METAL3 [pG,pB] (dx, 0.0);
edge eHC = METAL3 [pH,pC] (dx, 0.0);

region rADEF = AIR {eAE,eEF,eDF,eCD,eBC,eAB} ;
region rGHBC = AIR {eGB,eBC,eHC,eGH} RECTANGLES;

coordinates x, y;
refine C = 1.0 ;
```

2.9   Electromagnetism

In this section we introduce Maxwell's equations and briefly discuss their meaning and properties. In Section 2.10 we derive the boundary conditions on the fields. The main purpose of this section is to describe the properties of the equations. In Section 2.10, we obtain the boundary conditions in a form suitable to show how they should be numerically implemented. Once this is done, we will have all the information we need to solve problems in electromagnetism.

2.9.1   Maxwell's Equations

The basic equations of electromagnetism are Maxwell's equations (or the Maxwell equations)

$$\nabla \times \mathbf{E} = -\frac{\partial \mathbf{B}}{\partial t} \qquad (2.122)$$

$$\nabla \times \mathbf{H} = \mathbf{J} + \frac{\partial \mathbf{D}}{\partial t} \qquad (2.123)$$

$$\nabla \cdot \mathbf{D} = \rho \qquad (2.124)$$

and

$$\nabla \cdot \mathbf{B} = 0. \qquad (2.125)$$

$\mathbf{J}$ is the current density per unit area and ρ is the free charge density per unit volume. $\mathbf{J}$ and ρ are the sources in these equations, which create the fields.

The fields also obey constitutive relations such as

$$\mathbf{D} = \epsilon \mathbf{E} \qquad (2.126)$$

and

$$\mathbf{B} = \mu \mathbf{H}. \qquad (2.127)$$

ϵ is the permittivity, ϵ_0 is the permittivity of vacuum and ϵ_r is the relative permittivity. Similarly, $\mu = \mu_r \mu_o$ is the permeability.

The fields are defined by the force they exert on a charged particle. The Lorentz force $\mathbf{F}$ on a charge q moving with velocity $\mathbf{v}$ is given by

$$\mathbf{F} = q(\mathbf{E} + \mathbf{v} \times \mathbf{B}). \qquad (2.128)$$

From this it can be seen that the electric field is the force on a (stationary) unit charge, or the force per unit charge. The magnetic field can be defined similarly in terms of the force it exerts on a moving charge.

2.9.2   Integral Form of Maxwell's Equations

We previously defined *div* and *curl*. The definitions led almost directly to integral relations 2.129 and 2.130. Integrating the divergence of a vector $\mathbf{A}$ yields:

$$\int_v \nabla \cdot \mathbf{A} \, dv = \oint_s \mathbf{A} \cdot d\mathbf{s} \qquad (2.129)$$

where the vector $d\mathbf{s}$ is formed from the outward unit normal $\mathbf{n}$ to the surface and the element of area ds; $d\mathbf{s} = \mathbf{n} ds$. The surface s encloses the entire volume v, in this case, like the skin of a potato enclosing the inside of the potato. Integrating the *curl* of a vector $\mathbf{A}$ yields:

$$\int_s (\nabla \times \mathbf{A}) \cdot d\mathbf{s} = \int_\ell \mathbf{A} \cdot d\mathbf{r} \tag{2.130}$$

for any vector $\mathbf{A}$. In this case the loop ℓ being integrated over can be thought of as being like a wire which has been twisted so that its ends touch. The surface being integrated over is like a soap bubble suspended across the loop of wire.

We can use Equations 2.129 and 2.130 to rewrite Maxwell's equations in integral form.

In Equation 2.129, the divergence is integrated over a volume. The surface integral in that same equation is over the entire surface surrounding the volume. The surface integral of $\mathbf{A}$ is often called the flux of $\mathbf{A}$ through that surface. If we apply this to $\nabla \cdot \mathbf{B} = 0$ we get

$$\oint_s \mathbf{B} \cdot d\mathbf{s} = 0 \tag{2.131}$$

so the flux of $\mathbf{B}$ out of a closed surface is zero. From $\nabla \cdot \mathbf{D} = \rho$ we get

$$\oint_s \mathbf{D} \cdot d\mathbf{s} = \int \rho \, dv = Q; \tag{2.132}$$

the flux of $\mathbf{D}$ out of a closed surface is equal to Q, the total free charge in the volume.

In the *curl* relation, the surface is not closed. The path of integration ℓ is all the way along the edge of the surface.

Using the *curl* relation and the equation for $\nabla \times \mathbf{E}$, it can be shown that

$$\int_s \nabla \times \mathbf{E} \cdot d\mathbf{s} = \oint_\ell \mathbf{E} \cdot d\mathbf{r} = -\int_s \frac{\partial \mathbf{B}}{\partial t} \cdot d\mathbf{s}. \tag{2.133}$$

It can be shown from this that the electromotive force, defined to be $\Sigma = \int_s \mathbf{E} \cdot d\mathbf{r}$ is

$$\Sigma = -\frac{d\Phi}{dt} \tag{2.134}$$

although it is necessary to use the Lorentz force equation to find the force on a moving conductor, when the time derivative is taken outside the integration. Φ in this equation is the flux of $\mathbf{B}$ through the surface s.

Finally, the last Maxwell equation in the static case with $\frac{\partial \mathbf{D}}{\partial t} = 0$ is $\nabla \times \mathbf{H} = \mathbf{J}$. Using this in the *curl* relation gives

$$\int_s \nabla \times \mathbf{H} \cdot d\mathbf{s} = \oint_\ell \mathbf{H} \cdot d\mathbf{r} = \int_s \mathbf{J} \cdot d\mathbf{s}. \tag{2.135}$$

The integral $\oint_\ell \mathbf{H} \cdot d\mathbf{r}$ is the line integral of $\mathbf{H}$ around a closed loop. The integral $\int_s \mathbf{J} \cdot d\mathbf{s}$ is the integral of the current density over the area inside the loop, which is equal to the current through the loop.

$$\oint_{\ell} \mathbf{H} \cdot d\mathbf{r} = I \tag{2.136}$$

which is Ampere's law.

Lines of the vectors $\mathbf{E}$, $\mathbf{D}$, $\mathbf{B}$ and $\mathbf{H}$ are often drawn, with the tangent to the line being in the direction of the vector at each point along the line. Moving in the direction the vector points will lead to tracing out a 'line' of the vector.

Typically the number of lines drawn is proportional to the flux of the vector. For instance, in the case of the field $\mathbf{D}$ we know that the flux of $\mathbf{D}$ leaving a volume v is given by

$$\oint_{v} \mathbf{D} \cdot d\mathbf{s} = Q \tag{2.137}$$

where Q is the charge in the volume. This means that Q creates the flux of $\mathbf{D}$. Field lines of $\mathbf{D}$ start on positive charges and end on negative charges. This is reasonable because $\mathbf{E}$ is usually proportional to $\mathbf{D}$, and $\mathbf{E}$ is the force on a unit positive charge. That force will tend to point away from positive charges and toward negative charges.

Lines of the magnetic field $\mathbf{B}$ never start or stop, because the flux of $\mathbf{B}$ leaving a volume V is

$$\oint_{v} \mathbf{B} \cdot d\mathbf{s} = 0. \tag{2.138}$$

No flux goes in or out of any volume, because the field lines that enter the volume all leave again.

2.9.3   Properties of Maxwell's Equations

The Maxwell equations are not all independent. This is frequently significant when they are implemented numerically, since it may save effort or it may make certain expressions we try to use redundant. For example, if we take the divergence of the *curl* equations and use the fact that $\nabla \cdot (\nabla \times \mathbf{C}) = 0$ for any vector $\mathbf{C}$, we get from Faraday's law, Equation 2.122,

$$\nabla \cdot \left(\frac{\partial \mathbf{B}}{\partial t} \right) = \frac{\partial}{\partial t} (\nabla \cdot \mathbf{B}) = 0. \tag{2.139}$$

If $\nabla \cdot \mathbf{B} = 0$ initially, this equation states that $\nabla \cdot \mathbf{B}$ will stay zero, so $\nabla \cdot \mathbf{B} = 0$ is partly contained in one of the other equations.

Similarly, from the divergence of the other *curl* Equation 2.123, which expresses Ampere's law,

$$\nabla \cdot \mathbf{J} + \nabla \cdot \left(\frac{\partial \mathbf{D}}{\partial t} \right) = \nabla \cdot \mathbf{J} + \frac{\partial}{\partial t} \nabla \cdot \mathbf{D} = 0. \tag{2.140}$$

Conservation of charge says that

$$\nabla \cdot \mathbf{J} + \frac{\partial \rho}{\partial t} = 0 \tag{2.141}$$

so we would deduce that $\nabla \cdot \mathbf{D} = \rho$, which is one of the Maxwell equations, from these.

It is worth explaining the conservation equation. If we take a small volume ΔV, then from the definition of div, $\Delta V \, (\nabla \cdot \mathbf{J})$ is the flux of $\mathbf{J}$, that is the current leaving ΔV. The current leaving ΔV is the rate at which the charge Q in ΔV decreases. Since $Q = \rho \Delta V$, and the rate Q decreases is $-\frac{dQ}{dt}$ (minus because this is the rate of decrease)

$$\Delta V \, (\nabla \cdot \mathbf{J}) = -\frac{\partial}{\partial t}(\rho \Delta V) \tag{2.142}$$

and as ΔV is constant we obtain the equation of conservation of charge in the above form:

$$\nabla \cdot \mathbf{J} = -\frac{\partial \rho}{\partial t}. \tag{2.143}$$

2.10   Boundary Conditions on the Electromagnetic Fields

Since Maxwell's equations are partial differential equations we need to apply boundary conditions if we want to solve them. We can derive the boundary conditions from the equations. The derivation is basically the same for both of the divergence equations, and a second derivation works for both the *curl* equations.

2.10.1   Divergence Equations

To derive boundary conditions using the divergence equations, we need to imagine a very small 'drum' shape which crosses the boundary between two regions where the boundary conditions apply. d is the depth of the drum and Δs is the area of each of the top and bottom faces. The total flux of a vector out of this drum can be made to be just the flux through the top and bottom by taking the sides to have zero areas, which happens when d gets very small.

The equation which says $\nabla \cdot \mathbf{B} = 0$ is equivalent to saying that the total flux out of the drum is zero, $\int \mathbf{B} \cdot d\mathbf{s} = 0$. The flux leaving the drum is

$$\int \mathbf{B} \cdot d\mathbf{s} = \int_{top} \mathbf{B}_2 \cdot \mathbf{n} \, ds - \int_{bottom} \mathbf{B}_1 \cdot \mathbf{n} \, ds = 0. \tag{2.144}$$

$\mathbf{n}$ is the normal to the boundary, pointing from region 1 (which we assume is below the boundary) to region 2 (which we assume is above the boundary).

The minus sign is because $\mathbf{B}_1$ is drawn pointing in to the drum. The faces are small, so the integrals just give

$$[\mathbf{B}_2 \cdot \mathbf{n} - \mathbf{B}_1 \cdot \mathbf{n}]\Delta s = 0 \tag{2.145}$$

and so

$$\mathbf{B}_2 \cdot \mathbf{n} = \mathbf{B}_1 \cdot \mathbf{n}. \tag{2.146}$$

This is often stated as: the normal (to the surface) component of $\mathbf{B}$ is continuous.

Now if we repeat this for the equation which states that $\nabla \cdot \mathbf{D} = \rho$, there is a complication, because the total flux of D leaving the drum is not always zero. From the definition of *div* ,

$$\int_v \nabla \cdot \mathbf{D} dv = \oint_s \mathbf{D} \cdot ds \tag{2.147}$$

but we also know from Maxwell's equations that $\nabla \cdot \mathbf{D} = \rho$, so

$$\int \nabla \cdot \mathbf{D} = \int \rho dv = Q \tag{2.148}$$

where Q is the charge in the drum. When we let d get very small, the only way for Q to avoid shrinking to zero is if a finite amount of charge is right on the surface. If ρ_s is the free charge per unit area (instead of per unit volume) on the boundary surface, then $Q = \rho_s \Delta s$. The flux of $\mathbf{D}$ leaving is then approximated, for very small thickness t and a drum of small area Δs such that the integral can be replaced by the integrand multiplied by Δs, as the left side of

$$[\mathbf{D}_2 \cdot \mathbf{n} - \mathbf{D}_1 \cdot \mathbf{n}]\Delta s = \rho_s \Delta s \tag{2.149}$$

where the right side is the approximate expression for Q. Both of these expressions will be exact for small enough t and Δs. Canceling the factor of Δs,

$$\mathbf{D}_2 \cdot \mathbf{n} - \mathbf{D}_1 \cdot \mathbf{n} = \rho_s. \tag{2.150}$$

This can be stated: the jump in the normal component of D is equal to the surface charge density.

2.10.2 Curl Equations

To derive boundary conditions from the *curl* equations we imagine a rectangular loop which lies across the boundary. Its length parallel to the boundary will be ℓ, and its width in the direction perpendicular to the boundary will be d. As before we will let the side perpendicular to the boundary shrink to zero; $d \to 0$. When we do this to the loop, it has essentially zero area.

The definition of *curl* involves the integral of a vector around the edge of a loop. When d goes to zero, the integral around this loop becomes the integral along the top and the integral in the opposite direction along the bottom. For instance, from the definition of *curl*,

$$\int_s \nabla \times \mathbf{E} \cdot ds = \oint \mathbf{E} \cdot d\mathbf{r} = [E_{2,tan} - E_{1,tan}]\ell. \tag{2.151}$$

We assume ℓ is short enough that $\mathbf{E}$ is constant along the length of the loop, and d is essentially zero so the integrals up the sides perpendicular to the boundary also give zero. E_{tan} is the tangential component of the electric field, parallel to the boundary.

From the Maxwell equation for $\nabla \times \mathbf{E}$ we can see that

$$\int_s \nabla \times \mathbf{E} \cdot d\mathbf{s} = -\int_s \frac{\partial \mathbf{B}}{\partial t} \cdot d\mathbf{s}. \tag{2.152}$$

As d shrinks and the area goes to zero, so does the right-hand side so

$$\int_s \nabla \times \mathbf{E} \cdot d\mathbf{s} = 0 \tag{2.153}$$

and

$$E_{2,tan} = E_{1,tan}. \tag{2.154}$$

The tangential component of $\mathbf{E}$ is continuous.

Now when we use the other *curl* equation there is again a complication. Although the integral around the loop of $\mathbf{H}$ is similar to the integral of $\mathbf{E}$,

$$\int \nabla \times \mathbf{H} \cdot d\mathbf{s} = \oint \mathbf{H} \cdot d\mathbf{r} = [H_{2,tan} - H_{1,tan}]\ell, \tag{2.155}$$

this is not always zero. This is because

$$\int \nabla \times \mathbf{H} \cdot d\mathbf{s} = \int \left(\mathbf{J} + \frac{\partial \mathbf{D}}{\partial t} \right) \cdot d\mathbf{s}. \tag{2.156}$$

The second term on the right-hand side gives zero as the area shrinks, but if there is a current flowing on the surface the first term involving $\mathbf{J}$ does not.

$\mathbf{K}_s$ is the current per unit length (instead of per unit area) flowing on the surface. The total current through the loop is $K_{sn}\ell$, where K_{sn} is the component of $\mathbf{K}_s$ normal to the loop. Then

$$[H_{2,tan} - H_{1,tan}]\ell = K_{sn}\ell \tag{2.157}$$

and

$$H_{2,tan} - H_{1,tan} = K_{sn}. \tag{2.158}$$

If we write the equation as a vector equation, with $\mathbf{n}$ the unit normal from region 1 to region 2,

$$\mathbf{n} \times [\mathbf{H}_2 - \mathbf{H}_1] = K_s. \tag{2.159}$$

The jump in the tangential component of $\mathbf{H}$ is equal in magnitude to the surface current density $\mathbf{K}$.

This completes the boundary conditions. Once we have seen how to use these boundary conditions in a numerical description of the fields, we can solve problems in electromagnetics. This is described in Section 2.10.

2.11   Boundary Conditions on a Mesh

Now that we have found the boundary conditions, we need to decide how to apply them in a numerical calculation of the fields. This involves choosing where to put the mesh points relative to the boundary. The quantities are known at their mesh points, so if the mesh points are at the boundary, we know the associated quantity at the boundary. If the mesh points are on either side of the boundary, we know the quantity on either side of the boundary but not at the boundary. Both of these arrangements of the mesh points are useful for different purposes. The mesh used to find the potential is described next, in Subsection 2.11.1. The mesh used for the fields is the topic of Section 2.11.2. Finally Section 2.11.3 on boundary conditions on the mesh describes boundary conditions for time-dependent electromagnetic problems.

2.11.1   The Potential Mesh

To describe the electrostatic potential V near a boundary, we suggest that V be defined on a mesh with some of the mesh points set up to exactly coincide with the boundary. (In some cases we shall instead suggest that mesh points be placed on either side of the boundary but not exactly on the boundary.)

If we try to apply the boundary conditions on the electric fields

$$E_{1,tan} = E_{2,tan} \tag{2.160}$$

and

$$D_{1,n} = D_{2,n} + \rho_s \tag{2.161}$$

we have to find these components from V. Since V is to be defined on the boundary the potential V immediately above the boundary and V immediately below are the same. They are both equal to V at the boundary. In what follows the i index will run along the boundary (horizontally) and the j index will run normal to the boundary (vertically).

If the tangential electric field at point i on the boundary is $E_x[i,j]$ and if we define $E_x[i,j]$ as

$$E_x[i,j] = -\frac{V[i+1,j] - V[i-1,j]}{2\Delta x} \tag{2.162}$$

we have the same $E_x[i]$ above and below the boundary as well. The first boundary condition is satisfied automatically, because mesh points for V were placed on the boundary and no distinction is made as to whether those points on the boundary are on one side or the other. They are treated as being equally representative of both sides of the boundary, although they are at the boundary.

To find D_n, we define D_n above the boundary as

$$D_n^{above}[i,j] = -\epsilon_2 \frac{V[i,j+1] - V[i,j]}{\Delta y} \qquad (2.163)$$

and below the boundary

$$D_n^{below}[i,j] = -\epsilon_1 \frac{V[i,j] - V[i,j-1]}{\Delta y}. \qquad (2.164)$$

The fields in this case are really defined halfway between the mesh points used to define V. (Depending on the rule we use for differentiation, derivatives may be defined at the same mesh points as the quantity differentiated or between those mesh points.) D_n^{above} is defined at $j+1/2$ and D_n^{below} is defined at $j-1/2$. When we set $D_n^{above} = D_n^{below} + \rho_s$ we get an equation for $V[j]$, the potential at the boundary, in terms of $V[j+1]$ and $V[j-1]$, the potentials on either side of the boundary.

2.11.2   The Field Mesh

The voltage mesh which we chose to work with has points precisely on the boundary, as well as a little way away from the boundary. The electric field found from these voltages, for example using an expression like $(V[i,j] - V[i,j+1])/\Delta y$, is defined halfway between points labeled by (i,j) and $(i,j+1)$. This suggests that when we work with the fields, we probably should define the field at points on either side of the boundary.

The fields at the mesh points on either side of the boundary are clearly distinguished from each other. Continuity of tangential **E** means that

$$E_x[j+1] = E_x[j] \qquad (2.165)$$

whereas the condition on normal D gives

$$D_y[j+1] = \epsilon_2 E_y[j+1] = D_y[j] + \rho_s = \epsilon_1 E_y[j] + \rho_s. \qquad (2.166)$$

The same sort of mesh is used for the magnetic fields. If the wrong meshes were used the boundary conditions would cause extra problems.

2.11.3   Electromagnetic Waves and Boundary Conditions

We now discuss two time-dependent problems and their associated boundary conditions. First we consider transmission lines, a topic which is important in understanding the wave phenomena which occur in VLSI circuits. Second, the effect of periodic boundary conditions on the waves which are allowed in a cavity is outlined (partly to help with understanding the behavior of electrons in a crystal.)

Transmission Lines

The simplest transmission line is probably a coaxial cable with a cylindrical inner conductor, around which there is an insulator, which in turn is surrounded by a cylindrical outer conductor. There are a capacitance and an inductance associated with this arrangement, and they are both distributed along the length of the line. The inner conductor can act as one plate of a capacitor, with the outer conductor being the other plate. A voltage $V(z)$ between the inner and outer plates can be supported by this capacitor. This voltage can vary with z, the distance along the axis of the cylinder. If C is the capacitance per unit length, then in a length Δz there is a capacitance $C\Delta z$ and a stored electrostatic energy $\frac{1}{2}C\Delta zV^2$.

To see where the inductance comes from, suppose a current $I(z)$ flows in the z-direction along the outer surface of the inner conductor. The magnetic field produced by this current wraps around the inner conductor. The field produces an equal and opposite current, on the inside of the outer conductor. (If the conductors are perfect conductors, the current which is set up will shield the interior of the conductors from the magnetic field. To prevent the field penetrating the outer conductor, the total current inside the outer conductor must be zero. Thus the current on its own inner surface must be equal and opposite to the current on the outside of the inner conductor.) If the inductance per unit length is L then in a length Δz there is inductance $L\Delta z$ and stored magnetic energy $\frac{1}{2}L\Delta zI^2$.

Transmission-Line Equations

The inductors and capacitors which are distributed along the length of the transmission line can support waves which travel along the length of the line. Consider a short section of the line, of length Δz. The voltage $V(z)$ is between the two conductors. The current $I(z)$ flows along the conductors, in the z-direction. We need to find how $V(z)$ and $I(z)$ change in the length Δz, in order to obtain the equations governing the electrical behavior of the transmission line. To do this we make use of a circuit model of the line, which in the length Δz has capacitance $C\Delta z$ between the lines, and has inductance $L\Delta z$ which the current has to flow through.

The capacitance allows a current to 'flow between the conductors'. In reality charge builds up on the capacitor plates, creating a displacement current between the plates. The effect of this is to remove in the length Δz a current equal to $(C\Delta z)\frac{\partial V}{\partial t}$ from the current flowing along the conductor. This is the amount by which the current $I(z)$ decreases in Δz, so the change in $I(z)$ is $\Delta I = -C\Delta z\frac{\partial V}{\partial t}$. If there is a conductance $G\Delta z$ from one conductor to the other in the length Δz (which would represent leakage through the insulator between the conductors), there is an extra current between the conductors, equal to $(G\Delta z)V$. The total decrease in current is thus

$$-\Delta I = \left(C\frac{\partial V}{\partial t} + GV\right)\Delta z \qquad (2.167)$$

and dividing by Δz in the limit as Δz goes to zero,

$$\frac{\partial I}{\partial z} = -C\frac{\partial V}{\partial t} - GV. \tag{2.168}$$

The inductance $L\Delta z$ in the length Δz causes a voltage drop when the current $I(z)$ flows through the inductance. $V(z)$ changes by $\Delta V = -(L\Delta z)\frac{\partial I}{\partial t}$. Any resistance $R\Delta z$ in the length of conductor Δz which opposes current flowing in the z-direction along the conductor causes a voltage drop $(R\Delta z)I$, so combining these two voltage drops (and using a negative sign because these are both drops in voltage)

$$-\Delta V = \left(L\frac{\partial I}{\partial t} + RI \right)\Delta z \tag{2.169}$$

and so again dividing by Δz in the limit when Δz goes to zero,

$$\frac{\partial V}{\partial z} = -L\frac{\partial I}{\partial t} - RI. \tag{2.170}$$

Solution of Transmission Line Equations

The waves on a finite-length transmission line are of interest, not only because of their own importance but also because the topic illustrates the behavior of waves in general and because the boundary conditions we use are illustrative.

To solve the wave equations it is usual to differentiate the equations above, one with respect to t and the other with respect to z, and to eliminate the second derivative which is common to both equations which are obtained. The result for the case $R = 0$, $G = 0$ is a single, second-order partial differential equation:

$$\frac{\partial^2 V}{\partial z^2} = LC\frac{\partial^2 V}{\partial t^2}. \tag{2.171}$$

If the equation which was differentiated with respect to t had been differentiated with respect to z, and vice versa, an identical equation would have been obtained, except that V would be replaced by I. (It is common to neglect the conductance G and the resistance R, as has been done here, since they are both usually small, and in the ideal transmission line would be zero.)

The finite-difference equivalent of this equation is

$$\frac{V[i, k + 1] - 2V[i, k] + V[i, k - 1]}{(\Delta t)^2} = \frac{1}{LC}\frac{\partial^2 V}{\partial z^2}. \tag{2.172}$$

The integer k corresponds to time, so $t = k\Delta t$, while $z = i\Delta z$.

The second derivative with respect to z is usually approximated as

$$\frac{\partial^2 V}{\partial z^2} \simeq \frac{V[i + 1, k] - 2V[i, k] + V[i - 1, k]}{(\Delta z)^2}. \tag{2.173}$$

Boundary conditions are specified at both ends of the line. Suppose that at $i = 0$, which is also $z = 0$, we specify the voltage due to an incoming wave. We frequently set $V[i = 0, k] = \cos(\omega k \Delta t)$. Another common condition would be to use a rectangular pulse as input.

At $z = z_{max}$ we have the load. If there are N_m mesh points in z, then, since the mesh starts at $i = 0$, the highest value of i is $L_m = N_m - 1$. The load provides the boundary condition at $i = L_m$.

If the load is just a resistor R_L, then $V[i = L_m] = I[i = L_m]R_L$. The problem with this boundary condition is that the finite-difference equation we obtained above and which we use to describe the wave on the line does not involve I. To get around this, we can differentiate our boundary condition with respect to time (for instance) to obtain

$$\frac{\partial V}{\partial t} = \frac{\partial I}{\partial t} R_L. \tag{2.174}$$

We have Equation 2.170 for $\frac{\partial I}{\partial t}$ in terms of the z-derivative of V, above, so we can eliminate I from this boundary condition, to obtain

$$\frac{1}{R_L} \frac{\partial V}{\partial t} = -\frac{1}{L} \frac{\partial V}{\partial z}, \tag{2.175}$$

both sides of this equation being equal to $\frac{\partial I}{\partial t}$, although in the case of the left side of the equation it is only equal to $\frac{\partial I}{\partial t}$ at the load. We then convert it to finite-difference form, which is done in the file tlpulse.sg.

If the load is an inductor L_L, the boundary condition is actually easier to apply, since the load current and voltage obey $V = L_L \frac{\partial I}{\partial t}$ while the voltage and current on the line obey Equation 2.170, $\frac{\partial V}{\partial z} = -L \frac{\partial I}{\partial t}$. At $i = L_m$ we can eliminate $\frac{\partial I}{\partial t}$ and obtain an equation which in finite-difference form is

$$V[i = L_m, k] = -\frac{L_L}{L} \frac{V[i, k] - V[i - 1, k]}{\Delta z}. \tag{2.176}$$

Finally, for a capacitor C_L at the load, $I = C_L \frac{\partial V}{\partial t}$. On the line, however, $\frac{\partial V}{\partial z} = -L \frac{\partial I}{\partial t}$, so substituting $C_L \frac{\partial V}{\partial t}$ for I, at the load, $\frac{\partial V}{\partial z} = -L \frac{\partial}{\partial t} \left(C_L \frac{\partial V}{\partial t} \right) = LC_L \frac{\partial^2 V}{\partial t^2}$. The time derivative is found as usual, using $V[L_m, k-2]$, $V[L_m, k-1]$ and $V[L_m, k]$, while $\frac{\partial V}{\partial z}$ is found as for the inductor.

The boundary conditions we found are not the only possible ways to implement the boundary conditions, when we use V as the dependent variable. If I had been used instead of V as the dependent variable, we could find boundary conditions in a similar way.

The input file which follows, tlpulse.sg, uses a pulse input at $z = 0$ and has a resistor as its load. The characteristic impedance mentioned in the constant definitions is defined as $Z_0 = \sqrt{L/C}$.

Input File tlpulse.sg

```
// FILE:  sgdir\ch02\tlpulse\tlpulse.sg
```

```
//
// Time Domain Simulation of Waves on a Transmission Line.
// Try RL = 0.0 also.
//
// This example uses a pulse signal. Sinusoidal inputs are also possible.
// For better resolution, use smaller dz and make dt smaller by same factor.
//


CONST TMAX = 2;                    // time simulation runs for
CONST UP = 1.0;                    // phase velocity
CONST LENGTH = TMAX*UP/4.0;        // LENGTH of simulation region
CONST N = 5 ;                      //
CONST PI = 3.14159;                //
CONST NZ = 2*N*TMAX;               // number of mesh points
CONST LZ = NZ-1;                   // index of the last mesh point
CONST DZ = LENGTH/LZ;              // mesh spacing
CONST DT = 0.09*DZ/UP;             // time step
CONST Z0 = 50.0;                   // characteristic impedance of line
CONST RL = 50.0;                   // load resistance
CONST L  = Z0*UP;

var t, tw, V[NZ,3];

begin main
  assign t=0.0 ;
  assign tw=0.0;
  assign V[i=all,k=all] = 0.0 ;
  assign V[i=0..1,k=2] = 1.0;
  while (t <= TMAX) begin
    if (t > TMAX/4.0)
      assign V[i=0..1,k=2] = 0.0;

    // explicit wave equation
    assign V[i=1..LZ-1,k=2] = 2.0*V[i,k-1] - V[i,k-2] + sq(DT) * sq(UP) *
        (V[i+1,k-1] - 2*V[i,k-1] + V[i-1,k-1])/sq(DZ);

    // boundary condition
    assign V[i=LZ,k=2] = -RL/L*(V[i,k-1] - V[i-1,k-1])/DZ*DT + V[i,k-1];

    // write every time solution
    if (tw > 10*DT) begin
      write;
      assign tw = 0.0;
    end

    // shift the current value and last value of V
    assign V[i=all,k=0..1] = V[i,k+1];
```

```
      assign t = t  + DT;
      assign tw= tw + DT;
   end
end
```

In this file the time index is k, and k has only three values. $k = 0$ is the past, $k = 1$ is now and $k = 2$ is the future, would be one way of stating how k is used. After each time step the value of V is written out, then $V[i, k]$ is set equal to $V[i, k + 1]$ to allow for the change in time.

The load R_L is said to be 'matched' to the line when $R_L = Z_0$. This should give no reflected wave. When $R_L = 0$ there is a noticable reflected wave, however.

The equation specification file given above used an 'explicit scheme' and solved the finite-difference equations using assignment statements. A more accurate scheme could be set up which was 'implicit' and used equation statements; see the exercises.

Waves and Periodic Boundary Conditions

This section is devoted to the use of periodic boundary conditions in problems involving wave propagation. It can be skipped without making the following material harder to read.

An example where, in principle, we need to solve both of the *div* and *curl* equations can be found in application of Maxwell's equations to the case of a plane electromagnetic wave. Some semiconductor devices, such as GaAs MESFETs, do operate in the range of frequencies where electromagnetic wave phenomena become important. The purpose of including this material here, however, is mainly to illustrate properties of waves, some of which will be useful in understanding the solution of the Schrödinger equation which describes carrier behavior in a semiconductor. In particular, the 'eigenmodes' and 'eigenstates' of the equation and of its boundary conditions, which correspond to the allowed waves in the solid, will be illustrated. In addition, we shall discuss how eigenmodes can be found using SGFramework.

We shall now consider the effect of periodic boundary conditions, how we handle them numerically, and their effect on a wave. The relevance of this topic to this book will mainly be that it shows how a periodic boundary condition leads to a restriction on the allowed values of the wavevector. This in turn means there are only a certain number of allowed states for the wave. In the case of a carrier in a semiconductor, the number of allowed states is proportional to the volume available in 'wavevector space'.

The equation for the wave is written below in the form

$$\nabla^2 V + k^2 V = 0. \tag{2.177}$$

To find the 'eigenvalues', which are the allowed values of k, it is necessary to make k an unknown in the SGFramework input file and allow the code to seek an appropriate value of k. V must also be an unknown, but it cannot be altogether unknown, because in the absence of any constraint on V except the periodic boundary condition on V and the above equation, the solution is likely to turn out to be $V(x, y) = 0$, which is correct but not useful. The value of V at some point must be held at a nonzero value. The values of V at all the other points, and the value of k, are the unknowns. The value of the k which SGFramework finds will then depend on the initial guess which is given to it.

To see how this type of problem can arise, we consider a time-dependent case. The equations we need to solve to find the electric field are

$$\nabla \times \mathbf{E} = -\frac{\partial \mathbf{B}}{\partial t} \tag{2.178}$$

and if $\rho = 0$

$$\nabla \cdot (\epsilon \mathbf{E}) = 0. \tag{2.179}$$

If there is no current, $\mathbf{J} = 0$ and

$$\nabla \times \mathbf{H} = \frac{\partial \mathbf{D}}{\partial t} \tag{2.180}$$

while

$$\nabla \cdot \mathbf{B} = 0 \tag{2.181}$$

is always true. From the *curl* equations, making use of the divergence equations to simplify the result and assuming a 'plane wave', we can obtain the wave equation and then the Helmholtz equation given below for $\mathbf{H}$, which we shall solve using periodic boundary conditions on $\mathbf{H}$.

Plane Waves

If we assume the fields vary like

$$\mathbf{B} = \mathbf{B}_0 \cos(\omega t - \mathbf{k} \cdot \mathbf{r}) \tag{2.182}$$

then the divergence equations are automatically satisfied, provided the fields are at right angles to the wavevector $\mathbf{k}$. This expression is introduced here to help define the quantities ω and $\mathbf{k}$. In this expression, $\mathbf{r}$ is a position vector which in two Cartesian variables is $\mathbf{r} = (x, y)$. $\mathbf{k}$ is called the wavevector, $\mathbf{k} = (k_x, k_y)$ and $k = |\mathbf{k}| = \frac{2\pi}{\lambda}$, where λ is the wavelength. This means that

$$\mathbf{k} \cdot \mathbf{r} = k_x x + k_y y. \tag{2.183}$$

If the example were in three dimensions, there would be an extra $k_z z$ on the right-hand side.

If our field $\mathbf{B}$ has a particular period which in this case is controlled by $\mathbf{k} = (k_x, k_y)$, it is helpful to choose the size of the simulation region to have sides L_x in x and L_y in y so that an integer number of periods of $\mathbf{B}$ fit into each side. This means that

$$k_x = \frac{2\pi m}{L_x}, \; k_y = \frac{2\pi n}{L_y} \tag{2.184}$$

where m and n are integers. If the sides L_x and L_y satisfy these equations then since we know that **E** should also be periodic with the same period as **B**, we can say

$$E_x(x = L_x, y) = E_x(x = 0, y); E_x(x, y = L_y) = E_x(x, y = 0) \tag{2.185}$$

and

$$E_y(x = L_x, y) = E_y(x = 0, y); E_y(x, y = L_y) = E_y(x, y = 0). \tag{2.186}$$

An example input file which uses a periodic boundary condition is given next. The conditions are placed on the magnetic field H. In the interior of the simulation region H is found from the Helmholtz equation

$$\nabla^2 H + k^2 H = 0 \tag{2.187}$$

which is obtained from the wave equation

$$\nabla^2 H - \frac{1}{c^2} \frac{\partial^2 H}{\partial t^2} = 0 \tag{2.188}$$

when we assume that the time variation is sinusoidal, so that

$$\frac{\partial^2 H}{\partial t^2} = -\omega^2 H. \tag{2.189}$$

To get from this equation to the Helmholtz equation given above we used $k^2 = \omega^2/c^2$.

Before we give the example input file, we should emphasize that one of the major reasons for looking at this case is to show how the periodic boundary conditions limit the possible solutions which can fit inside the solution region. As explained above, a carrier inside a semiconductor behaves like a wave subject to a periodic boundary condition. The number Δn of allowed values of n in a range Δk_x of the variable k_x is $\Delta n = \frac{L_x}{\pi} \Delta k_x$. Similarly, $\Delta m = \frac{L_y}{\pi} \Delta k_y$ and if $k_z = \frac{p\pi}{L_z}$ then $\Delta p = \frac{L_z}{\pi} \Delta k_z$.

In the three 'dimensions' of **k**, each allowed value of **k** must correspond to an allowed value (that is an integer value) of (m, n, p). The number of states in the range $\Delta k_x \Delta k_y \Delta k_z$ is $\Delta m \Delta n \Delta p = \frac{L_x L_y L_z}{\pi^3} \Delta k_x \Delta k_y \Delta k_z$. This expression provides the number of states available to an electron or hole, which is proportional to the volume in wavevector space, $\Delta k_x \Delta k_y \Delta k_z$.

We now return to our example, which calculates the field due to a current sheet using a periodic boundary condition.

Input File isheet.sg

```
// FILE:   sgdir/ch02/isheet/isheet.sg
//
// isheet.sg ; Simulation of Fields due to Current Sheet
// Description: This simulation determines the magnetic and electric field
//              caused by a current sheet.

const NX  = 41;                 // number of x-mesh points
const NY  = 41;                 // number of y-mesh points
const LX  = NX - 1;             // index of last x-mesh point
const LY  = NY - 1;             // index of last y-mesh point
const DX  = 1.0;                // mesh spacing

const Mx  = 1.0;                // number of modes in x
const My  = 1.0;                // number of modes in y
const PI  = 3.14159;            // PI
const EPS = 8.85e-12;           // permittivity
const MU  = PI*4e-7;            // permeability
const Kx  = Mx*2.0*PI/LX;       // wavevector in x
const Ky  = My*2.0*PI/(LY-1);   // wavevector in y
const K   = sqrt(sq(Kx)+sq(Ky)); // wavevector
const ETA = sqrt(MU/EPS);       // intrinsic impedance

var H[NX,NY];
var Ex[LX,LY];

// Helmholtz equation
equ H[i=1..LX-1,j=1..LY-1] ->
 {H[i-1,j] + H[i,j-1] - 4*H[i,j] + H[i+1,j] + H[i,j+1]} / sq(DX) =
 -sq(K)*H[i,j];

// periodic boundary conditions
equ H[i=LX,j=all]   -> H[i,j] = H[0,j];
equ H[i=1..LX,j=0]  -> H[i,j] = H[i,LY-1];
equ H[i=1..LX,j=LY] -> H[i,j] = H[i,1];

begin main
  assign H[i=all,j=all] = 0.0;
  assign H[i=0,  j=all] = cos(j*Ky);
  solve;
  assign Ex[i=all,j=all] = (H[i,j+1] - H[i,j]) / DX / (K*ETA);
end
```

Properties of Plane Waves

While we are investigating this wave problem, it is worth examining some of the other properties of the plane electromagnetic wave. The divergence of B is always zero,

$$\nabla \cdot \mathbf{B} = 0 \tag{2.190}$$

which becomes

$$\frac{\partial B_x}{\partial x} + \frac{\partial B_y}{\partial y} = 0 \tag{2.191}$$

in two dimensions. This can be used to show that $\mathbf{k} \cdot \mathbf{B} = 0$. If $\mathbf{B}$ is written out in full as

$$\mathbf{B} = \mathbf{a}_x B_{x0} \cos(\omega t - k_x x - k_y y) + \mathbf{a}_y B_{y0} \cos(\omega t - k_x x - k_y y) \tag{2.192}$$

then

$$\frac{\partial B_x}{\partial x} = B_{x0} k_x \sin(\omega t - k_x x - k_y y) \tag{2.193}$$

while

$$\frac{\partial B_y}{\partial y} = B_{y0} k_y \sin(\omega t - k_x x - k_y y). \tag{2.194}$$

Since these add up to zero, we must have

$$B_{x0} k_x + B_{y0} k_y = 0 \tag{2.195}$$

which is equivalent to

$$\mathbf{k} \cdot \mathbf{B}_0 = 0 \tag{2.196}$$

where $\mathbf{k} = \mathbf{a}_x k_x + \mathbf{a}_y k_y$ and $\mathbf{B}_0 = \mathbf{a}_x B_{x0} + \mathbf{a}_y B_{y0}$. If $\mathbf{k} \cdot \mathbf{B}_0 = 0$, then $\mathbf{B}$ is perpendicular to $\mathbf{k}$.

If $\mathbf{k} \cdot \mathbf{B} = 0$, there is no variation in $\mathbf{B}$ if we move in the direction that $\mathbf{B}$ points in. Any box we draw will have the same flux leaving as entering. If a vector $\mathbf{C}$ varies as we move along the direction of $\mathbf{C}$, we have a nonzero divergence of $\mathbf{C}$.

We can illustrate some of these ideas by setting up an example file to evaluate *div* and *curl* of different vectors.

The field in the example does not give $\nabla \cdot \mathbf{B} = 0$. One might choose B_{x0} so that $\nabla \cdot \mathbf{B} = 0$ and plot $(\nabla \times \mathbf{B})_z$. Note that B_x is differentiated with respect to x in the first equation, and with respect to y in the second.

2.12   Propagators in Electromagnetism

In electrostatics the potential V at a point $\mathbf{r}$ due to a unit charge at a point $\mathbf{r}'$ is

$$V(\mathbf{r}) = [4\pi\epsilon|\mathbf{r} - \mathbf{r}'|]^{-1} . \tag{2.197}$$

Input File exvec.sg

```
// FILE:  sgdir\ch02\exvec\exvec.sg

const PI  = 3.14159265358979323846;   // pi
const DIM = 101;                       // number of mesh points
const DX  = 1.0;                       // horizontal mesh spacing
const DY  = 1.0;                       // vertical mesh spacing
const M   = 1;                         // number of horizontal modes
const N   = 2;                         // number of vertical modes
const KX  = 2*PI*M/(DIM-1)/DX;         //
const KY  = 2*PI*N/(DIM-1)/DY;         //
const BX0 = 5.0;                       // amplitude
const BY0 = 1.0;                       // amplitude

var x[DIM], y[DIM];
var Bx[DIM,DIM], By[DIM,DIM], divB[DIM,DIM], curlB[DIM,DIM];

equ divB [i=1..DIM-2,j=1..DIM-2]->
  divB[i,j] = (Bx[i+1,j]-Bx[i-1,j])/(2.0*DX)+(By[i,j+1]-By[i,j-1])/(2.0*DY);

equ curlB[i=1..DIM-2,j=1..DIM-2]->
  curlB[i,j] = (By[i+1,j]-By[i-1,j])/(2.0*DX)-(Bx[i,j+1]-Bx[i,j-1])/(2.0*DY);

begin main
  assign x[i=all] = i*DX;
  assign y[j=all] = j*DY;
  assign Bx[i=all,j=all] = BX0*cos(KX*i*DX+KY*j*DY);
  assign By[i=all,j=all] = BY0*cos(KX*i*DX+KY*j*DY);
  solve;
  write;
end
```

This is one of the best known and simplest examples of a propagator. A propagator, as the term is used here, is a mathematical expression which describes a 'response' of a 'system' to a 'unit input'. (Propagators are called by a variety of names, one of which is Green's functions.) The terminology of input and response is appropriate for an application such as signal processing. In the situations examined here, the input might be better described as a 'driving force'. The electrostatic potential $V(\mathbf{r})$ is the response to the unit charge, which is the unit 'driving force' in the electrostatic example, so the expression for $V(\mathbf{r})$ is the electrostatic 'propagator'.

It is useful to know what the response is to a unit input. In a linear system the 'unit input response' can be used to find the response to a more general input. In a linear system, if we double the input we double the response, and so on. If the charge is q_1 coulombs at a point $\mathbf{r}'_1$, then the voltage is

$$V(\mathbf{r}) = \frac{q_1}{4\pi\epsilon|\mathbf{r} - \mathbf{r}'_1|}. \tag{2.198}$$

This expression for $V(\mathbf{r})$ only depends on $\mathbf{r}$ and $\mathbf{r}'$ through the difference between them.

If a second charge q_2 is added at another position $\mathbf{r}'_2$, then the linearity property means that the total voltage is the sum of the voltages due to q_1 and q_2, so

$$V(\mathbf{r}) = \frac{1}{4\pi\epsilon} \left(\frac{q_1}{|\mathbf{r} - \mathbf{r}'_1|} + \frac{q_2}{|\mathbf{r} - \mathbf{r}'_2|} \right). \tag{2.199}$$

If we have N charges, labeled by an index i where i runs from 1 to N, then

$$V(\mathbf{r}) = \sum_{i=1}^{N} V_i(\mathbf{r}) \tag{2.200}$$

where $V_i(\mathbf{r})$ is the potential due to the ith charge.

$$V_i(\mathbf{r}) = \frac{q_i}{4\pi\epsilon|\mathbf{r} - \mathbf{r}'_i|}. \tag{2.201}$$

Then the overall potential is

$$V(\mathbf{r}) = \frac{1}{4\pi\epsilon} \sum_{i=1}^{N} \frac{q_i}{|\mathbf{r} - \mathbf{r}'_i|}. \tag{2.202}$$

In this way we can find the potential due to many charges, using the propagator.

If the charge is spread out over a volume, then the sum over charges is converted to an integral over the volume. The charge at $\mathbf{r}'$ is then written as $q = \rho(\mathbf{r}')\,dv'$, where $\rho(\mathbf{r}')$ is the charge density per unit volume at $\mathbf{r}'$ and dv' is a small volume containing the charge q.

$$V(\mathbf{r}) = \frac{1}{4\pi\epsilon} \int_{v'} \frac{\rho(\mathbf{r}')\,dv'}{|\mathbf{r} - \mathbf{r}'|}. \tag{2.203}$$

This integral is an example of a 'convolution integral', where the charge density $\rho(\mathbf{r}')$ is convoluted with the propagator which depends on the difference between $\mathbf{r}$ and $\mathbf{r}'$. The integral is over v', which means that $\mathbf{r}'$ is the quantity which varies during the integration.

A simple example is given in the example file 'emprop.sg'. This example uses a 'propagator' approach to find the electric field $\mathbf{E}$ at a point (x_1, y_1, z_1) due to a circular ring of charge with radius r_0 centered on the point (x_0, y_0, z_0). A similar file and results from it are given in the problems at the end of the chapter. The file in the problems is designed to find the electrostatic potential V in a plane.

Input File emprop.sg

```
// FILE: sgdir\ch02\emprop\emprop.sg

// simulation parameters
const NX  = 42;              // number of points on source
const LX  = NX - 1;          // index of last mesh point
const dth = 2.0 * 3.1415/NX ;  // angle between points on source
const r0  = 0.2 ;            // radius of ring source
const x0  = 0.1 ;            // x-coord of center of ring
const y0  = 0.1 ;            // y-coord of center of ring
const z0  = 0.1 ;            // z-coord of center of ring
const x1  = 0.1 ;            // x-coord of field point
const y1  = 0.4 ;            // y-coord of field point
const z1  = 0.2 ;            // z-coord of field point


//  component of E field
func E1(x,y,z)
  const pre = 9.0e+9;
  assign R = sqrt(x*x+y*y+z*z);
  assign Ex = pre*x/(R*R*R);
  return Ex;


var  dx[NX],  dy[NX], dz[NX];
var  Ex[NX],  Ey[NX], Ez[NX];

//  sum contributions to components of E
equ Ex[i=1..LX] -> Ex[i] = Ex[i-1] + E1(dx[i],dy[i],dz[i]);
equ Ey[i=1..LX] -> Ey[i] = Ey[i-1] + E1(dy[i],dz[i],dx[i]);
equ Ez[i=1..LX] -> Ez[i] = Ez[i-1] + E1(dz[i],dx[i],dy[i]);

begin main
  assign dx[i=0..LX] = x1 - r0 * cos(i*dth) - x0 ;
  assign dy[i=0..LX] = y1 - r0 * sin(i*dth) - y0 ;
  assign dz[i=0..LX] = z1 - z0 ;
  assign Ex[i=0] = E1(dx[i],dy[i],dz[i]);
  assign Ey[i=0] = E1(dy[i],dz[i],dx[i]);
  assign Ez[i=0] = E1(dz[i],dx[i],dy[i]);
  solve;
  write;
end
```

2.13   Solution of Problems in Electrostatics

In this section we will discuss a number of issues related to solving eletrostatic problems. We will also solve several electrostatic problems using the methods described above. First in Section 2.13.1 we shall reiterate the version of the Maxwell equations when there is no time dependence, and simplify them. Section 2.13.2 is on boundary conditions on a perfect conductor. This is an important case in semiconductor simulation. The function lsum (which performs a Label Sum) is introduced (see Section 15.3.4). Then we can begin to discuss how to use the the equations in some interesting applications. Section 2.13.3 discusses the displacement current. Understanding the displacement current will help explain the behavior we see in time-dependent electrostic problems, such as a capacitor with an alternating voltage applied to it. The conditions under which a 'slowly varying' field may be called static are outlined. Section 2.13.4 is on numerical examples of solution of Laplace's equation, 2.13.5 discusses problems involving conductors in an applied electric field, 2.13.6 solves a problem with a dielectric in an electric field and 2.13.7 presents a simulation of a microelectromechanical system.

2.13.1   Equations for Static Electric Fields

In electrostatics, the Maxwell equation

$$\nabla \times \mathbf{E} = -\frac{\partial \mathbf{B}}{\partial t} \tag{2.204}$$

reduces to $\nabla \times \mathbf{E} = 0$ because the time derivative is zero. When $\nabla \times \mathbf{E} = 0$ the electric field $\mathbf{E}$ can be written as the gradient of a scalar:

$$\mathbf{E} = -\nabla V. \tag{2.205}$$

This $\mathbf{E}$ is a conservative field, which means the work done in pushing a charge around a loop and back to where it started is zero,

$$\oint_{loop} q\mathbf{E} \cdot d\mathbf{r} = 0 \tag{2.206}$$

because

$$\oint_{loop} \mathbf{E} \cdot d\mathbf{r} = \int \nabla \times \mathbf{E} \cdot d\mathbf{s} \tag{2.207}$$

and $\nabla \times \mathbf{E} = 0$.

The divergence equation $\nabla \cdot \mathbf{D} = \rho$ can be used along with $\mathbf{E} = -\nabla V$ to get Poisson's equation:

$$\nabla \cdot (-\epsilon \nabla V) = \rho \tag{2.208}$$

which means

$$\nabla^2 V = -\frac{\rho}{\epsilon}. \tag{2.209}$$

The first examples we will look at have $\rho = 0$, so we get Laplace's equation,

$$\nabla^2 V = 0. \tag{2.210}$$

An example where $\nabla^2 V = 0$ was solved with V specified along the edges of a rectangular box was given in Section 2.8. A rectangular capacitor in air, inside a grounded rectangular box, is described by the example cap.sg, which is given in Section 2.13.4. Suppose the plates are each held at a specified potential $\pm V_p$. To find the capacitance C we need to know the charge Q on one of the plates, since $C = Q/V$ where $V = 2V_p$ is the potential difference between the plates. To find Q we need to apply one of the boundary conditions derived above, the condition on normal **D**, in the special case where one side of the boundary is a perfect conductor. This is an important case, and we describe it next.

2.13.2   Boundary Conditions on a Perfect Conductor

This section provides material we need before we can solve problems in electrostatics with conductors present.

We can calculate the charge on a conductor Q using the boundary condition which relates the normal component of **D** to the charge on the surface. There is no electric field inside the conductor, because we assume it is a perfect conductor. This is because the current density **J** is related to the electric field by Ohm's law

$$\mathbf{J} = \sigma \mathbf{E} \tag{2.211}$$

with σ the conductivity and $\sigma = \infty$ in a perfect conductor. As $\sigma = \infty$, **E** has to be zero inside the conductor or else **J** would also be infinite. Physically, electrons in a perfect conductor are free to move around, and they do so until they shield out any electric field.

Since **E** is zero inside the perfect conductor, the boundary condition on the normal component of **D** becomes

$$D_n^{outside} - D_n^{inside} = D_n^{outside} = \epsilon_0 E_n^{outside} = \rho_s. \tag{2.212}$$

If we draw field lines of **D**, the lines all start on the charge on the surface when ρ_s is positive (and end on the charge on the surface when ρ_s is negative).

If we know E_n at each point on the surface, we can find Q by summing the surface charge in each cell,

$$Q = \sum_{cells} \epsilon_0 \left(-\frac{\partial V}{\partial n} \right) \Delta s. \tag{2.213}$$

Δs is the area of each cell and $E_n = -\frac{\partial V}{\partial n}$. The normal component of ∇V has been written as $\frac{\partial V}{\partial n}$.

The sum over a region which is needed to find the charge is facilitated by the function lsum, which is introduced in Section 15.3.4. For an example of its use, see Section 7.3.1.

2.13.3 Displacement Current

In this section we describe the Displacement Current Density, which will help in understanding time-dependent electrostatic problems, such as the simulation of a capacitor with an applied ac voltage. We also outline a few related problems, which illustrate the situation in the capacitor, without solving them. These problems could be solved using the tools provided so far.

If charge flows into a region and if the charge builds up there, then an electric field also builds up. This situation can sometimes still be considered static, even though there is time variation, provided that the variation is not fast enough to make $\frac{\partial \mathbf{B}}{\partial t}$ large in the equation $\nabla \times \mathbf{E} = -\frac{\partial \mathbf{B}}{\partial t}$. (A test for whether a variation can be called static is that the frequency is low compared to the frequency of an electromagnetic wave whose wavelength is about the same size as a typical dimension in the problem.)

Suppose charge is flowing from one vicinity to another. A current starts at the place the charge leaves, so a decreasing charge in one region is associated with a current leaving that region. The current flows to the place to which the charge goes and the current stops there. An increasing charge in a region is associated with a current flowing in. The electric field $\mathbf{E}$ and the field $\mathbf{D}$ point from the positive charge to the negative charge. If the positive charge in one region is getting more positive and the negative charge in the other region is getting more negative, then $\mathbf{E}$ and $\mathbf{D}$ are increasing.

The rate of change of $\mathbf{D}$ at a point is called the displacement-current density,

$$\mathbf{J}_D \equiv \frac{\partial \mathbf{D}}{\partial t}. \tag{2.214}$$

In this discussion, the conduction current runs from one point to another, rather than going round a circuit back to where it started. Because the conduction current moves charge around, it creates a change in $\mathbf{D}$ and this means we have a displacement current. The displacement current starts where the conduction current ends. When the two currents are added, the overall current is continuous.

A one dimensional example of the calculation of the displacement current will now be discussed. In one dimension x suppose the charge density per unit length $\rho(x)$ is initially zero, and suppose it changes with time in different ways in each of two regions.

$$\frac{\partial \rho}{\partial t} = -A, \ 0.05 < x \le 0.1 \tag{2.215}$$

and

$$\frac{\partial \rho}{\partial t} = A, \ 0.25 < x \le 0.35. \tag{2.216}$$

The charge leaves to the left to $x = 0$, flows through an external circuit and is reinjected on the right at $x = 0.4$.

We might want to find (1) the conduction current, (2) the field $\mathbf{D}$ and (3) the displacement current, and compare (1) and (3). This problem can be solved analytically or by setting up a suitable 1D input file. A second question we might ask is: How would the answer change if the conduction current flowed to the right between the two regions?

A second example in two dimensions (x, y), has a charge density per unit area $\rho(x, y)$ which exists in two regions.

$$\rho(x, y) = B, \ -0.2 < x \le 0.15, y_0 < y \le y_0 + 0.2 \tag{2.217}$$

and

$$\rho(x, y) = -B, \ 0.15 \le x < 0.2, y_0 < y \le y_0 + 0.2. \tag{2.218}$$

The position of the lower edge in y is $y_0 = 10 + 10t$, so the charge is moving to higher y-values. The edge of the region is a conducting box at a potential $V = 0$, with sides at $x = \pm 0.25, y = 0$ and $y = 0.5$. A suggested approach to this is to find the static field using SGFramework when $y_0 = 10$. One can then deduce from this what will happen when y_0 is allowed to change. Alternatively one can find the static field again when $y_0 = 10 + 10\Delta t$ and subtract the two results for $\mathbf{D}$ to find $\frac{\partial \mathbf{D}}{\partial t}$. One could then find (1) the conduction current density $\mathbf{J}_c$, (2) the field $\mathbf{D}$, (3) the displacement current density $\frac{\partial \mathbf{D}}{\partial t}$ and (4) the total current density $\mathbf{J} = \mathbf{J}_c + \mathbf{J}_D$ and $\nabla \cdot \mathbf{J}$.

In the next section the displacement current is found by using Laplace's equation to obtain the potential near a capacitor with an alternating potential applied to the plates.

2.13.4   Solution of Laplace's Equation - Numerical Examples

In this section we will be using Laplace's equation

$$\nabla^2 V = 0 \tag{2.219}$$

and suitable boundary conditions to find the potential V. The finite difference form of this in Cartesian coordinates (x, y), with corresponding indices (i, j) and uniform mesh spacings $(\Delta x, \Delta y)$ is

$$\frac{V[i+1, j] + V[i-1, j] - 2V[i, j]}{(\Delta x)^2} + \frac{V[i, j+1] + V[i, j-1] - 2V[i, j]}{(\Delta y)^2} = 0. \tag{2.220}$$

This was derived earlier in this chapter.

Potential Inside a Conducting Box

A problem that is very easy to solve using finite-differences in SGFramework is finding the potential inside a rectangular conducting box with the walls held at fixed potentials. This was done in Section 2.7.2. This can also be solved analytically, using separation of variables and Fourier analysis to handle Laplace's equation. These techniques are very important and useful but they take a long time to explain.

They are sufficiently complex in fact that most undergraduate textbooks do not explain properly how the Fourier analysis handles the boundary conditions.

The situation considered in Section 2.7.2 was when one wall of a conducting box was at 10 volts and all the others were at zero volts. The boundary conditions used in that example can be implemented as follows.

The left-hand wall at $x = 0$ corresponds to $i = 0$ for all values of j, so for all j the $V[i = 0, j]$ are all set to 10. V on all the other walls is zero, so in terms of the mesh indices:

Along the bottom, at $y = 0$ we have $j = 0$, so $V[i, j = 0] = 0$ for all i.

Along the top, at $y = L_2$ we have $j = N_2 - 1$ so $V[i, j = N_2 - 1] = 0$ for all i.

Along the right-hand wall, at $x = L_1$ we have $i = N_1 - 1$, so $V[i = N_1 - 1, j] = 0$ for all j.

The values of V at all the interior points were found by solving Laplace's equation, as given in the equation above in finite-difference form.

In the next example we shall give, the walls are all at zero potential, which can be handled either as just described or by never setting the potential on the boundary to any other value. The code assumes the value of a quantity is zero until it is set to something else. The values of the indices corresponding to the boundary were also indicated above. A capacitor is included in the interior of the box. The capacitor plates have a sinusoidally oscillating voltage applied to them.

Input File cap.sg

```
// FILE:  sgdir\ch02\cap\cap.sg
// linear capacitor in a shielded box .
// Purpose: To show how Laplace's equation is used to find
//          the potential V and to illustrate the Displacement Current .
//

const NX = 21;               // number of x-mesh points
const NY = 21;               // number of y-mesh points
const LX = NX - 1;           // index of last x-mesh point
const LY = NY - 1;           // index of last y-mesh point
const X1 = NX / 2 - 5;       // x-location of first plate
const X2 = NX / 2 + 5;       // x-location of second plate
const Y1 = NY / 4;           // y-location of plates
const Y2 = 3 * NY / 4;       // y-location of plates
const DX = 1.0;              // x-mesh spacing
const DY = DX ;              // y-mesh spacing
const Va = 10.0;             // applied voltage
const omega = 6.0;           // angular frequency
const tmax  = 0.5;           // maximum time. Set tmax=0 for static case.
const dt = 0.05;             // time step.
const eps0 = 8.8e-12 ;       // permittivity of vacuum

var t,V[NX,NY],V0[NX,NY],Dx[LX,LY],Dy[LX,LY]; //Why is the dimension of D less?
```

```
// See where D is calculated. t is time; V is electrostatic potential,VO=old V.
// Dx and Dy are the components of the "electric displacement", D.

var DxO[LX,LY],DyO[LX,LY],dVdt[NX,NY],dDxdt[LX,LY],dDydt[LX,LY];
// dVdt is the time derivative of V. Other quantities are similar.

// all voltages are unknown except for the mesh points on the
// perimeter and the capacitor plates
unknown V[1..LX-1,1..LY-1];
known   V[X1,Y1..Y2], V[X2,Y1..Y2];

// Laplace's equation, with dx set equal to dy.
// We use this even when V varies with time, because if the frequency
// is small enough the equations are "nearly" the static equations.

equ V[i=all,j=all] -> {V[i-1,j] + V[i,j-1] - 4*V[i,j] +
                       V[i+1,j] + V[i,j+1]} / sq(DX) = 0;

begin main
  assign t= 0.0 ;

  // Initialize variables needed to find time derivatives.
  assign V[i=X1,j=Y1..Y2] = + Va ;
  assign V[i=X2,j=Y1..Y2] = - Va ;
  solve ;
  assign VO[i=all,j=all] = V[i,j];
  assign DxO[i=all,j=all]= - eps0 * (VO[i+1,j] - VO[i,j])/DX ;
  assign DyO[i=all,j=all]= - eps0 * (VO[i,j+1] - VO[i,j])/DY ;

  // Start varying t.
  while (t <= tmax) begin
    assign V[i=all,j=all] = 0.0;
    assign V[i=X1,j=Y1..Y2] = +Va*cos(omega*t); //oscillating voltage on plate
    assign V[i=X2,j=Y1..Y2] = -Va*cos(omega*t); //opposite voltage, other plate
    solve ;
    assign Dx[i=all,j=all] = - eps0 * (V[i+1,j] - V[i,j])/DX ;    // Find Dx.
    assign Dy[i=all,j=all] = - eps0 * (V[i,j+1] - V[i,j])/DY ;    // Find Dy.
    assign dVdt[i=all,j=all] = (V[i,j] - VO[i,j])/dt ;            // Find dVdt.
    assign dDxdt[i=all,j=all]= (Dx[i,j]- DxO[i,j])/dt;            // Find dDxdt.
    assign dDydt[i=all,j=all]= (Dy[i,j]- DyO[i,j])/dt;            // Find dDydt.
    assign VO[i=all,j=all] = V[i,j] ;                            // Update old V
    assign DxO[i=all,j=all]= Dx[i,j] ;                           // Update old Dx
    assign DyO[i=all,j=all]= Dy[i,j] ;                           // Update old Dy
    write;
    assign t = t + dt ;
  end
end
```

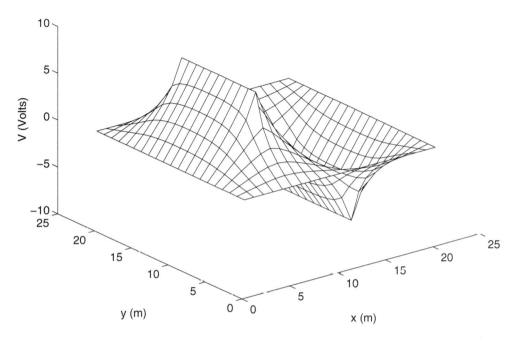

Figure 2.13. The electrostatic potential at some moment in time for a capacitor which, as shown in Figure 2.13, is driven with an alternating voltage. The input file is given in the text.

2.13.5   Conductors in an Applied Electric Field

In this section we discuss several problems which should be straightforward to implement, by modifying the files used in earlier examples. The example which follows later, of a dielectric in an external field, may also be useful.

A Rectangular Conductor in an External Field

Now we want to find the potential when a conductor is put into an external electric field. We can solve Laplace's equation as usual outside the conductor. The conductor is at a fixed potential - suppose we put a grounded rectangular conductor into the field.

If the conductor runs from $i = i_1$ to $i = i_2$ in the x-direction and from $j = j_1$ to $j = j_2$ in the y-direction, then $V[i, j] = 0$ for the region which has both $i_1 \leq i < i_2$ and $j_1 \leq j < j_2$.

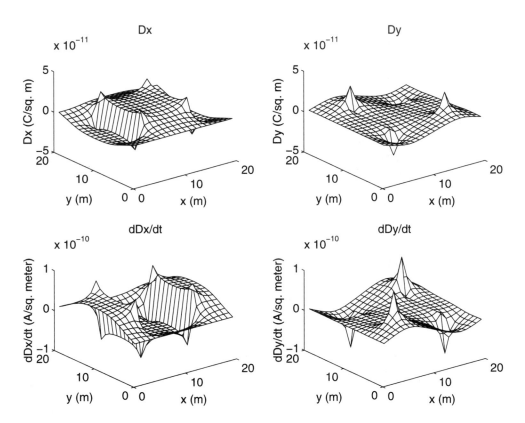

Figure 2.14. Various components of the electric field, and their time derivatives, for a capacitor driven by an alternating voltage.

The one other piece of information we need is the potential at the boundary of the solution region, which we assume is far enough from the conductor to be unaffected by it. If we want the field far away from the conductor to be $\mathbf{E}_{ext} = \mathbf{E}_0\mathbf{a}_x$, we can use $V_{ext} = -E_0(x - x_0)$, where x_0 is in the center of the simulation. V along the boundary is set equal to this potential V_{ext}.

The more usual textbook problem has a circular conductor. A circular conductor does not fit exactly on our rectangular mesh, but we can approximate a circle with a combination of rectangles. More general meshes which are not limited to rectangular cells are discussed and used earlier in this chapter and in Chapter 5.

Potential Near a Hole in a Conducting Plate

As a second example consider a thin grounded metal plate at $y = 0$ with a hole from $x = -0.1$ to $x = 0.1$, which separates a region with no electric field as $y \to -\infty$ from a region where the electric field goes to $\mathbf{E} = 10\mathbf{a}_y$ V/m as $y \to \infty$. The problem is to find the potential everywhere.

We will set the potential on the boundary to satisfy the conditions on the electric field. In the upper half we use $V = -10y$ on the boundary, so that $\mathbf{E} = -\nabla V = 10\mathbf{a}_y$. In the lower half we must use $V = 0$ on the boundary. A line of cells for $y = 0$ and (say) $-0.2 \le x < -0.1$ and another line for $y = 0$ and $0.1 < x \le 0.2$ are all held at zero potential to represent the metal plate. As usual, we solve $\nabla^2 V = 0$ in the interior.

2.13.6   A Dielectric in an Electric Field

Suppose we have an electric field which is the same everywhere, except for a small region where there is a piece of dielectric. See Figure 2.15. We can study what happens near the dielectric by assuming that the dielectric only has a significant effect on the electric field close to the dielectric.

To find the electric field we must solve the equations of electrostatics, using the boundary conditions on $\mathbf{E}$ and $\mathbf{D}$ as well. There is no free charge in the region we are looking at, so we can set $\rho = 0$, and all time derivatives are zero, so we can again use Laplace's equation, $\nabla^2 V = 0$.

To solve Laplace's equation on a mesh, it is convenient to have the rectangular piece of dielectric have its sides parallel to the mesh lines. This allows us to easily put on the boundary conditions at the surface of the dielectric. The next subsection discusses the boundary conditions in detail. The file diel.sg which implements the solution is also given.

Numerical Form of the Boundary Conditions on the Potential

The two boundary conditions we derived which apply to $\mathbf{E}$ and $\mathbf{D}$ are:

1. $\mathbf{E}_{tan}$ is continuous.

2. D_n is continuous.

The first of these is satisfied automatically, given the way we shall set up the problem. First we need to show where to put the boundary on the mesh, then we can explain why $\mathbf{E}_{tan}$ is automatically continuous. Two possibilities are

(a) to put mesh points right at the boundary (which is what we shall do) so that the values at those points represent conditions at the boundary, which effectively means they determine values very close to the boundary and on both sides of the boundary; or

(b) to put the mesh points a little way away from the boundary on either side of it.

In either case, V is defined at the mesh points. $\mathbf{E}_{tan}$ is found from the differences in V that are found as we move from point to point parallel to the surface. If we have V defined at mesh points at the boundary then when we find $\mathbf{E}_{tan}$ we find it at the boundary, not on one side or the other but on both sides, so it is forced to be the same on both sides. If the mesh points were not on the boundary, the values of $\mathbf{E}_{tan}$ could be different on either side. Assuming we have mesh points on the boundary so $\mathbf{E}_{tan}$ is continuous, then at the next row of points away from the boundary $\mathbf{E}_{tan}$ will be slightly different. As we approach the boundary from several rows away, $\mathbf{E}_{tan}$ on either side of the boundary goes to the same value.

Having a set of mesh points at the boundary allows us to define E_n on both sides of the boundary. Let the ith cell be on the boundary, with $i+1$ above and $i-1$ below. Above the boundary $E_n = -\frac{V[i+1]-V[i]}{\Delta x}$ and below the boundary $E_n = -\frac{V[i]-V[i-1]}{\Delta x}$. This placement of the mesh points is thus convenient to handle both of the boundary conditions we mentioned.

Using the other type of mesh, with V defined not on the boundary but on either side we can only find E_n on average near the boundary, from $V[above] - V[below]$. We need to distinguish between E_n above and below the boundary and this mesh does not allow us to do this. This means that the mesh with points on the boundary is necessary.

The second condition, D_n is continuous, is not automatic. Since $\mathbf{D} = \epsilon\mathbf{E}$, we take the expressions for $\mathbf{E}_n$ we found previously and set the values for $\epsilon\mathbf{E}_n$ above and below the boundary equal to each other. If air is above the boundary and dielectric is below,

$$\epsilon_{air}E_{n,air} = -\epsilon_{air}\frac{V[i+1]-V[i]}{\Delta x} = \epsilon_{diel}E_{n,diel} = -\epsilon_{diel}\frac{V[i]-V[i-1]}{\Delta x}. \quad (2.221)$$

These equations for V are used instead of $\nabla^2 V = 0$, right at the boundary.

The corners of the dielectric slab are not described well by the finite-difference form of the boundary conditions that we have given here. A more accurate and more general version of the boundary condition could be set up easily, using the 'nsum' function, on a mesh generated from a skeleton file.

The one remaining boundary condition is at the outside edge of the mesh. We find V from the known $\mathbf{E}$ there. For example, if $\mathbf{E}$ is $\mathbf{E} = E_0\mathbf{a}_x$ then $V = -E_0x$ can be used all along the outside boundary. A similar boundary condition is imposed

in the example which follows by initially setting V to a linearly varying set of
values everywhere. The potential on the boundary is then assumed to be "known",
since only the points away from the boundary are included in the declaration which
states where V is "unknown".   When the "solve" command is given and the
equation for V is solved it is only done for the locations where V is unknown. The
'equ' statement controlling which values of i and j are involved says "all", but this
is interpreted as meaning all the points where V is unknown and where previous
equations do not apply.

Now that we know all the boundary conditions we can solve $\nabla^2 V = 0$, the
numerical form of which has been given several times above.

Input File diel.sg

```
//   FILE:  sgdir\ch02\diel\diel.sg
//
//   Purpose: This example illustrates the solution of Laplace's equation,
//      and the use of the boundary conditions that D-normal is continuous,
//      and E-tangential is continuous. Look for continuity of Dx at a
//      boundary normal to x and continuity of Ey at the same boundary, etc.
//      Continuity of tangential E is 'automatic' the way the mesh is set up,
//      but continuity of normal D has to be put in explicitly.
//
//      The dielectric is sitting in an external field which is determined
//      by the potential V which is specified on the outside edge of the region.

const NX = 41;                     // number of x-mesh points
const NY = 41;                     // number of y-mesh points
const LX = NX - 1;                 // index of last x-mesh point
const LY = NY - 1;                 // index of last y-mesh point
const X1 = NX / 2 - 5;             // x-location of first corner
const X2 = NX / 2 + 5;             // x-location of second corner
const Y1 = NY / 4;                 // y-location of corner
const Y2 = 3 * NY / 4;             // y-location of corner
const DX =   1.0;                  // mesh spacing
const A  = -0.1;                   // constants which specify
const B  = -0.2;                   // potential on the edge of the region
const EPSr= 4.0;                   // relative permittivity
const EPSo= 8.854e-12;             // permittivity of vacuum (approximately)

var V[NX,NY],Ex[LX,LY],Ey[LX,LY],Dx[LX,LY],Dy[LX,LY];
var temp[LX,LY],tempv[NX,NY];


// The V array is bigger by 1 along each axis than the E and D
// arrays, since when the differentiation of V is done to find
// E and D we cannot go all the way to the edge. See below.
// All voltages are unknown except for the mesh points on the
// perimeter.
```

```
unknown V[1..LX-1,1..LY-1];

// Use D normal continuous. epsilon-zero and DX cancel out of the equations.
equ V[i=X1,j=Y1..Y2] -> -(V[i,j]-V[i-1,j]) = - EPSr*(V[i+1,j]-V[i,j]) ;
equ V[i=X2,j=Y1..Y2] -> -(V[i+1,j]-V[i,j]) = - EPSr*(V[i,j]-V[i-1,j]) ;
equ V[i=X1..X2,j=Y1] -> -(V[i,j]-V[i,j-1]) = - EPSr*(V[i,j+1]-V[i,j]) ;
equ V[i=X1..X2,j=Y2] -> -(V[i,j+1]-V[i,j]) = - EPSr*(V[i,j]-V[i,j-1]) ;

// Laplace's equation
equ V[i=all,j=all] ->
  {V[i+1,j] + V[i,j+1] -4.0*V[i,j] + V[i-1,j] + V[i,j-1]}/sq(DX) = 0.0;

begin main

  assign V[i=all,j=all]= A*j + B*i;
  solve;

  // Find E and D from the potential V.
  assign Ex[i=all,j=all] = - (V[i+1,j] - V[i,j])/DX ;  // Ex = - dV/DX
  assign Ey[i=all,j=all] = - (V[i,j+1] - V[i,j])/DX ;  // Ey = - dV/dy
  assign Dx[i=all,j=all]          = EPSo*Ex[i,j] ;      // Dx = EPSo*Ex in air
  assign Dx[i=X1..X2-1,j=Y1..Y2] = EPSr*EPSo*Ex[i,j] ; // Dx = EPSr*Ex in diel
  assign Dy[i=all,j=all]          = EPSo*Ey[i,j] ;      // Dy = EPSo*Ey in air
  assign Dy[i=X1..X2,j=Y1..Y2-1] = EPSr*EPSo*Ey[i,j] ; // Dy = EPSr*Ey in diel

  // transpose the values in the matrices for the purposes of plotting
  assign tempv[i=all,j=all] = V[j,i] ;
  assign V[i=all,j=all] = tempv[i,j] ;
  assign temp[i=all,j=all] = Ex[j,i] ;
  assign Ex[i=all,j=all] = temp[i,j] ;
  assign temp[i=all,j=all] = Ey[j,i] ;
  assign Ey[i=all,j=all] = temp[i,j] ;
  assign temp[i=all,j=all] = Dx[j,i] ;
  assign Dx[i=all,j=all] = temp[i,j] ;
  assign temp[i=all,j=all] = Dy[j,i] ;
  assign Dy[i=all,j=all] = temp[i,j] ;
end
```

2.13.7   Use of Complex Geometries: a MEMS Example

One of the most important features of the software we will be using is that it handles very complex meshes in a relatively straightforward fashion. To show how this works in a context which is physically simple but at the same time interesting and important in the microelectronics industry, and specifically for MEMS (Microelectromechanical Systems), we can solve an electrostatic problem for a component of a micromachine.

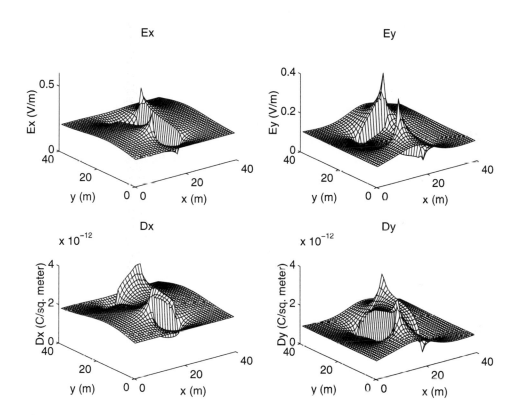

Figure 2.15. The fields around a slab of dielectric in a constant external electric field.

The micromachine consists in part of two interlocking metal combs. Different voltages are applied to each comb. The force they exert on each other is used to move one of the combs in the direction toward or away from the other. To describe the potential between the combs we need to set up a mesh, and this we do in turn by means of a mesh skeleton.

The mesh skeleton is built up from a list of points, defined by the coordinates of the points. From pairs of points we can construct a list of edges. By joining edges we can define regions. The edges and regions are given labels, so that different equations can be solved in different regions.

In many cases the mesh skeleton can be typed in by hand. In the example we are going to be using, the skeleton is simple but highly repetitive, so the skeleton file was written by a simple computer program. The skeleton file mem.sk is fairly simple to understand (if rather long). Many of the lines in the file have been deleted, the deletions being marked with blank lines.

The first part of the file defines points by their coordinates.

The next part of the file defines edges. Some edges are defined to belong to regions such as METC. The definition of the edge in each case is the pair of points it is made up from. These points are specified in square brackets. Finally in parentheses is given an initial spacing of mesh points to be used along the edge, and how much that spacing varies along the edge. Most sets of parentheses given below read something like $(5.00e - 2, 0.0)$ denoting a spacing of $5.00e - 2$ units which is constant all along that edge.

Finally, regions are defined. Each region is said to belong to a particular physical region, which in this case means one of the metal regions, such as META or METC, or free space FREE. Then the edges which go around the perimeter of the region are listed in curly brackets. Finally, the word RECTANGLES indicates that the region should be divided into rectangles (which implies that the region itself is a rectangle).

Skeleton File mem.sk

```
point p0000 = (1.000000e+00, 2.100000e+00);
point p0002 = (1.100000e+00, 2.100000e+00);
point p0004 = (1.200000e+00, 2.100000e+00);
point p0006 = (1.300000e+00, 2.100000e+00);
point p0008 = (1.400000e+00, 2.100000e+00);
point p000a = (1.500000e+00, 2.100000e+00);
point p000c = (1.600000e+00, 2.100000e+00);
point p000e = (1.700000e+00, 2.100000e+00);
point p001e = (1.000000e+00, 2.000000e+00);
point p0020 = (1.100000e+00, 2.000000e+00);

and so on....

point p0190 = (1.500000e+00, 8.000000e-01);
point p0192 = (1.600000e+00, 8.000000e-01);
```

```
point p0194 = (1.700000e+00, 8.000000e-01);

edge e0001 = METC [p0000, p0002] (5.00e-2, 0.0);
edge e0003 = METC [p0002, p0004] (5.00e-2, 0.0);
edge e0005 = METC [p0004, p0006] (5.00e-2, 0.0);
edge e0007 = METC [p0006, p0008] (5.00e-2, 0.0);

and so on....

edge e0193 = METC [p0192, p0194] (5.00e-2, 0.0);
edge e0011 = [p0002, p0020] (5.00e-2, 0.0);
edge e0013 = [p0004, p0022] (5.00e-2, 0.0);
edge e0015 = [p0006, p0024] (5.00e-2, 0.0);
edge e0017 = [p0008, p0026] (5.00e-2, 0.0);
edge e0019 = [p000a, p0028] (5.00e-2, 0.0);

edge e0041 = [p0040, p0042] (5.00e-2, 0.0);
edge e0043 = FREEMETA [p0042, p0044] (5.00e-2, 0.0);
edge e0045 = FREEMETA [p0044, p0046] (5.00e-2, 0.0);
edge e0047 = [p0046, p0048] (5.00e-2, 0.0);
edge e0049 = [p0048, p004a] (5.00e-2, 0.0);
edge e004d = [p003e, p005c] (5.00e-2, 0.0);
edge e004f = [p0040, p005e] (5.00e-2, 0.0);
edge e0051 = [p0042, p0060] (5.00e-2, 0.0);
edge e0053 = [p0044, p0062] (5.00e-2, 0.0);
edge e0055 = FREEMETA [p0046, p0064] (5.00e-2, 0.0);
edge e0057 = FREEMETA [p0048, p0066] (5.00e-2, 0.0);
edge e005b = [p005a, p005c] (5.00e-2, 0.0);
edge e005d = FREEMETB [p005c, p005e] (5.00e-2, 0.0);
edge e005f = FREEMETB [p005e, p0060] (5.00e-2, 0.0);
edge e0061 = FREEMETB [p0060, p0062] (5.00e-2, 0.0);

edge e0181 = [p0172, p0190] (5.00e-2, 0.0);
edge e0183 = [p0174, p0192] (5.00e-2, 0.0);

region r0010 = FREE {e000f, e001f, e0011, e0001} RECTANGLES;
region r0012 = FREE {e0011, e0021, e0013, e0003} RECTANGLES;
region r0014 = FREE {e0013, e0023, e0015, e0005} RECTANGLES;
region r0016 = FREE {e0015, e0025, e0017, e0007} RECTANGLES;
region r0018 = FREE {e0017, e0027, e0019, e0009} RECTANGLES;

region r006a = FREE {e0069, e0079, e006b, e005b} RECTANGLES;
region r006c = METB {e006b, e007b, e006d, e005d} RECTANGLES;
region r006e = METB {e006d, e007d, e006f, e005f} RECTANGLES;
region r0070 = METB {e006f, e007f, e0071, e0061} RECTANGLES;

region r0180 = FREE {e017f, e018f, e0181, e0171} RECTANGLES;
region r0182 = FREE {e0181, e0191, e0183, e0173} RECTANGLES;
```

```
region r0184 = FREE {e0183, e0193, e0185, e0175} RECTANGLES;


coordinates x, y;
const xmax = 10.0;
const PI = 3.1415927;

set minimum length = 0;
set maximum length = 1.0e-00;
set minimum length = 5.0e-04;
set minimum divisions = 0;
set maximum divisions = 0;
```

The input file which uses this mesh skeleton is mem.sg. This file

1. reads a mesh file, mem.sk and

2. uses equation specifications which refer to regions, instead of ranges of indices. For example, the equation for V in each metal region, META, METB or METC, is just that V is equal to a constant. In the free-space region, FREE, the equation for V is Laplace's equation.

Laplace's equation is written using 'nsum'. The command nsum means neighbor sum; for a given node it does a sum over the neighboring nodes to which the original node is attached. It is used to implement the Laplacian, ∇^2 in this example file. This is done starting from the fact that $\nabla^2 = \nabla \cdot \nabla$, that is to say, first we need to find the gradient of our scalar and then we take the divergence of that gradient.

The 'box integration' method is covered in detail in Chapter 5. The component of ∇V along the line between any two nodes is approximated in the usual way as being the difference in the values of V at the nodes, divided by the distance between the nodes. Within the nsum, the quantity $(V[node(i,j)] - V[i])/elen(i,j)$ is this component; $elen(i,j)$ is the distance between node i and its jth neighbor. On the other hand, $ilen(i,j)$ is the length of the perpendicular bisector of the line between those nodes. To find the divergence of the vector, we take its flux through the surface enclosing the central node and divide by the volume around the central node. The surface is made up of the perpendicular bisectors (since this is a two-dimensional problem) and the volume is then the area enclosed by the perpendicular bisectors. The flux through each perpendicular bisector is grad, found as was just described, multiplied by the length of the bisector. The approximation to *grad* given above is thus multiplied by $ilen(i,j)$, and the sum is run over all the neighbors. To find the divergence we should then divide by the volume (the area, in our example). This is done in the file below, although since the right hand side of $\nabla^2 V = 0$ is zero, we could multiply through by that volume and have it vanish from the equation.

Input File mem.sg

// FILE: sgdir\ch02\mem\mem.sg

Commands to Set Up mem.msh

```
mesh mem.sk
sggrid mem.xsk
sgbuild ref mem_ref
mem_ref
```

Commands to Run mem.sg

```
sgxlat -p16 mem.sg
order mem.top mem.prm
sgbuild sim mem
mem
```

```
mesh "mem.msh";

var V[NODES];
unknown V[FREE];
known V[META], V[METB], V[METC];

set NEWTON DAMPING    = 8;                    // maximum damping 2^(-Damping)
set NEWTON ITERATIONS = 100;
set NEWTON ACCURACY   = 1e-10;
set LINSOL ALGORITHM  = GAUSSELIM;
set LINSOL FILL       = INFINITY;

equ V[i=FREE] ->
  nsum(i,j,ALL,{(V[node(i,j)]-V[i])/elen(i,j)}*ilen(i,j))/area(i) = 0.0;

begin main
  assign V[i=META] = +1.0;
  assign V[i=METB] = -1.0;
  assign V[i=METC] =  0.0;
  solve;
  write;
end
```

The skeleton file can be used to set up a mesh, using the commands given in the table, and the simulation itself is run using the commands in the second table.

A postcript file called mem.ps can be made using SGFramework's plotting capabilities. After running the program, one can specify

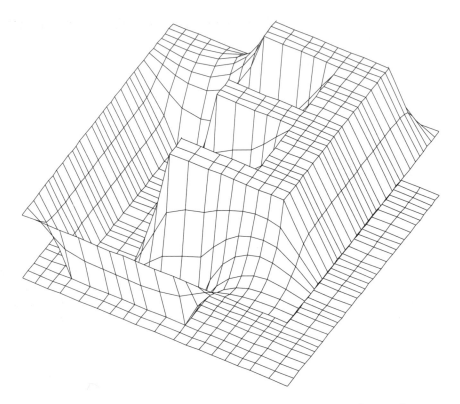

Figure 2.16. The potential around a simple interlocking-comb metal structure.

```
triplot mem V -wf -ps
```

If the -wf were left out, a color file would be produced which would be easier to see than the black-and-white wire frame plot we used here. Leaving out the -ps would plot the picture on the screen. The potential found is shown in Figure 2.16.

2.14   Summary

This chapter has reviewed several major conceptual items:

1. Vector calculus; gradient, divergence and curl; with emphasis on their physical meaning, which is essential for both electromagnetism and for transport theory.

2. Electromagnetism; its equations and boundary conditions on the electromagnetic fields; and

3. The numerical implementation of all of the above and solution of complex problems in electromagnetism by numerical means.

The Chapter 3 continues to present the overview of the physics of semiconductor devices, concentrating on the way in which particle transport is described.

PROBLEMS

1. Using the input file ex1.sg, numerically differentiate the function $\cos(0.2i)$ with respect to x. What is $0.2i$ equal to, given that $\Delta x = 0.1$ in this file? Plot i.) the original function, ii.) the result of the differentiation and iii.) the result you would get by doing the differentiation analytically, all on the same axes.

2. In this problem we vary the mesh size in our solution of Laplace's equation in a rectangular box and examine the effects.

   When changing the mesh size Δx, alter it by a factor of two on each occasion. Keep $\Delta y = \Delta x$.

   First make Δx smaller. (Record how long the run takes, each time.) If the solution changes significantly, keep on making Δx smaller until the solution does not change significantly with a change in Δx. (The meaning of significantly is a matter of judgment, but a change of more than a few percent might be considered significant.)

   If the solution does not change as the mesh size Δx gets smaller, next make Δx larger until the solution changes significantly. Finally, make Δx smaller until the solution takes 'too long' (which probably means that it takes more than ten minutes, although for a research problem 'too long' might mean it takes days). Plot the time to obtain a solution versus Δx, with both variables plotted on logarithmic scales. Explain the variation you observed.

3. The example file which follows is designed to evaluate the electrostatic potential due to a ring of charge. Examine the file and check that it is correctly set up. Where is the ring of charge located? In what plane is the potential calculated?

   Input File emprop2.sg

```
// sgdir\ch02\emprop2\emprop2.sg

// simulation parameters
const NS  = 42;              // number of points on source
const LS  = NS - 1;          // index of last mesh point on source
const NX  = 20;              // number of x-mesh points
const LX  = NX - 1;          // index of last x-mesh point
const NY  = 20;              // number of y-mesh points
const LY  = NY - 1;          // index of last y-mesh point
const dth = 2.0 * 3.1415/NS ; // angle between points on source
```

```
const r0  = 0.07;              // radius of ring source
const x0  = 0.1 ;              // x-coord of center of ring
const y0  = 0.1 ;              // y-coord of center of ring
const z0  = 0.1 ;              // z-coord of center of ring
const dx1 = 0.01;              // x-spacing of field mesh
const dy1 = 0.01;              // y-spacing of field mesh
const z1  = 0.12;              // z-coord of field point


// Potential due to point charge
func VP(x,y,z)
   const  pre = 9.0e+9;
   assign R = sqrt(x*x+y*y+z*z);
   assign V1 = pre/R;
   return V1;


var  dx[NX,NY,NS], dy[NX,NY,NS], dz[NX,NY,NS];
var  VT[NX,NY,NS], V[NX,NY];

//   sum contributions to V
equ VT[i=all,j=all,k=1..LS] ->
    VT[i,j,k] = VT[i,j,k-1] + VP(dx[i,j,k],dy[i,j,k],dz[i,j,k]);

begin main
   assign dx[i=all,j=all,k=all] = i*dx1 - (r0 * cos(k*dth) + x0) ;
   assign dy[i=all,j=all,k=all] = j*dy1 - (r0 * sin(k*dth) + y0) ;
   assign dz[i=all,j=all,k=all] = z1 - z0 ;
   assign VT[i=all,j=all,k=0] = VP(dx[i,j,k],dy[i,j,k],dz[i,j,k]);
   solve;
   assign V[i=all,j=all] = VT[i,j,LS] ;
   write;
end
```

The potential obtained is shown in Figure 2.17. Is this potential a reasonable result, in both shape and magnitude?

Add an extra variable, E_x, and, using an 'assign' statement, evaluate and plot the x-component of the electrostatic field from V, in the same plane. Is the field you obtained consistent with the potential? Explain.

4. Modify the file 'emprop2.sg' so that it describes a circle of charge of the same shape, but the half of the circle with $y < y_0$ now has the opposite charge from the half of the circle with $y > y_0$. Plot V and E_x in the same plane as before, and comment on whether your results are as expected.

5. The next few questions can be done using the Lamé functions (or scale factors) introduced in the text. Write out the Lamé functions for (a) Cartesian coordinates, (b) cylindrical coordinates and (c) spherical coordinates. Use your results to obtain

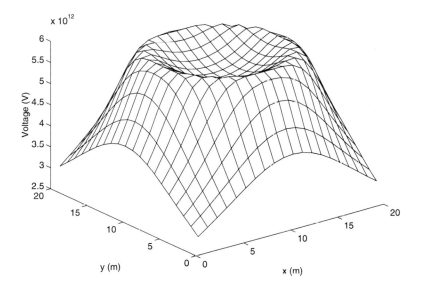

Figure 2.17. The results of using a propagator method to find the potential due to a ring of charge.

expressions for *grad, div* and *curl* in each case.

6. The example in input file ex3.sg implements Laplace's equation in one Cartesian coordinate. Modify this file so as to work in the cylindrical coordinate r, that is, the cylindrical radius. Compare the results of the simulation to the results for the Cartesian case.

7. The example files exgrad.sg, exdiv.sg and excurl.sg implement *grad, div* and *curl* in two Cartesian coordinates. Modify these files to work in the cylindrical coordinates (r, z). Are the functions which were given to be differentiated appropriate for cylindrical coordinates? If so, plot the results of using these files in cylindrical coordinates and compare to the equivalent Cartesian cases.

8. In the text, Laplace's equation was solved numerically inside a rectangular box, using two Cartesian coordinates. Suppose the problem was changed to one in which the cylindrical variables (r, z) were used. Suppose the radial variable ran from $r = 2$ to $r = 3$, and z ran from $z = 0$ to $z = 1$. $V(r = 2) = 1.0$, while all the other boundaries are held at zero potential. How do you expect the solution to change compared to the Cartesian case? Modify the input file given, so that it corresponds to the problem described, and solve Laplace's equation. Plot the results and show the difference between this and the Cartesian result. A second file in the text describes a capacitor subject to an alternating voltage, in a conducting box, in Cartesian coordinates. Modify this file to use the coordinates (r, z), in a box of the same dimensions as given above, run the file and comment on the results as compared to those in the Cartesian case.

9. Helmholtz's equation was solved subject to periodic boundary conditions, next to a current sheet, in the text. Modify the input file to work in the cylindrical coordinates (r, θ). Make the periodicity be in the right coordinate! Note that up until now in these problems we have used cylindrical coordinates (r, z). Run the simulation, plot and explain the results.

10. The example file diel.sg describes a dielectric in an external electric field. Modify the file so that the field is parallel to one axis. Run the simulation, for a long thin piece of dielectric aligned (a) parallel to the E-field and (b) perpendicular to the E-field. Plot and comment on the results.

11. Write a simple computer program which follows field lines, and try to find a method to plot the results. Use the analytic expression for the electrostatic potential due to a point charge, to find the potential due to an electric dipole and an electric quadrupole. Follow a few representative field lines in each case.

12. Find the electric field due to a point charge in a perfectly conducting square box, by solving Poisson's equation numerically. Since a true point charge cannot be represented on a mesh, let the charge density be nonzero in one cell at the center of a rectangular mesh. Set the potential on the outer boundary, which is the perfect conductor, to zero.

13. Compare the potential you obtain by solving Poisson's equation in the box in the previous question to the analytic expression for the potential due to a point charge at

the center of the box. Make sure the numerical solution and the propagator use the same charge. The charge in the numerical solution is the charge density multiplied by the cell volume. The two solutions should not be exactly the same. (Why not? Look at the boundary for clues.)

14. This question is intended to illustrate the basis of the method of images. The potential at the edges of the box in the previous examples was set to zero. Suppose we want to find the potential due to a point charge, but we want to make the potential along a line near the the point charge be zero because there is a grounded conductor there. We can use the analytic expression for the potential due to a point charge to make the potential along the line be zero. We do this by putting the actual charge in its real position and putting an equal and opposite charge at an equal perpendicular distance on the other side of the line. Plot the potential due to a unit charge at a position $(x = 0, y = 1)$ above a grounded plane along the y-axis at $x = 0$, by putting a unit positive charge at $(x = 0, y = 1)$ and a unit negative charge at $(x = 0, y = -1)$. Use the analytic expression for the potential due to a point charge. The region below $y = 0$ is really inside the metal, so the potential there is zero in fact. The solution obtained using the two point charges is only valid above $y = 0$. The negative charge is the image charge, and this is the simplest example of the use of the method of images.

Use the method of images at all four sides of the box in the first problem given on a point charge in a conducting box, with one image charge behind each side, to find the approximate potential inside the box. Plot the solution. Why is this solution only approximate?

15. The potential difference between two infinitely long concentric conducting cylinders is 10 V. Solve Laplace's equation in one dimension in the region between the cylinders, (a) analytically and (b) using the finite-difference method, if the inner cylinder has radius 2 meters and the outer cylinder has radius 3 meters. Find the electric field immediately outside the inner cylinder. From this find the total charge per unit length on the outside of the inner cylinder. Repeat the calculation of the field and charge, at the inside of the outer cylinder. Repeat this for concentric spheres, all the other information being unchanged.

16. A parallel-plate capacitor of horizontal width 1 meter, vertical spacing 0.1 meter and infinite length has a slab of dielectric with $\epsilon_r = 4$, inserted between the plates so that it fills the vertical gap and extends horizontally halfway across the gap. What are the appropriate boundary conditions on V at the edges of the dielectric slab?

If the voltage on the top plate is $V = 1$ volt and the bottom plate is grounded, find the potential V between the plates using a finite-difference scheme. (Make some simplifying assumption at the outer horizontal edge of the plates, such as that the potential varies linearly from top to bottom of the gap.) Calculate the potential approximately, analytically, and compare the two sets of results.

Repeat this for the case when the dielectric extends all the way horizontally but only halfway vertically.

17. This question involves the solution of Laplace's equation, in cylindrical coordinates (r, θ). In the first part the problem will be independent of θ, but θ-dependence will

be introduced later, so it might be best to solve the whole problem in two dimensions.

Consider two concentric conducting cylinders, the inner cylinder having outer radius $r = a$ and the outer cylinder having inner radius $r = b$. The inner cylinder is covered in a dielectric layer out to a radius $r = d$, where $d < b$. What is the appropriate boundary condition on the electrostatic potential, at each side of the dielectric layer?

If the relative dielectric constant of the dielectric is $\epsilon_r = 4$, and $a = 1.0$ while $b = 10.0$ meters, calculate the potential V for various values of d. Plot the electric field E as a function of r in each case. Where is the peak electric field in each case? Does the peak field ever change its location suddenly as d is varied?

The outer cylinder is now modified so that from $\theta = 0$ to $\theta = \pi$ its inner radius is $0.8b$ while for the rest of the range of θ its radius is unchanged. We shall choose $d < 0.8b$ now. How are the boundary conditions altered?

Now suppose the inner cylinder is modified so that its outer radius is $r = 2a$ for θ from 0 to π, but is unchanged elsewhere. The outer cylinder is restored to its original shape. The dielectric still covers the inner cylinder; its outer radius is $d > 2a$. How are the boundary conditions altered in this case?

Solve whichever of these two two-dimensional problems is easiest to set up. Where is the peak electric field when $d = 0.5b$?

18. The next two questions involve transport behavior, but since the transport is the flow of current they belong more in the present chapter.

Consider a medium having a current density given by $\mathbf{J} = \sigma \mathbf{E}$. The conservation of charge means that in steady state $\nabla \cdot \mathbf{J} = 0$. This in turn means that

$$\nabla \cdot (\sigma \mathbf{E}) = -\nabla \cdot (\sigma \nabla V) = 0 \tag{2.222}$$

where σ is the conductivity, $\mathbf{E}$ is the electric field and V is the electrostatic potential.

Suppose that the space between a pair of parallel-plate electrodes, of which one is at $x = 0$ and the other is at $x = L$, is filled with a material with conductivity $\sigma = \sigma_0 \left(1 + \frac{x}{L}\right)$, where $\sigma_0 = 1\,S/m$, and $L = 0.1$ meters. Suppose the plate at $x = 0$ is grounded and the plate at $x = L$ is at 10 volts.

Analytically and by means of a finite-difference scheme calculate the following, and check that the results agree: (a) $V(x)$, (b) $J_x(x)$, (c) the charge density (per unit volume) inside the material $\rho(x)$ and (d) the charge density (per unit area) on each plate. What is the resistance per unit area, $R = \frac{V}{J_x}$? Find an equivalent circuit (which in this case will be a combination of one or more resistors and capacitors) for the material between this pair of electrodes. Specify the layout of the circuit elements and their values.

19. Repeat the previous problem, but in cylindrical geometry, with $\sigma = \frac{0.1}{r}\sigma_0$, with a grounded cylindrical electrode at $r = 0.1$ meters and a second cylindrical electrode at 10 volts at $r = 0.2$ meters. (Replace quantities per unit area from the previous problem with quantities per unit length, as necessary, in your answers.) Compare your solution with that for $\sigma = \sigma_0$, and explain the differences.

20. This question is intended to illustrate the different types of vectors (and in particular basis vectors) used in general curvilinear coordinate systems. The two types of basis

vectors we usually define, when we use a general coordinate system (u^1, u^2, u^3) are tangent basis vectors, $\mathbf{e}_1, \mathbf{e}_2, \mathbf{e}_3$ and reciprocal basis vectors, $\mathbf{e}^1, \mathbf{e}^2, \mathbf{e}^3$. If $\mathbf{r}$ is the position vector, then varying (say) u^1 while keeping u^2 and u^3 constant causes $\mathbf{r}$ to move along a curve which is tangent to the surface in which u^2 and u^3 are constant. The tangent vector $\mathbf{e}_1 \equiv \partial \mathbf{r}/\partial u^1$, and similarly for the other two coordinates. The reciprocal basis vectors, on the other hand, are simply the gradients of the coordinates: $\mathbf{e}^1 \equiv \nabla u^1$, and so on [7]. The importance of these basis vectors is that they obey an orthogonality relation, $\mathbf{e}^i \cdot \mathbf{e}_j = \delta^i_j$. The orthogonality relation allows vectors to be decomposed into components in a general coordinate system.

Plot the basis vectors of each kind, for a.) spherical coordinates and b.) toroidal coordinates. Is the orthogonality relation valid in each case?

21. Modify the file tlpulse.sg to use an implicit scheme to evaluate the spatial derivatives at the current time. This will require the finite-difference expressions to be evaluated using 'equation' statements. Refer to Chapter 3 for more information on implicit schemes, if necessary. Compare the results to those obtained from the file tlpulse.sg.

22. The Bessel equation is

$$\frac{d^2\psi}{dx^2} + \frac{1}{x}\frac{d\psi}{dx} + \left(1 - \frac{n^2}{x^2}\right)\psi = 0. \tag{2.223}$$

The Legendre equation is

$$\left(1 - x^2\right)\frac{d^2\psi}{dx^2} - 2x\frac{d\psi}{dx} + l(l+1)\psi = 0. \tag{2.224}$$

What physical situations give rise to these equations? Consider Laplace's equation in cylindrical and spherical coordinates. In one case it is the 'radial part' of the equation obtained by separation of variables which is needed, $x = k\rho$, and in the other case $x = \cos\theta$. See ref. [6]. (a) Set up a finite-difference scheme to solve the Bessel equation. Solve the equation for an integer value of n (which could be zero) and for appropriate boundary conditions on $\psi(x = 0)$ and $\psi(x = 1)$, corresponding to a real physical problem of your choice. (b) Repeat part (a) for the Legendre equation with a suitable choice of the 'angular eigenvalue' l. Comment on the results.

Chapter 3

Transport Phenomena and their Numerical Analysis

The purpose of this chapter is (1) to describe the basic equations of transport theory, and (2) to show how they are solved numerically. The conservation of particles is described by a continuity equation, which makes use of the divergence operator; the fluxes are often proportional to gradients, so the vector operators and their numerical forms are essential here as well as in electromagnetism. The boundary conditions needed for different transport equations are explained. Then, once again, a number of numerical examples are given, some of which illustrate basic topics from numerical analysis such as accuracy and stability of numerical schemes.

This chapter is concerned with transport processes, particularly those which can occur in semiconductor devices. Chapter 2 provided background material on electromagnetism, some of which is needed to model semiconductors. This chapter completes the discussion of the basic physical processes which is needed to describe the device and the way it functions. As before, the numerical implementation of all the equations is emphasized throughout.

In most of this document, the main transport considered involves the fluid flow of the electrons and holes in a semiconductor device. Fluid transport will be introduced here. Transport must, in some cases, be studied using more detailed models, for instance when the mean free path is not very small compared to other typical lengths, such as the device dimensions. This 'long mean free path' transport is discussed in Chapter 13.

In order to introduce fluid transport models, we begin with simpler sets of fluid equations than those which describe carrier flow. Indeed, a logical approach to thinking about semiconductors might begin with the way in which semiconductors are made. The processes used in some of the stages of semiconductor fabrication are usually described using fluid transport equations similar to those which describe the flow of carriers. The example of dopant diffusion into a semiconductor is an important instance where this is true; it is used in much of this chapter both because it is a good illustration of diffusion processes, and because of its inherent importance.

122

In Chapter 13, on kinetic transport models, we shall consider other processes such as ion implantation and how they may be described.

Section 3.1 introduces the fluid transport equations and reviews the behavior of systems governed by diffusion equations. Finite-difference schemes will be described in Section 3.2 and used in several classes of examples that follow. The treatment given here is meant to be straightforward. A detailed description of more complicated finite-difference schemes is given in Chapter 5, which discusses numerical methods. After discussing various aspects of finite-difference methods, and applying them to dopant diffusion (in Subsection 3.2.3), we then turn to an alternative way of describing transport, based on 'propagators' or Green's functions, in Section 3.3. The performance of propagator methods and finite-difference methods are compared, to see which is most accurate and efficient.

3.1 Conservation Equations

The simple diffusion equation is obtained from the equation of conservation of particles, which is

$$\frac{\partial n}{\partial t} + \nabla \cdot \mathbf{\Gamma} = S. \tag{3.1}$$

n is the number density of particles (usually called simply the density). S is the production rate of particles per unit volume per second and $\mathbf{\Gamma}$ is the particle flux per unit area.

The key to understanding this equation is to recall the meaning of the divergence of a vector. The divergence of a vector quantity is defined to be the flux of the quantity leaving a very small volume Δv, after that flux has been divided by Δv. Because we find the divergence of the vector by taking the flux leaving Δv and then dividing that flux by Δv, it is helpful to take the conservation equation and multiply by that small volume Δv. The three parts of the equation then have a clear physical meaning.

The first term becomes the rate of change in the number of particles in the small volume.

$$\frac{\partial n}{\partial t}\Delta v = \frac{\partial(n\Delta v)}{\partial t} = \frac{\partial N}{\partial t} \tag{3.2}$$

where $N = n\Delta v$ is the number of particles in Δv.

The second (divergence) term when multiplied by Δv becomes the rate at which particles leave the volume through the surface, since by definition of the divergence, for a small enough Δv

$$\nabla \cdot \mathbf{\Gamma} = \frac{\text{flux out}}{\Delta v}. \tag{3.3}$$

The right-hand side of Equation 3.1 becomes $S\Delta v$, which is the rate of production of particles in the small volume.

If particles are conserved, then all those particles produced in a volume per second must be added to the number in the volume (in that second) or must leave through the surface (in the second). This equation makes what is therefore a reasonable statement, that the increase in the number of particles in the volume per second, $\frac{dN}{dt}$, plus the number leaving the surface of the volume per second (the flux of $\mathbf{\Gamma}$ out of the whole surface of the volume) equals the number produced in the volume in a second.

If $\mathbf{\Gamma} = -D\nabla n$, with D being the diffusion coefficient, then the particle-conservation equation reads

$$\frac{\partial n}{\partial t} - \nabla \cdot (D\nabla n) = S. \tag{3.4}$$

When D is independent of position, this is

$$\frac{\partial n}{\partial t} - D\nabla^2 n = S. \tag{3.5}$$

As was described in Chapter 2, a negative curvature (e.g. due to an initial spike in the density) makes the right side negative and so n decreases - as it should.

3.1.1   Boundary Conditions for Conservation Equations

In this book, the transport equations being solved are conservation equations, which are usually diffusion equations, either time-dependent or steady-state. The boundary conditions for these cases will be described next. In some cases, the transport is not primarily diffusive, however. The density is still described by a conservation equation, but the flux may be mainly due to a flow with a drift velocity rather than diffusion. This case needs a different boundary condition, as we shall see.

The conservation equation, in general, has a time-dependent solution. An initial condition must be specified, which is the value of the density at all points inside the region where the density is to be calculated, at the start of the time period of interest. The equation is solved numerically by stepping it forward in time, advancing by an amount of time Δt at each step.

The density at the edges of the region must also be provided externally to the numerical calculation. If the solution is being advanced in time, the density at the edges must be input to the calculation at every time step. The specification of the density at the edges is the boundary condition in space; the initial condition is the boundary condition in time.

Boundary Conditions for Diffusion

Physically it is reasonable to specify the initial condition for the diffusion equation, as opposed to a final condition in time. The diffusion equation describes processes which decay exponentially and which 'forget' the initial condition as time progresses. We cannot run a diffusion equation backward from a final condition and expect to recover the initial solution exactly, because it has 'forgotten' the details and will not be able to reverse itself. In other words, many initial conditions will give the same final state. A diffusion equation run backward cannot choose the actual initial condition from all the initial conditions which could have given the same final condition.

The spatial boundary condition used on a diffusion equation is the specification of the density on all the edges. This is an example of a Dirichlet boundary condition. (Other common types are described in Chapter 4, on the semiconductor equations.) In steady state, the diffusion equation is of the same basic form as the equation for the electrostatic potential, and so it has the same type of boundary condition.

Boundary Conditions for Convection

In some situations the flux is due to a drift of the fluid with a drift velocity $\mathbf{v}_{dr}$, so that $\mathbf{\Gamma} = n\mathbf{v}_{dr}$. In this case information is carried along in the direction of the flow, so boundary conditions specified downstream cannot effect the density upstream. In one dimension, the steady-state conservation equation in the absence of a source of particles becomes

$$\nabla \cdot \mathbf{\Gamma} = \frac{\partial}{\partial x}\left(nv_x\right) = S = 0. \tag{3.6}$$

Solving this gives $nv_x = \Gamma_0$ which is constant. The appropriate boundary condition is the specification of the density, at one point at the upstream end of the simulation region.

In two or more dimensions the flow traces out lines. The boundary condition for this equation should now be the specification of the density where the lines enter the simulation region. Specifying the density at more than one point on any given line will usually lead to a contradiction.

If diffusion is present, but the flow is dominant, we will have flow lines as described above which determine the direction the fluid travels. Boundary conditions will need to be applied upstream, not downstream. The extra spatial derivative in the flux makes the continuity equation a second-order PDE, and so, it needs an extra boundary condition. The density and its normal derivative at the boundary where the flow lines enter should now be given. In a numerical scheme this is equivalent to specifying the density on the first two rows of mesh points, at the boundary where the flow enters the simulation region.

3.2 Finite-Difference Solution of Transport Equations

In a problem which is one-dimensional in space, the density is to be calculated at discrete positions in space such as $x_i = i\Delta x$ and at discrete instants in time such as $t_k = k\Delta t$. The density is then $n(x_i, t_k)$ or equivalently $n(i, k)$. The derivatives with respect to time and space are typically approximated in finite-difference (FD) form as

$$\left(\frac{\partial n}{\partial t}\right)_{(i,k)} \simeq \frac{n(i, k+1) - n(i, k)}{\Delta t} \tag{3.7}$$

and

$$\left(\frac{\partial n}{\partial x}\right)_{(i,k)} \simeq \frac{n(i+1,k) - n(i-1,k)}{2\Delta x}. \tag{3.8}$$

As explained in Chapter 2, the second derivative with respect to space can be derived from the first derivative in the usual way

$$\left(\frac{\partial^2 n}{\partial x^2}\right)_{(i,k)} \simeq \frac{\left(\frac{\partial n}{\partial x}\right)_{(i+1,k)} - \left(\frac{\partial n}{\partial x}\right)_{(i-1,k)}}{2\Delta x}. \tag{3.9}$$

When i or $i \pm 1$ appear in the argument of n or as the subscript on the partial derivative, they are indicating where n or the partial derivative is to be evaluated: at the point labeled by i or $i \pm 1$.

The first derivatives at $i \pm 1$ can be found by shifting the above expression for the first derivative by ± 1.

$$\left(\frac{\partial n}{\partial x}\right)_{(i+1,k)} \simeq \frac{n(i+2,k) - n(i,k)}{2\Delta x} \tag{3.10}$$

and

$$\left(\frac{\partial n}{\partial x}\right)_{(i-1,k)} \simeq \frac{n(i,k) - n(i-2,k)}{2\Delta x}. \tag{3.11}$$

Using these in the approximation to the second derivative gives

$$\left(\frac{\partial^2 n}{\partial x^2}\right)_{(i,k)} \simeq \frac{n(i+2,k) - 2n(i,k) + n(i-2,k)}{(2\Delta x)^2}. \tag{3.12}$$

Since this expression uses the values of the density at the positions $x = i\Delta x$ and $x = i\Delta x \pm 2\Delta x$, it must also be possible to do the estimate using the densities at $x = i\Delta x$ and $x = i\Delta x \pm \Delta x$. This amounts to a rescaling in x by a factor of 2, so the factor of $2\Delta x$ is replaced by Δx. The second derivative is then approximated as

$$\frac{\partial^2 n}{\partial x^2} \simeq \frac{n(i+1,k) - 2n(i,k) + n(i-1,k)}{(\Delta x)^2}. \tag{3.13}$$

These discrete approximations can be put into the diffusion equation. Using a short-hand form in which the subscript indices on the derivative indicate the finite-difference approximation to the derivative, this becomes

$$\left(\frac{\partial n}{\partial t}\right)_{(i,k)} - D\left(\frac{\partial^2 n}{\partial x^2}\right)_{(i,k)} = S(i,k). \tag{3.14}$$

The new value of n at position i and time $k+1$ appears only in the time derivative. If the density n is known at all locations i at the old time k, then the discrete approximation gives a set of equations which can be solved easily for the density at each position i at the new time $k + 1$. An explicit expression for $n(i, k + 1)$ can be found using the expression for $\frac{\partial n}{\partial t}$.

$$\frac{n(i, k + 1) - n(i, k)}{\Delta t} = D \left(\frac{\partial^2 n}{\partial x^2} \right)_{(i,k)} + S(i, k) \qquad (3.15)$$

so that the new density is given by

$$n(i, k + 1) = n(i, k) + \Delta t \left[D \left(\frac{\partial^2 n}{\partial x^2} \right)_{(i,k)} + S(i, k) \right]. \qquad (3.16)$$

Because we can write $n(i, k + 1)$ explicitly in terms of known quantities, this is called an explicit finite-difference scheme. This kind of explicit scheme is usually sufficiently accurate, if a short enough time step is taken. In semiconductor applications, a more accurate scheme is almost always called for, which can be more difficult to implement. The difficulty of implementation is taken care of by the SGFramework so there is no reason not to use the more accurate scheme in what follows.

The inaccuracy in the explicit scheme follows, in part, from the approximation used to estimate $\frac{\partial n}{\partial t}$. This derivative is found using the values of n at times k and $k + 1$. It is more properly the derivative at time $t = (k + \frac{1}{2})\Delta t$. But if the left side of the equation is found at time $(k + \frac{1}{2})\Delta t$, then the right-hand side should also be found at time $(k + \frac{1}{2})\Delta t$. This is done by averaging the right side over the two times, which, using the subscript notation, reads

$$\left(\frac{\partial n}{\partial t} \right)_{(i,k+1/2)} \simeq \frac{1}{2} \left(D \left[\left(\frac{\partial^2 n}{\partial x^2} \right)_{(i,k)} + \left(\frac{\partial^2 n}{\partial x^2} \right)_{(i,k+1)} \right] + S(i, k) + S(i, k + 1) \right).$$
$$(3.17)$$

This approximation to the diffusion equation cannot be used to obtain an explicit expression for $n(i, k + 1)$ in terms of the old quantities at time k. The right side involves not only $n(i, k + 1)$ but also $n(i \pm 1, k + 1)$. It can only be solved to find the new densities by solving all the equations for all values of i simultaneously. The new density $n(i, k + 1)$ is given *implicitly* by this discrete equation, so this is called an implicit finite-difference scheme.

One advantage of this implicit scheme is that it has a beneficial feedback in it which the explicit scheme does not. If a large Δt is used in the explicit scheme, the change in n always varies in proportion with Δt, regardless of how big Δt or n become. In the implicit scheme, as Δt increases, the density appearing in $\frac{\partial^2 n}{\partial x^2}$ can change as well. This can prevent the change in n from becoming arbitrarily large. If $n(i, k + 1) - n(i, k)$ does not grow as fast as Δt, then when Δt is very large, the time derivative $(n(i, k + 1) - n(i, k))/\Delta t$ becomes small. The right-hand side must also become small when this, the left-hand side, becomes small. For a very large step Δt, the density which is the solution of the implicit equation is the steady-state density which is found by setting the right-hand side to zero:

$$D \left(\frac{\partial^2 n}{\partial x^2} \right)_{(i,k)} + S(i,k) = 0. \tag{3.18}$$

(The steady-state solution at times k and $k+1$ should be the same, so the later time was dropped from this equation.)

3.2.1 Accuracy of Finite-Difference Schemes

There is a limit on the time step which can be used in a finite-difference description of transport. If the expressions for the finite-difference form of the conservation equation are examined, the density in each cell is given in terms of the old densities in that cell, and in the neighboring cells. Suppose that the time step is so big that particles could, in reality, get into a cell from several cells away, in that time step. The FD expression is not accurate in this situation, because it only uses the density in the neighboring cells $i \pm 1$ (as well as in cell i) to find the density in cell i.

The two physical processes with which we usually are concerned which move particles around are diffusion and flow. We need to know how long it takes a particle to diffuse across a cell, and how long it takes a particle to flow across a cell, because if either of these times is less than the time step, the FD scheme will be inaccurate and may be unstable.

The time to flow across a cell width Δx is just

$$\Delta t_{Flow} = \frac{\Delta x}{v} \tag{3.19}$$

where v is the velocity of the motion. For diffusive processes, particles travel a distance Δx in a time of about

$$\Delta t_{Diff} = \frac{(\Delta x)^2}{2D}. \tag{3.20}$$

The time step Δt of the numerical scheme is required to be smaller than these; in practice, less than about a tenth of both of these 'transport times'. In an explicit scheme violation of this rule will usually lead to an instability of the solution, whereas in an implicit scheme it usually leads to inaccuracy. (If the implicit scheme is following time dependence to get to a steady-state solution, the inaccuracy may not matter to the steady-state.)

3.2.2 Conservative Schemes

A useful and potentially important feature of the type of scheme given by these discrete approximations, is that the scheme is conservative. This means that it adds no new particles except when it is physically appropriate to do so, because the numerical approximations conserve particles exactly. This is possible because the divergence of the flux is found from the difference in the fluxes entering and leaving a cell. The divergences at neighboring cells are found from differences in the same fluxes, so as long as the exact same expressions for the fluxes are used at the neighboring cells, the exact same flux removed from one cell is added at another. To be precise, let the flux going from cell A to cell B be $\Gamma_{A \to B}$. The expressions for the divergence at cell A and the divergence at cell B should contain the exact same expressions for $\Gamma_{A \to B}$ if the scheme is to be conservative. This issue will be important in the box integration Scheme which is introduced in Chapter 5.

In the example which follows, we show that the flux due to diffusion is implemented in such a way that it only leads to particles being moved around, with the exact number taken from one location being put back in some other location or locations.

The change in density due to diffusion is proportional to $\frac{\partial^2 n}{\partial x^2}$, which is, in turn, proportional to

$$n(i+1,k) - 2n(i,k) + n(i-1,k)$$

for some time k. (This assumes uniform mesh spacing so that $x_i = i\Delta x$.) At the adjacent location $i+1$ the equivalent expression in the diffusion term is

$$n(i+2) - 2n(i+1,k) + n(i,k).$$

At the other adjacent location $i-1$ the expression is

$$n(i,k) - 2n(i-1,k) + n(i-2,k).$$

Adding these,

$$
\begin{array}{llll}
n(i+2,k) - 2n(i+1,k) & +n(i,k) & & \\
+n(i+1,k) & -2n(i,k) & +n(i-1,k) & \\
& +n(i,k) & -2n(i-1,k) + n(i-2,k) & \\
= n(i+2,k) - n(i+1,k) & & -n(i-1,k) + n(i-2,k) &
\end{array}
$$

This shows the cancellation of terms involving $n(i,k)$. If we added the expressions from $i \pm 2$, the terms involving $n(i \pm 1, k)$ would also cancel, and so on until we reached the edges. At the edges, the boundary conditions determine whether a flux leaves the region or not.

Conservative schemes for other types of flux are straightforward to construct, provided the finite-differences are performed systematically.

3.2.3   Diffusion of Dopants

The diffusion of dopants is an extremely important issue in semiconductor modeling. From the point of view of modeling, it can be described straightforwardly, using the methods described above, provided we know the diffusion coefficients and can specify the boundary conditions. The issue of ion implantation, which should ideally come before diffusion, is probably not described well by the classes of equations we are treating here. We will return to it later in Chapter 13.

The two most common ways to think of the process in which ions diffuse into a semiconductor, are with an inexhaustible supply of dopants at the surface of the semiconductor, and with a finite supply of dopants which start near the surface. In the first case we have what is called a Dirichlet boundary condition, namely, that the density of dopants is known at the surface through which they enter the semiconductor. This density does not change with time in this case. At the other boundaries, we do not expect much of the dopant to arrive during the course of our simulation, so we might set the dopant density to zero. If we had symmetry about that boundary, we might use a zero-flux condition, and set the normal derivative of the density to zero.

In the second case which is frequently examined, the only difference is at the boundary where dopants are introduced. Since the supply is finite, no more dopants enter or leave through the surface, so we have a zero-flux condition and the normal derivative of the density is zero.

In some circumstances the effects of an electric field set up by the charged particles must be included in the description. The inclusion of the drift due to an electric field in the transport of charged particles is described below. The electric field, in such a case, is found by assuming the particles obey Boltzmann statistics and that the material is quasi-neutral. Quasi-neutrality is discussed in later chapters, beginning with Chapter 4. It implies that the charge density is near to zero. The charge can be approximated as zero for most purposes. (The one exception is in Poisson's equation. The fields produced by what little charge there is in a quasi-neutral region may still be significant.) In particular, if N_A is the number density of acceptor impurities, N_D is the number density of donor impurities and all the impurities are fully ionized, then $N_A - N_D$ is the net density of negative ions. In this case quasi-neutrality implies that this plus $p - n$, the density of holes minus the density of (conduction) electrons, must add up to zero. Since p and n are known in terms of the potential V through a Boltzmann relation, then the potential V can be found by setting the charge density equal to zero.

In the illustrations which follow next the electric field will be neglected. Including the electric field is fairly simple, if the mobility is known, and E is found as discussed above. The mobility can be found from the diffusion coefficient, using the Einstein relation $D = \frac{kT}{e}\mu$. (See Chapter 4.) The extension to the finite-difference equation is given in the next section. The diffusion coefficient is quite complex and is usually assumed to have an Arrhenius form

$$D = D_o \exp\left(-\frac{E_{act}}{k_B T}\right) \tag{3.21}$$

where E_{act} is an activation energy. Values for D_o and E_{act} for various dopants are given in [9]. Unfortunately data is not well known for many of the relevant processes. To give an example, for phosphorus (diffusing through neutral vacancies) in silicon D_o is 3.85 cm^2/s and E_{act} is 3.66 eV. If T were room temperature, $k_B T$ would correspond to about 0.03eV and the exponential would be about e^{-122}. The need to drive in the dopants at high temperature is clear.

Fixed Surface Density of Dopants

The first simulation given is for the case of a fixed dopant density at the semiconductor surface. The result of the simulation is shown in Figure 3.1. Among the points to note about this example,

1. One of the boundary conditions is that the density is fixed along part of the boundary. This implies that there are so many dopants on the surface that they are not depleted significantly during the diffusion.

2. The normal derivative of the density is zero elsewhere, implying no diffusive flux through those boundaries. The top boundary does indeed have no flux through it, because there is nowhere for the dopants to go (except along that part of the top boundary where the high density of dopants is placed, since extra dopants can come in from there). The boundary condition is not very realistic at the other boundaries unless they are very far away, or unless the problem is symmetric about that boundary so that there is no flux.

The diffusion coefficient has been set to an arbitrary, small value in cm^2/s, which if expressed in m^2/s, would be 10^{-4} times this value. Changing the value of D does not change the shape of the density profile at all, but it changes the time it takes for the dopants to diffuse into the silicon. It is a useful exercise to calculate the real value of D and find the time it would actually take for the entire diffusion which is being described here, assuming a temperature of the order of a thousand degrees Kelvin. If the time seems unrealistic, it should be possible to estimate what the temperature must be to have the diffusion take a more reasonable length of time. (A typical diffusion process would take tens of hours.) Once the diffusion coefficient is known, it is possible to estimate what time step is needed in the simulation.

Input File dif1.sg

```
// FILE:   sgdir\ch03\dif1\dif1.sg
// Dopant diffusion simulation, for fixed dopant concentration at the surface.

// constants
const NX    = 20;                   // number of mesh points, x-direction
const NY    = 20;                   // number of mesh points, y-direction
const L     = 10.0e-4;              // distance scaling (cm)
const dx    = L/NX;
const dy    = L/NY;
const T     = 300.0;                // operating temperature
const D     = 1.0e-2;               // diffusion constant
const dt    = 0.1*dx*dx/(2*D);      // time step -- obeys Courant condition
const Xmin  = 8;
const Xmax  = 12;

const q     = 1.602e-19;            // electron charge (C)
const kb    = 1.381e-23;            // Boltzmann's constant (J/K)
```

```
const e0     = 8.854e-14;          // permittivity of vacuum (F/cm)
const eSi    = 11.8;               // relative permittivity of Si
const eSiO2  = 4.2;                // relative permittivity of SiO2
const Ni     = 1.25e10;            // intrinsic concentration of Si (cm^-3)


// declare variables and specify which one are unknowns
var itime;
var iwrite;
var x[NX];
var y[NY];
var N[NX,NY], Nold[NX,NY];
unknown N[all,all];
known N[Xmin..Xmax,0];


// Diffusion equation
equ N[i=1..NX-2,j=1..NY-2] ->
D*(N[i+1,j  ]-2*N[i,j]+N[i-1,j  ])/(dx*dx) +
D*(N[i  ,j+1]-2*N[i,j]+N[i  ,j-1])/(dy*dy) -
(N[i,j]-Nold[i,j])/dt = 0.0;

//boundary at i=NX-1
equ N[i=NX-1,j=1..NY-2] ->
N[i,j] -N[i-1,j]= 0.0;

// boundary at i=0
equ N[i=0,j=1..NY-2] ->
N[i,j] - N[i+1,j] = 0.0;

// boundary at j=NY-1
equ N[i=1..NX-2,j=NY-1] ->
N[i,j] - N[i,j-1] = 0.0;

// first boundary at j=0
equ N[i=1..Xmin-1,j=0] ->
N[i,j] - N[i,j+1] = 0.0;

// second boundary at j=0
equ N[i=Xmax+1..NX-2,j=0] ->
N[i,j] - N[i,j+1] = 0.0;

// corner boundary conditions:
equ N[i=0,j=0] ->
N[i,j] - N[i+1,j+1] = 0.0;

equ N[i=0,j=NY-1]->
N[i,j] - N[i+1,j-1] = 0.0;
```

```
equ N[i=NX-1,j=0]->
N[i,j] - N[i-1,j+1] = 0.0;

equ N[i=NX-1,j=NY-1]->
N[i,j] - N[i-1,j-1] = 0.0;


// set the numerical algorithm parameters
set NEWTON DAMPING     = 3;
set NEWTON ACCURACY    = 1.0e+4;
set NEWTON ITERATIONS  = 100;
set LINSOL ALGORITHM   = GAUSSELIM;
set LINSOL FILL        = INFINITY;


begin InitVars
  assign x[i=all] = i*dx;
  assign y[i=all] = i*dy;
  assign N[i=5..15,j=0] = 1.0e16;
  assign Nold[i=all,j=all] = N[i,j];
end

begin main
  assign itime = 0;
  call InitVars;
  while (itime < 2000) begin
    assign Nold[i=all,j=all] = N[i,j];
    solve;
    assign itime = itime + 1;
    assign iwrite = iwrite + 1;
    if (iwrite == 100) begin
      write;
      assign iwrite = 0;
    end
  end
end
```

Fixed Number of Dopants

The equivalent calculation with a fixed number of dopants injected near the surface is shown in Figure 3.2. The 'fixed number' of dopants are free to move, so the dopant concentration at the surface can be depleted. In this case, the boundary condition at the surface is a density gradient of zero normal to the surface, like all the other surfaces of the simulation region. A boundary condition of no density gradient means that there is no (diffusive) flux through the boundary. In the pre-

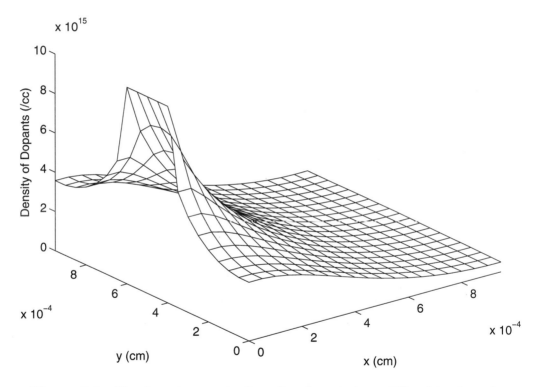

Figure 3.1. The dopant concentration after dopants have diffused in through a surface. The dopant concentration at the surface is so large that it effectively does not decrease due to loss of dopants during the diffusion.

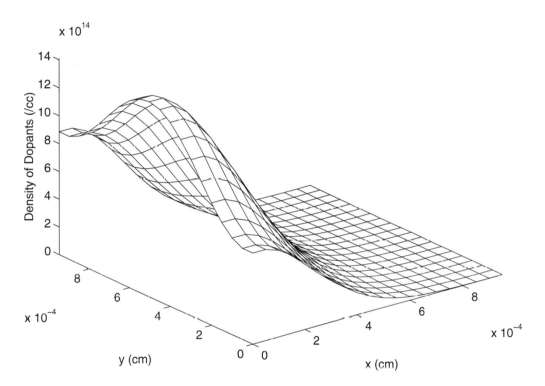

Figure 3.2. Dopant density after a diffusion. A fixed amount of dopant was used at the surface, and this density was depleted as dopants diffused away from the surface.

vious example, the density itself was held fixed along that part of the top surface, instead of its normal derivative. In addition, in the present example, there is an initial condition that the initial density is nonzero along a part of the top surface.

The input file is essentially the same as before, except as noted, so it will not be included here. The file used is dif2.sg. Figure 3.2 is a snapshot of the density profile after only one-tenth of the run which is simulated.

The analytic treatment of these two cases will be given in the remainder of this section. The reason why an analytic treatment is helpful is that it provides expressions, which can be compared to numerical results in order to test them. It also gives profiles for the dopant density which can be used in device simulations, as an alternative to a numerical calculation of the dopant density. The numerical calculation will be applicable in more general cases, since the analytic expressions are usually valid only in one dimension and for rather special cases which are not realistic.

Analytic Treatment of Dopant Diffusion

The two special cases of the boundary conditions, which were examined above, are now treated analytically, in one spatial dimension. The mathematical treatment in a case where a fixed amount of dopant is allowed to diffuse into the silicon is also discussed in some detail in a later section on the subject of 'propagators'. A propagator (also known as a Green's function) is the mathematical expression for the result produced by a unit 'driving force'. In the case of a diffusion equation, adding particles is the 'driving force' which provides a density. Adding one unit of density at a point in space at some time, causes there to be a Gaussian density centered at that point at later times. Adding more density elsewhere and at other times creates other Gaussian contributions to the density. The other Gaussians have different widths which depend on the time since the density was added, and each Gaussian is centered on the point where its density was added. The total density can be found by summing the contributions of all the individual Gaussians. Finding the density in this way is sometimes convenient, and this is the reason why 'propagators' are useful in diffusion problems. For details of the way this is handled mathematically, see the section on propagators.

The other 'classic' doping profile is produced when the amount of dopant is not taken to be fixed, as it was above. Instead, an inexhaustible supply is available at the surface of the semiconductor. The solution, in this case, is most easily introduced by starting with the answer and showing why it works. In one dimension the solution of the diffusion equation which satisfies this boundary condition is

$$N(x,t) = N_{surf} \left(1 - erf \left[\frac{x}{\sqrt{4Dt}} \right] \right). \tag{3.22}$$

Here D is the diffusion coefficient, and t is the time for which diffusion has been taking place, so Dt is not a time step. The function 'erf' is called the error function. One minus the error function is called the complementary error function, or '$erfc$'.

The error function is defined to be

$$erf(x) = \frac{2}{\sqrt{\pi}} \int_0^x \exp(-s^2)\, ds \tag{3.23}$$

so in this case, where the argument is $\frac{x}{\sqrt{4Dt}}$, the error function is

$$erf\left(\frac{x}{\sqrt{4Dt}}\right) = \frac{2}{\sqrt{\pi}} \int_0^{x/\sqrt{4Dt}} \exp(-s^2)\, ds. \tag{3.24}$$

To show that this is a solution of the diffusion equation, we shall differentiate it and substitute into the equation. When differentiating an integral such as

$$I = \int_{a(x,t)}^{b(x,t)} f(s)\, ds \tag{3.25}$$

with respect to x the result is

$$\frac{dI}{dx} = \frac{db}{dx} f(b) - \frac{da}{dx} f(a). \tag{3.26}$$

Since $a = 0$ this means that $\frac{dI}{dx} = f(b)\frac{db}{dx}$ and similarly $\frac{dI}{dt} = f(b)\frac{db}{dt}$.

Let A be the partial derivative of the above error function with respect to x; then

$$A = \frac{2}{\sqrt{\pi}} \cdot \frac{1}{\sqrt{4Dt}} \exp\left(\frac{-x^2}{4Dt}\right). \tag{3.27}$$

The next derivative with respect to x is then

$$B = \frac{\partial A}{\partial x} = \frac{1}{\sqrt{\pi Dt}} \left(-\frac{2x}{4Dt}\right) \exp\left(\frac{-x^2}{4Dt}\right). \tag{3.28}$$

The derivative of the error function with respect to time is

$$C = \frac{2}{\sqrt{\pi}} \frac{x}{\sqrt{4D}} \left(-\frac{t^{-3/2}}{2}\right) \exp\left(\frac{-x^2}{4Dt}\right). \tag{3.29}$$

These results show that $-D\,B = \frac{x}{2\sqrt{\pi}Dt^{3/2}} \exp\left(\frac{-x^2}{4Dt}\right)$, and C is equal to minus this, so the diffusion equation is satisfied by this error function. The diffusion equation is also satisfied by a constant, so the expression given for $N(x,t)$ satisfies the diffusion equation.

Satisfying the equation is only part of solving a partial differential equation. Satisfying the boundary conditions, including initial conditions when appropriate, is also necessary. The boundary condition in this problem is that $N(x = 0, t) = N_{surf}$ for all time. The error function is clearly zero at $x = 0$, so the boundary condition is satisfied.

3.2.4   Diffusion of Charged Particles in an Electric Field

If an electric field is present, then charged particles experience a force which causes them to drift with velocity $\mathbf{v}_{dr}$, given by their mobility μ multiplied by the electric field $\mathbf{E}$, $\mathbf{v}_{dr} = \mu \mathbf{E}$. The equation of conservation of particles has an extra term due to the flux $\mathbf{\Gamma}_E = n\mathbf{v}_{dr}$ leading to $\nabla \cdot \mathbf{\Gamma}_E$ appearing in the divergence of the flux. This means there is an extra term $\nabla \cdot (n\mu(-\nabla V))$, which we need to 'difference', in order to solve the conservation equation numerically.

In one dimension and using a simple central differencing scheme we find for constant mobility μ that

$$\frac{\partial}{\partial x}(n\mu(-\frac{\partial V}{\partial x})) \simeq -\frac{\mu}{2\Delta x}\left[n(i+1)\left(\frac{\partial V}{\partial x}\right)_{i+1} - n(i-1)\left(\frac{\partial V}{\partial x}\right)_{i-1}\right]. \qquad (3.30)$$

The right side of this can be written in a more useful form so that

$$\nabla \cdot \mathbf{\Gamma}_E \simeq -\mu\left[n(i+1)[V(i+2) - V(i)] - n(i-1)[V(i) - V(i-2)]\right]/(2\Delta x)^2. \qquad (3.31)$$

This expression can be added to the rest of the divergence of the flux in the finite-difference form of the diffusion equation given above. It is in conservative form. If the electric field is constant, then it simplifies to

$$\mu E \,\frac{n(i+1) - n(i-1)}{2\Delta x}.$$

3.2.5   Steady-State Diffusion

Steady-state diffusion is a straightforward case of the more general time-dependent situation. It will be examined separately, primarily because the examples which will be given are conveniently classified as steady-state problems.

Finite-Difference Solution of the Steady-State Diffusion Equation

The steady-state diffusion equation is simply the time-dependent equation with the time derivative set to zero. From the point of view of solving the finite-difference equation the absence of the time derivative may actually make the problem harder, however. Advancing the solution in time can be done using an explicit scheme, if the time step is small enough. The steady-state equation, as normally written, requires simultaneous solution of a system of equations if it is to be solved directly. An alternative to solving a system of equations is to put back the time dependence and advance the diffusion equation in time until it reaches steady-state. Even Laplace's equation is sometimes solved by introducing an artificial time evolution into it, converting it into a diffusion equation in the process. Thinking of the way in which the time-dependent version of the problem evolves over time can be a useful guide to the way in which some iterative schemes (which find a solution by a process akin to evolving the problem in time) converge or fail to converge. Using the equ statement in SGFramework leads to a relatively direct solution of the finite-difference equation, so the time-stepping issue need not concern us here. (If the equations are linear, they may be solved directly via Gaussian elimination, or iteratively via SOR or a PCGM. Nonlinear problems need an iterative method. SGFramework's default Numerical Algorithm Module implements a damped Newton's method which calls a linear solver, which may be (sparse) Gaussian elimination, Gaussian elimination is direct, or SOR or PCGM, both of which are iterative.)

The boundary conditions on the steady-state diffusion equation are essentially the same as the transient problem, except that the initial condition is not needed. If an arbitrary initial condition is used in a linear system and the solution is advanced in time, then, provided it converges, the equation will still reach the correct steady-state. The more complex problems we shall examine later are nonlinear, however, and the final state does depend on the initial condition in some of these cases. The diffusion equation cannot be run backwards in time, because information about the initial state is not transmitted into the final steady-state density which is obtained. The system forgets its initial state, so if time is reversed the results will be meaningless.

Steady-State Density with Fixed Boundary Conditions and with a Fixed Electrostatic Potential

This section will be devoted to a few examples which in some cases are difficult to solve analytically but are straightforward to solve numerically using SGFramework. The first example of a fixed potential which should be considered is one where there is no potential. In this case the steady-state diffusion equation

$$-D\nabla^2 n = S \qquad (3.32)$$

is mathematically identical to Poisson's equation,

$$\nabla^2 V = -\frac{\rho}{\epsilon} \qquad (3.33)$$

if there is a source S of particles present, and to Laplace's equation $\nabla^2 V = 0$ if there is no source, in which case $-D\nabla^2 n = 0$.

All of these problems can be solved with an input file such as cont.sg. cont.sg is set up to solve the steady-state continuity equation, with no source, but with a fixed electric field E. This file is the result of a simple modification of a file which was used to solve Laplace's equation.

3.2.6 Steady-State Convection

A two-dimensional, steady-state continuity equation in which the flux is convective,

$$\mathbf{\Gamma} = n\mathbf{v}_{dr}, \qquad (3.34)$$

and $\mathbf{v}_{dr}$ is the drift or flow velocity, can be solved solved using a file (rain.sg) given below. This problem demonstrates several important points about modeling transport and about convection, which are discussed in this section.

Input File cont.sg

```
// FILE:  sgdir\ch03\cont\cont.sg
// demonstration of solution of continuity equation

const NX = 21;                  // number of x-mesh points
const NY = 21;                  // number of y-mesh points
const LX = NX - 1;              // index of last x-mesh point
const LY = NY - 1;              // index of last y-mesh point
const DX = 1.0;                 // mesh spacing
const D  = 1.0;                 // diffusion coefficient
const mu = 1;                   // mobility
const E  = 1;                   // electric field

var n[NX,NY],x[NX],y[NY];

// Continuity equation
equ n[i=1..LX-1,j=1..LY-1]  ->
  - D * {n[i-1,j] + n[i,j-1] - 4*n[i,j] + n[i+1,j] + n[i,j+1]} / sq(DX)
  + mu * E * {n[i+1,j] - n[i-1,j]} / (2.0*DX) = 0.0;

begin main
  assign x[i=all] = i*DX;
  assign y[j=all] = j*DX;
  assign n[i=all,j=all] = 1.0*i/NX;
  solve;
end
```

First, the boundary conditions in a problem where flow dominates the transport are all specified 'upstream'. The flow velocity can be used to define 'flow lines' in the same way that any vector field can be integrated to generate lines (see Chapter 2.) The upstream boundary condition is put on where 'each' line enters the simulation region. Boundary conditions put on anywhere downstream will usually contradict the information carried along by the flow from the upstream point, and so downstream boundary conditions are not appropriate.

If the drift velocity $\mathbf{v}_{dr}$ is independent of the density, then the divergence of the flux gives a first-order partial differential equation. In this case, one boundary condition, the value of the density at the upstream end of the flow line, is used. There are often reasons why the flow velocity should depend on the density, however. If a second derivative of the density appears in the divergence of the flux, then a second boundary condition is needed at the upstream end. The density and its derivative normal to the boundary are normally used as boundary conditions. If, for example, the boundary is at $y = 0$, corresponding to the index j being zero, then the normal derivative is

$$\frac{\partial n}{\partial \ell} \simeq (n[i,1] - n[i,0])/\Delta y. \tag{3.35}$$

Specifying $n[i,0]$ and $n[i,1]$ is thus equivalent to specifying n and $\frac{\partial n}{\partial \ell}$.

The file which follows is intended to be a simple representation of rain water running off a hillside. The flow velocity $\mathbf{v}_{dr}$ was taken to be proportional to minus the gradient of the height h of the hill.

$$\mathbf{v}_{dr} = -\mu \nabla h. \tag{3.36}$$

As a second part of the problem, h was allowed to vary with time according to

$$\frac{\partial h}{\partial t} = -\beta \mathbf{v}_{dr}^2. \tag{3.37}$$

However, if allowing h to change with time allows it to develop a minimum, this drift velocity will be inwards towards the minimum from all sides, and the density will become infinite there. This model for the erosion rate suffers from several problems. It could be argued that the erosion rate should depend on the amount of water flowing, not just its speed. In addition, whichever model we use, the simulation has a tendency to become unstable as the erosion progresses. For this reason, β has been set to such a low value, in the file given, that the erosion is negligible. It might be interesting to change β, and to change the erosion model, and to see what happens.

When a nonphysical behavior such as this occurs, it is necessary to go back to the physical basis for the model to find the reason why it failed. Now in a self-consistent problem where the flow velocity depends on the density, there will usually be a negative feedback which pushes particles away from a maximum in the density. In a system of charged particles, the electric field set up by an accumulation of particles of one charge will push them apart. This is the sort of feedback we will expect to encounter in most of the work in this book. If the hill in the present problem has a minimum of h, in other words it has a hole in it, the minimum/hole will fill up with water until the water surface is flat and then water will stop flowing in. To attempt to model this we set the flow velocity to

$$\mathbf{v}_{dr} = -\mu \nabla(h + n) \tag{3.38}$$

where n is the water 'density' measured in units where the density is equal to the depth of the water.

The inclusion of the depth in this way provides the negative feedback we need. As the depth increases, the flow will eventually be away from the hole, once $h + n$ has a maximum there. The extra term in the flow velocity is mathematically equivalent to the contribution of diffusion and leads to a need for a second (upstream) boundary condition.

The input file which follows, rain.sg, makes use of a simple 'upwind' or 'upsteam' differencing, to handle the flow which carries information in the flow direction only.

Input File rain.sg

```
// FILE:  sgdir\ch03\rain\rain.sg
// Demonstration of fluid transport by flow, as opposed to diffusion
// Simulation of 'water flow' and an
// attempt at calculating the erosion rate.
const NX = 21;              // Number of x-mesh points
const NY = 21;              // Number of y-mesh points
const LX = NX - 1;          // Max. index on x-mesh
const LY = NY - 1;          // Max. index on y-mesh
const i0 = LX/2;            // Reference point on x-mesh
const j0 = LY;              // Reference point on y-mesh
const n0 = 1.0e+1;          // Reference value for density/depth
const H0 = 1.0e+10;         // Reference value for height
const S0 = 0.0e+18;         // Max. source rate
const beta = 1.0e-25;       // Sets rate of erosion
const mu = 1.0e+2;          // Sets flow velocity
const DX = 1.0e+03;         // Mesh spacing in x
const DY = 1.5e+03;         // Mesh spacing in y
const DT = 1.0e+6;          // Time step
const x0 = i0*DX;           // Reference position in x-direction
const y0 = (j0+5.0)*DY;     // Reference position in y-direction
const tmax = 40.0*DT;       // Run time.


SET NEWTON ACCURACY = 1.0e-9;


var time, n[NX,NY],ho[NX,NY],hn[NX,NY],Source[NX,NY],X[NX],Y[NY];
unknown n[0..LX-2,2..LY];


//
// continuity equation for flux of water
//
equ n[i=0..LX-2,j=2..j0] -> - mu * (
    (n[i+1,j] - n[i  ,j])*
    (ho[i+1,j]-ho[i  ,j] + n[i+1,j] - n[i  ,j])/sq(DX)
  +  (n[i,j ] - n[i,j-1])*
    (ho[i,j ]-ho[i,j-1] + n[i,j ] - n[i,j-1])/sq(DY)

  +  n[i,j] * ( (ho[i  ,j] + ho[i+2,j] - 2*ho[i+1,j]
                +n[i  ,j] +  n[i+2,j] - 2* n[i+1,j] )/sq(DX)
              +(ho[i,j-2] + ho[i,j ] - 2*ho[i,j-1]
                +n[i,j-2] +  n[i,j ] - 2* n[i,j-1] )/sq(DY)  )
              ) = Source[i,j];

// The next few lines allow the flow to come in through
// the other side in y as well as at y=0, if needed.
// To use these, j0 must be set to (say) LY/2 .
```

```
//
//equ n[i=0..LX-2,j=j0..LY-2] -> - mu * (
//      (n[i+1,j] - n[i  ,j])*
//      (ho[i+1,j]-ho[i  ,j] + n[i+1,j] - n[i  ,j])/sq(DX)
//   +  (n[i,j+1] - n[i,j ])*
//      (ho[i,j+1]-ho[i,j ] + n[i,j+1] - n[i,j ])/sq(DY)
//
//   +  n[i,j] * ( (ho[i  ,j] + ho[i+2,j] - 2*ho[i+1,j]
//                  +n[i  ,j] +  n[i+2,j] - 2* n[i+1,j] )/sq(DX)
//                +(ho[i,j ] + ho[i,j+2] - 2*ho[i,j+1]
//                  +n[i,j ] +  n[i,j+2] - 2* n[i,j+1] )/sq(DY)  )
//                ) = Source[i,j];
//
//

begin main
   assign time = 0.0;
   assign ho[i=all,j=all] = H0 * (sq(i*DX+x0)        // Initialize the height
                        +   4.0* sq(j*DY-y0) ); //
   assign hn[i=all,j=all] = ho[i,j] ;               // h(new) = h(old)
   assign n[i=all,j=all] = n0/sqrt(sq(i*DX + x0)+16.0*sq(j*DY-y0)) ; //Set n
           // n will be held at this value on the edges where flow comes in.
   assign Source[i=1..LX-1,j=1..LY-1] = S0; // Can set a source rate

   while (time <= tmax) begin
     solve;
     assign hn[i=0..LX-2,j=2..j0] = ho[i,j] - DT * beta * sq(mu)
              *( sq( (ho[i+1,j]-ho[i,j])/DX )        // Erosion rate
               + sq( (ho[i,j]-ho[i,j-1])/DY ) );     // of the height
     assign ho[i=all,j=all] = hn[i,j] ;             // Update h(old)
     assign time = time + DT;
     write;
   end
end
```

3.3 Propagators in Transport Theory

The ideas of propagators, which were introduced in Chapter 2, on electrostatics, are also very useful in studying transport. For example, the propagator for a simple time-dependent diffusion equation is a Gaussian distribution of density. The initial condition is a delta function of the density, which means a single particle at a point $\mathbf{r}'$ at the time t_o. At a later time t the density due to this initial 'spike' of density is

$$n(\mathbf{r}, t) = \frac{1}{\left[2\sqrt{\pi D(t - t_o)}\right]^{n_d}} \exp\left(-\frac{(\mathbf{r} - \mathbf{r}')^2}{2n_d D(t - t_o)}\right) \qquad (3.39)$$

where the number of dimensions is n_d, and we expect $n_d = 1, 2$ or 3 although the expression would work for other integer values of n [6] . This density is the propagator, which is also called a Green's function,

$$G \equiv G(\mathbf{r} - \mathbf{r}', t - t_o). \tag{3.40}$$

When a constant flow velocity $\mathbf{v}_{dr}$ is included, it effectively changes the initial position $\mathbf{r}'$ in the propagator; the same expression holds but with $\mathbf{r}'$ replaced by $\mathbf{r}' + \mathbf{v}_{dr}(t - t_o)$. (See also the exercises in Chapter 4.)

(The steady-state diffusion equation is essentially the same as Poisson's equation. The steady-state transport propagator in the absence of a drift velocity is thus the same as the electrostatic propagator.)

If the rate of production of particles per unit volume and per unit time is $S(\mathbf{r}, t)$, then the number of particles added in a time dt at time t' and in a volume dv' at position $\mathbf{r}'$ is

$$dN = S(\mathbf{r}', t') \, dt \, dv'. \tag{3.41}$$

The density profile this turns into, at a later time t, at some other position $\mathbf{r}$, is just

$$n(\mathbf{r}, t) = G(\mathbf{r} - \mathbf{r}', t - t') \, S(\mathbf{r}', t') \, dt dv'. \tag{3.42}$$

To find the density due to a source spread over time and space we need to sum the effects of all the contributions from the source. Often this would be written as an integral over the times and places where the particles were produced. If the expression is to be implemented numerically it will actually be done as a sum, however.

$$n(\mathbf{r}, t) = \sum_{(\mathbf{r}', t')} G(\mathbf{r} - \mathbf{r}', t - t') S(\mathbf{r}', t') \, \Delta t \, \Delta v' \tag{3.43}$$

where $\Delta v'$ is the volume of the cell on the mesh where the particles were added and Δt is the length of the time step in which they were added. This expression could conveniently be summed using lsum; see Section 15.3.5. These expressions are again convolution integrals, or in some cases convolution sums.

The propagator we have introduced here applies to any initial density which can be represented as a sum of 'point' densities. The densities in cells are considered to be nearly points, one point in each cell. The density added by a source S is just one way to produce that initial density. If instead of adding a density $S\Delta t$ to a cell we already had a density n in the cell, the density at later times due to this initial density would be given by the same expression, but with $S\Delta t$ replaced by n in the propagator integral. It is more usual to find the density by following the evolution of the density one time step at a time, than to try to find the density at a later time directly from S, as in the propagator integral over the sources at all times.

If we follow the density one step at a time, then we know the old density from the previous step, and we need to add to it the density added by the source in the next step and let the total density diffuse. The propagator allows for the diffusion. In general if the density in the cell at some time t' is $n(\mathbf{r}', t')$ and extra density is added in Δt equal to $S(\mathbf{r}', t')\Delta t$, then the density in the cell before diffusion is allowed for is $n(\mathbf{r}', t') + S(\mathbf{r}', t')\Delta t$. This combined density is used in the propagator integral. If the integral is being used to step the density through one time step Δt, then the 'new' time minus the 'old' time is $t - t' = \Delta t$, and since we are only allowing for one 'old' time, we do not integrate over time but only over space. The contributions from older times are already contained in the density at time t'.

$$n(\mathbf{r}, t) = \sum_{\mathbf{r}'} G(\mathbf{r} - \mathbf{r}', \Delta t) \left[n(\mathbf{r}', t') + S(\mathbf{r}', t')\Delta t \right]. \qquad (3.44)$$

3.3.1 Numerical Implementation of the Transport Propagator

When a propagator such as, for example, the Gaussian propagator is used in a numerical scheme to describe a transport process, the finite size of the mesh means that the 'point' initial densities are not quite points and the summation is not exactly equivalent to the analytic integral that it is supposed to correspond to. The finite mesh size can lead to inaccuracy. Avoiding some of the more serious types of errors is the main topic of this section.

It is equally important to conserve particles when using a propagator as when using an FD scheme, as was described in the section on conservative FD schemes. The way this is achieved is by normalizing the propagator. The Gaussian propagator is normalized so that when it is integrated analytically in two dimensions the answer is unity. When is is used on a mesh the answer is not likely to be exactly one, because the Gaussian is only sampled on the mesh. That is, we do not integrate over each cell. We evaluate an expression at the center of each cell and multiply by the cell volume. The cell-center value of the expression is supposed to be repesentative of the expression throughout the cell. The expression multiplied by the volume should then be a good approximation to the integral of the expression over the cell volume, but only an approximation.

If the Gaussian is evaluated at the cell centers, then multiplied by the corresponding cell volumes and the answers are summed, the sum $\sum_{Norm}$, which is supposed to be exactly one, will usually be close to one if a suitable mesh and time step Δt were used. If particles are to be conserved we need the propagator sum to be exactly one. This is accomplished by dividing the propagator for particles starting in each initial cell by the sum $\sum_{Norm}(\mathbf{r}')$ associated with that initial cell.

The other issue to be aware of is that although $G(\mathbf{r} - \mathbf{r}', t - t')$ is an exact solution of the transport equation for an initial delta-function density, the density in a finite cell is never a delta function. If Δt is very small, G might be so large that $G\Delta v \gg 1$ when G is evaluated at the cell center. Although the normalization will make sure that particles are conserved, the results will not be accurate unless particles have time to diffuse out of the cell. Otherwise G will still be peaked at the cell center. The time to diffuse a distance Δx is roughly $(\Delta x)^2/2D$. If Δt is much larger than this, then the exponential which appears in G can be seen to vary slowly across the cell as it should.

Simplified Numerical Form of Propagator

To illustrate how this works, we will begin with a simpler propagator than the Gaussian. This propagator takes the density in the initial cell, puts one-tenth of it in each adjacent cell and leaves two-tenths in the initial cell. The most obvious implementation of this is given in the file which follows, which is set up like the 'Game of Life' example, but with a different rule for how to advance the solution in time. It is worth noticing that the boundary condition on this simulation is that any density which leaves the edge can do so — nothing comes in through the edge.

Plots of the results may be extracted, exactly as was done for the Game of Life (but with the names of the files changed.)

Input File simpdiff.sg

```
// FILE:  sgdir\ch03\simpdiff\simpdiff.sg
// This example is intended to show how propagators
// can be used in a simplified model of diffusion, to step
//

CONST NX = 19  ;              // number of x-mesh points
CONST LX = NX-1;              // index of the last x-mesh point
CONST NY = 19  ;              // number of y-mesh points
CONST LY = NY-1;              // index of the last y-mesh point
CONST dt   = 1;               // time step
CONST tmax = 40;              // maximum time
CONST is   = 6 ;              // index of stop light
CONST js   = 8 ;              // index of stop light
CONST ig   = 8 ;              // initial index of glider
CONST jg   = 15;              // initial index of glider

var t,A[NX,NY],AO[NX,NY],PROP[3,3],X[NX],Y[NY];

begin main
  assign X[i=all] = 1.0*i ;
  assign Y[j=all] = 1.0*j ;
  assign PROP[i=0..2,j=0..2] = 0.1 ;  // Set up simple
  assign PROP[i=1,j=1]       = 0.2 ;  // propagator

  assign t=0.0 ;
  assign AO[i=is,j=js+1]    = 1 ;        // launch stop light
  assign AO[i=is-1..is+1,j=js] = 1 ;  //
  assign AO[i=ig,j=jg+2]    = 1 ;        // launch glider
  assign AO[i=ig+1,j=jg+1] = 1 ;      //
  assign AO[i=ig-1..ig+1,j=jg] = 1 ;  //

  assign A[i=all,j=all] = AO[i,j] ;

  while (t <= tmax) begin
```

```
   assign A[i=1..LX-1,j=1..LY-1] =
     PROP[0,0]*AO[i-1,j-1] + PROP[0,1]*AO[i-1,j] + PROP[0,2]*AO[i-1,j+1] +
     PROP[1,0]*AO[i  ,j-1] + PROP[1,1]*AO[i  ,j] + PROP[1,2]*AO[i  ,j+1] +
     PROP[2,0]*AO[i+1,j-1] + PROP[2,1]*AO[i+1,j] + PROP[2,2]*AO[i+1,j+1];
   write;
   assign AO[i=all,j=all] = A[i,j] ;
   assign t = t+dt ;
  end
end
```

An alternative way to do the same thing is given next. This may seem more complicated, but it allows us to be more careful about the normalization. In the present case, the propagator has nine values which clearly add up to one. Sometimes our propagator only adds up to approximately one. In this case we might want to be able to add up the numbers of particles which came from the initial cell (i, j) to make sure the number we were putting back was exactly equal to the number we started with in (i, j). To be able to do this, we need to take all the particles from (i, j) all at once and check we have the right number. If we remove them a few at a time, it is not convenient to check we have the right total number. In more complicated examples we might want to conserve other quantities, such as momentum. This is only really practical if we relate everything back to the initial cell, taking each initial cell at a time.

Some pieces of the second file which are identical to the first have been omitted, to try to clarify the differences, and the omissions pointed out.

Input File simpdif2.sg

```
// FILE:  sgdir\ch03\simpdif2\simpdif2.sg
// This example is intended to show how propagators
// can be used in a simplified model of diffusion, to step
// forward in time.

CONST NX = 19  ;              // number of x mesh points
CONST LX = NX-1;              // index of the last x mesh point
CONST NY = 19  ;              // number of y mesh points
CONST LY = NY-1;              // index of the last y mesh point
CONST dt   = 1;               // time step
CONST tmax = 40;              // maximum time
CONST is   = 6 ;              // index of stop light
CONST js   = 8 ;              // index of stop light
CONST ig   = 8 ;              // initial index of glider
CONST jg   = 15;              // initial index of glider

var t,A[NX,NY],AO[NX,NY],PROP[9],ADD[NX,NY,9],X[NX],Y[NY];

begin main
```

```
      assign X[i=all] = 1.0*i ;
      assign Y[j=all] = 1.0*j ;
      assign PROP[i=0..8] = 0.1 ;      // Set up simple
      assign PROP[i=4]    = 0.2 ;      // propagator

      assign t=0.0 ;
      assign AO[i=is,j=js+1]   = 1 ;       // launch stop light
      assign AO[i=is-1..is+1,j=js] = 1 ;   //
      assign AO[i=ig,j=jg+2]   = 1 ;       // launch glider
      assign AO[i=ig+1,j=jg+1] = 1 ;       //
      assign AO[i=ig-1..ig+1,j=jg] = 1 ;   //
      assign A[i=all,j=all] = AO[i,j] ;
      while (t <= tmax) begin
        assign ADD[i=1..LX-1,j=1..LY-1,k=0..8] =
                 PROP[k]*AO[i,j] ;

// AT THIS POINT WE COULD SUM ADD[i,j,k] OVER ALL k.
// THE ANSWER SHOULD EQUAL AO[i,j].
// IF IT DOES, PROP WAS PROPERLY NORMALIZED

        assign A[i=1..LX-1,j=1..LY-1] =
               ADD[i-1,j-1,0] + ADD[i-1,j,1] + ADD[i-1,j+1,2]
             + ADD[i  ,j-1,3] + ADD[i  ,j,4] + ADD[i  ,j+1,5]
             + ADD[i+1,j-1,6] + ADD[i+1,j,7] + ADD[i+1,j+1,8] ;
        write;
        assign AO[i=all,j=all] = A[i,j] ;
        assign t = t+dt ;
      end
end
```

The density after one, ten, twenty and thirty-nine steps is shown in Figure 3.3. The vertical axis has been rescaled from plot to plot. The density tends toward the characteristic shape which is the eigenfunction of the diffusion equation with the longest decay time.

Numerical Implementation of Gaussian Propagator

Another example, where the propagator is actually found from the Gaussian, is given next. The main point of this example is that the normalization is actually needed here, partly because the analytic normalizing factor has been left out, since it will be taken care of automatically by the numerical normalization. The main reason why we have to normalize, is that the values found from the Gaussian will not add up to one. The cell area was used in the propagator, to show where it belongs. It, too, would be included by the normalization in this example. When we do the normalization we divide a number which is sometimes zero by another number which may be zero if the number on top is zero. To avoid dividing zero by

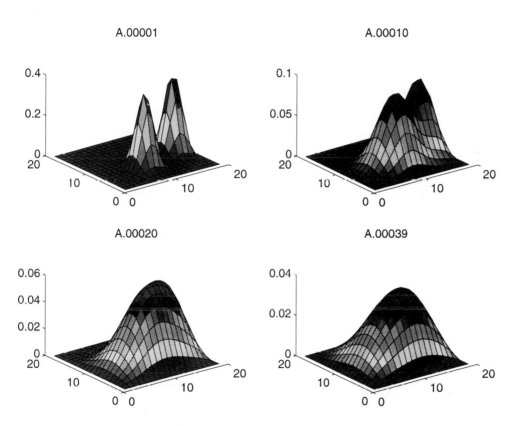

Figure 3.3. Density as calculated using a simple propagator method, shown at various times. The plots are not all on the same vertical scale.

zero, we set the divisor to a nonzero value whenever it is zero. SGFramework uses logical statments which are zero or one depending on the truth or falsehood of a statement; for instance, $(1.0 > 0.0)$ evaluates to 1.0 whereas $(1.0 < 0.0)$ would give 0.0; see below.

Input File gdf.sg

```
// FILE:  sgdir\ch03\gdf\gdf.sg
// This example shows how a more accurate propagator
// can be used in a model of diffusion, to step
// forward in time.
// The propagator is still not very accurate
// because the time step is too short and too few cells
// are used around the original cell.
// The numerical normalization is needed here -
// and since we normalize numerically, the constant
// outside the exponential is omitted.

CONST NX = 19  ;            // number of x-mesh points
CONST LX = NX-1;            // index of the last x-mesh point
CONST NY = 19  ;            // number of y-mesh points
CONST LY = NY-1;            // index of the last y-mesh point
CONST DX =     1;          // mesh spacing
CONST D  =     1;          // diffusion coefficient
CONST dt   =   1;          // time step
CONST R2   = sqrt(2.0) ;   // square root of 2.
CONST tmax = 41;           // maximum time
CONST is  = 6 ;            // index of stop light
CONST js  = 8 ;            // index of stop light
CONST ig  = 8 ;            // initial index of glider
CONST jg  = 15;            // initial index of glider


var t,A[NX,NY],AO[NX,NY],PROP[9],ADD[NX,NY,9],NORM[NX,NY];
var CELLAREA,EXPON,x[NX],y[NY];

begin main
  assign x[i=all] = i*DX;
  assign y[j=all] = j*DX;
  assign CELLAREA = DX * DX;
  assign EXPON = CELLAREA/(4.0*D*dt);
  assign PROP[i=all] = CELLAREA*exp(-R2*EXPON);    //
  assign PROP[i=1]   = CELLAREA*exp(-EXPON);       // Set up Gaussian
  assign PROP[i=4]   = CELLAREA;                   // propagator
  assign PROP[i=3]   = PROP[1];                    // THIS PROPAGATOR
  assign PROP[i=5]   = PROP[1];                    // IS NOT EVEN CLOSE
  assign PROP[i=7]   = PROP[1];                    // TO NORMALIZED
  assign t=0.0 ;
```

```
      assign AO[i=is,j=js+1]   = 1 ;      // launch stop light
      assign AO[i=is-1..is+1,j=js] = 1 ;  //
      assign AO[i=ig,j=jg+2]   = 1 ;      // launch glider
      assign AO[i=ig+1,j=jg+1] = 1 ;      //
      assign AO[i=ig-1..ig+1,j=jg] = 1 ;  //
      assign A[i=all,j=all] = AO[i,j] ;


      while (t <= tmax) begin

        assign ADD[i=1..LX-1,j=1..LY-1,k=0..8] =
                PROP[k]*AO[i,j] ;

// NORMALIZE!!
// ADD[i,j,k] summed over k should equal AO[i,j] - make it do so!

        assign NORM[i=1..LX-1,j=1..LY-1] = ADD[i,j,0]+ADD[i,j,1]+ADD[i,j,2]
                                          +ADD[i,j,3]+ADD[i,j,4]+ADD[i,j,5]
                                          +ADD[i,j,6]+ADD[i,j,7]+ADD[i,j,8] ;

// If NORM is zero, it is not needed, since AO is zero.
// But NORM must be set to a nonzero value since
// we will divide AO by it.

        assign NORM[i=all,j=all] = (NORM[i,j]> 0.0)*NORM[i,j] +
                                   (NORM[i,j]<=0.0)*10.0;

        assign ADD[i=1..LX-1,j=1..LY-1,k=all] = ADD[i,j,k]*AO[i,j]/NORM[i,j] ;

        assign A[i=1..LX-1,j=1..LY-1] =
                ADD[i-1,j-1,0] + ADD[i-1,j,1] + ADD[i-1,j+1,2]
              + ADD[i  ,j-1,3] + ADD[i  ,j,4] + ADD[i  ,j+1,5]
              + ADD[i+1,j-1,6] + ADD[i+1,j,7] + ADD[i+1,j+1,8] ;

      write;
      assign AO[i=all,j=all] = A[i,j] ;
      assign t = t+dt ;
    end
  end
```

The results of this simulation look the same as the previous one when plotted, so they are not shown separately.

The propagator used in this file could have been normalized at the start of the calculation. It was normalized each time it was used, to show how the normalization would be done if the propagator was different for each initial cell. The sum of the nine entries which go into the normalization is $\sum_{Norm}$, which is made up of 1 (from the center cell) $+4 \times 0.75$ (from the four cells closest to the center, at twelve o'clock, three o'clock, six o'clock and nine o'clock) $+4 \times 0.5$ (from the four corners) $\simeq 6$. The propagator evaluates to about 0.3 in the next nearest cells which were included, so an error of about 0.3/6 or about 0.05 is caused by omitting each of the next nearest cells. This is not as accurate as one might wish. More final cells further from the initial cell should be included to improve the accuracy. The extra final cells could be included by putting an extra 16 cells (the next set of cells which is one space further out from the center cell) into the propagator and in all the expressions in gdf.sg. This would actually be more convenient to do in a conventional programming language. It simply requires a little more typing in SGFramework, which is designed to handle FD or similar schemes, where there are relatively few final cells. (Alternatively, one could define a mesh of the desired shape, which in this case is rectangular, in a file called 'rect.sk'. Then lsum could be used to sum over all the initial cells in the region. lsum will work with assignment statements.)

Compared to an explicit FD scheme this propagator takes a very long time step. The Courant criterion for a diffusive problem, which applies to explicit FD schemes if they are to be stable and accurate (and to implicit FD schemes if they are to be accurate, in general), says that the time step Δt must obey

$$\Delta t < 0.1 \frac{(\Delta x)^2}{2D}. \tag{3.45}$$

But in the propagator scheme (which is itself an explicit scheme of a sort, in that the new values are given explicitly in terms of the old) $\Delta t = 1$, while $(\Delta x)^2/2D = 1/2$ so Δt is 20 times larger than would be allowed by the Courant criterion in an FD scheme.

The longer time step used in the propagator method makes the propagator much more efficient, but it also means that there is much less numerical diffusion in the overall simulation than in an FD scheme. At each time step some particles leave their initial cell and enter adjacent cells. Their location in the adjacent cell is not known - they are somewhere in the cell, and they are treated as being uniformly spread throughout the cell in the FD scheme and the propagator method. At the next time step some of these particles will be able to get into the next adjacent cell. This is because they were assumed to be uniformly spread through the cell they were put into, rather than close to the face of the cell that they crossed to get into the cell. This increased rate of spreading out of the particles is called numerical diffusion, and it happens at every time step. Taking fewer steps leads to less numerical diffusion, and this is hence an advantage of the propagator scheme.

Propagator for Diffusion with Flow

Including a drift velocity $\mathbf{v}_{dr}$ is relatively easy. The distance drifted in Δt is $v_{dr}\Delta t$. The diffusion is superimposed on the drifting, and appears to be outwards from a point moving with the flow, at the flow velocity. Instead of using the quantity

Input File ddf.sg

```
// FILE:  sgdir\ch03\ddf\ddf.sg
// This example is identical to gdf.sg except that
// the flow moves more rapidly to one side -
// particles move one cell further in a time step.

// SAME AS gdf.sg UNTIL
    assign A[i=2..LX-1,j=1..LY-1] =
            ADD[i-2,j-1,0] + ADD[i-2,j,1] + ADD[i-2,j+1,2]
          + ADD[i-1,j-1,3] + ADD[i-1,j,4] + ADD[i-1,j+1,5]
          + ADD[i  ,j-1,6] + ADD[i  ,j,7] + ADD[i  ,j+1,8] ;
    write;
    assign AO[i=all,j=all] = A[i,j] ;
    assign t = t+dt ;
  end
end
```

$(\mathbf{r} - \mathbf{r}')^2$ in the exponent we should instead use $(\mathbf{r} - (\mathbf{r}' + \mathbf{v}_{dr}\Delta t))^2$. If $v_{dr}\Delta t$ is not small compared to the cell size, then we must use more final cells on the side to which the flow carries us. Provided we choose the final cells to surround the point $(\mathbf{r}' + \mathbf{v}_{dr}\Delta t)$ we can use a large value of $v_{dr}\Delta t$, however.

A case in which $v_{dr}\Delta t = \Delta x$ and the flow is in the x-direction is especially easy to set up, since the propagator is the same as for $v_{dr} = 0$ except that the final cells are shifted by one space in the x-direction.

The file ddf.sg is identical to gdf.sg, except for the line where density is added back. Notice that density is being taken from initial locations which are more negative than before, because the flow is in the positive x-direction. The lower limit on x had to be changed, because of this.

Figure 3.4 only shows a few early frames, because the density all leaves quickly. In this plot the x-direction is upward (because of the SGFramework array convention).

3.3.2   Numerical Diffusion and Instability: a Comparison of the Propagator and an Explicit Scheme

To illustrate the concepts of numerical diffusion and numerical instability, we shall compare the propagator scheme given above to an explicit FD scheme. The input file for the propagator was modified only to change the initial density so that it was all at a point. The results of the run are shown in Figure 3.5.

Next, an equivalent finite-difference scheme was applied to this problem. The input file is fd.sg. The results are shown in Figure 3.6, and Figure 3.7.

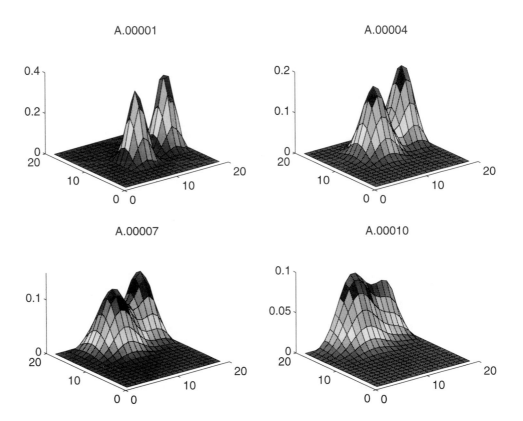

Figure 3.4. Diffusion calculated using a propagator method, when the flow carries particles a 'long way' in each time step.

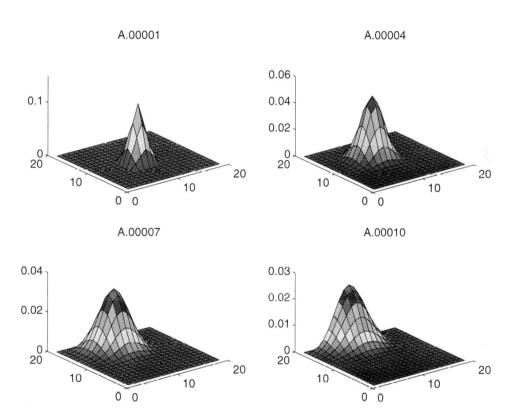

Figure 3.5. A propagator calculation of diffusion, for comparison to a finite-difference solution of the same problem (shown in Figures 3.6 and 3.7) which illustrates the effects of 'numerical diffusion'.

Input File fd.sg

```
// FILE: sgdir\ch03\fd\fd.sg
// This example shows how finite-differences
// can be used in a model of diffusion, to step
// forward in time.

CONST NX = 19  ;              // number of x-mesh points
CONST LX = NX-1;              // index of the last x-mesh point
CONST NY = 19  ;              // number of y-mesh points
CONST LY = NY-1;              // index of the last y-mesh point
CONST DX =     1;             // mesh spacing
CONST D  =     1;             // diffusion coefficient
CONST MU =     1;             // mobility
CONST E  =     1;             // electric field
CONST dt  = 0.2;             // time step; dt = 0.3 gives instability
CONST tmax = 11;             // maximum time
CONST is   = 6 ;             // index of initial cell
CONST js   = 8 ;             // index of initial cell


var t,A[NX,NY],AO[NX,NY],X[NX],Y[NY];

begin main
  assign X[i=all] = i*DX ;
  assign Y[j=all] = j*DX ;
  assign t=0.0 ;
  assign AO[i=is,j=js]   = 1 ;        //  Initialize array
  assign A[i=all,j=all] = AO[i,j] ;
  while (t <= tmax) begin

    assign A[i=1..LX-1,j=1..LY-1] = AO[i,j] + dt * D * (
                              + AO[i-1,j]
            + AO[i  ,j-1]- 4.0 * AO[i  ,j] + AO[i  ,j+1]
                              + AO[i+1,j]               )/sq(DX)
          - dt * MU * E * (AO[i+1,j] - AO[i-1,j])/(2.0*DX)          ;

    write;
    assign AO[i=all,j=all] = A[i,j] ;
    assign t = t+dt ;
  end
end
```

The first run of this file was done with the same time step as the propagator, of $dt = 1$. The expected result did indeed occur, as shown in Figure 3.6.

The finite-difference scheme gave oscillating values — oscillating in both space and time, from one cell to its neighbor and from one time step to the next. The FD scheme given in the input file makes it clear why this can happen. If Δt is

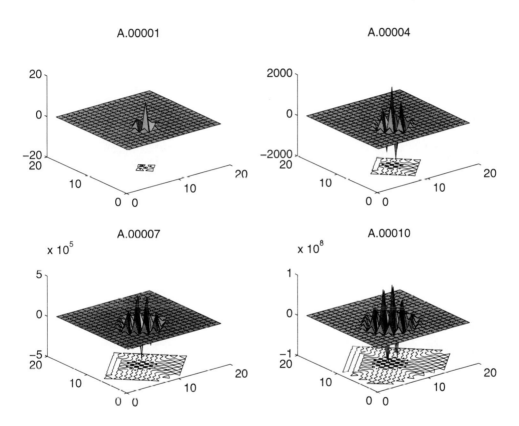

Figure 3.6. The results of using a large time step in an 'explicit' finite-difference scheme when the time step exceeds the Courant limit are usually instability, as shown. Note the vertical scale in each of these plots.

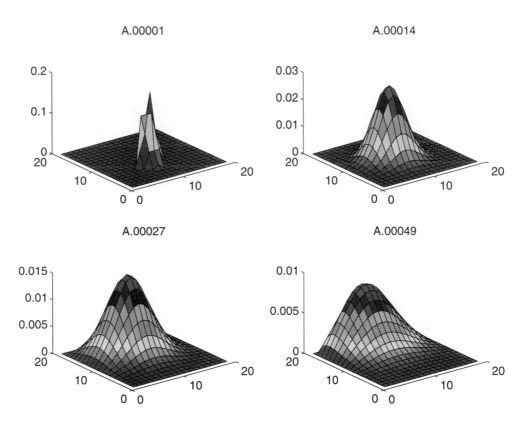

Figure 3.7. Results from an explicit finite-difference scheme with a smaller time step. Because of the small time step, more steps were taken in the same total elapsed time and more numerical diffusion is seen.

large, the change in the density at any point on the mesh can be larger than the original density in that cell (and if it is negative, the overall density in that cell goes negative). The propagator expression given above is always smooth and cannot go negative. The explicit FD scheme is an approximation to the propagator which fails for 'large' time steps, when the Courant/CFL criterion is violated. A time step less than 0.3 was needed to avoid instability. The results for a step of 0.2 are shown in Figure 3.7.

The last frame of this run corresponds to roughly the same time as the last frame from the propagator run. More time steps are needed because of the shorter time step. Because of the extra time steps, the FD run shows a wider and much lower peak, due to numerical diffusion. (The vertical scales are different.)

3.4 Summary

The purpose of this chapter was to review concepts of transport theory, such as the meaning of the conservation equation and of the various terms in it. The numerical representation of this equation, its solution and its use in relatively simple but important situations, such as dopant diffusion, were described.

PROBLEMS

1. Use the equation specification file wavtee.sg and the mesh specification file tee.sk from Chapter 2 to solve a diffusion problem. The longest boundary is to be held at a 'density' of one unit, the other boundaries are to be held at a density of zero and the rest (the interior) of the region is to have an initial density of zero. Step the equation forward in time until the solution (almost) reaches steady-state. Plot the density at various times during the diffusion, to show how the shape evolves with time. Estimate analytically how long it should take for the density to diffuse across the interior. Compare this to the time it took for the simulation to settle down.

2. This question is intended to illustrate basic ideas of diffusion in a 'direct', microscopic way. Write a Monte Carlo program which allows particles to move up and down in one dimension. (This is not to be done using SGFramework.) Give each particle an equal chance of traveling in either direction at each time step. Start all the particles at the same point. Let each step have the same magnitude but be either positive or negative. Follow a moderately large number $N \simeq 10^3$ of particles, one time step at a time, and plot a histogram of their locations after each time step. Calculate the mean position of the particles and the   standard deviation of their positions and plot both of these as functions of time. Comment on the mean and the standard deviation, and how they evolve in time, as well as the shape of the 'tail' of the distribution. The mean should not change on average; the variance should grow linearly with time, but again, only on average. Estimate the rate of growth of the variance.

3. Repeat the previous question, but with an absorbing boundary at 10 to 20 step sizes away from the starting point.

4. Repeat both of the previous questions, but instead of using a Monte Carlo method, use a propagator which simply splits the particles in each cell between the two neighboring cells, at each time step.

5. For a simple parabolic equation

$$\frac{\partial^2 \psi}{\partial x^2} = \frac{\partial \psi}{\partial t} \qquad (3.46)$$

with boundary conditions $\psi(x = 0) = 0$ and $\psi(x = 1) = 0$, show that the general term in the solution which is obtained using separation of variables is

$$\psi(x, t) = \sin(n\pi x)\exp(-n^2\pi^2 t). \qquad (3.47)$$

Choose an initial condition at $t = 0$ made up of contributions having several different n values, and plot the analytic solution for various times. Solve the equation using a finite-difference scheme, and compare the results at the same values of t.

6. Use the scale factors (or Lamé functions) (h_1, h_2, h_3) introduced in Chapter 2 to derive the finite-difference form of $-D\nabla \cdot (\nabla n)$, for use in solving the continuity equation in orthogonal coordinates (u^1, u^2, u^3). Check that your scheme is conservative.

7. Set up a one-dimensional finite-difference scheme to solve the time-dependent diffusion equation

$$\frac{\partial n}{\partial t} - D\nabla^2 n = 0 \qquad (3.48)$$

the one dimension being the cylindrical radius r. Use boundary conditions $n(r = a, t) = 1$, $n(r = b, t) = 0$, with $a = 0.1$ m and $b = 1.0$ m, and an initial condition $n(r, t = 0) = 0$. Advance the solution in time until steady-state is reached, using $D = 0.1$ m^2/s. What is the time scale $\tau = L^2/2D$ on which the density should be expected to settle down? Plot the density at $t = 0.1\tau, 0.3\tau, \tau, 3\tau$ and 10τ. Repeat with the boundary conditions at $r = a$ and $r = b$ reversed.

8. Solve the diffusion equation

$$\frac{\partial n}{\partial t} - D\nabla^2 n = S \qquad (3.49)$$

in two Cartesian coordinates (x, y), with an initial density of zero and a boundary condition $n(x, y, t) = 0$ on all the boundaries. Take the boundary to be a one meter square centered at $(x = 0, y = 0)$. The particle production rate $S(x, y)$ is nonzero only in the region between two squares centered at $(x = 0, y = 0)$. One of these squares has side 0.4 and the other has side 0.5 meters. Between the squares $S = 10^{10}$ m^{-2} s^{-1}.

[Define an array $S[i, j] = 10^{10} * ((sq(i * dx - x0) < 0.25) * (sq(j * dy - y0) < 0.25) - (sq(i * dx - x0) < 0.16) * (sq(j * dy - y0) < 0.16)).$]

Plot $n(x, y)$ at several times during the time evolution, and plot $n(x, y = 0)$ in steady-state. Explain the steady-state shape of $n(x, y = 0)$ and calculate (approximately, analytically) the peak density and the value of $\frac{dn}{dx}$ for all x at $y = 0$. Compare your analytic answer to the numerical answer obtained earlier.

9. Solve the time-dependent diffusion equation in two cylindrical variables (r, z) in the region $r < a$, $0 < z < b$, with $a = 1$ and $b = 3$ meters. The density is zero on the

boundaries at $r = a$, $z = 0$ and $z = b$. What is the appropriate boundary condition at $r = 0$? The source $S(r, z) = S_0$ in a rectangular region. The sides of the region are at $r = 0.2$ and $r = 0.3$ meters and $z = 1$ and $z = 2$ meters, and zero elsewhere. Plot the solution as a function of time, starting with an initial density of zero. Plot the steady-state solution $n(r, z = 1.5)$. Calculate this density approximately, analytically, and compare it to the numerical solution.

10. In the previous two examples the diffusion coefficient D was not specified. Why is the value of D not necessary to solve these problems as they were stated? The mobility μ of charged particles is often related to D by the relation $D = \frac{kT}{q}\mu$. (At room temperature, $\frac{kT}{q} \simeq 0.025 - 0.03$ volts.) The time to diffuse a distance L can be estimated from $\tau \simeq \frac{L^2}{2D}$. If an electric field is applied, the time to drift the distance L is $\tau_E = \frac{L}{\mu E}$. If these are about equal, then

$$\frac{L^2}{2D} \simeq \frac{L}{\mu E} = \frac{L}{DE}\frac{kT}{q}. \tag{3.50}$$

This implies that the electric field needed to make these times be about equal is $E = \frac{kT}{2qL}$. This is roughly the 'ambipolar' electric field. The drift velocity in this case is $v_z^{dr} = \mu E = 2D/L$.

Repeat the previous problem, with an electric field along the z-axis of the magnitude just derived.

11. This problem is concerned with transport of a fluid on the surface of a sphere of radius a. This is a two-dimensional flow, the two independent variables being the spherical coordinates (θ, ϕ). If the fluid motion is in steady-state, and fluid is not being produced or removed, then conservation of mass requires that $\nabla \cdot \mathbf{\Gamma} = 0$. $\mathbf{\Gamma}$ is the flux per unit area per second. If the transport is all due to a flow at velocity $\mathbf{v}$, then $\mathbf{\Gamma} = n\mathbf{v}$ where n is the (number) density. Suppose n is constant. What is the relationship between the components of the velocity, (v_θ, v_ϕ) ? If $v_\theta = 0$, show that n_ϕ can only be a function of θ.

What are the boundary conditions on n at $\phi = 2\pi$, and at $\theta = 0$ or $\theta = \pi$?

Suppose $v_\phi = v_0 \sin\theta$ and $v_0 = wa$ with $w \simeq \pi/12$ per day or about $3 \times 10^{-6} r/s$. If the fluid rotates in a pattern of cells, the rotation being superimposed on the flow described by v_ϕ, the rotation can mix the fluid and lead to an effective diffusion. Let the cell radius be r_c and the rotation period be τ_c. The diffusion coefficient is roughly $D_c \simeq r_c^2/(2\tau_c)$. If $r_c \simeq 0.1a$ and $\tau_c \simeq 5$ days $\simeq 4 \times 10^5$ seconds, find D_c.

Traces of a second fluid, a gas, with the same transport rates, are introduced in a range of θ values from $\theta = 50$ to 51 degrees, and for ϕ from 169 to 170 degrees. Using a finite-difference scheme, calculate the relative density of the gas at $(\theta = 40^0, \phi = 120^0)$, in steady-state, with $a = 10^5$ meters. Assume the sphere absorbs the gas with a time constant $\tau_a = 4\tau_c$. (Why is it necessary to include a mechanism for removing the gas?)

12. A gas is diffusing from a tank where it is kept at a high concentration, through a narrow tube into a region where the concentration is low. Model this by solving the diffusion equation in cylindrical coordinates (r, z), inside a pipe consisting of three

cylinders. The first cylinder has radius r_1 and runs from $z = 0$ to $z = z_1$. The second has radius r_2 and runs from $z = z_1$ to $z = z_2$, while the third has radius r_3 and runs from $z = z_2$ to $z = z_3$. At $z = 0$ the density is held at n_0, and at $z = z_3$ the density is $0.01n_0$. At all the other boundaries there is no flux normal to the boundary, so the normal derivative of n is zero, and this is the boundary condition which must be used there. This includes the radial boundaries, but also the surfaces at constant z where sections of pipe meet, for example at $z = z_1$ between the radii r_1 and r_2.

Set $r_1 = 1.0$, $r_2 = 0.5$ and $r_3 = 2.0$ meters, while $z_1 = 0.5$, $z_2 = 2.5$ and $z_3 = 4.0$ meters. Why are the values of D and n_0 not critical to the solution of the problem?

If a section of pipe absorbs a fraction γ of the particles striking it, the boundary condition at that wall is modified. The net flux to the wall is $\Gamma = -D\frac{\partial n}{\partial r}$ if the wall has its normal pointing radially inward. The number of particles striking the wall per unit area per second is $\frac{1}{4}nv_{th}$, where v_{th} is the 'thermal' speed of the gas particles. The number being absorbed per unit area per second is thus $\frac{\gamma}{4}nv_{th} = -D\frac{\partial n}{\partial r}$ so $\frac{\partial n}{\partial r} = -\alpha n$ where $\alpha \equiv \frac{\gamma v_{th}}{4D}$. Now $D \simeq \frac{1}{2}v_{th}\lambda$ so $\alpha = \frac{\gamma}{2\lambda}$. Modify the boundary condition on the middle section of pipe so that $\gamma = 0.1$ and $\lambda = 0.1$ meters and repeat the calculation.

13. The excess density of carriers (in this case holes) in a constant electric field E obeys the equation

$$-D\frac{\partial^2 p}{\partial x^2} + \mu E\frac{\partial p}{\partial x} = -\frac{p}{\tau_p} \tag{3.51}$$

where τ_p is the hole recombination time and the other symbols have their usual meanings. If the solution has the form $\exp(-\lambda x)$, find the expression for the two possible values of λ, which are known as the upstream and downstream solutions. If the units of length and the electric field are chosen so that $D\tau_p = 1$ and $\mu E\tau_p = 1$, set up a finite-difference scheme and a mesh to solve for p on either side of a point at $x = 0$ where $p(x = 0) = 1$. Let $p = 0$ at the edges of the simulation region, which must be several times the appropriate value of λ on each side of $x = 0$. Compare the solution to the exponential solutions by plotting them side by side.

14. This question and several which follow are intended to illustrate the motion of minority carriers in a pn junction. The first few questions set up the physical basis for the electric field which is used and the transport coefficients D and μ. Suppose the electric field in a one-dimensional semiconductor is $\mathbf{E} = \mathbf{a}_x \frac{\alpha V_T}{W}(W - |x|)$, for $|x| < W$, and zero otherwise. $x = 0$ is the center of the junction region between two differently doped regions and W is the width of the junction on either side of $x = 0$, so the junction region extends from $x = -W$ to $x = W$. V_T is the thermal voltage, $V_T \equiv \frac{kT_e}{e}$, and α is a constant which is of order one. Plot the electric field.

15. (*Continued from the previous question.*) The electric field given repels majority carriers from the junction and pushes minority carriers across the junction. As the minority carriers are pushed away from the junction at $x = 0$, the ionized impurities are exposed. Plot the charge density which should be found by evaluating $\rho = \frac{dD}{dx}$.

16. Explain the shape of the electric field, in terms of field lines starting on positive charges and ending on negative charges. Are electrons the majority or minority carriers at $x > W$, based on the charge density?

17. Electrons will be pushed to the left by this field. However, electrons produced outside the junction region will not enter the junction unless they diffuse into it. Let the electron diffusion coefficient be $D = 1$ cm^2/s. Find the electron mobility from the Einstein relation. Choose a time step which satisfies the the CFL criteria for diffusion and flow. Specify the time step in terms of the mesh spacing Δx which will be determined later.

18. Let electrons be produced at a rate of 10^{10} cm^{-3} in the region $x > W$. Assume the device has a cross-sectional area of 1 cm^2 so that 10^{10} particles are produced per centimeter per second. Use a one-dimensional finite-difference diffusion equation, similar to the two-dimensional diffusion equation used in the text, allowing for both diffusion and drift, to advance the electron density in time. Start with zero density. Consider only the region from $x = 0$ to $x = L = 10\,\mu$m with $W = 1\,\mu$m. Set the density to zero at the ends. Choose a mesh size such that about 50 cells fit into this region. Run the simulation for several hundred time steps or until it seems to be close to steady-state. Plot the resulting density of electrons.

19. The boundary conditions at $x = 0$ and $x = L$ in the previous example were intended to be that the density go to zero at the ends. In reality, the value at $x = L$ would be better approximated by the equilibrium density. Assume the doping density at $x = L$ is 10^{19} cm^{-3}. The type of doping can be determined from the fact that electrons are minority carriers in this region. Calculate $n(x = L)$. At $x = 0$ the flux per unit area Γ of electrons should not vary with position. Γ is mainly due to the electric field and not to diffusion at $x = 0$. If the flux is mobility dominated, then $\Gamma = n\mu E$. Set $\frac{d\Gamma}{dx} = 0$ at $x = 0$ and find $\frac{dn}{dx}$ at $x = 0$. This provides the other boundary condition that we should have used in the previous problem. Repeat the previous problem, using these boundary conditions.

20. Find realistic values of D and μ for silicon. What should the time step have been in the previous problem for a realistic value of D? Estimate the dielectric relaxation time τ_d. Why was τ_d not relevant in the problems above?

21. Solve the steady-state diffusion problem using the file provided in the text, and then repeat the exercise with the sign of the electric field reversed. Explain the differences. Add a source of particles in the center of the simulation region and repeat.

22. Solve the one-dimensional diffusion equation numerically for the two classic cases which are used to describe dopant diffusion; (a) with a fixed number of dopants in the silicon and (b) with a fixed density of dopants at the surface. Compare the results to the analytic expressions for the density in these two cases.

23. This question is intended to illustrate the relation between the transport equations and the heat equations, and the numerical solution of the full set of those equations.

    The equations for the temperature T and density ρ in a sound wave are [6]

    $$\frac{\partial T}{\partial t} = \alpha \frac{\partial \rho}{\partial t} + \beta \nabla^2 T \tag{3.52}$$

    and

    $$\frac{\partial^2 \rho}{\partial t^2} = \gamma \nabla^2 \rho + \epsilon \nabla^2 T. \tag{3.53}$$

The constants in these equations are $\alpha = \frac{\partial T_0}{\partial \rho_0}|_S$, $\beta = \frac{\kappa}{c_p \rho_0}$, κ being the thermal conductivity and c_p the specific heat at constant pressure; $\gamma = \frac{\partial p_0}{\partial \rho_0}|_T$ and $\epsilon = \frac{\partial p_0}{\partial T_0}|_\rho$. Choose 'reasonable' values for these quantities and for the time step on which the equations can be integrated.

Set up a simulation in one spatial dimension to solve these equations. Choose an initial perturbation which consists of a sinusoidal variation in the density, modulated by a slower variation so that the amplitude goes to zero after about ten wavelengths. Integrate the wave in time. Plot T and ρ versus position at various times.

Part II

Numerical Methods and Physical Models for Semiconductor Simulation

Chapter 4

The Semiconductor Equations

The purpose of Part II of this book is to extend the material of Part I to the level needed for semiconductor simulation. Chapter 4 reviews the semiconductor equations and the mathematical models of the semiconductor parameters. Chapter 5 discusses numerical methods, in general, and the methods that are needed to solve the semiconductor equations, in particular. Chapter 6 describes how the software employed in this work enables these numerical schemes and these models to be set up. After these chapters, the focus is switched to the simulation of particular semiconductor devices in Part III.

4.1 Introduction

This chapter presents the equations describing carrier behavior in semiconductors. The main emphasis is on the exact form of the transport equations and their boundary conditions. Following the presentation of the semiconductor equations is a compilation of the parameters needed in the semiconductor equations. Finally, there is a review of more advanced topics in semiconductor carrier behavior.

This chapter has four major sections. Section 4.2 outlines the equilibrium description of carriers in a semiconductor. Section 4.3 presents the transport equations used to describe particle and heat conservation, and discusses them and their boundary conditions. Section 4.4 describes the mathematical models of the semiconductor parameters. Section 4.5 considers momentum conservation equations for carriers. (Section 4.6 makes some concluding comments and summarizes this chapter.) Two appendices follow; one, 4.7.1 on use of the momentum equations. Section 4.7.2 addresses the behavior of implanted ions, injected at high energy into a semiconductor.

4.2 Equilibrium Carrier Behavior in Semiconductors

In this section the equilibrium behavior of carriers in a semiconductor will be discussed. In much of this document we shall be concerned with transport of carriers. (Transport is used here to indicate the means by which particles move around.) By the definition of equilibrium, there is no net flux of any kind in equilibrium. Since

167

there is no net flux in equilibrium, transport (fluxes) must involve nonequilibrium processes. The transport is described by a set of partial differential equations derived from the continuity equation. The transport equations are introduced in the next section. The purpose of this section is largely to define or to introduce concepts, such as the Fermi level and expressions for carrier densities, which are used in analytic analysis and in approximate models. The nonequilibrium models will use the transport equations.

The equilibrium densities of electrons and holes are found by summing the numbers of each type of carrier over all the energy levels available to them. The available states are uniformly spread out with respect to the particle wavevector. A uniform plane wave's amplitude is proportional to $\cos(\mathbf{k} \cdot \mathbf{r} - \omega t + \phi)$, where ϕ is a constant phase, where $\mathbf{k} \equiv (k_x, k_y, k_z)$, $\mathbf{r} \equiv (x, y, z)$, and $\mathbf{k \cdot r}$ is given by $\mathbf{k \cdot r} = k_x x + k_y y + k_z z$. The vector $\mathbf{k}$ is the wavevector, which points in the direction the wave is traveling. The magnitude of the wavevector is equal to $2\pi/\lambda$. Suppose we look at a 'snapshot' of the wave at some time t, and move from one point on the snapshot that we are looking at through one full oscillation of the wave to another point on the wave where the phase of the wave is the same. Then we have moved by one wavelength, λ, along the direction $\pm\mathbf{k}$ (and an unknown distance perpendicular to $\mathbf{k}$). The phase of the wave must change by $k\lambda$. The change in phase must also equal 2π in order for there to be one oscillation in one wavelength, so $k = 2\pi/\lambda$.

The number of states in a range of wavevectors k_x to $k_x + dk_x$, k_y to $k_y + dk_y$ and k_z to $k_z + dk_z$ is proportional to $dk_x dk_y dk_z$, which is the volume in wavevector space. This was discussed in more detail in Section 2.11.3. This condition is derived by saying that the wave function is required to be periodic in each direction within the crystal. That is, it must repeat an integer number of times on going across the crystal. This periodicity means that there is an allowed value of k_x every time k_x is equal to an integer multiple of $2\pi/L_x$, where L_x is the length of the crystal in the x-direction. $k_x = 2\pi m_x/L_x$ and similarly $k_y = 2\pi m_y/L_y$ and $k_z = 2\pi m_z/L_z$. (m_x, m_y, m_z) are all integers. This means that the allowed values of $\mathbf{k} \equiv (k_x, k_y, k_z)$ are uniformly spread out in k-space. If k is the length of the wavevector, then the volume in wavevector space Δv_k between the sphere of radius k and the sphere of radius $k + \Delta k$ is

$$\Delta v_k = \frac{4}{3}\pi[(k + \Delta k)^3 - k^3] \simeq 4\pi k^2 \Delta k. \tag{4.1}$$

So far we have used $\mathbf{k}$ without relating it to the particle speed or energy. However, the de Broglie relation $\mathbf{p} = \hbar\mathbf{k}$ states that the momentum $\mathbf{p}$ of a free particle is proportional to its wavevector. Its kinetic energy is then $\frac{p^2}{2m} = \frac{\hbar^2 k^2}{2m}$. For a carrier inside the semiconductor lattice, which is not too far from the bottom of the band, a similar result is approximately true. The energy of the carrier, measured from the band edge, is

$$E - E_C \simeq \frac{\hbar^2 k^2}{2m_e^*} \tag{4.2}$$

for electrons, with E_C being the bottom of the conduction-band and m_e^* the electron 'effective mass'. A similar expression applies to the holes. Writing k^2 in terms of the energy,

$$k^2 = \frac{2m_e^*(E - E_C)}{\hbar^2} \tag{4.3}$$

so

$$\Delta v_k = 4\pi k^2 \Delta k = 2\pi(2m_e^*/\hbar^2)^{3/2}(E - E_C)^{1/2}\Delta E = A(E - E_C)^{1/2}\Delta E \tag{4.4}$$

where A is a constant.

As was explained above, the number of states in Δv_k is proportional to Δv_k. The fraction of these states that are full is given by the Fermi-Dirac distribution, $f(E)$, which is the probability of any given state at energy E being occupied.

$$f(E) = \left[1 + \exp\left(\frac{E - E_F}{k_B T}\right)\right]^{-1}. \tag{4.5}$$

The number of electrons Δn in the range of energies ΔE is proportional to the number of states in the range of energy ΔE. The number of states in the range of energy ΔE is proportional to the volume in k-space Δv_k, multiplied by the fraction $f(E)$ of those states which are occupied.

$$f(E)\Delta v_k = f(E)A(E - E_C)^{1/2}\Delta E. \tag{4.6}$$

The total number of electrons per unit volume is obtained by summing over all energies. Converting the sum to an integral, (which is reasonable because the energy levels are very close together compared to typical particle energies,)

$$n_e(x) = A' \int_{E=E_C}^{\infty} f(E)(E - E_C)^{1/2}\,dE \tag{4.7}$$

where A' is (another) constant.

This integral may be approximated when the Fermi level lies a few times $k_B T$ below the conduction-band. The Fermi-Dirac distribution is then approximated as

$$f(E) = \exp\left(\frac{E_F - E}{k_B T}\right) \tag{4.8}$$

and the integral can be analytically evaluted. The density of electrons is then written

$$n = N_C \exp\left(\frac{E_F - E_C}{k_B T}\right). \tag{4.9}$$

Similarly the density of holes is written

$$p = N_V \exp\left(\frac{E_V - E_F}{k_B T}\right). \tag{4.10}$$

N_C (N_V) is the effective density of states at the bottom of the conduction-band (top of the valence-band) and $E_C(E_V)$ is the energy of the bottom of the conduction-band (top of the valence-band).

When these expressions are multiplied together, they give

$$np = N_C N_V \exp\left(\frac{E_V - E_C}{k_B T}\right). \qquad (4.11)$$

The right-hand side of this expression is a function of temperature only, for any given material, so the product np does not depend on the doping (unless $E_V - E_C$ is altered by it). In the intrinsic semiconductor, which is undoped, $n = p = n_i$, the intrinsic density, and $np = n_i^2$. According to the above result np is a constant, so the constant must be n_i^2. (At high doping levels, however, the assumptions used to calculate n and p break down.)

The value of the Fermi level which makes each of n and p equal to n_i, which occurs in an undoped semiconductor, is called the *intrinsic level* E_i. E_i is between E_C and E_V and is inside the band-gap. (The band-gap is the range of energies between the top of the valence-band and the bottom of the conduction-band; the width of the band-gap is $E_g \equiv E_C - E_V$.) If the potential $V(x)$ is altered by an amount ΔV, then the energies of electrons change by $-e\Delta V$, so E_C, E_V and E_i all change by $-e\Delta V$.

Since E_i differs from E_C by a constant, then we may replace E_C with E_i in Equation 4.9, provided N_C is multiplied by another constant which is equal to $\exp(\frac{E_C - E_i}{k_B T})$.

$$n = n_i \exp\left(\frac{E_F - E_i}{k_B T}\right) \qquad (4.12)$$

gives $n = n_i$ when $E_F = E_i$ so this is the correct replacement for N_C. Similarly in the expression for the hole density

$$p = n_i \exp\left(\frac{E_i - E_F}{k_B T}\right) \qquad (4.13)$$

The difference between E_i and E_F in the bulk semiconductor where quasi-neutrality holds, is measured by a quantity $\phi_F = \frac{1}{e}(E_i(quasi) - E_F)$.

Suppose the material is p-type, so $E_i > E_F$ and $\phi_F > 0$ is required to make $p > n_i$. In the bulk, the hole density is at its quasi-neutral value, $p = p(quasi)$. This is the hole density which is needed to make the bulk semiconductor be (quasi)neutral. As V is increased, however, we can decrease p and increase n. It is interesting to find the potential required to create inversion, which occurs when the minority-carrier density has the same value that the majority-carrier density would have in the quasi-neutral bulk. To achieve inversion the electron density must be $n = p(quasi)$, so

$$n = n_i \exp\left(\frac{E_F - E_i(inv)}{k_B T}\right) = p(quasi) = n_i \exp\left(\frac{E_i(quasi) - E_F}{k_B T}\right). \qquad (4.14)$$

Because the exponents must be equal, $E_F - E_i(inv) = E_i(quasi) - E_F$ or

$$E_i(inv) = -E_i(quasi) + 2E_F. \tag{4.15}$$

We said above that a change in potential ΔV changed E_i by the amount $-e\Delta V$. If V is zero in the quasi-neutral bulk, then the change in E_i between the bulk and a point x is

$$E_i(x) - E_i(quasi) = -e\Delta V = -eV(x). \tag{4.16}$$

This means that at inversion the potential is $V(inv)$, where

$$V(inv) = -\frac{1}{e}[E_i(inv) - E_i(quasi)] = 2\phi_F. \tag{4.17}$$

Quasi-Fermi Levels

When the semiconductor is not in equilibrium, it is still sometimes useful to try to use equilibrium concepts to describe the carrier behavior. The quasi-Fermi levels are defined, by replacing the Fermi level E_F in eq 4.12 for the electron density, and eq 4.13 for the hole density, with the electron quasi-Fermi level, E_{Fn} in the expression for n; and with the hole quasi-Fermi level E_{Fp} in the expression for p;

$$n = n_i \exp\left(\frac{E_{Fn} - E_i}{k_B T}\right) \tag{4.18}$$

and

$$p = n_i \exp\left(\frac{E_i - E_{Fp}}{k_B T}\right). \tag{4.19}$$

The usual context in which the concept of the quasi-Fermi levels arises is when excess carriers are generated in a semiconductor. The Fermi level is the energy at which a state has a probability of $\frac{1}{2}$ of being occupied. (In a sense it is like the 'water level' for the electrons.) If excess carriers are generated 'suddenly', for instance by illumination of the semiconductor, both kinds are created in equal numbers. This pushes the energy at which we expect to find holes (with probability of $\frac{1}{2}$) down, and the level where we expect to find electrons up. The Fermi levels for the two species split apart, and become E_{Fn} and E_{Fp}.

A second context where we find the quasi-Fermi levels is when we are studying the effects of spatial gradients. The band-edge energies and the intrinsic level E_i all depend on the applied voltage V. When V changes they all change by an amount $-eV$. This means that if we differentiate Eq (4.18) with respect to position, using the ∇ operator, we get

$$\nabla n = n\frac{1}{k_B T}(\nabla E_{Fn} + e\nabla V). \tag{4.20}$$

Rearranging this,

$$\nabla E_{Fn} = -e\nabla V + \frac{k_B T}{n}\nabla n. \tag{4.21}$$

Using the Einstein relation which relates the diffusion coefficient to the mobility, $D = \frac{k_B T}{e} \mu$, one can write the usual expression for the flux per unit area $\boldsymbol{\Gamma} = n\mu \mathbf{E} - D\nabla n$, in the case of electrons as

$$\boldsymbol{\Gamma}_n = +n\mu_n \nabla V - \frac{k_B T}{e}\mu_n \nabla n. \tag{4.22}$$

The electron mobility used here is $-\mu_n$, so that μ_n is a positive number, whereas the hole mobility μ_p is positive. The electron flux per unit area can be seen to be equal to

$$\boldsymbol{\Gamma}_n = -n\mu_n \nabla E_{Fn}. \tag{4.23}$$

Similarly, for the holes,

$$\boldsymbol{\Gamma}_p = p\mu_p \nabla E_{Fp}. \tag{4.24}$$

The Debye Length

The Debye length λ_D is the scale length in which the electric field is expected to decay (i.e. the field might vary like $\exp(-x/\lambda_D)$), inside a region filled with mobile charges. The equilibrium expressions given above show how λ_D arises.

Suppose that in some region there are 'free' electrons and a background of positive charges (which in a semiconductor would most likely be ionized donors). The charge density is expected to be of the form which was given above. The charge is due to a 'fixed' density of charges on the ionized dopants and an electron density which varies exponentially with the potential V:

$$\rho = e \left(N_D - n_0 \exp\left[eV/k_B T_e\right] \right). \tag{4.25}$$

N_D is the number density of donors, n_0 is a reference density of electrons (at the point where the potential $V = 0$) and the other symbols have their usual meanings. To derive λ_D we have to assume $\frac{eV}{k_B T_e}$ is small, so that the exponential can be expanded as $\exp(x) \simeq 1 + x$. Further, if $n_0 = N_D$, then $\rho \simeq -\frac{e^2 N_D}{k_B T_e}V$.

Poisson's equation, $\nabla^2 V = -\rho/\epsilon$, is then, in one dimension,

$$\frac{d^2 V}{dx^2} = -\frac{e^2 N_D}{k_B T_e \epsilon}V = -\frac{V}{\lambda_D^2} \tag{4.26}$$

where λ_D is defined to be

$$\lambda_D = \sqrt{\frac{k_B T_e \epsilon}{N_D e^2}}. \tag{4.27}$$

λ_D is small if the shielding of the field due to the ions, which is done by the electrons, is strong. If there are more electrons there is stronger shielding; this explains the factor of N_D in the denominator. As T_e increases the kinetic energy of the electrons increases. Hence the electrons are more likely to have enough energy to escape the potential well, in which they are trapped by their attraction to the positive charge. In that case they shield less; this explains the factor of T_e in the numerator.

4.3   Transport Equations for Semiconductors

In this section the semiconductor equations are presented. The section is divided into two subsections. The basic semiconductor equations are presented in Subsection 4.3.1. These equations consist of three coupled, nonlinear PDEs: Poisson's equation, an electron continuity equation and a hole continuity equation. Subsection 4.3.2 discusses the heat-transport equation.

4.3.1   Particle-Conservation Equations for Semiconductors

The basic semiconductor equations consist of three coupled, nonlinear partial differential equations: Poisson's equation (4.28), an electron continuity equation (4.29) and a hole continuity equation (4.30). Poisson's equation is implied by Maxwell's equations. It follows from (1) the electric field being the gradient of the electrostatic potential, $\mathbf{E} = -\nabla V$, and (2) the flux of $\mathbf{E}$ leaving a volume being proportional to the free charge inside the volume. The flux of $\mathbf{D} = \epsilon \mathbf{E}$ produced per unit volume is $\nabla \cdot \mathbf{D}$ (which is the definition of what the divergence of a vector means), and this is equal to the free-charge density ρ.

The continuity equations ensure that electrons and holes are conserved, locally in space. They account for the different ways that the carriers can be created and removed in a volume, or lost from the surface of the volume. The fluxes of particles, which appear in these equations, are written using mobilities and diffusion coefficients.

Suppose the flux per unit area of some quantity is $\mathbf{\Gamma}$. The divergence is defined, so as to make the amount of that quantity which is leaving a unit volume per second equal to $\nabla \cdot \mathbf{\Gamma}$. The particle-generation and recombination rates are all given in terms of rate coefficients. All of these coefficients are estimated using results from solid-state physics and by means of simplifying assumptions [9].

The physical interpretation of these equations is as follows. Poisson's equation states that electric flux emanates from positively charged particles (holes and ionized donor impurities) and terminates on negatively charged particles (electrons and ionized acceptor impurities). Furthermore, Poisson's equation implies that a unit flux of $\mathbf{D}$ ($\mathbf{D} = \epsilon \mathbf{E}$) is created by a unit charge.

The continuity equations state that the net source (the source minus the sink) of each type of carrier, which is to say all the carriers of that type produced per second in a unit volume, can be used in two ways. They can add to the density of those carriers in the volume. Alternatively, they can flow out of the surface of the volume. In general some of the new carriers will be added to the density. That is, some fraction of those carriers produced per second will appear in the rate of increase of the density per second. (Since the number produced was measured per unit volume, the number which are produced per unit volume per second which do not leave is equal to the increase in density per second. This is despite the fact that the volume we are looking at may be a lot smaller than a unit volume.)

Some of the new carriers flow out of the surface. Then some fraction of those carriers produced per second will appear in the flux leaving the surface. The flux leaving the surface of a very small volume is, by definition of the divergence, (approximately) equal to the divergence of the flux times the volume. That is, the divergence is the flux leaving the surface of a small volume divided by that small volume. The reason for defining divergence is that it measures that rate at which flux is produced per unit volume, when that flux leaves the surface of the volume. The continuity equation, therefore, uses the divergence of the flux to measure the flux leaving the volume. This is a conservation equation of the type discussed in Chapters 2 and Chapter 3.

$$\nabla \cdot (\epsilon \mathbf{E}) - e\,(p - n + C) = 0 \tag{4.28}$$

$$+\nabla \cdot \mathbf{J}_n - eR = e\frac{\partial n}{\partial t} \tag{4.29}$$

$$-\nabla \cdot \mathbf{J}_p - eR = e\frac{\partial p}{\partial t}. \tag{4.30}$$

The net concentration of positive ions C is given by $C = N_D - N_A$, assuming all the dopants are ionized.

Each of the basic semiconductor equations is of the form $\nabla \cdot \mathbf{F} = S$, where $\mathbf{F}$ is a flux and S is a source or sink. The flux in Poisson's equation is the electric flux density $\mathbf{D}$, while the fluxes in the electron and hole continuity equations are the electron current density $\mathbf{J}_n$ and the hole current density $\mathbf{J}_p$, respectively. The electric field intensity is given by Equation (4.31), while the electron and hole current densities are given by Equations (4.32) and (4.33), respectively. A derivation of the current density expressions for the semiconductor equations from kinetic theory is a rather lengthy task, and one which cannot be done entirely convincingly from first principles. The reader is referred to Selberherr [9] for an outline of as much of the derivation as can be justified, the assumptions used in the derivation, and a list of references to more advanced texts which discuss the finer points of the derivation.

In the following equations, V is the (scalar) electrostatic potential, n is the electron concentration and p is the hole concentration. V, n and p are also the dependent variables of the semiconductor equations. Mathematical models for the dielectric constant ϵ, carrier mobilities μ_n and μ_p, diffusion coefficients D_n and D_p and recombination/generation mechanisms which contribute to the net recombination rate per unit volume R will be presented in the next section.

$$\mathbf{E} = -\nabla V \tag{4.31}$$

$$\mathbf{J}_n = en\mu_n\mathbf{E} + eD_n\nabla n \tag{4.32}$$

$$\mathbf{J}_p = ep\mu_p\mathbf{E} - eD_p\nabla p. \tag{4.33}$$

These expressions are used in the more advanced examples given throughout this book. To explain the need for modification of these equations when the doping density becomes high we shall briefly consider the equilibrium treatment of the carrier densities again. As described in Section 4.2, the electron and hole concentrations in equilibrium are assumed to be defined by Fermi-Dirac distributions and a parabolic density of states. In order to determine the electron concentration, the electron energy-level occupancy probability (the Fermi function) is multiplied by an electron density-of-state function and integrated. Unfortunately, no closed-form expression for this integral exists. A similar procedure is used to obtain the hole concentration. In the case of a nondegenerately doped semiconductor, an approximation can be made by assuming the electrons and holes obey Boltzmann statistics. This non-degenerate case is the situation for which the nonequilibrium transport equations apply. The modeling of degenerately doped semiconductors will be discussed in Subsection 4.4.8. For simulation of devices at low temperature where the impurities are not completely ionized, degeneracy factors [10] and Fermi-Dirac statistics must be used and, in addition, the fixed carrier concentration of Poisson's equation will be temperature-dependent [11].

4.3.2 Heat-Conservation Equation

In order to simulate power devices under such conditions that the interaction of electrical and thermal phenomena is important, one cannot assume an isothermal temperature profile throughout the device. The lattice temperature and its distribution in power devices can significantly affect the devices' electrical characteristics. One of the most important thermal effects is thermal runaway. Thermal runaway is a condition in which the electrical dissipation of energy causes an increase in temperature which, in turn, causes a greater dissipation of energy. Unless the heat can be adequately removed from the device, this positive-feedback mechanism will ultimately lead to device failure. Other important effects, which need to be included, are discussed in the next section.

In order to account for thermal effects in semiconductor devices, the heat-conservation or heat-transport Equation (4.34) has to be solved in conjunction with the semiconductor equations. In the heat-transport equation, κ is the thermal conductivity, T is the temperature, H is the local heat-generation rate per unit volume, ρ is the specific mass density and c is the specific heat. The temperature T is the dependent variable of the heat transport equation. Models for the parameters κ, H, ρ and c will be given in Section 4.4. As with the semiconductor equations, the heat transport is of the form $\nabla \cdot \mathbf{F} = S$, where F is the flux of power and the S is the net generation of heat. Physically, the heat-transport equation states that the rate of change in the temperature multiplied by the specific heat per unit volume (which is the rate of increase in the thermal energy per unit volume), is equal to the net rate of generation of heat plus the rate at which heat is brought into the unit volume by transport. The rate at which heat leaves a unit volume is the divergence of the heat flux (in the limit $\Delta V \to 0$). The heat flux is $-\kappa(T)\nabla T$, so the divergence of $\kappa(T)\nabla T$ is the rate at which heat comes into the unit volume.

$$\nabla \cdot (\kappa(T) \nabla T) + H = \rho c \frac{\partial T}{\partial t}. \tag{4.34}$$

Heat-Conservation Equations for Semiconductors

We now go into more detail of the form of the energy (or heat) conservation equation, and the form it takes in a semiconductor. As explained above, accounting for temperature variations can be very important in power devices. It is also critical in some very small devices used in VLSI. The conventional drift-diffusion model of charge transport (the electron and hole continuity Equations (4.29) and (4.30)) neglect 'nonlocal' transport effects such as velocity overshoot and energy-dependent impact ionization. While these effects are minor in large devices, they are often significant in submicron devices. Neglecting 'nonlocal' effects in submicron device simulations can produce substantial errors in both the terminal currents and breakdown voltages of these devices.

To account for 'nonlocal' effects, additional electron and hole heat-conservation equations (Equations (4.35) and (4.36)) are solved in addition to the basic semiconductor equations. In these equations, e is the charge of an electron, k_B is Boltzman's constant, n and p are the electron and hole concentrations, $\mathbf{S}_n$ and $\mathbf{S}_p$ are the electron and hole energy flux densities, $\mathbf{J}_n$ and $\mathbf{J}_p$ are the electron and hole current densities, V is the electrostatic potential, T_n, T_p and T_0 are the electron, hole and lattice temperatures, κ_n and κ_p are the thermal conductivities for electrons and holes, and $\tau_{\epsilon n}$ and $\tau_{\epsilon p}$ are the electron energy and hole energy relaxation times.

$$\nabla \cdot \mathbf{S}_n = -\mathbf{J}_n \cdot \mathbf{E} - \frac{3}{2} n k_B \frac{T_n - T_0}{\tau_{\epsilon n}} - \frac{3}{2} k_B \frac{\partial (n T_n)}{\partial t} \tag{4.35}$$

$$\nabla \cdot \mathbf{S}_p = -\mathbf{J}_p \cdot \mathbf{E} - \frac{3}{2} p k_B \frac{T_p - T_0}{\tau_{\epsilon p}} - \frac{3}{2} k_B \frac{\partial (p T_p)}{\partial t}. \tag{4.36}$$

The electron and hole energy flux densities $\mathbf{S}_n$ and $\mathbf{S}_p$ are calculated according to Equations (4.37) and (4.38), respectively. In addition to drift and diffusion currents, there is an additional current due to thermal gradients in the carrier temperatures. To account for this current, the electron and hole current densities $\mathbf{J}_n$ and $\mathbf{J}_p$ are modified according to Equations (4.39) and (4.40). According to the Wiedeman-Franz relation, κ_n and κ_p are defined by Equations (4.41) and (4.42). δ_n, δ_p, Δ_n and Δ_p in Equations (4.37), (4.38), (4.41) and (4.42) are transport coefficients which are generally dependent on the electron and hole temperatures. Finally, ξ_n and ξ_p in Equations (4.39) and (4.40) are defined according to Stratton [12] and given by Equations (4.43) and (4.44).

$$\mathbf{S}_n = -\kappa_n (T_n) \nabla T_n - \frac{k_B \delta_n}{q} \mathbf{J}_n T_n \tag{4.37}$$

$$\mathbf{S}_p = -\kappa_p (T_p) \nabla T_p + \frac{k_B \delta_p}{q} \mathbf{J}_p T_p \tag{4.38}$$

$$\mathbf{J}_n = e n \mu_n \mathbf{E} + e D_n \nabla n + n \mu_n k_B (1 + \xi_n) \nabla T_n \tag{4.39}$$

$$\mathbf{J}_p = e p \mu_p \mathbf{E} - e D_p \nabla p + p \mu_p k_B (1 + \xi_p) \nabla T_p \tag{4.40}$$

$$\kappa_n\left(T_n\right) = en\mu_n\left(T_n\right)\left(\frac{k_B}{q}\right)^2 \Delta_n T_n \tag{4.41}$$

$$\kappa_p\left(T_p\right) = ep\mu_p\left(T_p\right)\left(\frac{k_B}{q}\right)^2 \Delta_p T_p \tag{4.42}$$

$$\xi_n = \frac{d\ln\mu_n}{d\ln T_n} \tag{4.43}$$

$$\xi_p = \frac{d\ln\mu_p}{d\ln T_p} \tag{4.44}$$

Equations (4.35)–(4.44) are known as the *energy balance model* (EBM) [12]. If ξ_n and ξ_p are set   to zero, the *Meinerzhagen's hydrodynamic model* (HDM) [13] is obtained. In the classic test case provided by the 'ballistic' diode, the HDM gives a large nonphysical velocity spike. The EBM, however, gives a more realistic velocity distribution that is in much better agreement with Monte Carlo simulations. In both of these models the transport parameters, such as mobility and impact-ionization coefficients, are functions of carrier temperatures instead of the local electric field. This is physically more plausible, because the microscopic scattering mechanisms, which control the transport, depend mainly on the energy of the carriers [14]. References to models for these parameters will be given later. Over the years, researchers have simulated semiconducting devices using the HDM [14, 13, 15, 16] and the EBM [12].

4.3.3 Boundary Conditions

The boundary conditions which are applied in solving a partial differential equation are one of the most important aspects of the problem specification. They control the nature of the solution that is obtained, but because they present difficulties at several levels they are often handled less accurately than they should be.

The list of potential reasons why it is not straightforward to use the exactly right set of boundary conditions includes physical reasons. For instance, the processes which occur at boundaries are less well known than the behavior in other regions. From the point of view of numerical implementation, schemes such as finite-elements are not always easy to use with all of the boundary conditions which are believed to be appropriate. In some circumstances fluid equations of the type which are used in most semiconductor modeling are oversensitive to the boundary conditions, and some sets of boundary conditions which seem reasonable will not allow a solution at all. It is not uncommon to encounter physical behavior in semiconductors that makes it difficult to choose the right boundary condition. An example of this would be a floating region inside a device where the potential of the region is determined in principle by current balance at a pn junction which encloses the region. The potential is not determined in this case by direct contact with an external electrode. In the real device the potential of this region can vary sensitively with conditions. In a simulation it may prevent convergence unless the potential at at least one point in the floating region is forced to take a value which reflects the built-in potential of the pn junction. Finally for now we will mention that the boundary conditions take up most of the effort in setting up simulations, as we shall see in the examples that follow.

The appropriate boundary conditions for sets of linear, second-order partial differential equations are discussed in standard texts on theoretical physics [6]. These are useful for the classification scheme that they provide, even if the simple equations considered do not directly apply. Three types of boundary condition which are commonly referred to are Dirichlet, Neumann and Cauchy conditions. Dirichlet boundary conditions specify the quantity of interest on the boundary. (A mnemonic which may be useful is that Dirichlet implies a Direct specification of the boundary condition. Knowing the quantity itself is a direct way to specify it.) Neumann boundary conditions specify the normal derivative of the quantity at the boundary. (The mnemonic this time is easy to anticipate; Neumann suggests Normal.) Cauchy boundary conditions use both. They specify the quantity and its normal derivative along part of the boundary. (A possible mnemonic here is that Cauchy means Complete. Knowing both the quantity and its derivative is a complete specification of the boundary conditions.)

The way this applies to the present topic is that if we specify the potential or the density at a boundary, we are using a Dirichlet condition. If we specify the flux through a boundary, we may be using the Neumann condition, since the flux will probably involve the normal derivative of the density through a term like $-D\nabla n$. This flux will be evaluated in the direction normal to the boundary. If the flux also involves a mobility term like $n\mu E = -n\mu\nabla V$ as well as the diffusion term $-D\nabla n$, then we have a mixed boundary condition involving n as well as the normal derivatives of n and V. This is not a Cauchy condition, since the quantities and their derivatives were not specified separately.

Physically the boundaries of the semiconductor are likely to be either contacts or insulating edges. At the contacts it is usual, if the contacts are Ohmic, to assume the carrier densities are equal to their equilibrium values and that the semiconductor is neutral. This is an example of the lack of an accurate physical boundary condition. Schottky contacts are another example where the physical model is in doubt, which will not be described here.

We also usually specify either the potential or the flux at the contact but not both. The other quantity, which was not specified, will be found by solving the semiconductor equations. This is motivated by physical reasoning. If we know the potential at the contacts of a device, we should be able to calculate the current, which is a flux (or a sum of two fluxes.) If a model of the external circuit is included, then the internal solution of the device equations and the model of the external circuit will probably have to be iterated until a self-consistent set of currents and voltages can be found, or the entire system (the semiconductor equations and circuit equations) have to be solved self-consistently. Examples of this are given in the mixed-mode simulations of Chapter 12.

On insulating surfaces the flux will be determined by the rate at which charges can recombine at the surface. If charge is not building up on the surface, then the fluxes of each type of carrier must be the same and equal to the surface recombination rate. If the surface recombination rate is infinite, the densities will take their equilibrium values, as was assumed at a conducting boundary. At an insulator, however, it is more common to assume the surface recombination rate is zero.

The normal component of the electric displacement is specified at surfaces other than good conductors. At the surface of a good conductor a charge is set up to prevent the interior from having an electric field in it. The lines of electric field either begin or terminate on surface charges, so if D is known, the charge can be found. At other surfaces the change in the normal D field on crossing the surface is equal to the charge per unit area on the surface. In this case, we assume the charge is known and fixed, so D can be found from it.

4.3.4   Alternate Sets of Independent Variables

There are several sets of variables which can be used instead of (n, p, V). The most commonly used sets exploit the exponential dependence of n and p on the potential. In doing so, there is an implication that the exponential variation is close to reality, since if it were not, the reason for using the alternative set of variables would not apply.

One set of variables can be constructed where the appropriate exponential is divided into n and p to get a new pair of variables. This factor is essentially an integrating factor for the current density, and allows the two terms to be combined into one. The very strong cancellation between the terms is then explicit in the expression. The problem with this approach is that it only works well for low voltages where the exponential of the voltage divided by $k_B T/e$ is not unmanageably large (and might lead to a precision overflow).

A more effective set of variables, based on the same ideas, involves the use of the quasi-Fermi levels E_{Fn} and E_{Fp}. The current density in terms of the quasi-Fermi level is

$$\mathbf{J}_n = -e\mu_n n \nabla E_{Fn} \tag{4.45}$$

for electrons, so the two terms in the current have again been combined. n is not a permissible variable in this set; it is a shorthand for $n_i \exp\left(\frac{e(V - E_{Fn})}{k_B T}\right)$ in this case.

It is not clear whether these variables are preferable to (n, p). (The voltage V is used in all of these cases.) Most numerical schemes seem to employ (n, p, V). The differences between schemes are most acute in cases where numerical accuracy is being stretched by very high doping and carrier densities. In these circumstances the schemes may lead to errors in somewhat different places. For example, if a problem is solved using (n, p, V), then the current computed from the density gradients and V involves a cancellation of two large numbers to get a much smaller one. (This could lead to a precision underflow.) If the quasi-Fermi level is known, then the current is found in a less error-prone fashion, using the above expression which involves no cancellation. On the other hand, the charge density which appears in

Poisson's equation is more naturally found from n and p. If n and p are the variables employed, then finding ρ is presumably more accurate than when working with the quasi-Fermi levels. In that case, the charge density is found from a small difference of two exponentials of large numbers.

It may be possible to construct more accurate schemes which would combine the best aspects of both sets of variables, for instance by iterating between approaches. However, this is beyond what is appropriate to speculate about in this book.

4.4   The Semiconductor Parameters

This section presents the mathematical models for the semiconductor parameters. Several models exist for each parameter. The models described in this section are the ones which are used in this book. References to other models will be given; in particular, see [9]. This section is organized as follows:

Subsection 4.4.1 describes the dielectric-constant models. Subsections 4.4.2, 4.4.3, 4.4.4 and 4.4.5 present the models for the carrier mobilities, surface-scattering, velocity saturation and diffusion coefficients. Subsection 4.4.6 describes models for several recombination and generation mechanisms. Subsection 4.4.7 discusses the minority-carrier lifetime models. Subsection 4.4.8 presents the model for band-gap narrowing which accounts for degeneracy in heavily doped semiconductors. Subsections 4.4.9 and 4.4.10 discuss the models for the specific mass density, the specific heat and the thermal conductivity. Subsection 4.4.11 presents the heat-generation mechanisms in semiconductors.

4.4.1   Permittivity

In the derivation of Poisson's equation, in Section 2.7, an assumption was made that $\mathbf{D} = \epsilon\mathbf{E}$, where ϵ is the permittivity. This relation is valid for materials which have a time-independent permittivity and for which polarization due to mechanical forces is negligible. In the materials used to make semiconducting devices, the above assumptions are valid. It should be noted that it is not possible to use Poisson's equation in the above form in the investigation of piezoelectric phenomena, ferroelectric phenomena or nonlinear optics. In general, ϵ is a tensor and may be a function of the applied field and the field's history. However, since most semiconductors are isotropic and nonmagnetic, ϵ is often a constant. Table 4.1 lists the relative permittivities ϵ_r of several materials used to make semiconducting devices. The absolute permittivities of the materials listed in the table are equal to the the relative permittivities ϵ_r multiplied by the permittivity of free space ϵ_0, which is equal to 8.854×10^{-14} F/cm.

4.4.2   Carrier Mobilities

Much research has been devoted to modeling carrier mobilities in silicon and other semiconductors. Many mechanisms reduce the carrier mobilities such as scattering

Material	Relative Permittivity
Si	11.8
SiO_2	3.9
Si_3N_4	7.2
GaAs	12.5
Ge	16.1

Table 4.1. Relative permittivities of common semiconductors and insulators.

by thermal lattice vibrations, ionized impurities, neutral impurities, vacancies, interstitials, dislocations, surfaces and by electrons and holes themselves [9]. Further reduction in mobility is possible, due to the saturation of the drift velocity. Many of these mechanisms are very complicated and difficult to model. To further complicate these issues, there is little agreement between researchers as to which models most accurately describe the actual physics. Even when there is consensus, the data for the models' parameters is scattered. From a computational view, simple empirical models are preferred. Sah et al have stated that more elaborate formulae based on complicated theoretical models do not justify the additional effort for the purpose of simulation [17]. Selberherr [9] agrees with this statement. Therefore, theoretical models for the various mechanisms that influence carrier mobilities and their interaction will not be discussed. This section will focus on empirical models of the dominant scattering mechanisms: lattice-scattering, impurity-scattering and carrier-carrier scattering. In addition, this section will discuss how the various scattering mechanisms can be combined to yield a single mobility. Finally, this section will present the surface-scattering models of Watt [18]. A good review of carrier mobilities can be found in the literature [19].

Lattice Scattering

Carriers may be scattered by thermal lattice vibrations known as phonons. A derivation of the lattice-scattering rate may be found in the literature [20]. Lattice-scattering is not a simple process since the intervalley band structure of acoustic and optical phonons in semiconductors gives rise to additional lattice-scattering mechanisms. For simulation purposes, lattice-scattering may be modeled by a simple power law [21] such as equations (4.46) and (4.47). Selberherr [9] has compiled a table of lattice mobility constants μ_n^0, μ_p^0, α_n and α_p for materials such as Si, GaAs and Ge. Sah et al [17] have proposed a different set of empirical models (Equations (4.48) and (4.49)) which account for scattering due to optical and intervalley phonons as well as acoustical phonons. Sah's models are useful over a broad temperature range of 4.2 K to 600 K. At room temperature, the two sets of models are in close agreement. At low temperatures, Sah's models predict higher mobilities than the simple power-law model; whereas at high temperatures the opposite is true.

$$\mu_n^L = \mu_n^0 \, (T/300)^{-\alpha_n} \tag{4.46}$$

$$\mu_p^L = \mu_p^0 \, (T/300)^{-\alpha_p} \tag{4.47}$$

$$\mu_n^L = \left\{ \left[4195.0 \, (T/300)^{-1.5} \right]^{-1} + \left[2153.0 \, (T/300)^{-3.13} \right]^{-1} \right\}^{-1} \tag{4.48}$$

$$\mu_p^L = \left\{ \left[2502.0 \, (T/300)^{-1.5} \right]^{-1} + \left[591.0 \, (T/300)^{-3.13} \right]^{-1} \right\}^{-1} . \tag{4.49}$$

Impurity Scattering

Impurity-scattering is another important mechanism that affects carrier mobilities in semiconductors. Conwell and Weiskopf [22] derived the first useful impurity-scattering model based on their theoretical investigations. The Conwell and Weiskopf model accounted for the screening of neighboring impurities when present in high concentrations. Subsequently, Brooks [23] proposed a refinement to the charge screening in the Conwell and Weiskopf model and also included the impurity-atom screening due to electrons and holes. A model which is useful for simulation purposes is that of Brooks [23] and is given by equation (4.50). The parameter T is the temperature and N is the effective doping concentration.

$$\mu_{n,p}^I = \frac{AT^{3/2}}{N} \left[\ln\left(1 + \frac{BT^2}{N}\right) - \frac{BT^2}{N + BT^2} \right]^{-1}. \qquad (4.50)$$

At low temperatures ($T < 77$ K), neutral impurity-scattering becomes important. Early theoretical work suggests that scattering from neutral impurity atoms is temperature independent [20, 24]. More recent investigations predict a weak temperature dependence [25]. At room temperature neutral impurity-scattering is usually ignored, since it is very difficult to determine the neutral impurity concentration in a real device and, more importantly, its effect is insignificant. A more thorough discussion of this scattering mechanism can be found in Selberherr [9].

Carrier-Carrier Scattering

The final important scattering mechanism is carrier-carrier scattering. This type of scattering is particularly important in conductivity-modulated power devices, such as pin diodes and insulated-gate bipolar transistors (IBGTs), where the on-state carrier concentrations exceed the doping concentration. Fletcher [26] noted that Chapman and Cowling's [27] research on the mutual diffusion of two groups of charged gaseous particles applies, following a classical approximation, to the quasi-neutral electron-hole plasma present in conductivity-modulated power devices under high-level injection. Choo [28] proposed a slightly modified expression (Equation (4.51)) relative to that of Fletcher whose range of validity as a function of injection level is greater than that of Fletcher's formula. At low-level injection, Choo's equation for the mobility evaluates to an extremely large value. When suitably combined with the other mobility expressions, it has a negligible effect on the overall mobility at low-level injection (see 4.4.2). This does not imply that no carrier-carrier scattering occurs at low-level injection, but rather that this scattering is a small perturbation of the other scattering mechanisms.

$$\mu_{n,p}^{ccs} = \frac{2 \times 10^{17}}{\sqrt{pn}} T^{3/2} \left[\ln\left(1 + \frac{8.28 \times 10^8}{\sqrt[3]{pn}} T^2\right) \right]^{-1}. \qquad (4.51)$$

Researchers have investigated like-carrier collisions at low-level injection [29, 30]. Their research suggests multiplicative corrections which reduce the lattice and impurity partial mobilities. Since the variation of these corrections is not known as a function of doping concentration, Dorkel and Leturcq [31] proposed correcting the impurity partial mobility, which is dominant at high doping, and not correcting the lattice partial mobility, which is dominant at low doping. Furthermore, Dorkel and Leturcq assumed that collisions between unlike carriers (electrons and holes) are identical in rate and effect on transport to like-carrier collisions.

Combining the Scattering Mechanisms

Since the scattering mechanisms are not independent of each other, they cannot be combined using the Mathiessen rule (added like resistors in parallel) to obtain an overall mobility. Based on theoretical considerations, Debye and Conwell [29] derived equation (4.52) for the combination of mobilities $\mu_{n,p}^L$ and $\mu_{n,p}^I$. Since the computational requirements of this equation are excessive, Dorkel and Leturcq [31] proposed a simple approximation (Equation (4.56)) which they claim is accurate over a large range of temperatures, doping concentrations and injection levels. Furthermore, since the relaxation times associated with $\mu_{n,p}^I$ and $\mu_{n,p}^{ccs}$ have similar energy dependence, they may be combined according to Mathiessen rule (Equation (4.57)). The combined mobility $\mu_{n,p}^{Iccs}$ may then be used in the Debye and Conwell expression in place of $\mu_{n,p}^I$. Other researchers have proposed different methods for computing the overall mobility. Many of these methods are discussed by Selberherr [9].

$$\mu_{n,p}^{LI} = \mu_{n,p}^L \left\{ 1 + x^2 \left[\text{Ci}\left(x\right)\cos\left(x\right) + \text{Si}\left(x\right)\cos\left(x\right) \right] \right\} \tag{4.52}$$

$$x = \sqrt{6\mu_{n,p}^L / \mu_{n,p}^I} \tag{4.53}$$

$$\text{Ci}\left(x\right) = -\int_x^\infty \frac{\cos\left(t\right)}{t} dt \tag{4.54}$$

$$\text{Si}\left(x\right) = -\int_x^\infty \frac{\sin\left(t\right)}{t} dt \tag{4.55}$$

$$\mu_{n,p}^{LI} = \mu_{n,p}^L \left[\frac{1.025}{1 + \left(x/1.68\right)^{1.43}} - 0.025 \right] \tag{4.56}$$

$$\mu_{n,p}^{Icss} = \left(\frac{1}{\mu_{n,p}^I} + \frac{1}{\mu_{n,p}^{ccs}} \right)^{-1}. \tag{4.57}$$

4.4.3 Carrier Velocity Saturation

When carriers are in a high electric field region of a semiconductor, their drift velocity saturates, resulting in a further reduction of their mobility. Shockley [32] derived the first useful equation for the effect of carrier heating on the drift velocity. A more commonly used expression for silicon devices is given by equation (4.58) where E is the electric field and β is a constant (usually 1 or 2), μ_{lf} is the low-field mobility and v_{sat} is the saturation velocity [33]. An empirical temperature-dependent model for the saturation velocity [34] is given by Equation (4.59). Models for carrier velocity saturation in GaAs can be found in the literature [35, 36].

$$\mu\left(E, T\right) = \frac{\mu_{lf}}{\left\{1 + \left[\mu_0 E / v_{sat}\left(T\right)\right]^\beta\right\}^{1/\beta}} \tag{4.58}$$

$$v_{sat}\left(T\right) = \frac{2.4 \times 10^7}{1 + 0.8 \exp\left(T/600\right)}. \tag{4.59}$$

4.4.4 Surface Scattering

In inversion-layer devices such as MOSFETs, surface-scattering becomes significant and surface mobilities may be substantially lower than bulk mobilities. Watt [18] derived a model which considers the following inversion-layer scattering mechanisms: phonon scattering between two-dimensional inversion-layer carriers and bulk phonons, surface-roughness scattering and charged-impurity scattering caused by the interaction between inversion-layer charge and surface charge. The first two mechanisms depend on the effective electric field and the last depends on the channel doping density. This model is given by Equations (4.60) and (4.61). The parameters for this model may be found in ref. [18].

$$\frac{1}{\mu_n^{eff}} = \frac{1}{\mu_{n1}}\left(\frac{10^6}{E_{eff}}\right)^{\alpha_{n1}} + \frac{1}{\mu_{n2}}\left(\frac{10^6}{E_{eff}}\right)^{\alpha_{n2}} + \frac{1}{\mu_{n3}}\left(\frac{N_B}{10^{18}}\right)\left(\frac{10^{12}}{N_i}\right)^{\alpha_{n3}} \tag{4.60}$$

$$\frac{1}{\mu_p^{eff}} = \frac{1}{\mu_{p1}}\left(\frac{10^6}{E_{eff}}\right)^{\alpha_{p1}} + \frac{1}{\mu_{p2}}\left(\frac{10^6}{E_{eff}}\right)^{\alpha_{p2}} + \frac{1}{\mu_{p3}}\left(\frac{N_B}{10^{18}}\right)\left(\frac{10^{12}}{N_i}\right)^{\alpha_{p3}}. \tag{4.61}$$

4.4.5 Carrier Diffusion Coefficients

For nondegenerately doped semiconductors where Boltzmann statistics are applicable, the carrier mobilities and diffusion coefficients obey the Einstein relationship and are given by Equations (4.62) and (4.63).

$$D_n = \frac{k_B T}{q}\mu_n \tag{4.62}$$

$$D_p = \frac{k_B T}{q}\mu_p. \tag{4.63}$$

4.4.6 Recombination and Generation

Generation is a process whereby electrons and holes are created. Conversely, recombination is a reverse process whereby electrons and holes are destroyed. Several recombination and generation (R-G) mechanisms occur in a semiconductor, and each of these mechanisms has been modeled by several mathematical expressions. The dominant mechanisms are photon transitions, phonon transitions, three-particle (Auger) transitions and impact ionization [9]. Other such mechanisms exist, but they are less important [37]. Models of each of the dominant recombination and generation mechanisms will be discussed. Focus is given to the models used in the simulations done in this book. References will be given to other models.

Photon Recombination/Generation

Photogeneration is one of the simplest R-G processes, involving a direct transition of an electron across the band-gap. Photogeneration occurs when sufficiently energetic photons impinge upon a semiconductor. If the photon's energy is greater than or equal to the semiconductor band-gap energy E_g, the photon may be absorbed. The absorbed energy excites a valence electron to the conduction-band, creating an electron-hole pair. The opposite process, radiative recombination, can also occur. Radiative recombination occurs when a conduction-band electron falls back to the valence-band and releases its energy in the form of light (a photon). Photogeneration and radiative recombination may be modeled by equations (4.64) and (4.65) assuming a photon capture rate C_c^{OPT} and a photon emission rate C_e^{OPT} [9].

$$R_{OPT} = C_c^{OPT} np \tag{4.64}$$

$$G_{OPT} = C_e^{OPT}. \tag{4.65}$$

The emission rate C_e^{OPT} is a function of the quantum efficiency for the conversion of photons into electron-hole pairs γ_{qe}, the absorption coefficient α, the reflectance r between the two media in question, the power of the impinging radiation P, the photon energy E_p and the distance from the interface surface y. The emission rate is modeled by equation (4.66). The photon energy E_p is given by equation (4.67), where h is Planck's constant, c is the speed of light in a vacuum and λ is the wavelength of the light. The absorption coefficient α is defined by equation (4.68), where n is the index of refraction. Finally the reflectance r is given by Equation (4.69), where n_1 and n_2 are the indices of refraction of the media in question.

$$C_e^{OPT} = \gamma_{qe} \alpha \left(1 - r\right) \frac{P}{E} \exp\left(-\alpha y\right) \tag{4.66}$$

$$E = hc/\lambda \tag{4.67}$$

$$\alpha = (4\pi/\lambda) \operatorname{Im}\{n\} \tag{4.68}$$

$$r = \frac{\left(\operatorname{Re}\{n_1\} - \operatorname{Re}\{n_2\}\right)^2 + \left(\operatorname{Im}\{n_1\} - \operatorname{Im}\{n_2\}\right)^2}{\left(\operatorname{Re}\{n_1\} + \operatorname{Re}\{n_2\}\right)^2 + \left(\operatorname{Im}\{n_1\} + \operatorname{Im}\{n_2\}\right)^2}. \tag{4.69}$$

The capture rate can be computed from the emission rate. In thermal equilibrium $R_{OPT} = G_{OPT}$. Equating equations (4.64) and (4.65) yields the desired result, Equation (4.70).

$$C_e^{OPT} = C_c^{OPT} n_i^2. \tag{4.70}$$

Phonon Recombination/Generation

Phonon transitions are primarily two-step processes and occur at special sites within the semiconductor. These sites are referred to as R-G centers or traps and have energies between the conduction-band and the valence-band. The two-step process is important, because a one-step process of electron transfer between the top of the valence-band and the bottom of the conduction-band would require a significant change of momentum in indirect semiconductors, such as silicon, making one-step photon transitions unlikely. Shockley, Read and Hall (SRH) [38, 39] established a theory of phonon recombination/generation, and Dhariwal, Kothari and Jain [40] generalized it. SRH R-G involves four processes: electron capture, hole capture, electron emission and hole emission. Electron capture occurs when a conduction-band electron is trapped by a vacant R-G center. Consequently, the R-G center becomes occupied. The captured electron can either fall to the valence-band and annihilate a hole (hole capture) or move back to the conduction-band (electron emission). In either event, the occupied R-G center becomes vacant. Similarly, hole emission occurs when a valence-band electron is trapped by a vacant R-G center. Consequently, the R-G center becomes occupied, and a hole is produced in the valence-band. This electron can either fall back to the valence-band and annihilate a hole (hole capture) or move to the conduction-band (electron emission). In either case, the occupied R-G center becomes vacant. If electron capture is followed by hole capture, then recombination occurs and an electron-hole pair is destroyed. If hole emission is followed by electron emission, then generation occurs and an electron hole pair is created. Electron capture followed by electron emission and hole emission followed by hole capture have no net effect on the number of valence-band electrons and conduction-band holes.

In order to construct a mathematical model of SRH R-G, the R-G centers are characterized by associating them with an energy level E_t, a concentration N_t and capture cross sections for electrons and holes. The fraction of occupied R-G centers in equilibrium can be determined by application of Fermi-Dirac statistics. With this information electron and hole capture and emission rates can be computed. Following the ideas of Shockley, Read and Hall, a model of SRH recombination can be derived (Equation (4.71)). The variable E_i is the intrinsic energy level of the semiconductor and is approximately located at the center of the band-gap, for most semiconductors. During the derivation one typically defines electron (τ_n) and hole (τ_p) minority-carrier lifetimes. The lifetimes are reciprocals of the corresponding capture rates. Models for the minority-carrier lifetimes will be presented in Subsection 4.4.7.

$$R_{SRH} = \frac{np - n_i^2}{\tau_p\,(n + n_1) + \tau_n\,(p + p_1)} \tag{4.71}$$

$$n_1 = n_i \exp\left(\frac{E_t - E_i}{k_B T}\right), \quad p_1 = n_i \exp\left(\frac{E_i - E_t}{k_B T}\right). \tag{4.72}$$

Additional recombination can occur at insulator-semiconductor interfaces. This surface recombination is described by a surface recombination velocity [41]. Surface recombination may be modeled by effective electron and hole lifetimes. Since it is very difficult to analytically solve the semiconductor equations, the equations are usually solved numerically on a suitable solution mesh. Each mesh point on the insulator-semiconductor interface will have associated effective carrier lifetimes. The effective carrier lifetimes are given by Equations (4.73) and (4.74), where s_n and s_p are the electron and hole recombination velocities, $d|_i$ is the length of the interface associated with node i and $A|_i$ is the semiconductor area associated with node i.

$$\tau_n^{eff}\big|_i = \left(\frac{s_n \, d|_i}{A|_i} + \frac{1}{\tau_n|_i} \right)^{-1} \tag{4.73}$$

$$\tau_p^{eff}\big|_i = \left(\frac{s_p \, d|_i}{A|_i} + \frac{1}{\tau_p|_i} \right)^{-1}. \tag{4.74}$$

In the derivation of the SRH R-G equation, it is assumed that the number of vacant R-G centers remains constant. In other words, the number of R-G centers is much larger than the number of carriers involved in the R-G process. It is also assumed that the trapped electrons may instantaneously move to the valence or conduction-band. Dhariwal, Kothari and Jain did not make these assumptions and derived equation (4.75). In this equation, δt_n and δt_p are the transit times for conduction and valence electrons to 'jump' to the R-G center's energy level, and $\delta t'_n$ and $\delta t'_p$ are the transit times for the reverse process. Their derivation does assume, however, that the conduction band electrons which transit to the R-G centers are close to the bottom of the conduction-band and, similarly, that the valence-band electrons which transit to the R-G centers are close to the top of the valence-band. If the transitions are instantaneous ($a = 1$, $b = 1$ and $c = 0$), then the SRH formula is recovered. In general, τ_n^* and τ_p^* are given by complex expressions which contain values which are not precisely known. Hence, experimental data is used to determine the values of the carrier lifetimes. From a simulation point of view, the extra computational effort in using the DKJ equation is not justified, and usually the SRH equation is used.

$$R_{DKJ} = \frac{np - n_i^2}{\tau_p^* \, (an + bn_1) + \tau_n^* \, (bp + ap_1) + c \, (np - n_i^2)} \tag{4.75}$$

$$a = 1 + \frac{\delta t_n}{\delta t'_n}, \; b = 1 + \frac{\delta t_p}{\delta t'_p}, \; c = \frac{\delta t_n + \delta t_p}{N_t}. \tag{4.76}$$

Recombination at Low Levels of the Excess Carrier Density

The excess carrier density is the carrier density in excess of the equilibrium density. If the equilibrium densities are n_0 and p_0, and the excess densities are Δn and Δp, then $n = n_0 + \Delta n$ and $p = p_0 + \Delta p$.

The expressions for the recombination rate can all be written in the form

$$R = \alpha(np - n_i^2), \tag{4.77}$$

although we will compare our results mainly to the SRH result, which is why we introduce this topic here. When the excess densities are small, and assuming they are equal to each other, $\Delta n = \Delta p$, we can simplify this equation since $np = (n_0 + \Delta n) \times (p_0 + \Delta p)$. $n_0 p_0 = n_i^2$ and neglecting the product of the excess densities,

$$R = \alpha(n_0 + p_0)\Delta n. \tag{4.78}$$

Setting this equal to $-\frac{d\Delta n}{dt}$ allows us to define a recombination rate for the excess minority-carriers, given by $\tau = [\alpha(n_0 + p_0)]^{-1}$.

If we take the SRH expression for the recombination rate, if we neglect n_1 and p_1 and if we set $\tau_n = \tau_p$, then we find

$$R = \frac{np - n_i^2}{\tau_n(n + p)}. \tag{4.79}$$

This shows that the above definition of τ_n and τ_p is consistent with that in the SRH expression.

τ_n or τ_p is the minority-carrier recombination time. The reason that this is called the minority-carrier recombination time is clearer if one assumes that the excess minority-carrier density is actually larger than the equilibrium minority density. In this case the rate appearing in eq. 4.78 can be written as $\alpha(n_0 + p_0)n$ and this can be set equal to $-\frac{dn}{dt}$ (assuming electrons are the minority, for now.) For the majority-carrier this replacement would not be possible.

If extra carriers are generated, for instance by illumination, with a generation rate G_{opt}, then we expect a steady state excess density of minority-carriers given by $\Delta n = G_{opt}\tau_n$. This was obtained by adding G_{opt} to the source rate, which is $-R = -\Delta n/\tau_n$, and setting this equal to $\frac{d\Delta n}{dt}$ which is zero.

Three-Particle (Auger) Recombination

As the carrier concentrations increase, another R-G mechanism becomes important. Auger recombination [42, 43, 44], also known as three-body or three-particle recombination, becomes dominant when the carrier concentration become very large as in the case of high-level current injection in power devices. Like SRH R-G, Auger R-G involves four processes: electron capture, hole capture, electron emission and hole emission. Electron capture occurs when a conduction band electron recombines with a hole in the valence-band and transmits its excess energy to another electron in the conduction-band. Hole capture is similar to electron capture, except the excess energy is transmitted to a hole in the valence-band instead of an electron in the conduction-band. Electron and hole emission are the opposite processes of electron and hole capture. An electron from the valence band moves to the conduction-band by absorbing the energy of an energetic conduction-band electron (electron emission) or a valence band hole (hole emission). The net result of electron and hole emission is the creation of an electron-hole pair.

A model for Auger recombination (Equation (4.80)) can be derived in a straightforward manner [9]. Researchers do not agree on the values of the Auger capture

Temperature [K]	C_n [cm^{6}s^{-1}]	C_p [cm^{6}s^{-1}]
77	2.3×10^{-31}	7.8×10^{-32}
300	2.8×10^{-31}	9.9×10^{-32}
400	2.8×10^{-31}	1.2×10^{-31}

Table 4.2. Auger recombination coefficients for silicon.

coefficients C_n and C_p. Selberherr [9] recommends the data of Dziewior and Schmid [44] (Table 4.2) and an extensive collection of Auger coefficients can be found in the literature [45]. The Auger R-G model represented in Equation (4.80) assumes that Auger recombination is a direct band-to-band process. Recent investigations have shown that trap-assisted Auger R-G has a higher probability of occurring than does direct band-to-band Auger R-G. Trap-assisted Auger R-G is considered by Fossum et al [42].

$$R_{Auger} = \left(pn - n_i^2\right)\left(C_n n + C_p p\right). \tag{4.80}$$

Impact Ionization

The last important R-G mechanism is impact ionization. Impact ionization is purely a generative process. As electrons and holes are accelerated by an electric field, they gain energy. As the carriers collide with the lattice atoms, they may impart some of their energy to the electrons of these atoms. If these carriers are allowed to gain sufficient energy between collisions, they may impart enough energy to cause a valence-band electron to undergo a transition to the conduction-band. If this occurs, an electron-hole pair is created. These newly created carriers may, in turn, be accelerated by the electric field, and, upon colliding with lattice atoms, create additional electron-hole pairs. Hence, impact ionization is a multiplicative process capable of generating huge numbers of carriers in the presence of high electric fields.

Microscopically, impact ionization is identical to the Auger generation process of electron and hole emission. An electron from the valence-band moves to the conduction-band by absorbing the energy of another carrier. The other carrier may be an electron in the conduction-band or a hole in the valence-band. In either case, the net result is the creation of an electron-hole pair. The differences between Auger generation and impact ionization are as follows. The rate of Auger generation is proportional to the carrier concentrations where the rate of impact ionization is proportional to the current density. Auger recombination can occur in regions with high carrier concentrations and negligible current flow. On the contrary, impact ionization can occur in regions with low carrier concentrations but requires nonnegligible current flow.

A commonly used model for impact ionization is given by equation (4.81). The ionization rates α_n and α_p are defined as the numbers of collisions which result

Parameter	$E \leq 4.000 \times 10^5$	$E > 4.000 \times 10^5$	units
α_n^{∞}	7.030×10^5	7.030×10^5	cm^{-1}
E_n^{crit}	1.231×10^6	1.231×10^6	V/cm
β_n	1	1	
α_p^{∞}	1.582×10^6	6.710×10^5	cm^{-1}
E_p^{crit}	2.036×10^6	1.693×10^6	V/cm
β_p	1	1	

Table 4.3. The ionization-rate parameters of Van Overstraeten and De Man.

in the generation of an electron-hole pair from a single electron or hole travelling one centimeter through the lattice. These rates do not take into consideration the subsequent collisions caused by the generated electrons and holes; they only consider the collisions of the original electron or hole. The ionization rates are functions of the electric field in the direction of current flow and are given by Equations (4.82) and (4.83).

$$G_{ii} = \alpha_n \frac{|\mathbf{J}_n|}{q} + \alpha_p \frac{|\mathbf{J}_p|}{q} \tag{4.81}$$

$$\alpha_n = \alpha_n^{\infty} \exp\left[- \left(\frac{E_n^{crit}}{\mathbf{E} \cdot \mathbf{J}_n / |\mathbf{J}_n|} \right)^{\beta_n} \right] \tag{4.82}$$

$$\alpha_p = \alpha_p^{\infty} \exp\left[- \left(\frac{E_p^{crit}}{\mathbf{E} \cdot \mathbf{J}_p / |\mathbf{J}_p|} \right)^{\beta_p} \right]. \tag{4.83}$$

There is much debate over the appropriate choices of ionization-rate parameters [46, 47, 48, 49, 50]. For the purpose of device simulation, several researchers use the values of Van Overstraeten and DeMan [51]. Table 4.3 summarizes their results.

4.4.7 Minority-Carrier Lifetimes

When the product of the electron and hole concentrations exceeds its equilibrium value, the excess carriers will recombine faster than they are generated. Some carriers will recombine very quickly and others will take a long time to recombine. The average time it takes a carrier to recombine is known as the carrier's lifetime. The electron and hole lifetimes are denoted by τ_n and τ_p, respectively. Physically, τ_n and τ_p are interpreted as the average times an excess minority-carrier will live in a sea of majority carriers [52]. Mathematically, τ_n and τ_p are defined as the reciprocals of the electron and hole capture rates as discussed in 4.4.6. Since electrons and holes are captured by R-G centers known as traps and experimental findings suggest that additional traps are created at high doping levels [53], the carrier lifetimes are functions of the doping concentration. Equations (4.84) and (4.85) are

Parameter	Value	Units
τ_{n0}	3.95×10^{-4}	s
τ_{p0}	3.52×10^{-5}	s
N_n^{ref}	7.10×10^{15}	cm^{-3}
N_p^{ref}	7.10×10^{15}	cm^{-3}

Table 4.4. Electron and hole lifetime parameters for silicon.

empirical formulae which fit experimental data. Typical values [54, 55] for τ_{n0}, τ_{p0}, N_n^{ref} and N_p^{ref} are listed in Table 4.4. Alternative values for the aforementioned parameters and other empirical expressions for the doping dependence of the carrier lifetimes can be found in the literature [56, 57]. The carrier lifetimes may also be influenced by various fabrication procedures, such as gettering, gold doping and electron irradiation. Unfortunately, it is very difficult to precisely determine the carrier lifetimes which result from these processing procedures. Subsidiary experimental measurements must usually be performed in order to determine the carrier lifetimes in a semiconductor sample which has been treated in one of these ways.

$$\tau_n = \frac{\tau_{n0}}{1 + (N_D + N_A)/N_n^{ref}} \tag{4.84}$$

$$\tau_p = \frac{\tau_{p0}}{1 + (N_D + N_A)/N_p^{ref}}. \tag{4.85}$$

Input File recomb.sg

The equation specification file recomb.sg, follows the evolution of the carrier density in '0D', allowing for the SRH and Auger recombination.

```
// FILE:  sgdir\ch04\recomb\recomb.sg
// physical and scaling constants
const Ni  = 1.25e10;          // intrinsic conc. of Si  (cm^-3)
const C   = 0.00e10;          // net doping conc.       (cm^-3)
const dno = +1.0e+3*Ni;       // initial excess electron conc. (cm^-3)
const dpo = +1.0e+3*Ni;       // initial excess hole conc.  (cm^-3)
const NT  = 1000;             // number of time steps


// equilibrium electron concentration of homogeneously doped silicon
func Neq(<N>)
  assign temp = [abs(N)+sqrt(sq(N)+4*sq(Ni))]/2.0;
  return (N>=0)*temp + (N<0)*sq(Ni)/temp;


// equilibrium hole concentration of homogeneously doped silicon
func Peq(<N>)
  assign temp = [abs(N)+sqrt(sq(N)+4*sq(Ni))]/2.0;
```

```
  return (N<=0)*temp + (N>0)*sq(Ni)/temp;

// electron lifetime
func LTn(<N>)
  const LT0  = 3.95e-4;
  const Nref = 7.1e15;
  return LT0 / (1.0 + N / Nref);

// hole lifetime
func LTp(<N>)
  const LT0  = 3.52e-5;
  const Nref = 7.1e15;
  return LT0 / (1.0 + N / Nref);

// recombination
func R(n,p,<tn>,<tp>)
  const Cn = 2.8e-31;
  const Cp = 1.2e-31;
  const pn = Ni*Ni;
  assign Rsrh = (n*p-pn)/[tp*(n+Ni)+tn*(p+Ni)];
  assign Raug = (n*p-pn)*(Cn*n+Cp*p);
  return Rsrh + Raug;

// declare variables
var dn[NT], dp[NT], neq, peq, tn, tp, dt;

// solve dn/dt = -R(n,p), dp/dt = -R(n,p) using implicit time stepping
equ dn[i=1..NT-1] -> (dn[i]-dn[i-1])/dt = -R(neq+dn[i],peq+dp[i],tn,tp);
equ dp[i=1..NT-1] -> (dp[i]-dp[i-1])/dt = -R(neq+dn[i],peq+dp[i],tn,tp);

begin main
  assign neq = Neq(C);
  assign peq = Peq(C);
  assign tn = LTn(abs(C));
  assign tp = LTp(abs(C));
  assign dn[i=0] = dno;
  assign dp[i=0] = dpo;
  assign dt = 0.1*max(tn,tp);
  solve;
  write;
end
```

recomb.sg can be run using the script recomb.bat.

4.4.8 Degeneracy and Band-Gap Narrowing

Many useful devices, such as power devices, are degenerately doped, and Boltzmann statistics do not apply. One method of dealing with degenerately doped semicon-

Table: Script File recomb.bat

```
echo off
cls
sgxlat -p15 recomb.sg
order recomb.top recomb.prm
call sgbuild sim recomb
recomb -vs2
extract recomb.res
del recomb.cpp
del recomb.h
del recomb.res
del recomb.top
del recomb.prm
del recomb.exe
```

ductors is to use Fermi-Dirac statistics and introduce degeneracy factors into the semiconductor equations [10]. An alternative method, which is attractive from a computational point of view, is to assume that the semiconductor's band-gap narrows as the dopant concentration increases. The justification for this method is as follows. At low concentrations, the dopant atoms do not interact and introduce a single allowable impurity energy level in the semiconductor's band gap. As the dopant concentration increases, the distance between neighboring dopant atoms decreases. At high dopant concentrations, the dopant atoms influence each other and, instead of introducing a single allowable energy level in the band-gap, they introduce a continuum of energies. These energies merge into the semiconductor's valence-band in the case of acceptors or the conduction-band in the case of donors. The net effect is a narrowing of the semiconductor's band-gap.

Band-gap narrowing [58] is modeled by a temperature- and dopant- dependent pn-product (an effective intrinsic concentration squared n_{ie}^2) which is given by equation (4.86). This model is valid for silicon. V_{g0} is the band-gap of silicon at 0 K, which is 1.206 eV. The spatial gradients of the pn-product also have an effect on the semiconductor transport equations. The electric field $\mathbf{E}$ in equations (4.32) and (4.33) as well as equations (4.35) and (4.36) is replaced by an effective electric field given by Equations (4.89) and (4.90).

$$pn = n_{ie}^2\left(N, T\right) = n_{i0}^2\left(T\right) \exp\left(\frac{q}{k_B T}\Delta V_{g0}\left(N\right)\right) \tag{4.86}$$

$$n_{i0}^2\left(T\right) = 9.61 \times 10^{32} T^3 \exp\left(-\frac{e V_{g0}}{k_B T}\right) \tag{4.87}$$

Material	ρ $[\mathrm{m^2s^{-2}K^{-1}}]$	c $[\mathrm{VAs^3m^{-5}}]$
Si	2328	703
SiO$_2$	2650	782
Si$_3$N$_4$	3440	787
GaAs	5316	351
Ge	5323	322

Table 4.5. Specific mass density and specific heat of common semiconductors and insulators.

$$\Delta V_{g0}\left(N\right) = 9.0 \times 10^{-3} \left\{ \ln\left(\frac{N}{10^{17}}\right) + \sqrt{\left[\ln\left(\frac{N}{10^{17}}\right)\right]^2 + \frac{1}{2}} \right\} \tag{4.88}$$

$$\mathbf{E}_n^{eff} = -\nabla\left(V + \frac{k_B T}{q}\ln\left(n_{ie}\right)\right) \tag{4.89}$$

$$\mathbf{E}_p^{eff} = -\nabla\left(V - \frac{k_B T}{q}\ln\left(n_{ic}\right)\right). \tag{4.90}$$

4.4.9 Specific Mass Density and Specific Heat

The specific mass density ρ and specific heat c of silicon, silicon dioxide, silicon nitride, gallium arsenide and germanium are listed in Table 4.5. The data is that of Selberherr [9] and is given for $T = 300\,\mathrm{K}$. The temperature dependences of the specific mass density and specific heat are negligible in the aforementioned materials and are seldom included in device simulations.

4.4.10 Thermal Conductivity

A model for the thermal conductivity κ of silicon has been developed by Glassenbrenner and Slack [59] and is given by equation (4.91). The units of κ are voltamperes per centimeter per degree Kelvin [$\mathrm{VAcm^{-1}K^{-1}}$]. Their model is in good agreement with theory and deviates by only 5 percent over the temperature range from 250 K to 1000 K. Measurements indicate that the thermal conductivity of a semiconductor may decrease as much as 30 percent in heavily doped semiconductors. The Glassenbrenner and Slack model does not take this into consideration.

$$\kappa\left(T\right) = \frac{1}{0.03 + \left(1.56 \times 10^{-3}\right)T + \left(1.65 \times 10^{-6}\right)T^2}. \tag{4.91}$$

4.4.11 Thermal Generation

A useful model for modeling thermal generation was suggested by Adler [56] and is given by Equation (4.92). Adler's model includes Ohmic losses, energy loss and gain through recombination and generation and accounts for the effects of band-gap narrowing caused by degenerate doping (see Section 4.4.8). For nondegenerate materials, equation (4.92) simplifies to Equation (4.93).

$$H = \frac{1}{q}\mathbf{J}_n \cdot \nabla E_c + \frac{1}{q}\mathbf{J}_p \cdot \nabla E_v + R\left(E_c - E_v\right) \tag{4.92}$$

$$H = \left(\mathbf{J}_n + \mathbf{J}_p\right) \cdot \mathbf{E} + RE_g. \tag{4.93}$$

4.4.12   Parameter Models for High-Field Transport

In this subsection, references for the carrier temperature-dependent mobility and impact-ionization models will be given. These transport parameters are functions of the carrier temperatures instead of the local electric field. As stated in Subsection 4.3.2, these models are more physically plausible because the microscopic mechanisms, which control transport, depend mainly on the energy of the carriers. Quade et al [14] have developed a method of determining electron and hole generation rates which are dependent on the average carrier temperature. Furthermore, they have demonstrated the influence of impact ionization as a cooling mechanism which reduces the mean energy and velocity of the carriers and have shown that the ionization coefficients α_n and α_p are spatially retarded with respect to the local electric-field models normally used, as discussed in Subsection 4.4.6. The spatial retardation results in larger multiplication factors. Cook [60] developed models which treat the carrier mobilities and diffusion-coefficients as functions of the average carrier energy. These models are valid over a larger range of electric fields, current densities and concentration gradients than the conventional models of Subsection 4.4.2.

4.5   The Momentum Conservation Equation

In situations where the force on a fluid changes rapidly in space or time, it is often not sufficient to calculate the fluid velocity from (say) a mobility. In that case the equation of conservation of momentum can be used to find the fluid velocity.

The equation of conservation of momentum which is used to describe the rate of change of the fluid velocity is just Newton's law, applied to the fluid.

$$\frac{d}{dt}\left(m_i n_i \mathbf{v}_i\right) = \mathbf{F}_i \tag{4.94}$$

where m_i is the mass of a particle of species i, n_i is the number per unit volume of species i, and $\mathbf{v}_i$ is the flow velocity of species i. The left side of the equation is therefore the rate of change of momentum of a unit volume of the fluid made up of species i. $\mathbf{F}_i$ is the force per unit volume on that fluid.

In a fluid simulation in one dimension r, for example, the momentum equation given above, after being divided by the mass, might become

$$\frac{\partial n_i v_i}{\partial t} = -\frac{1}{r^\alpha}\frac{\partial}{\partial r}\left(r^\alpha n_i v_i v_i\right) - e\frac{n_i}{m_i}\frac{\partial V}{\partial r} - \frac{1}{m_i}\frac{\partial n_i}{\partial r}n_i k_B T_i - n_i v_i \nu. \tag{4.95}$$

The parameter α is 0 for Cartesian coordinates, 1 for cylindricals and 2 for sphericals. ν is a collision frequency which determines the frictional drag.

A simulation based on these equations is given in Section 4.7.1.

4.6 Summary

In this chapter, the semiconductor equations and mathematical models of the semiconductor parameters have been reviewed. Emphasis has been placed on models which are suitable for numerical computer simulation, and references were given for those seeking derivations and further information. The equations and parameters reviewed in this chapter will be extensively referenced in the following chapters.

4.7 Appendix

Two appendices to the material in this chapter are given next. The first contains an example of the use of the momentum equations. The second contains parameters for ion stopping, which is discussed in Chapter 13.

4.7.1 Solution of the Momentum Equation

In this section we discuss a simulation based on solving the momentum equation. This material can be regarded as an appendix to the section on the momentum equation. In simulations of ion implantation from a plasma, this equation is solved along with a source-free continuity equation

$$\frac{\partial n_i}{\partial t} + \frac{1}{r^\alpha}\frac{\partial}{\partial r}\left(r^\alpha n_i v_i\right) = 0 \tag{4.96}$$

and Poisson's equation. The second term is the one-dimensional divergence of the flux, see Section 2.6.

The purpose of the simulation psii.sg is to describe the implantation of ions from a plasma into a semiconductor. The fluid being followed is thus the ions. The electrons are assumed to have a Boltzmann distribution of density with respect to the electrostatic potential V. The time evolution of the ion density as the ions are electrostatically drawn into the target is calculated. This allows the electric field to be found, and in turn the ion energies with which the ions implant are obtained. (From the ion energies, the depth profile of the implanted ions can be calculated.) The file which follows is a 1D example of the use of the ion momentum equation in 'Plasma Source Ion Implantation' (PSII).

This example makes use of 'upwind'/'upstream' finite-differencing. This is complicated by the fact that the upwind direction can potentially change during the run. Even when we think we know the upwind direction, if the solution method could find the direction to be reversed momentarily during the iterations, then the code needs to be able to handle that reversal. In this case the flow direction is determined by the electric field, which is in turn found from Poisson's equation,

and this can indeed change direction temporarily during the iterations. This leads
to some complications in setting up the file.

In the file, the force on the fluid is referred to as a 'source', since it is the source
of momentum which appears on the right-hand side of the momentum equation.

Input File psii.sg

```
// FILE:  sgdir\ch04\psii\psii.sg
//  This file specifies a simulation of
//  Plasma Source Ion Implantation (PSII) which repeatedly solves
//  a nonlinear poisson's equation, ion-momentum equation, and
//  an ion continuity equation.  The electrons are assumed to
//  have a Boltzmann distribution, and are assigned a temperature.
//  They are only in the calculation as part of the poisson
//  equation.
//
//  The simulation is in 1 dimension and can be set up as being
//  a cartesian, cylindrical or spherical geometry, depending
//  on the constant alpha.
//
//  The target is considered to be on the left side of the
//  1D mesh (i=0), and the right boundary (i=NR) can be specified
//  to be either plasma or an outer wall, with the constant rmode.

//CONSTANTS
const NR = 101;                         // number of mesh points
const LR = NR - 1;                      // index of last mesh point
const PI = 3.1415927;
const C0 = 1.0e10;                      // density scaling constant
const R0 = 1.0e14;                      // ionization scaling constant
const ITMAX = 5000;                     // number of time steps
const IOUTPUT = 1000;                   // output every IOUTPUT iterations
const Rmax = 15.0;                      // maximum mesh radius, cm
const Rmin = 1.0;                       // minimum mesh radius, cm
const alpha = 0;                        // alpha = 0, cartesian   geometry
                                        // alpha = 1, cylindrical geometry
                                        // alpha = 2, spherical   geometry
const rmode = 0;                        // rmode = 0, right boundary = plasma
                                        // rmode = 1, right boundary = wall
const kb = 1.381e-23;                   // Boltzmann's constant
const q  = 1.602e-19;                   // charge of an electron, C
const e0 = 8.854e-14;                   // permitivity of free space, F/cm.
const L0 = e0/q;                        // Poisson's equation factor
const Te = 3.0;                         // electron temperature, eV
const TeK = Te*11600;                   // electron temperature, K
const Ti = 100.0;                       // ion temperature, K
const dtMax = 1.0e-9;                   // maximum time step, sec
const miKG = 1.0e-26;                   // ion mass, kg
```

```
const nuCon = 0.0;                   // ion-neutral collision freq.(sec^-1)
const InitVMax = 20.0;               // initial maximum potential (v)
const InitVwall = 0.0;               // initial wall potential (v)
const VtargetFinal = 5000.0;         // final target potential (v)
const Vrate = 500.0/dtMax;           // target potential ramping rate (v/sec)


// VARIABLES
var it;           // time index variable
var iout;         // output index variable
var sinfac;       // used in InitVars
var BohmVel;
var dV;
var Vfinal;
var VMaxFinal;
var MaxVel;
var dt;
// Array magnitude constants
var NiMax,NeMax,RiMax,VMax,Vwall;

// mesh, geometry constant variables
var dr,dc;

// array variables
var  r[NR],  h[LR];                  // mesh variables, cell
var  rb[NR+1], hb[NR+1];             // mesh variables, face

var  V[NR], Ni[NR], Ne[NR];          // potential, ions, electrons
var oldNi[NR], Ri[NR];               // old ions, ion source rate

var flTerm[NR];                      // flux term, momentum equation
var SoTerm[NR];                      // source term, momentum equation

var nu[NR];                          // ion-neutral collision frequency
var Ex[NR+1];                        // electric field (cell faces)

var Velc[NR];                        // ion velocity at cell centers
var Vel[NR+1];                       // ion velocity at cell faces

var fluxc[NR];                       // ion flux at cell centers
var flux[NR+1];                      // ion flux at cell faces

var velpos[NR+1];                    // upwind velocity array, positive
var velneg[NR+1];                    // upwind velocity array, negative
var testr[NR+1];


// Specify the nonlinear Poisson's equation.
// variable alpha determines geometry:
```

```
//      alpha = 0:      cartesian
//      alpha = 1:      cylindrical
//      alpha = 2:      spherical
//
equ V[i=1..NR-2] ->
  L0*{ (1/pow(r[i],alpha)) * (
        pow((r[i] + 0.5*h[i  ]),alpha)*(V[i+1]-V[i])/h[i] -
        pow((r[i] - 0.5*h[i-1]),alpha)*(V[i]-V[i-1])/h[i-1]) /
        (0.5*h[i]+0.5*h[i-1]) } =
        ( NeMax*exp(q*(V[i]-VMax)/(kb*TeK)) - Ni[i]);

// specify the right-hand boundary.  If rmode = 0 (plasma boundary), then
//      apply Neumann condition(dV/dr = 0).  If rmode = 1 (wall boundary), then
//      apply Dirichlet condition (V = Vwall).
equ V[i=NR-1] ->
        V[i] = (rmode==0)*V[i-1] + (rmode==1)*Vwall;

// set the numerical algorithm parameters
set NEWTON DAMPING    = 3;
set NEWTON ACCURACY   = 5.0e-10;
set NEWTON ITERATIONS = 100;
set LINSOL ALGORITHM  = GAUSSELIM;
set LINSOL FILL       = INFINITY;


// Function:    InitVars
// Description: This function initializes all the variables.
begin InitVars
// physical constants:
  assign Vfinal = VtargetFinal;
  assign VMaxFinal = InitVMax + VtargetFinal;
  assign dV = dt*Vrate;
  assign dt = dtMax;

// set up mesh
  assign dr = (Rmax-Rmin)/LR;
  assign r[i=0..NR-1] = Rmin + (dr*i);
  assign h[i=0..NR-2] = r[i+1]-r[i];
  assign rb[i=1..NR-1] = r[i]-0.5*h[i-1];
  assign hb[i=1..NR-1] = rb[i+1]-rb[i];
  assign rb[i=0] = r[i]-0.5*h[i];
  assign hb[i=0] = rb[i+1]-rb[i];
  assign rb[i=NR]= rb[i-1]+hb[i-1];

// set up variables
  assign NiMax = C0;
  assign NeMax = C0;
  assign BohmVel = sqrt(kb*TeK/miKG)*1.0e-2;     // Bohm velocity , cm/s
```

```
  assign VMax =  InitVMax;
  assign Vwall = InitVwall;
  assign sinfac = (rmode==0)*(0.5*PI) + (rmode==1)*(PI);

// set up initial arrays:
  assign testr[i=all] = 0.0;
  assign testr[i=0..NR-1] = (r[i]-r[0])/(r[NR-1]-r[0]);
  assign Ni[i=1..NR-1] = NiMax*sin(sinfac*(testr[i]));
  assign oldNi[i=all] = Ni[i];
  assign V[i=1..NR-1] = VMax*sin(sinfac*(testr[i]));
  assign nu[i=all] = nuCon;
end


// Function:    UpDatePot
// Description: This function updates the electrostatic potential
//              while the system is still in the ramping phase. It
//              changes the target potential boundary condition.
begin UpDatePot;
        assign dV = 0.0 + (VMax < VMaxFinal)*dt*Vrate;
        assign Vwall = Vwall + dV;
        assign VMax  = VMax + dV;
        assign V[i=1..NR-1] = V[i] + dV;
end


// Function:    NeSolve
// Description: This function calculates the electron density
//              assuming a Boltzmann distribution.  Array Ne
//              is not used in any calculation - it is only calculated
//              for output.
begin NeSolve
        assign Ne[i=all] = NeMax*exp(q*(V[i]-VMax)/(kb*TeK));
end


// Function:    ESolve
// Description: This function calculates the electric field using
//              a central difference.  It is mostly used as a
//              diagnostic.
begin ESolve
        assign Ex[i=0] = (V[i+1]-V[i])/h[i];
        assign Ex[i=1..NR-2] = (V[i+1]-V[i-1])/(h[i-1]+h[i]);
        assign Ex[i=NR-1] = (V[i]-V[i-1])/h[i-1];
end


// Function:    InitVel
// Description: This function initializes the ion velocities
begin InitVel
        assign Vel[i=all] = (-Ex[i]>0)*(abs(BohmVel)) +
                            (-Ex[i]<=0)*(-abs(BohmVel));
```

```
            assign Velc[i=all] = Vel[i];
end


// Function:    SetUpwindArrays
// Description: This function assigns the velocity array to the
//              upwind velocity arrays velpos and velneg.  The
//              velpos array will contain velocities >= 0, and
//              the velneg array will contain velocities <= 0.
begin SetUpwindArrays

// calculate upwind velocities  (if Vel> 0, velpos = Vel, velneg = 0. )
//                              (if Vel< 0, velpos = 0.,  velneg = Vel)
        assign velpos[i=all] = 0.5*(Vel[i]+abs(Vel[i]));
        assign velneg[i=all] = 0.5*(Vel[i]-abs(Vel[i]));

// calculate upwind fluxes
        assign flux[i=1..NR-1] = Ni[i-1]*velpos[i] + Ni[i]*velneg[i];
        assign flux[i=0] = Ni[0]*velneg[0];
        assign flux[i=NR] = -BohmVel*Ni[i-1];
end


// Function:    SetDt
// Description: This function sets the time step so that the Courant
//              limit will be obeyed.
begin SetDt
        assign MaxVel = 2.0*abs(Vel[0]);
        assign dt = dr/MaxVel;
        assign dt = (dt > dtMax)*dtMax + (dt <= dtMax)*dt;
end

// Function:    VelSolve
// Description: This function solves the momentum equation to obtain
//              the ion velocity at the cell centers.  It then linearly
//              interpolates to get the ion velocity at the cell faces.
//              The differencing is upwind.
begin VelSolve

// Calculate SoTerm, the source term(s):
// the 1.0e4 is to ensure that SoTerm will be in terms of cm, not m.
        assign SoTerm[i=1..NR-2] = -(q/miKG)*Ni[i]*(1.0e4)*
                    (V[i+1]-V[i-1])/(h[i-1]+h[i])
                    -nu[i]*Velc[i]*Ni[i]
                    -(1.0e4)*miKG*Ti*kb*(Ni[i+1]-Ni[i-1])/(h[i-1]+h[i]);

// Calculate flTerm, the flux term:
        assign flTerm[i=1..NR-2] =
```

```
                   (pow(rb[i+1],alpha)*(velneg[i+1]*fluxc[i+1]+
                                        velpos[i+1]*fluxc[i ])
                 - pow(rb[i ],alpha)*(velneg[i ]*fluxc[i ]+
                                        velpos[i ]*fluxc[i-1]))/
                   (hb[i]*pow(r[i],alpha));

// Sum to get updated momentum, update fluxc (flux at cell center)
        assign fluxc[i=1..NR-2] = fluxc[i] + dt*(-flTerm[i] + SoTerm[i]);
        assign fluxc[i=0] = fluxc[1] - h[0]*(fluxc[2]-fluxc[1])/h[1];

// Get cell center velocity:
        assign Velc[i=1..NR-2] = (Ni[i] > 0.0) * fluxc[i]/Ni[i];

// Set Velc, fluxc at the plasma boundary (at the right side)
//      if rmode = 0, then Rmax boundary is a plasma boundary, and
//      therefore should have a dVel/dr = 0 BC.
//      if rmode = 1, then Rmax is a wall boundary, and
//      therefore should have a dVel/dr = extrapolation

        assign Velc[i=NR-1] - Velc[i-1] +
                       (rmode==1)*(h[i-1]*(Velc[i-1]-Velc[i-2])/h[i-2]);
        assign fluxc[i=NR-1]= Velc[i]*Ni[i];

// set Velc at left (target) side.
        assign Velc[i=0] = Velc[i+1] - h[i]*(Velc[i+2]-Velc[i+1])/h[i+1];

// Get cell face velocity:
        assign Vel[i=1..NR-1] = 0.5*(Velc[i] + Velc[i-1]);

// set left boundary Vel to be an interpolation. This is the velocity at
//      the target face.
        assign Vel[i=0] = Vel[i+1] - hb[i]*(Vel[i+1]-Vel[i+1])/hb[i+1];
end


// Function:    ContSolve
// Description: This function solves the continuity equation, using
// the ion velocity at cell faces calulated in VelSolve
// to obtain the ion density at cell centers.
// The differencing is upwind.
begin ContSolve

// Solve continuity equation:
        assign Ni[i=1..NR-2] =
                oldNi[i] - dt* (
        pow(rb[i+1],alpha)*(velpos[i+1]*oldNi[i ]+velneg[i+1]*oldNi[i+1])-
        pow(rb[i ],alpha)*(velpos[i ]*oldNi[i-1]+velneg[i ]*oldNi[i ]))/
                (hb[i]*pow(r[i],alpha));
```

```
// Add in ionization:
        assign Ni[i=1..NR-1] = Ni[i] + dt*Ri[i];

// set up Ni at the target boundary. (interpolate)
        assign Ni[i=0]=Ni[i+1] - h[i]*(Ni[i+2]-Ni[i+1])/h[i+1];

// set up Ni at the left (target) boundary.
        assign Ni[i=0] =
                 (abs(velneg[i]) > abs(velpos[i])) *
                 (
                 oldNi[i] - dt* (
         pow(rb[i+1],alpha)*(velneg[i+1]*oldNi[i+1])-
         pow(rb[i  ],alpha)*(velneg[i  ]*oldNi[i  ]))/
                 (hb[i]*pow(r[i],alpha))
                 )
                 //  (pow(rb[i+1],alpha)*velneg[i+1]*oldNi[i+1])/
                 //  (pow(rb[i],alpha)*velneg[i])
                 +
                   (abs(velneg[i]) <= abs(velpos[i]))*Ni[i+1];


// set up Ni at right boundary (plasma boundary).
//      assign Ni[i=NR-1] =
//              (abs(velpos[i+1])>abs(velneg[i+1]))*
//              (pow(rb[i],alpha)*velpos[i]*oldNi[i-1])/
//              (pow(rb[i+1  ],alpha)*velpos[i+1  ])
//              +
//              (abs(velpos[i+1])<=abs(velneg[i+1]))*
//                       Ni[i-1];
        assign Ni[i=NR-1] = Ni[i-1];


// check for negative densities.
        assign Ni[i=0..NR-1] = (Ni[i]<=0) + (Ni[i]>0)*Ni[i];

end

// Function:    UpDateOldNi
// Description:
begin UpDateOldNi
        assign oldNi[i=all] = Ni[i];
end

// Function:    main
// Description: This is the main function of the simulation.  It contains
// the main time-stepping loop, and calls to all the other
// functions.
begin main
```

```
   call InitVars;
   call ESolve;
   call InitVel;

   assign it = 0;
// begin time loop
      while (it <= ITMAX) begin
         assign iout = 0;
         while (iout <= IOUTPUT) begin
           if(it <= ITMAX) begin
           call UpDatePot;       // update target potential
           solve;                // solve Poisson's equation
            call NeSolve;        // calculate electron density
            call ESolve;         // calculate electric field
            call SetUpwindArrays;// assign the upwind arrays
            call SetDt;          // set time step to obey courant criterion
            call VelSolve;       // solve the ion-momentum equation
            call ContSolve;      // solve the ion-continuity equation
            call UpDateOldNi;    // set oldNi = Ni
           end
           assign it = it+1;     // update the time index
           assign iout = iout+1; // update the output index
         end
         write;                  // output to results file
      end
// end time loops
end
```

4.7.2   Models for Ion Stopping

In Chapter 13 we shall consider the processes which take place during ion implantation, and how they may be included in an overall model of ion implantation. The processes which slow the ions down and stop them are reviewed briefly here. There is considerable uncertainty about much of this information, it should be noted. In our experience, the electronic stopping data is particularly likely to be in error. This section begins with the effects of collisions with nuclei on fast implanted ions. The second part deals with interactions of fast ions with electrons.

Nuclear Scattering of Ions

We shall describe a model of nuclear scattering due to Lindhard et al. [61] and Firsov [62],[63],[64], where the nuclear-scattering cross section may be written using an approximate one-parameter expression. They first define reduced energy and length parameters

$$\varepsilon = \varepsilon_1 E = \left(\frac{M_2}{M_1 + M_2} \cdot \frac{a}{Z_1 Z_2 e^2} \right) E \qquad (4.97)$$

$$\rho = \rho_1 x = [N\pi a^2 \gamma]x \tag{4.98}$$

where E is the initial ion energy, M_2 is the target mass, Z_1 (Z_2) is the ion (target) atomic number, e is the electron charge, N is the number density of the target, $\gamma = 4M_1 M_2/(M_1 + M_2)^2$, a is the screening radius,

$$a = 0.8853a_0(Z_1^{2/3} + Z_2^{2/3})^{-1/2} \tag{4.99}$$

and a_0 is the Bohr radius. They introduce the parameter

$$t = \frac{T}{\gamma E}\varepsilon^2 \tag{4.100}$$

where T is the energy transferred from the ion to the target. The nuclear-scattering cross section may then be written in the approximate, one-parameter version

$$Nd\sigma(T) = \frac{\rho_1}{\gamma}\frac{dt}{2t^{3/2}}h(t^{1/2}) \tag{4.101}$$

where the function h is of the form

$$h(t^{1/2}) = \lambda_0 t^{1/2-m}[1 + (2\lambda_0 t^{1-m})^1]^{-1/q} \tag{4.102}$$

with $\lambda_0 = 2.54$, $m = 0.25$, and $e = 0.475$.

For each group of ions with energy $E \pm dE/2$, we need the probability of a final energy after the atomic scatter: in our simulations we use the probability of going to a range of final energies, and we denote the probability as $T_{col}(E, E')$. The probability of an ion with initial energy E scattering to a range of final energy $E' \pm dE/2$ is proportional to the cross section:

$$T_{col}(E, E') \propto \int_{T_0}^{T_1} d\sigma(T) \tag{4.103}$$

where $T_{0,1} = E' - (E \pm dE/2)$.

Unfortunately, as T_0 approaches zero, T_{col} goes to infinity. This represents very small angle scattering. A lower limit $T_0^{\min}$ is imposed to avoid this problem. Since the energy transferred is related to the scattering angle Θ by

$$T = \gamma E \sin^2(\Theta/2) \tag{4.104}$$

this is equivalent to resolving nuclear-scattering angles above a certain limit. Since the algorithm described here has a fixed angular resolution, a typical angular cell width is used in the above equation to find the minimum energy transferred. Typical angular widths give $T/E \sim 2\%$.

For each initial energy E, the equation is integrated over each valid final energy cell $E' \pm dE/2$ to find each component of $T_{col}(E, E')$. The 'total' cross section for large-angle nuclear scattering at energy E is then

$$\Sigma(E) \equiv \int_{T_0^{\min}}^{E} d\sigma(T). \tag{4.105}$$

The constant of proportionality for T_{col} is then simply $1/\Sigma(E)$. The small angle nuclear-scattering which we have neglected is accounted for as a viscous drag which does not change the ions' direction, only their speed (see below).

Electron Drag on Fast Ions

Large-angle nuclear-scattering is not the only way for an ion to lose its kinetic energy. In addition, as the ion travels through the target, the electron gas can result in a viscous drag. We now describe this process and how the probability T_{ele} (which is again the probability of going to a range of final energies from a given initial energy) is constructed. We then describe how small angle nuclear-scattering (which can also be thought of as a viscous drag) is included with this.

We start with expressions similar to those of Lindhard [61] and Firsov [62],[63],[64]. The expressions used to calculate the energy loss due to electronic stopping of ions are

$$dE_e = dL \, N \, S_e \tag{4.106}$$

and

$$S_e(E) = S_L = kE^q \tag{4.107}$$

where dL is the path length, S_e is called the stopping power, $q = 1/2$ and k is given by

$$k = \frac{1.212 Z_1^{7/6} Z_2}{\left(Z_1^{2/3} + Z_2^{2/3}\right)^{3/2} M_1^{1/2}}. \tag{4.108}$$

The expression can be integrated to give the change of energy ΔE, given that the ion traveled a distance L.

$T_{ele}(c, c', E, E')$ gives the probability that, due to electron drag, the ion will drop down to a final energy $E \pm dE/2$ in cell c, given that it last scattered in cell c', with energy E'. Since we know the initial and final Cartesian cells, we know the average length traveled l and therefore we can find the final energy.

As described previously, small-angle nuclear-scattering can also be considered to be simply a viscous drag term. The energy loss due to traveling a short distance dl is given by

$$dE = N \, dl \int_0^{T_0^{\min}} T \, d\sigma(T) \tag{4.109}$$

For each beam the above equation is numerically integrated from each initial cell to each final cell. We note that as the integration is carried out along the trajectory, the energy of the ion is changing not only due to small-angle nuclear-scattering, but also due to electronic drag. These energy losses change the lower bound of the integral. It is important to include this effect while one is computing the energy loss due to small angle nuclear-scattering. The corresponding energy loss is simply added to that included in T_{ele}.

PROBLEMS

1. The following questions refer to equilibrium behavior in a semiconductor.

   a.) A silicon sample has a donor density of 10^{15} /cm^3. What is the hole density at $300K$, using the simplest approximation for the electron density?

   b.) Repeat part a.), but impose local charge neutrality.

   c.) If the Fermi level in a piece of silicon is above E_i, what is the dominant type of doping? If $E_F - E_i = 0.1k_BT_e$, and there is only one type of dopant, what is the density of dopant atoms?

   d.) If the sample is doped with acceptors, which create energy levels at an energy $0.1k_BT_e$ below E_F, what is the chance of any acceptor level being filled?

2. This question examines the variation of carrier density with time due to generation and recombination for an infinite, homogeneous semiconductor. Write the expressions for the main contributions to the net generation rate G in a sample of silicon. Assume a doping density of 10^{18} cm^{-3} of donors. The equation for the time evolution of the density has no space-dependent terms in this case. It is just $\frac{dn}{dt} = G$. Set up an input file (perhaps based on recomb.sg) to step this equation forward in time.

   If we evaluate this equation and the equivalent equation for holes consistently with each other, advancing in time, what can we expect the steady-state solution to be? Now suppose the sample is illuminated with light so that 10^{22} electron-hole pairs are generated per cubic centimeter per second. The illumination is maintained for much longer than the recombination time, then turned off. Solve the self-consistent problem for both species, for SRH recombination. Repeat the calculation analytically, assuming the excess density can be treated as being low, and compare the results.

3. Equation 3.39 gives the density after a time t has elapsed, which results from an initial 'spike' in particle density. Suppose the particles were recombining, with a time constant τ. How would Eq. 3.39 be modified? Choose appropriate values for the physical parameters, to describe minority-carriers in silicon, and plot the density as a function of position, at various times, using the expression you obtained.

4. Find the expression given in the text for the electron mobility as a function of the electric field. Evaluate it for intrinsic silicon and plot the drift velocity versus electric-field strength, for a range of electric fields which are appropriate for VLSI devices. (It may be helpful to extract the expression for the mobility from one of the equation specification files, where it is given within a function. Remember that this function is normalized, so it is necessary to 'undo' the normalization.)

5. Plot the collision frequency for an electron in intrinsic silicon as a function of its speed. How fast must it be traveling to have on average one collision in the time it takes to go a tenth of a micrometer? Repeat the exercise for a doping density of 10^{19} cm^{-3}.

6. Estimate the mean free path λ of a 'thermal' electron in intrinsic silicon. What is the collision frequency ν that corresponds to this? From λ and ν find the diffusion coefficient D and the mobility μ.

7. Assume $N_D = 10^{14}$ /cm^3. Plot μ_n versus T for $T = 250K$ to $400K$.

   Now assume $T = 300K$. Plot μ_n versus N_D for $N_D = 10^{14}$ to 10^{19} /cm^3.

   Finally, assume $N_D = 10^{14}$ /cm^3 and $T = 300K$. Plot μ_n versus pn for $pn = 10^{20}$ to 10^{38} /cm^6.

   Comment on the results, i.e. when does the independent variable start to significantly affect the mobility?

8. Write out the equations which are being solved in the input file psii.sg. Pay particular attention to (a) the boundary conditions and (b) the form of the finite-difference equations which is used for the ion continuity equation. Explain both of these. What is this form of FD equation called?

9. Plot the rate of loss of energy due to each of nuclear-scattering and electronic-scattering, for an injected ion of (a) boron and (b) phosphorus, for a range of energies from 1 keV to 100 keV.

Chapter 5

Numerical Solution of PDEs

This is the second chapter of Part II of this book. It provides a more advanced treatment of the numerical methods which were introduced, as needed along the way, in chapters 2 and 3, in Part I of the book. The next chapter will show how we can easily implement these more sophisticated numerical techniques, presented in this chapter, using SGFramework.

5.1 Introduction

In general, it is not possible to find an exact analytic solution to a system of coupled, nonlinear PDEs. The following reasons may contribute to the failure of classical approaches [65, 66]:

1. The PDE is not linear and cannot be linearized without seriously affecting the result.

2. The solution region is complex.

3. The boundary conditions are of mixed types.

4. The boundary conditions are time dependent.

5. The medium is inhomogeneous or anisotropic.

When it is not possible to find an analytic solution, numerical techniques must be used to find approximate solutions to such systems of equations. The purpose of this chapter is to discuss the most common numerical methods used to solve systems of coupled, nonlinear PDEs. In addition, this chapter will go into detail in describing particular methods which are needed for semiconductor simulation.

As is emphasized at various points in this text, there are many reasons why a simulation may fail. Some of the most common are the use of:

1. an inappropriate model;

2. a simulation mesh (or other discrete approximation such as the size of the time step) which does not provide adequate spatial (or temporal) resolution; and

210

3. a simulation which leads to too large a computational problem for the available resources.

The first and last of these items are discussed in this chapter. The setting up of a mesh is described in the next chapter.

This chapter is divided into seven main sections. Section 5.2 discusses two major classes of numerical techniques used to solve systems of PDEs. To illustrate one of the numerical techniques which is employed throughout this work, the semiconductor equations are discretized via the finite-difference method in Section 5.3. Since most numerical techniques reduce the system of partial differential equations to large systems of algebraic equations, Sections 5.4 and 5.5 describe numerical techniques used to solve systems of nonlinear and sparse linear algebraic equations, respectively. Section 5.6 reviews several software packages designed for the numerical solution of PDEs. Section 5.7 discusses the measures we can and should employ to attempt to check whether our simulations are trustworthy. Finally, Section 5.8 summarizes this chapter.

5.2   Numerical Techniques for Solving PDEs

Several numerical techniques have been invented to solve systems of PDEs approximately. In this section, two major numerical techniques will be described. Subsection 5.2.1 describes the finite-difference method. Subsections 5.2.2, on Difference Operators, and 5.2.3 on Box Integration, both expand on the theme of finite-differences. Subsection 5.2.4 discusses the finite-element method. Finally, other methods will be mentioned in Subsection 5.2.5.

5.2.1   Finite-Difference Methods

The finite-difference method (FDM) was developed during the late 1800s and early 1900s by several researchers. Richardson, a meteorologist, published a paper in 1910 and Thom [67] published a paper in the 1920s titled 'the method of squares' to solve nonlinear hydrodynamic equations. Since then, the method has found applications in many areas of science and engineering. The goal of finite-difference methods is to approximate differential equations by a system of difference equations which are algebraic in form. This process involves replacing derivatives with differences between quantities evaluated at discrete locations, and is known as discretization. In general, there are three basic steps involved in a finite-difference solution of PDEs [66].

1. Dividing the solution region into a grid of nodes.

2. Approximating the given PDEs by finite-difference equivalents. The finite-differencing procedure relates the dependent variables at points in the solution region to the values of the same variables at the neighboring points.

3. Solving the difference equations subject to prescribed boundary conditions and initial conditions.

The amount of literature devoted to the subject of finite-difference methods is very large [66, 68, 69, 70]. A thorough discussion of finite-difference methods is beyond the scope of this book. The purpose of this section is to present two methods of discretizing PDEs via finite-difference methods: direct substitution of difference operators for differential operators and the box integration method. Several important topics such as convergence, consistency, stability and accuracy of finite-difference schemes will not be discussed in detail here. Interested readers are referred to Strikwerda's book [68].

5.2.2 Substitution of Difference Operators for Differential Operators

One method of deriving finite-difference equations from partial differential equations is to substitute difference operators for differential operators. (See Section 2.1.) Difference operators may be derived using Taylor's theorem, which gives both the form and degree of accuracy of the approximation [71]. For instance, to derive the forward first-difference operator, one expands $F(x + h_+)$ about $F(x)$, as in Equation (5.1). The forward first-difference operator δ_+ is obtained by solving Equation (5.1) for $F'(x)$, as shown in Equation (5.2).

$$F(x + h_+) = F(x) + h_+ F'(x) + \frac{h_+^2}{2} F''(x) + \frac{h_+^3}{6} F'''(x) + O(h_+^4) \qquad (5.1)$$

$$\delta_+ F \equiv F'(x) = \frac{F(x + h_+) - F(x)}{h_+} + O(h_+). \qquad (5.2)$$

In a similar fashion the backward first-difference operator is derived by expanding $F(x - h_-)$ about $F(x)$, as in Equation (5.3). The backward first-difference operator δ_- is obtained by solving Equation (5.3) for $F'(x)$, as shown in Equation (5.4). The forward (δ_+) and backward (δ_-) first-difference operators are first-order accurate with respect to mesh spacing h_+ and h_-, respectively.

$$F(x - h_-) = F(x) - h_- F'(x) + \frac{h_-^2}{2} F''(x) - \frac{h_-^3}{6} F'''(x) + O(h_-^4) \qquad (5.3)$$

$$\delta_- F \equiv F'(x) = \frac{F(x) - F(x - h_-)}{h_-} + O(h_-). \qquad (5.4)$$

The central first-difference operator δ_0 is obtained by subtracting Equation (5.3) from Equation (5.1) and solving for $F'(x)$, as in equation (5.5). In general, the central first-difference operator is first order accurate. If, however, $h = h_+ = h_-$, then equation (5.5) simplifies to Equation (5.6), and the central first-difference operator is second-order accurate.

$$\delta_0 F \equiv F'(x) = \frac{F(x+h_+) - F(x-h_-)}{h_+ + h_-} + O(h_- - h_+) \qquad (5.5)$$

$$\delta_0 F \equiv F'(x) = \frac{F(x+h) - F(x-h)}{2h} + O(h^2). \qquad (5.6)$$

The last difference operator we will consider is the central second-difference operator δ_0^2. The derivation is similar to that of the central first-difference operator and proceeds as follows. We first divide Equation (5.1) by h_+ and equation (5.3) by h_-. Next we add the two resulting equations and solve for $F''(x)$. The result is equation (5.7). In general, δ_0^2 is first order accurate. Again, if $h = h_+ = h_-$, then equation (5.7) simplifies to equation (5.8) and the central second difference operator is second-order accurate.

$$\delta_0^2 F \equiv F''(x) = \frac{\frac{F(x+h_+)}{h_+} - \frac{F(x)}{h_+} - \frac{F(x)}{h_-} + \frac{F(x-h_-)}{h_-}}{\frac{h_+ + h_-}{2}} + O(h_+ - h_-) \qquad (5.7)$$

$$\delta_0^2 F \equiv F''(x) = \frac{F(x+h) - 2F(x) + F(x-h)}{h^2} + O(h^2). \qquad (5.8)$$

To illustrate this method, we will discretize Laplace's equation $(\nabla^2 V = 0)$ on a two-dimensional Cartesian mesh with uniform spacing h using central differences. We start by noting the ∇^2 operator is equivalent to $\partial^2/\partial x^2 + \partial^2/\partial y^2$ in Cartesian coordinates, as in Equation (5.9). Next we substitute the differential operators for difference operators, as in equation (5.10). The operators $\delta_{0,x}^2$ and $\delta_{0,y}^2$ are the central second difference operators for the x and y partial derivatives, respectively. Expanding the difference operators and simplifying (Equation 5.11) yields the central-differenced form of Laplace's equation. For convenience, we use the notation $V_{i,j} \equiv V(ih, jh)$. Equation (5.11) is second-order accurate.

$$\nabla^2 V = \frac{\partial^2 V}{\partial x^2} + \frac{\partial^2 V}{\partial y^2} = 0 \qquad (5.9)$$

$$\delta_{0,x}^2 V + \delta_{0,y}^2 V = 0 \qquad (5.10)$$

$$\frac{V_{i+1,j} + V_{i,j+1} - 4V_{i,j} + V_{i-1,j} + V_{i,j-1}}{h^2} + O(h^2) = 0 \qquad (5.11)$$

This subsection introduced difference operators which are first- and second-order accurate. In general, it is possible to derive higher-order differences [68]. In addition, this section also showed, by means of an example, how PDEs may be discretized by substituting difference operators for differential operators. The next subsection describes an alternative method of discretizing certain PDEs by finite-difference methods, the box integration method.

5.2.3   Box Integration Method

The box integration method is a subset of finite-difference methods, but it is very important in the 2D semiconductor simulations which are presented in Part III of this book. Box integration provides a convenient method of discretizing a large class of PDEs, (especially PDEs where a divergence of a flux appears) on both regular and irregular meshes. We will limit the scope of this presentation to PDEs of the form given by equation (5.12) below. The symbol $\mathbf{F}$ represents a flux per unit area and S a source per unit volume. (The discussion below is essentially two-dimensional, so we shall assume there is no variation with the third coordinate and that $\mathbf{F}$ and S are integrated over that coordinate. The same sort of technique can be used for a fully 3D problem.)

To discretize a PDE of this form, one constructs a 'box' around each node in such a way that the mesh elements and 'boxes' span the same simulation domain. A convenient way to construct the boxes is to connect perpendicular bisecting segments. These are the edges obtained by drawing the perpendicular bisectors of the edges that connect the mesh nodes, as shown in Figure 5.1. However, the bisectors are only allowed to extend until they cut another bisector. The short lines so formed are the 'bisecting segments'. The edges of the box are shown as dashed lines and the mesh edges are shown as solid lines. On the exterior boundaries, the box edges and mesh edges coincide. The notation used in this figure is as follows. A is the area of the integration box surrounding node 2, ∂A is the boundary of the integration box (not labeled), F_i is the flux along the edge that connects node 2 to its ith neighbor and L_i is the length of the corresponding perpendicular bisecting segment.

Once the box is constructed, the PDE is integrated over the area of the box, and Green's theorem is then applied to convert the area integral on the left side of the equality to a line integral (Equation (5.13)). If

$$\nabla \cdot \mathbf{F} = S \qquad (5.12)$$

then

$$\oint_{\partial A} (\mathbf{F} \cdot \mathbf{n}) \, d\mathbf{l} = \iint_A S \, ds \qquad (5.13)$$

which is implemented as

$$\sum_i F_i L_i = SA \qquad (5.14)$$

where i labels the perpendicular bisecting segments (integration edges) surrounding a node; F_i is the flux per unit length normal to the integration edge i and L_i is the length of the integration edge. Since the mesh quantities are known only at the mesh points, the flux can be determined only along a mesh edge. This being the case, Equation (5.13) is approximated by Equation (5.14). The area integral is approximated by the product of the source per unit volume at the node and the area of the integration box. (Since one integrates around the box boundary ∂A, and the box boundary consists of perpendicular bisecting segments, the perpendicular bisecting segments are sometimes referred to as integration edges.)

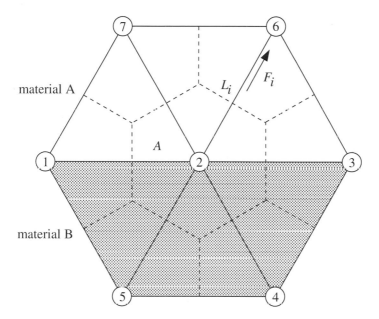

Figure 5.1. A simple triangular mesh. The solid lines are the mesh edges while the dashed lines are the edges of the integration 'box'. A is the area of the box surrounding node 2, F_i is the flux along the edge that connects node 2 to its ith neighbor and L_i is the length of the corresponding integration edge.

This scheme can be seen to be a conservative scheme, since (1) the scheme is based on using fluxes flowing between cells to find the divergence, and (2) the flux through a face between cells is exactly the same when evaluated for use in the divergence in each of the cells adjacent to the face.

Box Integration versus a Simple Finite-Difference Scheme

This short section compares box integration to the results of a 'straightforward' finite-differencing. It should be skipped unless one is interested in technical details of how Box Integration works.

To make contact between the Box Integration method and conventional finite-differences, suppose that we take a regular rectangular mesh and distort it so that the vertical lines of the original mesh are now at an angle θ to the vertical. In other words, if we keep the index i fixed and change the index j by one, we change y by Δy but we also change x by $\alpha \Delta x$, where $\tan \theta$ will be equal to the horizontal distance moved divided by the vertical distance moved; $\tan \theta = \alpha \frac{\Delta x}{\Delta y}$. The spacing of the horizontal and vertical mesh points is unchanged, but the origin of each horizontal row is offset. The equations for the x and y coordinates of each point (i, j) are now

$$x_{i,j} = i\Delta x + j\alpha \Delta x \;=\; (i + j\alpha)\Delta x; \quad y_j = j\Delta y. \tag{5.15}$$

Since the horizontal rows are simply displaced sideways, within each row the derivative $\frac{\partial V}{\partial x}$ (assuming our dependent variable is a voltage, for example) can be evaluated as usual;

$$\frac{\partial V}{\partial x} \simeq \frac{V[i + 1, j] - V[i - 1.j]}{2\Delta x}. \tag{5.16}$$

However, the derivative $\frac{\partial V}{\partial y}$ is altered by the distortion of the mesh.

To find an approximation for the derivative at (i, j) we can interpolate the voltage between mesh points. Looking 'up' from (i, j) to a point a distance Δy above, that point is not at a mesh point but between the points $(i - 1, j + 1)$ and $(i, j + 1)$, provided alpha is positive. In fact, it is a distance $\alpha \Delta x$ to the left of the point (i, j). This suggests an interpolation scheme, which would find the value of V at this point from a weighted average of $V[i, j + 1]$ and $V[i - 1, j + 1]$. The interpolated voltage would be

$$V_{up}^{int}[i, j] = (1 - \alpha)V[i, j + 1] + \alpha V[i - 1, j + 1]. \tag{5.17}$$

This scheme returns the correct value when α is zero. It also implies that the derivative will be approximated as

$$\begin{aligned} \frac{\partial V}{\partial y} \simeq\; & [\,(\,(1 - \alpha)V[i, j + 1] + \alpha V[i - 1, j + 1]\,) \\ & - \;(\,(1 - \alpha)V[i, j - 1] + \alpha V[i + 1, j - 1]\,)\,]/2\Delta y \end{aligned}$$

where a similar interpolation has been used when looking 'down',

$$V_{down}^{int}[i, j] = (1 - \alpha)V[i, j - 1] + \alpha V[i + 1, j - 1] \tag{5.18}$$

a distance Δy to the line below.

To derive the second derivatives, we can argue in the same way that $\frac{\partial^2 V}{\partial x^2}$ is unchanged. Likewise, in the 'usual' expression

$$\frac{\partial^2 V}{\partial y^2} \simeq \left(V[i, j+1] - 2V[i, j] + V[i, j-1] \right) / (\Delta y)^2 \tag{5.19}$$

we need to replace the values on the lines 'above' and 'below' so that one possible approximation that could be suggested is

$$\frac{\partial^2 V}{\partial y^2} \simeq \left(V_{up}^{int}[i, j] - 2V[i, j] + V_{down}^{int}[i, j] \right) / (\Delta y)^2 \tag{5.20}$$

where the interpolated values are given by the expressions above. However, there is a problem with this method of interpolating. If the mesh has the shape shown in Figure 5.1, so that it forms a set of equilateral triangles, then $0.5 * (V[i-1, j+1] + V[i, j-1])$ can be interpreted as an interpolated value of V to the left of (i, j) and $0.5 * (V[i, j+1] + V[i+1, j-1])$ is an interpolated value to the right of (i, j). If these two interpolations are combined and $2V[i, j]$ subtracted, this would appear to give an answer proportional to the second derivative with respect to x. In fact, if we examine the approximation we set up for $\frac{\partial^2 V}{\partial y^2}$, we see that it contains information about both second derivatives.

To get a clearer idea of what is happening, we can make use of an expansion of V about the point $(x_{i,j} = (i + j\alpha)\Delta x, y_j = j\Delta y)$, where all distances will be measured from $(x_{i,j}, y_j)$.

$$V(x, y) = V_0 - E_x x - E_y y + \gamma x^2 + \delta y^2 + \epsilon x y. \tag{5.21}$$

For now, however, we set $\epsilon = 0$; see the exercises at the end of the chapter for more on this. Using this, we can find three combinations of voltages, which we shall call V_A, V_B and V_C where

$$V_A = V[i+1, j] + V[i-1, j] - 2V[i, j] \tag{5.22}$$

$$V_B = V[i-1, j+1] + V[i+1, j-1] - 2V[i, j] \tag{5.23}$$

and

$$V_C = V[i, j+1] + V[i, j-1] - 2V[i, j]. \tag{5.24}$$

Each of these is proportional to the second derivative along a particular line. Evaluated using the power series,

$$V_A = 2\gamma(\Delta x)^2 \tag{5.25}$$

$$V_B = 2\gamma(1 - \alpha)^2(\Delta x)^2 + 2\delta(\Delta y)^2 \tag{5.26}$$

and

$$V_C = 2\gamma\alpha^2(\Delta x)^2 + 2\delta(\Delta y)^2. \tag{5.27}$$

We need to take a combination of these which gives the correct Laplacian in two dimensions. By differentiating the expression for V given by the power series we find the Laplacian to be equal to

$$\nabla^2 V = 2\gamma + 2\delta. \tag{5.28}$$

Now the problem with using $(V_B + V_C)/2(\Delta y)^2$ to represent $\frac{\partial^2 V}{\partial y^2}$ was that this combination turned out to have a part proportional to $\frac{\partial^2 V}{\partial x^2}$. This combination should be equal to 2δ if the interpolation worked as intended. Instead it has a term proportional to γ in addition. The contribution of the curvature along the x-direction (which should be just 2γ) will be overcounted if we use the interpolation as if it were exactly equal to $\frac{\partial^2 V}{\partial y^2}$.

Because of this, we can try to find $\nabla^2 V$ from

$$\nabla^2 V = \nu \frac{V_A}{(\Delta x)^2} + \frac{(V_B + V_C)/2}{(\Delta y)^2} \tag{5.29}$$

where ν is an unknown to be determined. Substituting the values of V_A, V_B and V_C gives

$$\nabla^2 V = 2\gamma \left[\nu + \frac{(1-\alpha)^2 + \alpha^2}{2} \frac{(\Delta x)^2}{(\Delta y)^2} \right] + 2\delta. \tag{5.30}$$

If we choose ν to be

$$\nu = 1 - \frac{(1-\alpha)^2 + \alpha^2}{2} \frac{(\Delta x)^2}{(\Delta y)^2} \tag{5.31}$$

then we get the right value for the Laplacian.

In the symmetric case shown in Figure 5.1, $(\Delta x)^2 = \frac{4}{3}(\Delta y)^2$, and $\alpha = 0.5$ so $\nu = \frac{2}{3}$. This implies that our Laplacian should be

$$\nabla^2 V = \frac{2}{3} \frac{V_A}{(\Delta x)^2} + \frac{V_B + V_C}{2(\Delta y)^2} = \frac{2}{3(\Delta x)^2} [V_A + V_B + V_C] \tag{5.32}$$

for the case of a mesh of equilateral triangles. Now we shall compare this to the box integration result, for this symmetric case.

The box integration method implies that we first find ∇V along each line from (i, j) to one of its neighbor mesh points. Along the line from (i, j) to $(i-1, j)$, for example, this would be $\frac{V[i-1,j]-V[i,j]}{L_p}$, where L_p is the distance between the points. Next, we multiply this by the length L_b of the perpendicular bisector of the line between the points, to find the flux through the bisector. Thus, the flux of ∇V between (i, j) and $(i-1, j)$ is

$$\Gamma = (V[i-1, j] - V[i, j]) \frac{L_b}{L_p}. \tag{5.33}$$

Input File lapt.sg

```
// FILE:  sgdir\ch05\lapt\lapt.sg
// demonstration of Laplace's equation
// on a 'tilted' mesh

const NX = 21;                  // number of x-mesh points
const NY = 21;                  // number of y-mesh points
const LX = NX - 1;              // index of last x-mesh point
const LY = NY - 1;              // index of last y-mesh point
const DX = 1.0;                 // mesh spacing

var V[NX,NY],x[NX],y[NY];

// Laplace's equation
equ V[i=1..LX-1,j=1..LY-1]  ->
    V[i-1,j+1] +   V[i,j+1] +
    V[i-1,j  ] - 6*V[i,j  ] +  V[i+1,j  ] +
                   V[i,j-1] +  V[i+1,j-1] = 0.0;

begin main
  assign x[i=all] = i*DX ;
  assign y[j=all] = j*DX ;
  assign V[i=all,j=all] = 0.0;
  assign V[i=0,  j=all] = 10.0;
  solve;
end
```

Repeating this for all six neighbor nodes shown in the figure, for the symmetric case where each node has the same L_p and L_b, simply gives the total flux

$$\Gamma_T = \frac{L_b}{L_p}(V[i-1,j]+V[i-1,j+1]+V[i,j+1]+V[i+1,j]+V[i+1,j-1]+V[i,j-1]-6V[i,j]).$$

$$(5.34)$$

This expression is clearly proportional to the combination $V_A + V_B + V_C$, so the correspondence between the two ways of doing the finite-differencing has been shown.

The example file 'laplace.sg' which was given in chapter 2 has been modified to use this scheme. The results can be plotted using Matlab, but since the plotter assumes the array should be treated as a square, instead of a highly flattened parallelepiped, the differences between this and the solution in a rectangular region are difficult to see. (See Figure 5.2.) Further comparison to the box integration method will be left to an exercise.

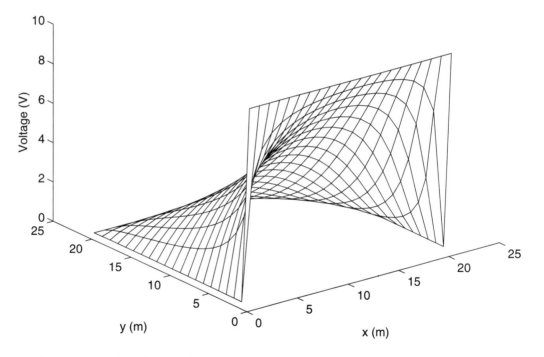

Figure 5.2. Result of solving Laplace's equation in a diamond-shaped region. Since this is plotted using a square region, the potential looks only slightly different from the example in the second chapter.

5.2.4 Finite-Element Methods

At present, finite-element methods are the most widely used methods for the solution of PDEs on irregular meshes [72, 73, 66]. Finite-element calculations usually begin by taking the set of PDEs to be solved and recasting them in 'variational form', where a 'variational quantity' is to be minimized in order to obtain the physically correct solution to the PDEs. The simplest example of this may be the solution of Laplace's equation $\nabla^2 \Psi = 0$ in a volume V. This can be recast as a statement that $\int_V E^2 dV = \int_V |\nabla \phi|^2 dV$ should be minimized, subject to the appropriate boundary conditions. In this case the variational quantity can be thought of as $W_e = (1/2)\,\epsilon \int E^2 dV$, the stored energy in the electrostatic field. The correct solution for ϕ is that which minimizes W_e. Strictly speaking, it is not necessary to cast the PDEs in variational form. Finite-element methods may be applied if the equations can be locally integrated around a point in the simulation domain.

It becomes necessary to choose a representation for the 'trial function', which will be varied to minimize the variational quantity. The trial function is usually written as a sum of a set of unknown coefficients a_i each multiplied by a corresponding known basis function V_i. When this form of the function is substituted into the variational quantity, all the operations involving the V_i should be capable of being evaluated, leaving an expression in terms of the unknown coefficients a_i only. This is then minimized with respect to the a_i. The resulting equations for the a_i usually must be solved by sparse-matrix techniques, and it is this task to which much of a 'finite-element package' is devoted.

Each of the basis functions used in finite-elements is usually nonzero only in its own small region, while most of the other trial functions are zero in that region. This is distinct from Fourier methods where the basis functions extend throughout all space. A simple finite-element basis set in one dimension might use triangle functions, the ith of which is of unit height at the ith node but drops linearly to zero at nodes $i - 1$ and $i + 1$ and is zero elsewhere.

5.2.5 Other Numerical Methods

Although the finite-difference and finite-element methods are the most popular methods for numerically solving PDEs, other numerical methods have been developed. The finite-box method [74] is a generalization of the classical rectangular-grid finite-difference method and, specifically, a generalization of the 'terminating-line' approach introduced by Adler [75]. The usually rectangular pattern of grid points is dropped and the mesh is built up from rectangular cells called boxes.

Fourier methods may also be used to numerically solve PDEs. With Fourier methods, one assumes that the dependent variables may be represented as Fourier series. The series are substituted into the PDEs and the equations obtained for the coefficients are solved numerically. Fast Fourier transforms (FFTs) may be used in this method to switch between the time and frequency domains. This approach has been applied to semiconductor simulation by Axelrad [76, 77].

5.3   Discretizing the Semiconductor Equations

The most popular numerical method for solving the semiconductor equations is the finite-difference method. In this section the semiconductor equations will be discretized using the box integration method. Not only will this presentation serve as an illustration of the finite-difference method, but the resulting difference equations will also be referenced by the following chapters. As mentioned in Section 4.3, the basic semiconductor equations consist of three coupled, nonlinear PDEs: Poisson's equation (4.28), an electron continuity equation (4.29) and a hole continuity equation (4.30).

This section consists of three subsections. Prior to numerically solving systems of equations on a finite-precision digital computer, the equations usually should be scaled. Subsection 5.3.1 discusses a suitable set of scaling constants and gives references to other choices of scaling constants. Subsection 5.3.2 discretizes the semiconductor equations using the box integration method with central differences and discusses the problems that result from discretizing the electron and hole continuity equations via central differences. Finally, Subsection 5.3.3 presents an alternative discretization for the electron and hole continuity equations which is known as the Scharfetter-Gummel method.

5.3.1   Scaling the Semiconductor Equations

As mentioned in Subsection 4.3.1, the electrostatic potential V, the electron concentration n and the hole concentration p are the dependent variables in the basic semiconductor Equations (5.35), (5.36) and (5.37). Since the dependent variables may vary by many orders of magnitude from one to another (as well as throughout the solution domain), the dependent variables should be appropriately scaled in order to aid in the analysis of semiconductor equations [9, 71]. In order for the scaling to be useful, the variables should be scaled according to their intrinsic or characteristic values. If the scaling constants are chosen correctly, then the dimensionless variables are typically of the order unity [71].

DeMari [78, 79] proposed a standard method of scaling which is attractive from a computational point of view. When the semiconductor equations are reformulated using DeMari's scaling, there are no operations involving constants. However, from a mathematical point of view, this scaling is not satisfactory, since the dependent variables are not at all the same order of magnitude and the scaled space charge in Poisson's equation $(p - n + C)$ may be very large [9]. According to Selberherr, an alternative method of scaling which is attractive from a mathematical point of view was introduced by Vasil'eva et al [80, 81] and further developed by Markowich et al [82, 83]. The scaling constants of Vasil'eva and Markowich et al are enumerated in Table 5.1.

Upon scaling the semiconductor variables and parameters by the constants listed in Table 5.1, the semiconductor equations (4.28), (4.29) and (4.30) transform into Equations (5.35), (5.36) and (5.37). The parameter λ^2 which multiplies the diver-

quantity	symbol	value
x, y	x_0	$\max |\mathbf{x} - \mathbf{y}|, \mathbf{x}, \mathbf{y} \in D$
V	V_0	$k_B T / e$
$n, p, C,$	C_0	$\max |C(\mathbf{x})|, \mathbf{x} \in D$
D_n, D_p	D_0	$\max(D_n(\mathbf{x}), D_p(\mathbf{x})), \mathbf{x} \in D$
μ_n, μ_p	μ_0	D_0 / V_0
R	R_0	$D_0 C_0 / x_0^2$
t	t_0	x_0^2 / D_0
E	E_0	V_0 / x_0
J_n, J_p	J_0	$x_0 / (e D_0 C_0)$

Table 5.1. Scaling factors for the semiconductor equations and parameters.

gence term in Poisson's equation is defined by Equation (5.38). λ^2 is a very small number with a typical value of 10^{-10}. Physically, λ represents the scaled minimum Debye length in the device. Since λ^2 is a very small number, Poisson's equation is singularly perturbed. The implications of the previous statement will become evident in the following sections. Since the space charge in Poisson's equation is scaled to unity, the residual of equation (5.35) may be directly used as a measure of computational accuracy. The same is true of the scaled continuity equations ((5.36) and (5.37)).

$$\lambda^2 \nabla \cdot (\epsilon_r \mathbf{E}) - (p - n + C) = 0 \tag{5.35}$$

$$\nabla \cdot \mathbf{J}_n - R = \frac{\partial n}{\partial t} \tag{5.36}$$

$$-\nabla \cdot \mathbf{J}_p - R = \frac{\partial p}{\partial t} \tag{5.37}$$

$$\lambda^2 = \frac{V_0 \epsilon_0}{e x_0^2 C_0} \tag{5.38}$$

The scaled electric-field and electron and hole current densities are given by Equations (5.39), (5.40) and (5.41).

$$\mathbf{E} = -\nabla V \tag{5.39}$$

$$\mathbf{J}_n = \mu_n n \mathbf{E} + D_n \nabla n \tag{5.40}$$

$$\mathbf{J}_p = \mu_p p \mathbf{E} - D_p \nabla p. \tag{5.41}$$

If the diffusion coefficients are related to the mobilities by the Einstein relation, then the scaled diffusion coefficients are equal to the scaled mobilities. This fact will be used in the derivation of the Scharfetter-Gummel discretization in Subsection 5.3.3.

5.3.2   Discretization of Semiconductor Equations

Since the scaled semiconductor equations are of the form $\nabla \cdot \mathbf{F} = S$, they may easily be discretized via the box integration method (see Subsection 5.2.3). To illustrate this procedure, Poisson's Equation (5.35) will be discretized. Consider an arbitrary interior node i. The first step in discretizing Poisson's equations at node i is to construct a suitable box around this node as described in Subsection 5.2.3. Next Poisson's equation is integrated over the box's area and Green's theorem is applied (Equation (5.42)).

$$\lambda^2 \oint_{\partial A} \epsilon_r \left(\mathbf{E} \cdot \mathbf{n} \right) dl = \iint_A (p - n + C) \, ds. \tag{5.42}$$

Since the electric-field intensity $\mathbf{E}$ is only known along the edges which connect the mesh nodes to their neighboring nodes, and the electron, hole and fixed dopant concentrations are only known at the nodes, the integrals in Equation (5.42) are approximated as shown in Equation (5.43). The notation used in equation (5.43) is as follows. Node i is an arbitrary interior node in the simulation domain. Index j represents the jth neighbor of node i. E_{mdpt} is the component of the electric-field intensity along the edge and at the midpoint of the edge which connects node i to its jth neighbor. n_i, p_i and C_i are the electron, hole and fixed dopant concentrations, respectively. L_j is the length of the integration edge (the edge of the box which perpendicularly bisects the edge which connects node i to its jth neighbor). Finally, A_i is the area of the integration box.

$$\lambda^2 \sum_j \left(\epsilon_r \left. E \right|_{\text{mdpt}} L_j \right) = (p_i - n_i + C_i)\, A_i. \qquad (5.43)$$

Every quantity in Equation (5.43) is either a constant or a dependent variable except the electric-field intensity $\left. E \right|_{\text{mdpt}}$. To derive an expression for E_{mdpt} one notes that $\mathbf{E} = -\nabla V$ is simply $E = -\frac{dV}{dx}$ in one dimension and applies Taylor's theorem, as in Subsection 5.2.2. The resulting expression is given by Equation (5.44) where V_i and V_j are the electrostatic potential at nodes i and the jth neighbor of i, respectively and $h_{i,j}$ is the length of the edge which connects node i to its jth neighbor. Equation (5.44) is the central difference of $\mathbf{E}$ since $\left. E \right|_{\text{mdpt}}$ is defined at the midpoint of the edge which connects node i to its jth neighbor and V_i and V_j are defined on either side of $\left. E \right|_{\text{mdpt}}$.

$$\left. E \right|_{\text{mdpt}} = - (V_j - V_i) / h_{i,j}. \qquad (5.44)$$

Equations (5.43) and (5.44) represent a generic central-difference approximation of Poisson's equation on a general two-dimensional mesh. A similar method may be used to discretize the time-independent electron and hole continuity equations (5.36) and (5.37). Equations (5.45) and (5.47) represent the finite-difference approximation to the time-independent electron continuity equation, while Equations (5.46) and (5.48) represent the finite-difference approximation to the time-independent hole continuity equation. $\left. J_n \right|_{\text{mdpt}}$ and $\left. J_p \right|_{\text{mdpt}}$ are the normal components of the electron and hole current densities at the midpoint of the edge that connects node i with its jth neighbor, and R_i is the net recombination minus generation at node i.

$$+ \sum_j \left(\left. J_n \right|_{\text{mdpt}} L_{i,j} \right) - R_i A_i = 0 \qquad (5.45)$$

$$- \sum_j \left(\left. J_p \right|_{\text{mdpt}} L_{i,j} \right) - R_i A_i = 0 \qquad (5.46)$$

$$\left. J_n \right|_{\text{mdpt}} = \mu_n \left(n_i + n_j \right) / 2 \cdot \left. E \right|_{\text{mdpt}} + \left. D_n \right|_{\text{mdpt}} \left(n_j - n_i \right) / h_{i,j} \qquad (5.47)$$

$$\left. J_p \right|_{\text{mdpt}} = \mu_p \left(p_i + p_j \right) / 2 \cdot \left. E \right|_{\text{mdpt}} - \left. D_p \right|_{\text{mdpt}} \left(p_j - p_i \right) / h_{i,j} \qquad (5.48)$$

Equations (5.47) and (5.48) are central-difference approximations to the electron and hole current densities. The derivatives dn/dx and dp/dx have been replaced

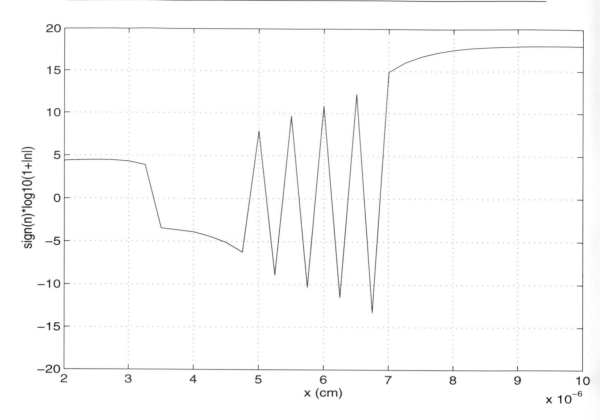

Figure 5.3. Nonphysical oscillations resulting from discretizing the semiconductor equations via central differencing.

with the differences $(n_j - n_i) / h_{i,j}$ and $(p_j - p_i) / h_{i,j}$, respectively. Furthermore, the electron and hole concentrations at the midpoint of the edge which connects node i to its jth neighbor have been computed by taking the average of the carrier concentrations at nodes i and its jth neighbor. Although these central-difference approximations are valid, provided the discrete points sample the density profile with sufficient frequency, in practice they are seldom used. If the mesh spacing is not sufficiently fine, the numerical solution of the electron and hole continuity equations will exhibit large nonphysical oscillations, as shown in Figure 5.3, since Poisson's equation is singularly perturbed [84]. This phenomenon is not unique to the semiconductor equations. It is well known that standard discretization methods may give spurious oscillations on a relatively coarse mesh in the case of singularly perturbed problems [85, 86]. The most popular method of overcoming this problem is to discretize the electron and hole current densities using the Scharfetter-Gummel method. This is the topic of the next subsection.

5.3.3   The Scharfetter-Gummel Discretization

In this subsection, we will derive the Scharfetter-Gummel form of the transport equations. The general form is given here. A simple derivation for simplified conditions is given in the examples at the end of the chapter. This and the associated examples may be useful in understanding what the purpose of the discretization is, since the discussion here gives the derivation but relatively little review of the physics involved. The simple treatment in the examples avoids difficulties which arise at zero electric-field, etc., by ignoring them. The general form given here handles them properly.

It is important to realize that while the Scharfetter-Gummel form of the semiconductor equations is very widely used and vital for much of the semiconductor modeling work which is done, it may suffer from inaccuracy. As we shall see below, it is derived in one dimension. In more than one dimension it is believed to introduce a potentially significant amount of what is known as 'numerical diffusion'. Numerical diffusion is an artificial increase in the rate of transport. Numerical diffusion is due to the representation of the physical quantities on a discrete mesh with finite spacing of the mesh points (as opposed to the mesh points being separated by infinitesimal distances). Attempts to improve this discretization by using it in combination with what are known as 'upwind' or 'upstream' finite-difference equations are being made, with only limited success to date.

The Scharfetter-Gummel discretization consists essentially of a discretization of the expression for the component of the electron current density along the edge, at the midpoint of the edge that connects nodes i and j. This current consists of a mobility contribution and a diffusion contribution, and these two contributions often cancel each other to a very great extent. Because their difference is a (very) 'small difference between (very) large numbers', it must be handled (very) carefully. The derivation for electron current density is shown. The derivation for the hole current density is not given since it is nearly identical.

The derivation starts with the one-dimensional normalized electron current density expression (Equation (5.49)). This will be used to obtain an expression for the density consisting of an exponential plus a constant term. The exponential by itself gives no flux. The constant is responsible for the flux and will be zero if there is no flux. This expression for the density will in turn be used to find a discrete approximation to the density at the center of the edge between points i and j. The discrete approximation to the density will then be substituted back into the expression for the current.

Equation (5.49) uses the fact that the normalized mobility is equal to the normalized diffusion coefficient. This assumes the Einstein relation is valid. As is common when finite-differencing, we assume that various quantities are constant between nodes. In this case we assume that the electron current density and the electric field are constant on the edge. With these assumptions, equation (5.49) can be rearranged so as to become the first-order ordinary differential equation given by Equation (5.50). Equation (5.50) can be solved by multiplying both sides of the equation by the integrating factor $\exp(Ex)$ (Equation (5.51)) and then integrating both sides (Equation (5.52)). The result of the integration is given by Equation (5.53).

$$J_n = \mu_n \left(n(x)E + \frac{d}{dx}n(x) \right) \tag{5.49}$$

$$\frac{d}{dx}n(x) + En(x) = C_0 \tag{5.50}$$

$$\frac{d}{dx}\left(n(x)\exp(Ex) \right) = C_0 \exp(Ex) \tag{5.51}$$

$$\int d\left(n(x)\exp(Ex) \right) = \int C_0 \exp(Ex)\,dx \tag{5.52}$$

$$n(x) = \frac{C_0}{E} + C_1 \exp(-Ex). \tag{5.53}$$

The constants C_0 and C_1 are determined by the boundary conditions. The boundary conditions are given by equations (5.54) and (5.55). The final solution is given by Equation (5.56) and is expressed in terms of the growth function $g(x, V)$ which is defined by equation (5.57). The growth function $g(x)$ consists of a combination of the same exponential which appears in the expression for the density (multiplied by a constant) plus a constant, so the density $n(x)$ can itself be made up out of a multiple of $g(x)$ plus a constant. In addition, $g(x)$ is chosen so as to be 0 at $x = x_i$ and to be 1 at $x = x_j$. The expression for $n(x)$ is written in terms of n_i, n_j and $g(x)$ and clearly gives $n = n_i$ at $x = x_i$ and $n = n_j$ at $x = x_j$.

The central-difference approximation, $E = (V_i - V_j)/h_{i,j}$, has been used in equation (5.57). The quantity U_t is the scaled thermal voltage. In this derivative it is equal to 1 since the electrostatic potential is scaled by the thermal voltage.

$$n(x = x_i) = n_i \tag{5.54}$$

$$n(x = x_j) = n_j \tag{5.55}$$

$$n\left(x \in [x_i, x_j] \right) = (1 - g(x, V))\,n_i + g(x, V)n_j \tag{5.56}$$

$$g(x, V) = \frac{1 - \exp\left(\frac{V_j - V_i}{U_t}\frac{x - x_i}{h_{i,j}} \right)}{1 - \exp\left(\frac{V_j - V_i}{U_t} \right)}. \tag{5.57}$$

In order to discretize Equation (5.49), we need to know two quantities: the value of the electron concentration, and its derivative, at the midpoint of the edge which connects node i and node j. Using (5.56) we find the values at the midpoint; see equations (5.60) and (5.61), which are expressed in terms of functions aux1 and aux2 which are defined by equations (5.58) and (5.59). Substituting Equations (5.60) and (5.61) into Equation (5.49) yields the Scharfetter-Gummel discretization of the electron current density (Equation 5.62). If the electric-field is zero along the edge, then Equation (5.62) degenerates to the central-difference expression given by Equation (5.47), since $\mathrm{aux1}(x) = 0$ and $\mathrm{aux2}(x) = 0.5$. The derivation of the Scharfetter-Gummel discretization of the hole current density is analogous, and the result is given by equations (5.63), (5.64) and (5.65).

$$\mathrm{aux1}(x) = \frac{x}{\sinh x} \tag{5.58}$$

$$\mathrm{aux2}(x) = \frac{1}{1 + \exp(x)} \tag{5.59}$$

$$n\big|_{\mathrm{mdpt}} = n_i\,\mathrm{aux2}\left(\frac{V_i - V_j}{2U_t}\right) + n_j\,\mathrm{aux2}\left(\frac{V_j - V_i}{2U_t}\right) \tag{5.60}$$

$$\frac{dn}{dx}\bigg|_{\mathrm{mdpt}} = \mathrm{aux1}\left(\frac{V_i - V_j}{2U_t}\right)\frac{n_j - n_i}{h_{i,j}} \tag{5.61}$$

$$J_n\big|_{\mathrm{mdpt}} = \mu_n\big|_{\mathrm{mdpt}}\left(n\big|_{\mathrm{mdpt}}\frac{V_i - V_j}{h_{i,j}} + \frac{dn}{dx}\bigg|_{\mathrm{mdpt}}\right) \tag{5.62}$$

$$p\big|_{\mathrm{mdpt}} = p_i\,\mathrm{aux2}\left(\frac{V_j - V_i}{2U_t}\right) + p_j\,\mathrm{aux2}\left(\frac{V_i - V_j}{2U_t}\right) \tag{5.63}$$

$$\frac{dp}{dx}\bigg|_{\mathrm{mdpt}} = \mathrm{aux1}\left(\frac{V_i - V_j}{2U_t}\right)\frac{p_j - p_i}{h_{i,j}} \tag{5.64}$$

$$J_p\big|_{\mathrm{mdpt}} = \mu_p\big|_{\mathrm{mdpt}}\left(p\big|_{\mathrm{mdpt}}\frac{V_i - V_j}{h_{i,j}} + \frac{dp}{dx}\bigg|_{\mathrm{mdpt}}\right). \tag{5.65}$$

SGFramework provides the internal functions aux1 and aux2; see Section 15.1.8.

5.4 The Solution of Nonlinear Systems of Algebraic Equations

Discretizing a system of PDEs on an appropriate solution mesh usually results in a massive system of algebraic equations whose solution should approximate the solution of the original system of PDEs. Thus, once the PDEs have been discretized using an appropriate discretization scheme, the next task is to solve the resulting system of algebraic equations. If the PDEs are nonlinear, then the discretized algebraic equations will also be nonlinear. In general, the algebraic equations are coupled, so it is not possible to solve any given equation independently of the rest of them. This section describes two methods of solving a nonlinear system of algebraic equations of the form $\mathbf{f}(\mathbf{x}) = \mathbf{0}$, which is shorthand for the vector Equation (5.66). Subsection 5.4.1 discusses fixed-point iterative methods and Subsection 5.4.2 discusses relaxation methods.

$$\begin{bmatrix} f_1(x_1, x_2, x_3, \ldots, x_n) \\ f_2(x_1, x_2, x_3, \ldots, x_n) \\ \vdots \\ f_n(x_1, x_2, x_3, \ldots, x_n) \end{bmatrix} = \begin{bmatrix} 0 \\ 0 \\ \vdots \\ 0 \end{bmatrix}. \tag{5.66}$$

Section (5.5) introduces schemes which make use of the sparseness of the matrices to offset the effects of their size. Nevertheless, the size of the problems to be solved can be a severe impediment. It is a useful exercise to set up a simple specimen problem and try to solve it by various methods (such as those available in SGFramework) as the size of the problem is increased. The solution will ultimately slow down to an extent which makes it impractical. One 'hits a wall' when the computer does not contain enough RAM to store the problem in memory. Once the virtual memory starts to be used, the computation time increases drastically. Virtual memory is memory implemented by storing the contents of physical memory to a device such as a disk drive. Since disk access is several orders of magnitude slower than RAM access, the simulation slows down. (This can occur during some of the large 2D simulations.) The question is, how large can it be, on a given computer, before this happens?

5.4.1   Fixed-Point Iterative Methods

The basic idea of fixed-point iterative methods is to rewrite the equation $\mathbf{f}(\xi) = \mathbf{0}$ in an equivalent equation of the form $\xi = \mathbf{g}(\xi)$. Then, starting from an initial guess $\mathbf{x}^{(0)}$, one generates the sequence $\mathbf{x}^{(m+1)} = \mathbf{g}(\mathbf{x}^{(m)})$ for $m = 0, 1, 2, \ldots$ Hopefully, the series converges to the fixed-point ξ of $\mathbf{g}$. This subsection will focus on one particular fixed-point iterative method, Newton's method. Only the algorithm of Newton's method will be discussed. Readers interested in a more thorough mathematical analysis of the method are referred to the literature [9, 87, 88].

The iteration function of Newton's method is given by equation (5.67), where $\mathbf{f}'(\mathbf{x})$ is the Jacobian matrix defined by Equation (5.68). Upon substituting equation (5.67) into $\mathbf{x}^{(m+1)} = \mathbf{g}(\mathbf{x}^{(m)})$, one recovers the classic formula for Newton's method (Equation (5.69)). One positive feature of Newton's method is its quadratic convergence, once the solution vector $\mathbf{x}^{(m)}$ is sufficiently near the actual solution.

$$\mathbf{g}(\mathbf{x}) = \mathbf{x} - \mathbf{f}'(\mathbf{x})^{-1}\mathbf{f}(\mathbf{x}) \tag{5.67}$$

$$\mathbf{f}'(\mathbf{x}) = \frac{\partial \mathbf{f}}{\partial \mathbf{x}} = \begin{bmatrix} \frac{\partial f_1}{\partial x_1} & \frac{\partial f_1}{\partial x_2} & \cdots & \frac{\partial f_1}{\partial x_n} \\ \frac{\partial f_2}{\partial x_1} & \frac{\partial f_2}{\partial x_2} & \cdots & \frac{\partial f_2}{\partial x_n} \\ \vdots & \vdots & \ddots & \vdots \\ \frac{\partial f_n}{\partial x_1} & \frac{\partial f_n}{\partial x_2} & \cdots & \frac{\partial f_n}{\partial x_n} \end{bmatrix} \tag{5.68}$$

$$\mathbf{x}^{(m+1)} = \mathbf{x}^{(m)} - \mathbf{f}'\left(\mathbf{x}^{(m)}\right)^{-1}\mathbf{f}\left(\mathbf{x}^{(m)}\right). \tag{5.69}$$

Since matrix inversion is computationally expensive, each Newton iteration is usually performed in two steps. First, the linear equation (5.70) is solved by either a direct or indirect method, as discussed in Section 5.5. Then, the solution vector $\mathbf{x}$ is updated, as shown in Equation (5.71).

$$\mathbf{f}'\left(\mathbf{x}^{(m)}\right)\Delta\mathbf{x}^{(m)} = \mathbf{f}\left(\mathbf{x}^{(m)}\right) \tag{5.70}$$

$$\mathbf{x}^{(m+1)} = \mathbf{x}^{(m)} - \Delta\mathbf{x}^{(m)} \tag{5.71}$$

The main problem with the classical Newton's method, as outlined above, is its tendency to overestimate the length of the actual correction step $\Delta\mathbf{x}^{(m)}$ if the solution vector $\mathbf{x}^{(m)}$ is not sufficiently close to the actual solution. This phenomenon is commonly known as overshoot. A common remedy is to limit the length of the correction step by introducing a damping factor η in the classical Newton's method such as equation (5.72). The damping factor η is within the interval $(0, 1]$ and is chosen so that the residual norm of the equations $\mathbf{f}(\mathbf{x})$ decreases with each iteration $(\|\mathbf{f}(\mathbf{x}^{(m+1)})\| < \|\mathbf{f}(\mathbf{x}^{(m)})\|)$. This method is known as a damped Newton's method. Although the damping can increase the robustness of the classical Newton's method by ensuring that the residual norm of the equations decreases with each iteration, it may unnecessarily hinder the rate of convergence [89]. Empirical tests have suggested that it may be advantageous, depending upon the particular problem, to limit the amount of damping, thus allowing the residual norm of the equations to temporarily increase in order to increase the overall rate of convergence.

$$\mathbf{x}^{(m+1)} = \mathbf{x}^{(m)} - \eta\mathbf{f}'\left(\mathbf{x}^{(m)}\right)^{-1}\mathbf{f}\left(\mathbf{x}^{(m)}\right). \tag{5.72}$$

5.4.2 Relaxation Methods

In their simplest form, relaxation methods assume it is possible to rearrange the system of equations $\mathbf{f}(\xi) = \mathbf{0}$ so that one can solve the ith equation for the ith variable $(\xi_i = g_i(\xi_1, \ldots, \xi_{i-1}, \xi_{i+1}, \ldots \xi_n))$. Then, starting with an initial guess $\mathbf{x}^{(0)}$, one attempts to improve the ith component $x_i^{(0)}$ by setting it equal to the function $g_i(x_1, \ldots, x_{i-1}, x_{i+1} \ldots x_n)$. One typically loops through each variable in a systematic fashion and uses new values in subsequent calculations. There are many variants of the basic relaxation method described above. They are discussed in the literature [87, 88].

5.5 The Solution of Sparse Linear Systems of Algebraic Equations

As mentioned in the previous section, discretizing a system of nonlinear PDEs results in a massive system of nonlinear algebraic equations. Often the nonlinear system of equations is solved via Newton's method. At each iteration, Newton's method solves the linear system given by Equation (5.70) in order to find a correction vector $\Delta\mathbf{x}$. In most numerical methods a significant fraction of the total CPU time required to solve a system of PDEs is spent solving linear systems of algebraic equations. Therefore, it is important to implement an efficient linear solution algorithm.

There are several methods for solving linear systems of algebraic equations $\mathbf{Ax} = \mathbf{b}$, where $\mathbf{A}$ is an $n \times n$ matrix and $\mathbf{x}$ and $\mathbf{b}$ are $n \times 1$ vectors. The coefficients of $\mathbf{A}$ and $\mathbf{b}$ are given and the coefficients of $\mathbf{x}$ are unknown. As mentioned in the previous paragraph, the resulting number of discretized equations is often very

large, hence the dimension of the matrix n is very large indeed. Fortunately, matrix
A is sparse for most discretization methods. A matrix is sparse if most of its
coefficients are zero. Sparsity is important, because its exploitation can lead to
enormous computational savings in terms of CPU time and memory. Although
there is no strict definition as to what percentage of the coefficients have to be zero
for a matrix to be sparse, a matrix is considered sparse if there is an advantage in
exploiting the fact that many of its elements are zero. Usually the nonzero elements
of a sparse matrix are stored in linked lists or other suitable data structures, since
the memory requirements of using a two-dimensional array are excessive and it is
not necessary or useful to explicitly store the zeroes.

In general, the methods of solving systems of linear equations may be divided
into two categories: direct methods and indirect methods. Consequently, this section is divided into parts. Subsection 5.5.1 discusses direct methods, while Subsection 5.5.2 discusses indirect methods.

Prior to solving a sparse linear system of algebraic equations with either direct
or indirect methods, it is often advantageous to permute the rows and columns
of the matrix by multiplying the matrix by permutation matrices **P** and **Q** in
order to reduce the matrix to block triangular form, as shown in equation (5.73).
This is usually done in two stages. First, rows and columns are permuted so that
the diagonal coefficients a_{ii} are nonzero. This is known as finding a transversal.
The second step uses symmetric permutations to find the matrix's block form.
Algorithms for performing the above steps can be found in the literature [90]. Once
a matrix is reduced to block triangular form, the problem of solving the entire system
is reduced to solving several smaller problems, with savings in both computational
work and memory storage. The algorithms presented in Subsections 5.5.1 and 5.5.2
assume that matrix **A** has a transversal (nonzero coefficients on its diagonal).

$$\mathbf{PAQ} = \begin{bmatrix} \mathbf{B}_{11} & & & & \\ \mathbf{B}_{21} & \mathbf{B}_{22} & & & \\ \mathbf{B}_{31} & \mathbf{B}_{32} & \mathbf{B}_{33} & & \\ \vdots & \vdots & \vdots & \ddots & \\ \mathbf{B}_{N1} & \mathbf{B}_{N2} & \mathbf{B}_{N3} & \cdots & \mathbf{B}_{NN} \end{bmatrix} \tag{5.73}$$

5.5.1 Direct Methods

Almost all direct methods for sparse matrices are based on variants of Gaussian-
elimination. The Gaussian-elimination algorithm [91] is an organized method for
solving systems of linear algebraic equations. The idea of the algorithm is to manip-
ulate the system of equations into a form that is more easily solved without changing
the actual solution of the original system. The operations which may be applied to
a linear system of algebraic equations without changing its solution are known as
elementary or Gaussian operations. There are three basic Gaussian operations: the
interchange operation, the scaling operation and the substitution operation. The
interchange operation swaps equations i and j. The scaling operation multiplies

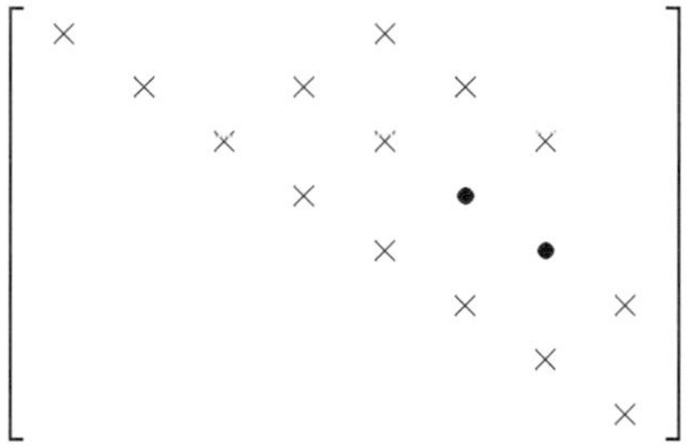

$$
\begin{bmatrix}
\times & & & \times & & & \\
& \times & & \times & & \times & \\
& & \times & & \times & & \times \\
& & & \times & & \bullet & \\
& & & \times & & \bullet & \\
& & & & \times & & \times \\
& & & & & \times & \\
& & & & & & \times
\end{bmatrix}
$$

Figure 5.4. An example of an upper-triangular matrix. The $\times$ symbols are the nonzero elements of the sparse matrix and the $\bullet$ symbols are the 'fill'.

equation i by a constant c which is not equal to zero. The substitution operation replaces equation j with the sum of equation j and a constant c times any other equation i.

Suppose one wished to solve the linear system of algebraic equations $\mathbf{Ax} = \mathbf{b}$ for $x = [x_1, x_2, \ldots, x_n]^T$. A description of the Gaussian-elimination algorithm for sparse systems of equations is outlined below.

For each row $i = 1, 2, 3, \ldots, n$, loop through the nonzero elements. Let j denote the column index of one of the nonzero elements. If i is greater than j then use the substitution operation to replace row i with the sum of row i and $-a_{ij}/a_{jj}$ times row j. This operation will make matrix coefficient a_{ij} equal to zero. After this portion of the algorithm has been completed, the matrix $\mathbf{A}$ will be in upper-triangular form, as shown in Figure 5.4. During this step, additional nonzero elements will be created as a result of the substitution operation. These nonzero elements are known as fill and are designated by the $\bullet$ symbol. The original nonzero elements (whose values may change in the process) are designated by the $\times$ symbol.

The second portion of the Gaussian-elimination algorithm is known as back substitution. This portion starts by computing the last component first, using $x_n = b_n/a_{nn}$. Then for each row $i = n - 1, n - 2, n - 3, \ldots 1$, the solution x_i is obtained by using equation (5.74).

$$
x_i = \left(b_i - \sum_{j>i} a_{ij} x_j \right) / a_{ii} \tag{5.74}
$$

The advantage of the Gaussian-elimination algorithm is that the solution vector $\mathbf{x}$ is found directly by applying the algorithm once. However, there are two problems. The first problem is the memory requirement. During the first portion of the Gaussian-elimination algorithm, additional nonzero elements (fills) are created. The number of fills can be quite large and often greatly exceeds the original number of nonzero elements. In order to reduce the number of fills, the linear system can be permuted by ordering routines. A discussion of these routines is beyond the scope of this work. Interested readers are directed to the literature [90, 92, 93, 94].

The second problem with the Gaussian-elimination algorithm is the accumulation of round-off errors. If the absolute value of a diagonal coefficient is much less than the absolute value of an off-diagonal coefficient on the same row ($|a_{ii}| \ll |a_{ij}|$ for $i \neq j$), then significant errors may be present in the solution. One method of overcoming this problem is to employ pivoting [90, 95]. Pivot strategies attempt to minimize the accumulation of round-off errors by permuting the matrix during the first portion of the Gaussian-elimination algorithm. Unfortunately, pivoting increases the complexity of the algorithm and often counteracts the effects of the fill-reducing ordering routines.

5.5.2   Indirect Methods

In contrast to direct methods which, in the absence of round-off errors, would exactly solve the linear system $\mathbf{Ax} = \mathbf{b}$ to machine precision in one pass or iteration of the solution algorithm, indirect methods are iterative methods which may require many passes of the solution algorithm to improve the accuracy of an initial guess. Hopefully, after a reasonably small number of iterations the solution will converge to a prescribed accuracy. In general, iterative methods do not add additional nonzero elements to the matrix $\mathbf{A}$. This property makes iterative methods very attractive for very large systems of linear equations which may not be feasibly solved with a direct method due to memory requirements. The disadvantage of iterative methods is that they may diverge. For instance, the Gauss-Seidel method discussed below can be shown to converge if the coefficient matrix $\mathbf{A}$ is strictly (row) diagonally dominant. It also converges if $\mathbf{A}$ is positive   definite (if A is real symmetric and for all nonzero vectors $\mathbf{y}$, $\mathbf{y}^T \mathbf{Ay} > 0$).

This subsection will focus on an important class of indirect methods, fixed-point iteration. The basic algorithm of these methods is as follows.

1. Pick an initial guess $\mathbf{x}^{(0)}$.

2. For $m = 0, 1, 2, \ldots$, iterate $\mathbf{x}^{(m+1)} = \mathbf{x}^{(m)} + \mathbf{C}\left(\mathbf{b} - \mathbf{Ax}^{(m)}\right)$ until the maximum update between iterations is sufficiently small.

The matrix $\mathbf{C}$ is chosen to have two desirable properties. First, it should be easy to calculate the matrix $\mathbf{C}$ times an arbitrary vector. Second, $\mathbf{C}$ should have the property $\|\mathbf{I} - \mathbf{CA}\| < 1$ in some matrix norm.

The two most common choices of the matrix $\mathbf{C}$ are $\mathbf{C} = \mathbf{D}^{-1}$ and $\mathbf{C} = (\mathbf{L} + \mathbf{D})^{-1}$. The matrices $\mathbf{D}$ and $\mathbf{L}$ are the diagonal and strictly lower-triangular parts of $\mathbf{A}$. The first choice leads to the Jacobi iteration (Equation (5.75)) and the second to the Gauss-Seidel iteration (Equation (5.76)). Successive overrelaxation (SOR) is a variant of the Gauss-Seidel iteration and is given by Equation (5.77). The SOR method may exhibit an increased rate of convergence if the relaxation parameter ω is appropriately chosen. A more detailed analysis of these iterative methods can be found in the literature [87].

$$x_i^{m+1} = x_i^m - \left(\sum_{j<i} a_{ij} x_j^m + \sum_{j\geq i} a_{ij} x_j^m - b_i \right) / a_{ii} \qquad (5.75)$$

$$x_i^{m+1} = x_i^m - \left(\sum_{j<i} a_{ij} x_j^{m+1} + \sum_{j\geq i} a_{ij} x_j^m - b_i \right) / a_{ii} \qquad (5.76)$$

$$x_i^{m+1} = x_i^m - \omega \left(\sum_{j<i} a_{ij} x_j^{m+1} + \sum_{j\geq i} a_{ij} x_j^m - b_i \right) / a_{ii}. \qquad (5.77)$$

Unfortunately, these iterative methods will diverge when solving semiconductor equations given in Chapter 4. Often, preconditioned conjugate gradient methods are employed for semiconductor simulation. Various preconditioned conjugate gradient methods such as BICG [96], GMRES [97] and CGS [98] may be found in the literature. The convergence of these methods is strongly dependent upon the preconditioner. Typical preconditioners are based upon incomplete factorization, such as ILU [98], ILLU [84] and ILUV [99]. The default NAM in SGFramework uses a damped Newton method with a choice of three linear solvers: (sparse) Gaussian-elimination; SOR; and preconditioned conjugate gradient.

5.6   Review of Software for the Solution of PDEs

We now turn to computer programs which are available for solution of PDEs. Recently, several computer programs have been introduced to solve systems of equations which arise from problems in applied physics. General-purpose mathematical packages are of limited use in modeling physical systems since they cannot directly solve PDEs. This section outlines the various software packages designed for the numerical solution of PDEs. The first half of this section will mention application-specific software while the last half will discuss general-purpose software. There has not been too much activity in the research community to develop powerful general-purpose software for the solution of PDEs. The majority of effort is spent on application specific software. Application-specific frameworks such as ATLAS [100] for semiconductor processing and device modeling and ANSYS [101] for non-linear structural, thermal, electromagnetics, acoustics and coupled-field analyses can solve a predetermined system of PDEs. If the problem of interest is supported, they can be very effective. The main disadvantage of these programs is their lack of flexibility. Application-specific programs also do not allow the user to investigate arbitrary physical models and change the underlying structure of the governing PDEs.

A broader class of problems can be solved using general-purpose finite-element packages. A code-generating package which uses symbolic manipulation to automate the labor-intensive steps involved in generating the input to a finite-element library has been described by Gates [102]. The same philosophy of automating repetitive and error-prone tasks also underlies the SGFramework which will be described in Chapter 6.

In recent years a few researchers have been working on computer frameworks that could solve relatively general classes of PDEs. Moore, Ozturan and Flaherty (MOF) have developed software which can handle systems of second-order PDEs in two space dimensions [103]. Although the MOF code can solve a large class of PDEs, there are many problems that cannot be solved using their framework. PDEase [104] is a relatively new product for solving PDEs numerically by finite-element analysis. PDEase solves a variety of nonlinear systems of PDEs of mixed elliptic-parabolic and hyperbolic types. It solves static, dynamic and eigenvalue problems with two space dimensions. PDEase defines its own element grid, performs error analyses and refines the grid as needed until error criteria are met. PDEase also takes care of most of the decisions needed for numerical solution of PDEs. PDEase is a very general and powerful tool; however, the automation of this package comes at the expense of generality. There are certain classes of problems that it was not designed to handle. For instance, it would be very difficult to model semiconductor devices using PDEase.

In this section, application-specific and general-purpose software packages for the numerical solution of PDEs were discussed. To summarize, application-specific packages are useful if one's problems are the type of problems for which the software was designed. General purpose PDE solvers, on the other hand, are more flexible but seem to lack the power necessary to perform extremely complicated problems such as semiconductor device modeling. The SGFramework described in the next chapter is designed to be both general and flexible and to overcome the deficiencies of these other approaches.

5.7 Validation of the Results of Simulations

One of the most important and difficult issues which needs to be addressed when using a simulation of a physical system is that of the accuracy of the simulation. In a complex problem such as that which we face when setting up a semiconductor device model it is unlikely that one can prove that the model is accurate. In fact, it is probable that any model we set up is at best accurate only for a limited set of circumstances. The uncertainty about how much the results of the model can be trusted should have a profound effect on how the task of simulation is done. The strategy we recommend is, as far as possible, to perform every useful and reasonable test that one can envision. Then it is necessary to attempt to make sure that the results of the test simulations are consistent with experiments, with each other and with intuitive expectations. To show that a model is consistent with experiment it is not sufficient to demonstrate that the trends are similar; a quantitative comparison is necessary. One common failing in modeling work involves the setting up of a crude model which is known to be at best marginally accurate, but which is easier to use than a more correct model. If this model can be shown to reproduce any features of actual experiments, there is a danger that it will be claimed that the model 'agrees with experiment'. This is then taken to be a justification to use the model in circumstances where there is no reason of any sort to believe that it is valid. The ultimate justification which is given is that such a model is 'better than nothing' — but this claim could easily be wrong.

There are many reasons why a simulation could fail to give the right answers. The main problems include that:

1. The wrong physical models, or incorrect data, may have been used.

2. The computer code which represents those models may have errors in its derivation from the models or in its implementation. (That is to say, it may have a 'bug'. One computer scientist once said that 'The definition of a nontrivial code is that it has a bug'. A better statement would be 'The definition of a trivial code is that it has no bugs'. This needs to be kept in mind; more will be said about this later.)

3. The models may be correct but they may be incomplete or only applicable in a limited set of circumstances.

4. The representation of the real problem by means of numerical approximations to the actual behavior may be significantly flawed. For example, if we are using a mesh, the mesh may not provide adequate resolution. This is one of the most common causes of inaccuracy, and also of failure of the model to converge, that we have encountered in semiconductor modeling. It is a common practice to vary the mesh spacing in simulations. If the solution varies when the mesh size is changed, the mesh is usually made smaller until there is no longer a variation as it is made smaller still. Making the mesh smaller may not ever give a solution which settles down - and even if it does, it may start changing again if the mesh is made smaller still.

5. A combination of several of these failings can occur. For instance, we may have two different ways in which we could represent a physical problem on a discrete mesh. Both may be equally accurate in principle. However, when implemented on the most accurate mesh we can set up, one may perform much better than the other. Use of the 'wrong' method is not strictly a 'bug', but it is not desirable either.

The ways we attempt to convince ourselves that these problems are not occurring, or at least are not causing fatal flaws in our work, are not always satisfactory. They will be discussed in many of the sections which follow, as we attempt to test whether the simulation results we obtain are believable.

Depending on the context, different ones of these problems may be the most critical. In semiconductor modeling, the standard fluid equations are relatively well established (that is to say, they are probably correct), provided the conditions are not too extreme. If the electric-fields are high, for instance, then various effects can occur which will lead to some or all of the carriers being heated to high energies. This in turn can lead to a breakdown in the fluid equations. In very small devices substantial voltages may exist across very narrow layers. A finite voltage across a narrow layer leads to a strong electric field. In addition if the mean free path λ is bigger than the layer thickness, electrons which cross the layer may be able to pick up an energy corresponding to the entire voltage dropped across the layer, because they have few if any collisions. Their acceleration in the electric-field will then create a beam of hot electrons, which is not described well by fluid equations. The beam's velocity is not determined only by the local electric-field, as in the usual mobility expression for the drift velocity, $\mathbf{v}_{dr} = \mu \mathbf{E}$. Instead, the electric-field along the entire trajectory of the beam is involved, when we calculate the electron's velocity. In this case a nonlocal description of the carriers is called for. Acceleration of the carriers by an electric-field can be described by the momentum conservation equation, and if all the carriers were a more or less coherent beam with the same velocity, a fluid approach could be set up to handle this. This fluid formulation would be much more complex than the usual semiconductor models and it would not work, even then, for a realistic device. A kinetic approach is called for in this case, as discussed in chapter 13.

Even if the fluid approximation is still valid as the carriers heat up, (which it may be if they only get hot, provided that they still have a Maxwellian distribution of energies), the correct expressions for the coefficients which appear in the fluid equations may not be known. In power semiconductors, for example, an important failure mode is caused by local heating in a channel carrying substantial current. This heating leads to an increase in the local conductivity. (In semiconductors the extra heat will usually generate extra carriers, so the conductivity goes up even though the mobility goes down due to the increased collision frequency.) The increase in the conductivity locally can lead to more of the current being carried in a narrower channel and to even more localized heating. The end result of this feedback can be a breakdown of the device, which takes place in the current-carrying channel. Under extreme conditions such as this, the coefficients describing the avalanching of electrons may not be known accurately.

One of the most difficult parts of the model to specify accurately is frequently the boundary conditions (bcs). There are many reasons why the bcs can be problematic. One common issue is that the location where the bcs should ideally be specified is far from the device. Unless the device is enclosed in a conducting can, the fields set up by the device do not conveniently limit themselves to a small region close to the device. It may be necessary to include a substantial volume around the device in the simulation. Use of a variable-sized mesh is helpful in allowing a large region outside the device to be included with relatively few, large cells in it. The specification and generation of such a mesh is fairly straightforward, using the software provided here, and is described in the next chapter. Even with a large exterior region included, however, the boundary condition may not be entirely convincing.

As was stated previously, setting up an appropriate mesh is frequently a cause of problems in semiconductor simulations. Experience with simulation packages which perform mesh generation automatically shows that the mesh obtained automatically sometimes performs poorly and fails to resolve the variations which take place across the mesh. Automatic mesh generation is difficult to do well, so the user must exercise caution.

The meshes used in semiconductor simulations are usually nonuniform. A simple finite-difference scheme is typically more accurate on a mesh with uniform spacing; see Section 2.1. However, the Scharfetter-Gummel (SG) scheme can be highly asymmetric in the way it includes adjacent mesh points. This means that even on a uniform mesh its accuracy is likely to be no better than on a nonuniform mesh. In addition, the SG scheme is believed to introduce other errors, commonly referred to as numerical diffusion, since its treatment of the electric-field is truly appropriate only in one spatial dimension.

5.8 Summary

In this chapter we described the finite-difference techniques to be used in later chapters. We also reviewed alternative discretizations. We then discussed techniques for solving nonlinear and linear systems of algebraic equations. We reviewed existing software for solving PDEs. Finally, we discussed the possible failings of numerical models and the ways in which models must be tested if we are to have any confidence in the usefulness of their predictions.

PROBLEMS

1. Suppose that the semiconductor equations (the continuity equations for each of the electrons and the holes, plus Poisson's equation) are to be solved by a finite-difference method. The mesh which is to be used has N_m nodes. This question is concerned with the maximum reasonable value of N_m.

   What are the usual methods of matrix inversion which are used to solve the discretized semiconductor equations? Do not include methods of matrix inversion which are unlikely to converge. Assess the computational resources which are required for each method. In other words, how do the requirements scale with N_m for each method? What is the most effective method, and how large can N_m be, for a commonly available modern computer? (One common problem occurs when the mesh is too large for the computer to handle. As the number of mesh points increases, the runs get dramatically slower until the solution is no longer practical.)

2. Solve Laplace's equation by converting it to a diffusion equation and advancing the solution in 'time' to steady state. Instead of solving $\nabla^2 V = 0$, solve the equation $\frac{\partial V}{\partial t} - \nabla^2 V = 0$. Why is this preferable to solving the original equation? Choose an appropriate time step, using a diffusion coefficient with a magnitude of one. Use a rectangular region, with $V = 0$ on three sides and $V = 10$ volts on the fourth side. Compare the answer to the direct solution, found for instance in Chapter 2. What does the use of a diffusion equation suggest about the convergence of this method and the role of the boundary conditions?

3. Write out the result of discretizing the Laplacian using the box integration method on a regular triangular mesh, with arbitrary lengths of the sides of the triangles but all the triangles being the same shape and size, and with one side of each triangle being chosen parallel to the x-axis. Compare your result to the result obtained using interpolation, in the text. Are the two schemes conservative?

4. Repeat the comparison of box integration and interpolation which was done in the text, but allow ϵ, which appears in the assumed form of the potential V, to have an arbitrary value.

5. Write out by hand what the elements of the Jacobian matrix are for Laplace's equation. Evaluate in general terms what are the other quantities appearing in the Newton's method. Perform a few representative calculations by hand.

6. A frequent cause for failure of Newton's method in a nonlinear problem is a poor initial guess. Suppose Laplace's equation were modified to read $V \nabla^2 V = 0$. A common initial guess is that $V = 0$. What will happen if this is used?

7. The next several problems are intended to show the need for the Scharfetter-Gummel (or some similar accurate) method for handling semiconductor equations, and to introduce the ideas behind the derivation of the method.

   We will now solve the steady-state continuity equation

$$\nabla \cdot \mathbf{J} = S, \qquad (5.78)$$

   in one dimension, for a constant (in space and time) electric-field, at first with the source rate S set to zero and on a uniform mesh. This problem will not use the Scharfetter-Gummel method, primarily to show why it is needed. It will be used in a subsequent problem. Let the region modeled be $L = 10\,\mu$m across. Suppose the electron density at one side, where $x = 0$ is $n(x = 0) = n_i$, the intrinsic density. Assume the electrons are in equilibrium so the electron density obeys a Boltzmann relation, $n(x) = n_0 \exp\left(\frac{eV}{k_B T_e}\right)$, where T_e is the electron temperature and $V^{th} \equiv k_B T_e / e \simeq 0.03$ V. The electric-field E_x pushes electrons towards $x = L$ and $E_x = -V_0/L$. What is $n(x = L)$ when $V_0 = 0.03$; 0.1; 1.0; 10 V? Suppose there are 100 cells of a mesh between $x = 0$ and $x = L$. For which of these values of V_0 will the mesh be able to resolve the density variation? ($\Delta n/n$ should be small, for each cell, where Δn is in this case the change in n across that cell.)

8. Set up a simulation using a simple differencing of the continuity equation, to solve the continuity equation under the conditions described in the previous problem. Impose the boundary conditions $n(x = 0) = n_i$, and since $V(x = 0) = 0$ and $V(x = L) = V_0$ we must have $n(x = L) = n_i \exp\left(\frac{V_0}{V^{th}}\right)$ in equilibrium. Solve the equation numerically for each V_0, starting with the smallest V_0, since beyond some value of V_0 the simulation will fail to converge. (Do not try to solve for values of V_0 greater than this.) Explain why the convergence fails at a particular value of V_0. Compare the results of the simulations to the analytic result. For a given value of the electric-field, change the density at $x = L$ slightly and calculate the current. A one-dimensional version of the file used in two dimensions in the last chapter (cont1.sg) is given below.

Input File cont1.sg

```
// FILE:  sgdir\ch05\cont1\cont1.sg
// demonstration of solution of continuity equation

const NX = 101;                  // number of x mesh points
const LX = NX - 1;               // index of last x mesh point
const DX = 1.0;                  // mesh spacing
const D  = 1.0;                  // Diffusion coefficient
const mu = 1;                    // mobility
const E  = 0.1;                  // Electric Field

var n[NX],x[NX];

// Continuity equation
equ n[i=1..LX-1]  -> - D * {n[i-1]  - 2*n[i] + n[i+1] } / sq(DX)
                    +  mu * E * {n[i+1] - n[i-1]} / (2.0*DX) = 0.0 ;


begin main
  assign x[i=all] = i*DX    ;
  assign n[i=all] = 1.0*i/NX;
  solve;
end
```

9. Repeat the previous question using a Scharfetter-Gummel form of the discretized continuity equation. To help with this, we shall rederive the Scharfetter-Gummel (SG) discretization, for a uniform mesh and a constant electric-field. The points to notice here are that we are finding a more accurate finite-difference form of the equation for the carrier density than we get by simple central-differencing; but our 'simple' SG expressions will not work in a general case (for instance if the electric-field goes to zero).

To evaluate $\nabla \cdot \Gamma$ at any point i we really need $(\Gamma_{i+1/2} - \Gamma_{i-1/2})/(x_{i+1/2} - x_{i-1/2})$. Here $\Gamma_{i+1/2}$ is the flux between points i and $i+1$, and so on. If the spacing of the mesh points is constant and equal to Δx, this is just $(\Gamma_{i+1/2} - \Gamma_{i-1/2})/\Delta x$.

Now $\Gamma = n\mu E - D\frac{dn}{dx} = \Gamma_0$, which is assumed to be constant between mesh points. If we define $\alpha \equiv |\mu|E/D$, then show that $\Gamma = -D\exp(\pm\alpha x)\frac{d}{dx}(n\exp(\mp\alpha x))$, where the sign on top is for particles which have a positive mobility. If the mobility is negative, the sign on α must be reversed. This equation becomes $\frac{d}{dx}(n\exp(\mp\alpha x)) = -\Gamma_0\exp(\mp\alpha x)/D$. This gives $n\exp(\mp\alpha x) = A\pm\frac{\Gamma_0}{\alpha D}\exp(\mp\alpha x)$, so that $n = A\exp(\pm\alpha x)\pm\Gamma_0/\alpha D$. But Γ_0 is the flux, and the variable we solve for is n, so we need to express Γ_0 in terms of n.

First we define $\epsilon = \exp(\pm\alpha\Delta x)$. The $\pm$ allows for the sign of μ; if μ is negative the negative sign must be used. Further, since $\alpha \equiv |\mu|E/D$, then $\Gamma_0/\alpha D = \Gamma_0/|\mu|E = n_0$

is a density (which may be negative). The density at the point x_i is $n(x_i) = A \pm n_0$, if we measure x from $x = x_i$. Similarly the density at point $x_{i+1} = x_i + \Delta x$ is $n(x_{i+1}) = A \exp(\pm \alpha \Delta x) \pm n_0 = A\epsilon \pm n_0$. Subtracting these, we get $n(x_{i+1}) - n(x_i) = A[\epsilon - 1]$, so that $A = (n(x_{i+1}) - n(x_i))/[\epsilon - 1]$. Now between points i and $i+1$ the flux we have been calling Γ_0 should more usefully be called $\Gamma_{i+1/2}$. In terms of $\Gamma_{i+1/2}$, show that the density at point x_i is $n(x_i) = A \pm \Gamma_{i+1/2}/\alpha D = \frac{n(x_{i+1}) - n(x_i)}{\epsilon - 1} \pm \frac{\Gamma_{i+1/2}}{\alpha D}$. From this, show that $\pm \frac{\Gamma_{i+1/2}}{\alpha D} = \frac{(\epsilon - 1)n(x_i) + n(x_i) - n(x_{i+1})}{\epsilon - 1} = \frac{\epsilon n(x_i) - n(x_{i+1})}{\epsilon - 1}$ and so obtain one of the main results we need:

$$\Gamma_{i+1/2} = \pm \frac{\alpha D}{\epsilon - 1}(\epsilon n(x_i) - n(x_{i+1})). \tag{5.79}$$

$\Gamma_{i-1/2}$ can be found by replacing x_i by x_{i-1} and replacing x_{i+1} by x_i in this expression.

Rederive this expression for the flux, from the Scharfetter-Gummel expression for the current given earlier in this chapter.

If we use this to evaluate $\frac{\partial \Gamma}{\partial x}$ at the point i, we find

$$\frac{\Gamma_{i+1/2} - \Gamma_{i-1/2}}{\Delta x} = \pm \frac{\alpha D}{(\epsilon - 1)\Delta x}[(\epsilon + 1)n(x_i) - (n(x_{i+1}) + \epsilon n(x_{i-1}))]. \tag{5.80}$$

Check to see if this expression for the divergence of the flux gives the usual discretization when the electric-field goes to zero. Why would $E = 0$ be a problem in a numerical implementation of this scheme?

In equilibrium $\Gamma = 0$ everywhere and the Boltzmann relation would apply so that $n(x_{i+1}) = \epsilon n(x_i) = \epsilon^2 n(x_{i-1})$. Check that if the density obeys these equations, then $\Gamma = 0$. On the other hand, if Γ is constant (in what follows we are measuring x so that $x_i = 0$), then show that $n(x_{i-1}) = A/\epsilon \pm \Gamma_0/\alpha D$, while $n(x_i) = A \pm \Gamma_0/\alpha D$ and $n(x_{i+1}) = A\epsilon \pm \Gamma_0/\alpha D$. Check that these expressions for the density give $\frac{\partial \Gamma}{\partial x} = 0$.

If there were no diffusion, the flux would all be due to the drift, which is at the velocity $v_{dr} = \mu E$. A constant density n_0 gives no diffusive flux, so the density n_0 leads to a total flux $\Gamma_0 = n_0 \mu E$. The other part of the density we found, $n_1 = A \exp(V/V^{th})$ (which is valid for electrons; for holes there would be a negative sign on the exponent), gives no net flux. Show this as follows. The diffusive flux $-D\frac{dn}{dx} = -\frac{DA}{V^{th}} \exp\left(\frac{V}{V^{th}}\right) \frac{dV}{dx}$. Now $E = -\frac{dV}{dx}$, so this equals $\frac{D}{V^{th}}En_1$. Because $D/V^{th} = -\mu$ (according to the Einstein relation, the minus sign being because these are negative particles and so μ is negative), this cancels the flux μEn_1 which is caused by the density n_1 being in the electric-field E.

The file which follows (scharf.sg) is supposed to allow a side-by-side comparison of a regular and an SG scheme. Note that the values of D and μ are not realistic, and the sign of μ would be wrong for electrons.

Plot the analytic (Boltzmann) solution and compare this to the numerical solutions as the number of mesh points is reduced. Start with 100 mesh points and reduce the number by about a quarter for each new simulation. How do you expect the solution to change as a result of decreasing the number of mesh points?

Input File scharf.sg

```
// FILE:  sgdir\ch05\scharf\scharf.sg
// demonstration of solution of continuity equation
// compared to a simplified SG equation.

const NX = 101;                    // number of x-mesh points
const LX = NX - 1;                 // index of last x-mesh point
const DX = 20.0/NX;                // mesh spacing
const D  = 1.0;                    // diffusion coefficient
const mu = 1;                      // mobility
const E  = 1;                      // electric field
const eps= exp(mu*E*DX/D) ;        // constant needed for SG scheme

var n[NX],nsg[NX],x[NX];

// Continuity equation
equ n[i=1..LX-1]  ->
 - D*{n[i-1] - 2*n[i] + n[i+1] } / sq(DX)
 + mu * E * {n[i+1] - n[i-1]}  / (2.0*DX) = 0.0;

// Scharfetter-Gummel equation, for E=const. and Dx = const.
// and with a multiplicative constant left out.
equ nsg[i=1..LX-1]->
  [(eps+1)*nsg[i] - (nsg[i+1] + eps*nsg[i-1])]/sq(DX) = 0.0 ;

begin main
  assign x[i=all] = i*DX;          // Coordinate Array
  assign n[i=all] = 1.0*i/NX;
  assign nsg[i=all] = n[i];
  solve;
end
```

10. The above form of the Scharfetter-Gummel discretization can be generalized to a situation where the electric-field is not constant, by evaluating the factor ϵ which appears in $\Gamma_{i+1/2}$ using the electric field which exists between points i and $i + 1$, and similarly for $\Gamma_{i-1/2}$. Then the expressions for these values of Γ can be subtracted from each other to give a new expression for $\frac{\partial \Gamma}{\partial x}$.

To make use of this we can find the electric-field from Poisson's equation and use this field to find the density by means of the SG discretization. The nonlinear Poisson's equation we use is discussed in more detail in Chapter 7, on pn-junctions. It assumes equilibrium densities of carriers. Why is this assumption appropriate and/or necessary? Check that the file which follows (pniv.sg) is 'correct' for electrons. The physical constants are not chosen accurately in these examples, since we are concerned with the numerical issues here. A similar file with roughly correct physical data is used Chapter 7, on pn-junctions, however. What values of the electric-field are likely to cause problems? What is the particle production rate in this example? What will the electron current be determined by? Find Γ by putting an 'assign' statement after the 'solve' statement and plot the results for V, n and Γ. Are the values of D and μ realistic?

11. The optimal solution scheme for an equation or for a system of equations will often be one which is designed so as to have 'built in' to it the features of the physical problem it is intended to represent. The problem which follows concerns rate equations   (also known as *Master equations* .)

The nucleation and growth of 'islands' of sputtered material, formed when adatoms (sputtered atoms on the surface) come together, is believed to be governed by rate equations such as

$$\frac{dN_i}{dt} = \alpha_{i-1} N_{i-1} - (\alpha_i + \beta_i) N_i + \beta_{i+1} N_{i+1}. \tag{5.81}$$

N_i is the number of islands per unit area with i atoms in the island. α_i and β_i are rate constants. What is the significance of each of the terms on the right-hand side of the equation? (Coalescence of islands with each other is not included here.)

There is usually a 'critical size' for the islands, $i = i_c$. Below the critical size the islands are most likely to decay, but above the critical size they are likely to grow. Suppose that an atom in an island of size i has energy E_i. For $i < i_c$ the energy increases with increasing i, whereas for $i > i_c$ the energy decreases with increasing i. In the range of sizes $i < i_c$ the density of islands is close to its equilibrium profile, $N_i^{eq} = N_0 \exp(-N_i E_i / k_B T)$, where T is the temperature and N_0 is a constant. Find a relationship between the α_i and the β_i which must hold if $\frac{dN_i}{dt} = 0$ when $N_i = N_i^{eq}$.

For $i > i_c$, suppose $\beta_i = 0$. Let $N_{i_c} = N_i^{eq}$ for all t, with the initial condition $N_i(i > i_c) = 0$ at $t = 0$. [Compare these conditions to the boundary conditions on the diffusion equation, given in Chapter 3.] Solve the rate equations for $i > i_c$ using a straightforward numerical integration, with $i_c = 20$ and for $i < 200$. Let $\alpha_i = \alpha_{i_c}[2 - i_c/i]$. Choose Δt to be a suitably small fraction of $\tau_\alpha \equiv [2\alpha_{i_c}]^{-1}$, and run the integration for about $100\tau_\alpha$.

Solve the rate equation analytically for $i = i_c + 1$. Repeat this for $i = i_c + 2$. Generalize the solution for $i = i_c + k$, where k is arbitrary. Write a computer program which makes use of the analytic solution to obtain $N_i(t)$ [105]. Compare the analytic solution to the result of the direct numerical solution. The analytic solution can form the basis for a more complex numerical solution which includes coalescence of islands with each other, and which is close to optimal in this case [105].

Input File pniv.sg

```
// FILE:  sgdir\ch05\pniv\pniv.sg
// demonstration of solution of 1D nonlinear Poisson's equation
// - applied to a linear pn junction.

const NX = 101;              // number of x-mesh points
const LX = NX - 1;           // index of last x-mesh point
const DX = 1.0e-6;           // mesh spacing
const VR = 1.0;              // voltage at the rhs
const VT = 0.03;             // thermal voltage
const N0 = 1.0e+20;          // doping density
const ec = 1.6e-19;          // electronic charge
const eps0=8.8e-12;          // permittivity of vacuum
const epsi=11.8;             // relative permittivity of Si
const eoeps = ec/eps0/epsi;  // e/epsilon
const mu  = 1.5;             // magnitude of mobility
const D   = VT * mu;         // diffusion coefficient
const ni = 1.0e17;           // intrinsic density

var V[NX],NA[NX],n[NX],E[NX],EP[NX],x[NX];
set NEWTON ACCURACY = 1.0e-8 ;

// Poisson's equation
equ V[i=1..LX-1] ->
  {V[i-1] - 2*V[i] + V[i+1]}/sq(DX) =
  eoeps * (NA[i] + N0*(exp((V[i]-VR)/VT) - exp(-V[i]/VT)));

// these two equations could be implemented as functions
equ E[i=0..LX-1]  -> E[i]  = - {V[i+1] - V[i]}/DX ;
equ EP[i=0..LX-1] -> EP[i] = exp(-E[i]*DX/VT)  ;

equ n[i=1..LX-1]  ->
  mu/DX * [{EP[i] *n[i]   - n[i+1]}*E[i]  /(EP[i]  -1.0) -
          {EP[i-1]*n[i-1] - n[i]  }*E[i-1]/(EP[i-1]-1.0)] = 0.0;

begin main
  assign x[i=all]       = i*DX;                 // coordinate Array
  assign NA[i=all]      = N0*(1.0 - 2.0*i/LX);  // acceptors on left.
  assign V[i=0]         = 0.0;
  assign V[i=LX]        = VR;
  assign n[i=LX/2..LX]  = N0;
  assign n[i=0]         = ni*ni/N0;             // eqbm. density
  assign E[i=all]       = 1.0;
  assign EP[i=all]      = exp(-E[i]*DX/VT);
  solve;
end
```

Chapter 6

The SGFramework

This is the final chapter of Part II, in which the SGFramework software tool is described. Some of the discussion concerns how the software was constructed and so it may not be of interest to most readers, who only need to be able to use the software. Sections 6.1 and Section 6.2 are essential to understanding Simgen, but Section 6.3 could be omitted if the details of the implementation of the software are not essential.

6.1 Introduction

This chapter describes the SGFramework. SGFramework is designed for the simulation of problems in applied physics and many other areas which are typically described by partial differential equations (PDEs). SGFramework allows users to concentrate on the physics and numerical analysis of their problem rather than on laborious computer implementation considerations. SGFramework performs many of the time-consuming and error-prone tasks involved in setting up simulations. It does not discretize the equations being solved, which are often a system of coupled, nonlinear PDEs. However, SGFramework can take a set of algebraic equations specified in an equation specification file (input file), which may be finite-difference equations and the appropriate discretized boundary conditions, and handle the rest of the steps required to obtain their solution. Since the equations are in general nonlinear, and in some cases they lead to highly ill-conditioned Jacobian matrices, this is a nontrivial exercise. The syntax of the equation specification file is designed to handle large sets of discretized equations efficiently. SGFramework automatically constructs a computer simulation by

1. parsing the equation specification file,

2. generating, by symbolic differentiation and manipulation, the functions which are needed to solve the equations specified in the input file, and

3. interfacing (compiling and linking) the computer-generated code to an appropriate numerical algorithm module (NAM) and/or user-written code.

246

Since SGFramework does not determine the scheme used to represent the solution of the PDEs, the choice of an accurate and appropriate numerical scheme is left to the user. This aspect of numerical analysis is beyond the scope of SGFramework. Software which chooses an optimal solution scheme for an arbitrary system of PDEs is unlikely to be successful in the near future. SGFramework is designed, instead, to solve the large systems of nonlinear algebraic equations, which arise from most numerical schemes.

The system of coupled algebraic equations to be solved is specified in an equation specification file. Typically, the discretization of a system of PDEs involves hundreds or thousands of equations which differ only in their mesh indices. Usually the equations involve array variables and are written for the 'general' array element in terms of indices such as $(i, j, k, \ldots)$, where the range of each index can also be specified. For example, the statement below is used to obtain a solution of Laplace's equation on a rectangular mesh with uniform spacing of DX in x and DY in y. The indices are specified in a natural manner. The expression between the arrow '$\rightarrow$' and the semicolon is the finite-difference form of Laplace's equation, $\nabla^2 V = 0$, on a Cartesian mesh with uniform spacing in the x- and y-directions. The expression to the left of the arrow is interpreted by SGFramework to mean 'Use this expression to calculate $V[i, j]$ for $i = 1$ to $LX - 1$ and for $j = 1$ to $LY - 1$.'

```
equ V[i=1..LX-1,j=1..LY-1] ->
  {V[i+1,j]-2*V[i,j]+V[i-1,j]}/sq(DX) +
  {V[i,j+1]-2*V[i,j]+V[i,j-1]}/sq(DY) = 0.0;
```

The mesh used in this example is rectangular and uniform. It is straightforward to implement a nonuniform rectangular mesh, for instance, by reading the mesh coordinates from an input file. Rectangular subregions, where different equations apply, are easily defined. Automatic generation of irregular two-dimensional (2D) meshes and pseudo three-dimensional meshes (with azimuthal symmetry) has also been implemented to allow more accurate representation of complex 2D geometries. One of the principal difficulties in solving a system of coupled, nonlinear partial differential equations (the PDEs) in complex geometries involves the setting up of a suitable computational mesh. If the system of PDEs has domains of smooth behavior and other domains of rapid variation, the problems of designing a suitable irregular mesh are compounded. It cannot be pretended that any simple method of mesh refinement, which is available at present, will always produce good results. However, mesh refinement according to user-specified criteria is a simple exercise using SGFramework.

In summary, the stages of the problem solution which are done by SGFramework are in some sense routine. The essential numerical analysis is still done by the user. There are real advantages to having these other steps automated. These advantages include:

1. The input file needed to specify a problem is very compact. This has consequent advantages, such as ease of understanding, ease of problem implemen-

tation and ease with which the file for solving that problem can be understood and modified by users other than the author.

2. If the user writes a code to perform the stages which are automated in SGFramework, it will be time-consuming to implement and prone to errors. It may also require specialized knowledge of techniques needed to solve the discretized equations which is not common among physicists and engineers.

3. SGFramework integrates almost all of the routine procedures of constructing a simulation in one package.

4. SGFramework supports the use of irregular meshes, in such a way as to make their use straightforward.

As was already indicated, this chapter is divided into twelve sections. Section 6.2 presents an overview of the SGFramework. A brief description of each program is given in this section. Following this section, each program in the SGFramework is described in detail. Section 6.3 describes the SGFramework translator, which parses and processes the equation specification file and generates C++ code which may be linked to NAMs and/or user-written code. Section 6.4 discusses the minimum-degree ordering program which is useful in reducing the memory requirement and execution time of simulations which use the Gaussian-elimination algorithm. Sections 6.5 and 6.6 describe the generation and refinement of irregular meshes. Sections 6.7 and 6.8 discuss the interfacing issues which occur when discretized equations and boundary conditions are specified on irregular meshes. Sections 6.9 and 6.10 describe the SGFramework build script and visualization tools, respectively. Finally, Section 6.11 summarizes this chapter.

6.2   An Overview of the SGFramework

A flowchart of the SGFramework is shown in Figure 6.1. [1]

The flowchart contains two types of symbols: data files and programs. Data files are represented by rectangular boxes with folded corners. The data files with 1's and 0's are binary files, while the files without the 1's and 0's are text data files. The asterisk character '*' stands for the simulation name. The text that follows the period '.' is the data file's extension. Programs and shell scripts are represented by circles. All of the programs are coded in the C++ programming language except for the mesh generator, which is coded in C.

The framework is divided into three main sections: mesh generation, simulation construction and results visualization. The programs and data files in each section

---

[1]Most of the figures in this chapter and some of the text are reprinted from *Computer Physics Communications,* Vol. 93, "Strategies for mesh-handling and model specification within a highly flexible simulation framework", 179-211, 1995, with permission from Elsevier Science — NL, Amsterdam, the Netherlands.

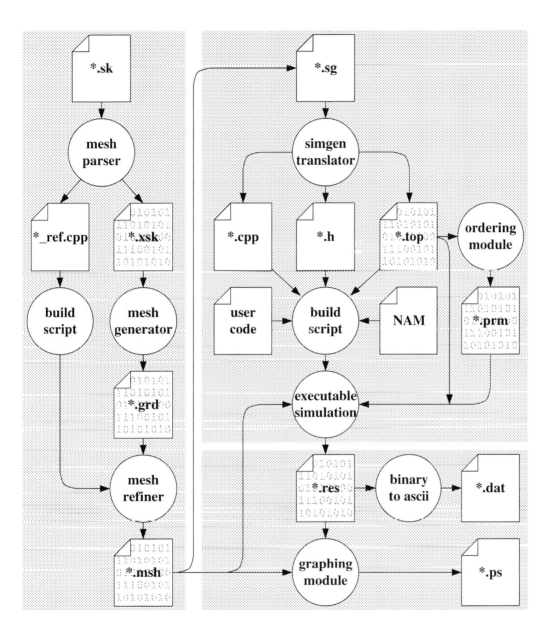

Figure 6.1. A flowchart of the SGFramework. The programs in the left gray box generate and refine the simulation mesh, the programs in the upper right box generate the executable simulation and the programs in the lower-right box visualize the simulation results.

are enclosed by a gray box. The mesh generation section contains the programs which generate and refine the mesh. This section is enclosed by the left box. The programs in this section parse the mesh specification file, generate an initial mesh and refine this mesh to construct an appropriate solution mesh. The simulation construction section contains the programs which build the simulation. This section is enclosed by the top right box. The programs in this section translate the equation specification file into C++ code and link this code with optional user-supplied code and a numerical algorithm module or NAM to generate an executable simulation. In addition, programs which reorder the unknowns in order to reduce memory requirements are included in this section. The data visualization section contains the programs which plot and organize the data. This section is enclosed by the bottom right box. The user-supplied specification files are shown at the top of the mesh generation and simulation construction sections.

6.3   The SGFramework Translator

(This section can be omitted by readers not concerned with the internal workings of SGFramework, who should proceed to Section 6.4).

The SGFramework translator parses the equation specification file and generates C++ code which contains the functions and procedures that initialize and control the simulation as well as evaluate the elements of the Jacobian matrix. The generated C++ code is compiled and linked to an appropriate numerical algorithm module (NAM) to create an executable program. If necessary, application-specific code may also be compiled and linked to the code generated by the SGFramework translator. Running the simulation executes the statements contained in the equation specification file's main procedure. Usually the equations specified in the input file are solved for one or more sets of boundary conditions.

The SGFramework translator is coded using the C++ programming language. It comprises eight major modules: the scanner, the parser, the error handler, the symbol table, the math expression module, the indices module, the statements module and the code generator. A flowchart of these modules is shown in Figure 6.2. The modules are represented by circles, and the arrows depict the flow of information (data) between the modules. For instance, the parser provides data to and receives data from the symbol table but only provides data to the indices modules. All of the modules are implemented as separate classes with the exception of the scanner and the parser which are implemented as a single class. The rectangular boxes are data files. The data files with 1's and 0's are binary files, while the files without the 1's and 0's are text data files. The asterisk character '*' stands for the simulation name. The text that follows the period '.' is the data file's extension. The box with the rounded corners represents the standard output device, which is usually a terminal (or a terminal window).

This material is handled in several subsections. Eight of the subsections describe the modules of the SGFramework translator. Subsections 6.3.1, 6.3.2 and 6.3.5 discuss the scanner, parser and error handler. Subsections 6.3.6, 6.3.7 and

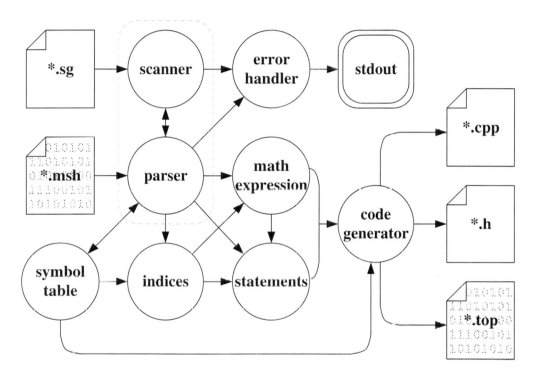

Figure 6.2. A flowchart of the SGFramework translator. The circles represent the program's modules. The arrows depict the flow of information and data between the modules.

6.3.8 describe the symbol table, math expression module and indices module. Subsections 6.3.9 and 6.3.10 discuss the statements module and code generator. Finally Subsection 6.3.10 summarizes the discussion of the translator.

In addition to the subsections listed above, there are three supplemental subsections. Subsection 6.3.1 discusses the scanner's finite-state table. Subsection 6.3.3 explains the grammar of the SGFramework language. Subsection 6.3.4 outlines the parser's finite-state table.

6.3.1   The Scanner

The equation specification file is a text file which the user creates using a text editor. The text file consists of a hierarchy of characters, tokens and statements. Fundamentally, the text file may be viewed as an array of characters. The characters are grouped together to form tokens and the tokens are grouped together to form statements. It is the responsibility of the scanner (also known as the lexical analyzer) to recognize and divide the input file into valid tokens.

Upon invoking the SGFramework translator, the program enters an initialization procedure. During this procedure the equation specification file is read into a character array. This array along with some additional variables is referred to as the input stream. After the initialization procedure is completed, the parser is called to process the equation specification file. During the parsing procedure, the parser frequently asks the scanner to return the next token in the input stream.

The scanner is implemented as a finite-state machine (FSM). The FSM table is shown in Subsection 6.3.1. When a token is requested from the scanner, the scanner first checks if a token is already available. If one has already been identified, it is returned to the parser. Otherwise, the scanner enters the FSM with an initial state of ST_START. The FSM examines the current character in order to determine the next state and the action(s) to take. Once a token is identified, the scanner exits from the FSM and returns the token to the parser, which is discussed in the next section.

The Scanner's Finite-State Table

This subsection lists the scanner's finite-state table. The scanner's finite-state machine (FSM) is implemented with 'if-then-else' type statements. The first column of the table lists the current state of the FSM. The second column lists the expected input characters. The third column lists the next state to which to move if the input character matches the expected character. The fourth column lists the appropriate action to be taken if a match is found. The possible choices are to keep the character, discard the character or temporarily store the character in a buffer. The first choice uses the current input character as the input character for the next state. The last two choices use the next character in the input stream as the input character for the next state. The fifth column represents the return token. A nonblank entry in this column means the scanner identified a valid token and should return it to the parser.

For example, suppose the scanner is in the start state 'ST_START' and the input character is a digit. This would correspond to a match in the second line of the finite-state table, hence the scanner's next state would be 'ST_NUM1' and the current input character would be stored in a buffer. The next character in the input stream would become the input character to state 'ST_NUM1'. This process continues until a token is recognized.

A sample of the table is shown below. For more information, see [106].

CURRENT STATE	CURRENT CHARACTER	NEXT STATE	CHARACTER ACTION	RETURN TOKEN(S)
ST_START	ALPHA	ST_ID	BUFFER	
ST_START	DIGIT	ST_NUM1	BUFFER	
ST_START	SPACE	ST_START	DISCARD	
ST_START	'/'	ST_COMM1	DISCARD	
ST_START	'"'	ST_STRING	DISCARD	
ST_START	'-'	ST_ARROW	DISCARD	
ST_START	'.'	ST_RANGE1	DISCARD	
ST_START	'<'	ST_LT	DISCARD	
ST_START	'>'	ST_GT	DISCARD	
ST_START	'='	ST_EQUAL	DISCARD	
ST_START	EOF	ST_EOF	KEEP	
ST_START	DEFAULT		DISCARD	CHARACTER
ST_ID	ALPHA	ST_ID	BUFFER	
ST_ID	DIGIT	ST_ID	BUFFER	
ST_ID	DEFAULT		KEEP	TOK_ID
ST_NUM1	DIGIT	ST_NUM1	BUFFER	
ST_NUM1	'.'	ST_RANGE2	BUFFER	
ST_NUM1	'e'|'E'	ST_NUM3	BUFFER	
ST_NUM1	DEFAULT		KEEP	TOK_INT

6.3.2 The Parser

The scanner and the parser have similar functions. As discussed in the previous section, the scanner recognizes and returns legal groups of characters known as tokens. It is the responsibility of the parser to recognize legal groups of tokens known as statements. Once a statement is identified, the parser calls an appropriate routine to process the statement.

In general, there are two commonly used approaches to parsing. Top-down methods look for a legal program by first looking for parts of a legal program and then looking at parts of parts, etc., until the pieces are small enough to match the input directly. Bottom-up methods put pieces of the input together in a structured way, making bigger and bigger pieces, until a legal program is constructed. In general, top-down methods are recursive, bottom-up methods are iterative; top-down methods are thought to be easier to implement, bottom-up methods are thought to be more efficient [107].

```
CURRENT      INPUT        NEXT         STACK        USER
STATE        TOKEN        STATE        ACTION       ACTION
----------   ----------   ----------   ----------   ----------
ST_EXPR      TOK_LEXPR    ST_EXPR      ACT_EXPR
ST_EXPR      TOK_EXPR     ST_EXPR1     ACT_PUSH
ST_EXPR      DEFAULT      ST_LEXPR     ACT_CALL
ST_EXPR1     TOK_AND      ST_EXPR2     ACT_PUSH
ST_EXPR1     TOK_OR       ST_EXPR2     ACT_PUSH
ST_EXPR1     TOK_XOR      ST_EXPR2     ACT_PUSH
ST_EXPR1     DEFAULT      ST_RETURN    ACT_POP
ST_CDECS     TOK_ID       ST_CDECS1    ACT_PUSH
ST_CDECS1    '='          ST_CDECS2    ACT_DROP
ST_CDECS2    TOK_EXPR     ST_CDECS3    ACT_PUSH      USR_DCONST
ST_CDECS2    DEFAULT      ST_EXPR      ACT_CALL      USR_EM1
ST_CDECS3    ','          ST_CDECS     ACT_DROP
ST_CDECS3    ';'          ST_CDECS4    ACT_DROP
ST_CDECS4    DEFAULT      ST_PROG
ST_PROG      TOK_CONST    ST_CDECS     ACT_DROP
```

Table 6.1. The constant-declaration portion of the SGFramework grammar finite-state table.

Since the SGFramework language is not necessarily finalized, the parser is implemented as a combination of the two methods described. Most of the parser is bottom-up with the exception of certain language entities such as mathematical expressions. In addition, the grammar is totally specified by a finite-state table. For instance, the constant-declaration portion of the SGFramework grammar is shown in table 6.1. The syntax of a constant declaration is the keyword 'CONST' followed by an identifier which is the constant's name, an equal sign and a mathematical expression which evaluates to a constant. The constant declaration is terminated by a semicolon and multiple constants can be declared in the same statement by concatenating their declarations with a comma separating each from the next.

The parsing of a constant declaration proceeds as follows. If the FSMs current state is 'ST_PROG' and the 'TOK_CONST' token is returned from the scanner, the parser drops (discards) the token and sets the state variable to 'ST_CDECS'. The state 'ST_CDECS' expects the next token to be an identifier. If the next token is not an identifier, the scanner reports a syntax error and enters a recovery procedure so that parsing may be resumed. Assuming the next token is an identifier, the FSM pushes this token on to the parsing stack and goes to the state 'ST_DECS1'. 'ST_DECS1' expects the next token to be an equal sign. If the token is not an equal sign, then a syntax error is reported. Otherwise, the FSM discards the token and goes to state 'ST_DECS2'. Up to this point the parser is using a bottom-up approach. However, the next state 'ST_DECS2' uses a top-down approach.

The state 'ST_CDES2' expects the next token to be a mathematical expression (the next token is expected to be 'TOK_EXPR'). Since the scanner returns individual tokens and not legal mathematical expressions, the next token will not be 'TOK_EXPR'. In the previous cases a syntax error was reported when the current token did not match the expected token. However, 'ST_CDECS2' contains a 'DEFAULT' entry which always matches the token returned by the scanner. The 'DEFAULT' entry informs the FSM that its next state will be 'ST_EXPR' and that it will enter this state via a call action ('ACT_CALL'). The call action stores the FSM current state on a state stack and then jumps to the next state 'ST_EXPR'. Once the FSM parses a mathematical expression using a top-down approach, it will generate an expression token and encounter a line whose next state is 'ST_RETURN'. This will cause the FSM to jump to the state which is on the top of the state stack. In our example, the top of the state stack is 'ST_CDECS2' so the FSM will go back to that state. However, unlike the first time the FSM was in 'ST_CDES2', the current token is now 'TOK_EXPR', and the user action 'USR_DCONST', which declares a constant, is called. This action validates the constant declaration and stores the declaration in the symbol table (see Subsection 6.3.6).

The combination of bottom-up and top-down approaches combines the advantages of both parsing methods. The majority of the grammar is parsed in an efficient bottom-up approach; however, the strategic use of the top-down approach allows the grammar to be easily modified. Finally, the ability to specify the grammar via a finite-state table produces a compact and efficient representation of the grammar.

6.3.3 The SGFramework Grammar

Programming languages are often described by a particular type of grammar called context-free grammar. SGFramework is no exception. The grammar is summarized in [106].

6.3.4 The Parser's Finite-State Table

The parser's finite-state table is stored in an array. The parser contains a parsing engine which uses the finite-state table to identify legal groups of tokens. Once a legal group of tokens is identified, the parsing engine calls an appropriate procedure to process the tokens.

The parser's finite-state table contains five columns. The first column is the current state. The next column contains the expected input tokens. If the FSM is in a state such that the input token does not match any of the expected input tokens, then a syntax message is generated by the error handler. If, on the other hand, the FSM state contains a 'DEFAULT' entry, a match occurs regardless of the input token and the appropriate actions specified by the default entry are taken. The third column specifies the next state to which to move if a match between the input and the expected token occurs. The fourth column instructs the parsing engine as to what action to take. For instance, 'ACT_PUSH' pushes the input token on the parsing stack while 'ACT_DROP' discards the input token. In both cases, a new input token is requested from the scanner. The last column contains a list of 'user' actions. If this column contains an entry, a special routine is called to perform some specific action such as declare a symbol, generate a binary math node object, etc. The finite-state table is summarized in [106].

6.3.5 The Error Handler

As the equation specification file is parsed, the SGFramework translator may detect errors and potential problems. When a problem occurs, the error handler is called to generate an error or warning message. These messages are not immediately reported. Instead, they are stored in an array and reported to the user prior to program termination. The syntax of the messages is '[X] (R,C): message', where 'X' is the type of message and 'R' and 'C' are the row and column indices of the location, where the error occurred in the equation specification file. Messages may be warnings, errors or fatal errors and are represented by the characters 'W', 'E' and 'F', respectively. When warnings and errors occur, the SGFramework translator attempts to recover and continue parsing the specification file. When a fatal error occurs, the SGFramework translator initiates program termination. The SGFramework translator recognizes and reports 50 types of warnings, errors and fatal errors. The next module to be discussed is the symbol table.

6.3.6 The Symbol Table

The SGFramework language allows users to declare constants, variables, arrays, functions and procedures. Since these declarations are often used by subsequent program statements, they must be stored in a symbol table for efficient retrieval. The SGFramework symbol table allows declarations to be retrieved both by name and type.

The symbol table is implemented as an object which is derived from another object, a hash table. The hash-table object has member functions which efficiently store, retrieve and delete hash node objects. Hash node objects store the declaration's name and type. In addition, each type of declaration has a corresponding object which is derived from the hash node object. These objects contain additional declaration-specific information. For instance, a constant symbol object stores the constant's value, while an array symbol object stores the dimensions of the array.

The modules described thus far are common to almost all compilers and translators. The next module to be described is the math expression module. This module gives the SGFramework translator its symbolic calculus and algebra capabilities which distinguish the SGFramework translator from typical compilers.

6.3.7 The Math Expression Module

The overall goal of the SGFramework is to solve problems in applied physics and engineering, thus the equation specification file typically contains many mathematical expressions such as the governing PDEs and the models of the physical parameters.

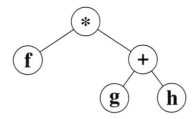

Figure 6.3. The expression tree of f * (g + h).

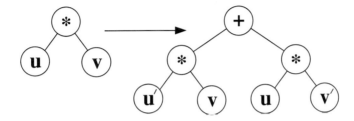

Figure 6.4. An illustration of the product rule of differentiation.

Internal to the SGFramework translator, mathematical expressions are represented by n-ary trees of math node objects. The math node objects contain several useful member functions. The most important function is the differentiate function, since the translator must symbolically differentiate each user-defined function with respect to its arguments and each equation with respect to the unknown variables; hence, symbolic differentiation is the first topic of this subsection. Following symbolic differentiation is a brief discussion of symbolic simplification. Finally, this subsection will conclude with a brief discussion of the precomputation of user-defined functions.

Symbolic Differentiation

Symbolic differentation is performed recursively. For instance, consider the expression $f * (g + h)$, where f, g and h are subexpressions. The n-ary tree corresponding to this tree is represented by Figure 6.3. To symbolically differentiate this expression, the differentiate member function is called from the root node. Since the root node is a multiplication node, the differentiation function employs the product rule of differentiation $(uv)' = u'v + uv'$, as shown in Figure 6.4. In order to implement the product rule, the differentiation function recursively calls itself to differentiate the left and right operands. Since the right operand is a sum of two expressions, the additive property of differentiation is used $((u + v)' = u' + v')$, as illustrated in Figure 6.5. The recursive process is terminated once the entire expression has been differentiated. The expression tree for the derivative of the example expression is represented by Figure 6.6.

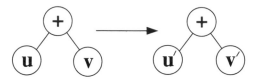

Figure 6.5. An illustration of the additive rule of differentiation.

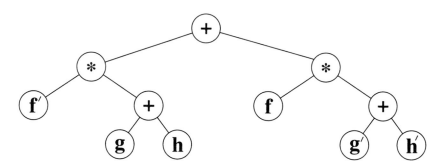

Figure 6.6. The expression tree which results from symbolically differentiating the expression f * (g + h).

Symbolic Simplification

Another important group of math node member functions are the ones that symbolically simplify expressions using rules such as $x + 0 = x$, $x * 0 = 0$, $x/1 = x$ and $-x * -y = x * y$. The symbolic simplification functions eliminate certain unnecessary computations which may result from the symbolic differentiation function. In addition to the symbolic differentiation and symbolic simplification function, math node objects have member functions that identify unknown variables, compare two expressions for equality, and determine if an expression is odd or even. The last operation is important in implementing the precomputation of user-defined functions, which is discussed next.

Precomputation of User-Defined Functions

Since a significant portion of simulation time is spent evaluating the elements of the Jacobian matrix, it is desirable to make this process as efficient as possible. One manner in which this may be accomplished is to eliminate redundant calculations by storing intermediate values in memory. For instance, a finite-difference discretization of Poisson's equation using the box integration method at node i requires the calculation of the electric-field intensity between the node i and a neighboring node j. Typically, the same electric-field intensity is needed (perhaps with a sign change) when we come to node j and look at its neighbors. The field can be computed once and stored in an array. Whenever the flux is needed, the array can be read and some simple logic can determine if the value should be multiplied by minus one.

The SGFramework translator allows user-defined functions such as the electric-field intensity between nodes to be precomputed and stored in arrays for quick retrieval. This feature results in significant increases in computational efficiency. However, the description given above is overly simplistic. In addition to storing the values of precomputed functions, the derivatives of these functions with respect to their arguments need to be stored as well. Consider the electric field intensity again. The electric-field intensity from node i to j is given by the expression $E_{ij} = -(V_j - V_i)/h$, where V_i and V_j are the electrostatic potentials at nodes i and j, respectively and h is the distance between nodes i and j. A typical declaration for this function is as follows.

```
func E(V1,V2,<h>)
  return -(V2 - V1)/h;
```

Suppose the electric-field intensity function above and the derivatives of this function with respect to its first two arguments are evaluated for every adjacent pair of nodes l and k and the results are stored in an array. To make the specification unambiguous, the function arguments V1 and V2 are chosen to be $V_{\min(l,k)}$ and $V_{\max(l,k)}$, respectively. In other words, V1 is the electrostatic potential of the node which appears first in the order (has the smaller index). Suppose the derivative of the electric field intensity function with respect to the first argument is needed from node j to i with $j > i$. As is the situation above, the value retrieved from the array needs to be multiplied by negative one, since the flux computed from i to j is the value that is stored. Now suppose that we request the derivative with respect to the first argument. The value returned needs to be multiplied by negative one, since the flux was computed from i to j not j to i. However, in addition to this, the first two arguments of this function are passed to the function in the opposite order from that in which the function and its derivatives were precomputed. Therefore, negative one times the precomputed value of the function with respect to its second argument needs to be returned rather than negative one times the precomputed value of the function with respect to its first argument.

The problem can even become more complicated, since the arguments of a precomputed function may be precomputed functions themselves. SGFramework transparently handles embedded precomputation of user-defined functions and provides support for the precomputation of user-defined functions which are computed at either the nodes or the edges which connect adjacent nodes. SGFramework automatically determines if arguments need to be swapped in order to properly determine derivative information and if the arguments themselves are even or odd functions with respect to the transposition of the mesh indices.

6.3.8   The Indices Module

The SGFramework translator is designed to efficiently handle large systems of equations resulting from the discretization of PDEs. Typically the discretization of a PDE generates hundreds or thousands of equations which differ only in their mesh indices. The SGFramework language allows such systems of equations to be represented by a single indexed equation. In order to generate the topology information for the Jacobian matrix and verify the bounds of array indices, the SGFramework translator frequently needs to loop through nested sets of indices. The indices objects provide this support. In addition, the indices objects also provide interfacing support between the mesh specification and equation specification files through the use of labels. These interfacing issues were the topic of Section 6.7. The next topic presented will be the statements module.

6.3.9   The Statements Module

The statements module is similar to the symbol table. Whereas the symbol tables store declarations of constants, variables, arrays, etc., the statements module stores procedure statements. Since the procedure statements need only to be retrieved in the order in which they were declared, the underlying structure of the statement module   is a dynamic array of pointers to generic statement objects. Just as particular declaration objects (symbols) are derived from a generic hash-node object, particular statement objects are derived from a generic statement object.

The statement module also stores a list of mathematical equations, which are usually discretized PDEs. This module, in conjunction with the mathematical expression module, has the facilities to identify the unknown variables and array elements within the equations and symbolically differentiate the equations with respect to these variables and array elements in order to obtain mathematical expressions for the elements of the Jacobian matrix. The merit of this method of evaluating the elements of the Jacobian matrix will be elaborated upon in the next subsection.

6.3.10 The Code Generator

In this subsection a brief description of code buffer objects will be given. Following this description, two methods of evaluating the elements of the Jacobian matrix will be described as will the approach used by the SGFramework translator. Finally, this subsection will conclude with a presentation of the construction of the Jacobian matrix itself.

Code Buffer Objects

The code generator's function is to produce the C++ code which contains the mathematical expressions that correspond to the elements of the Jacobian matrix as well as the statements which control the execution of the simulation. The code generator works in conjunction with the other modules to accomplish this task. Instead of generating the C++ source and header files directly, the code generator passes all of the code it generates to a code buffer object. The code buffer object formats and writes the C++ code to text files.

Evaluating the Elements of the Jacobian Matrix

We now turn to the general approach employed to evaluate the elements of the Jacobian matrix. In principle, there are two possible methods of calculating the elements of the Jacobian matrix: analytically or numerically. Analytic calculation of the Jacobian elements is advantageous in that the calculated partial differentials may be implemented in a straightforward manner without any additional overhead. The disadvantages of this method are that the equations must be known *a priori* and that the user does not have the freedom to investigate structurally different models. Analytic differentiation can also be error-prone. However, numerical calculation of the Jacobian elements does incur a substantial penalty in terms of coding complexity and computational efficiency. The finite-box semiconductor device simulation program of Franz et al [74] uses a hybrid analytical/numerical scheme. Their program hard codes the semiconductor PDEs but allows the user to specify arbitrary models for the physical parameters. This flexibility leads to a penalty in computational efficiency.

SGFramework combines the advantages of both the analytic and numerical methods. The SGFramework translator analytically differentiates each equation with respect to each unknown and generates a compact, efficient computer code for all the elements of the Jacobian matrix. Since the analytic differentiation is done by the SGFramework translator, and not by the user, it is not prone to errors.

Construction of the Jacobian matrix

We now discuss the details of how the Jacobian matrix is set up efficiently. The traditional approach to automatically constructing the Jacobian matrix is to symbolically differentiate each equation $f_i(x_1, x_2, x_3, \ldots, x_n)$ with respect to each independent variable x_n to obtain the Jacobian matrix element j_{in}. However, if each equation is only a function of a few independent variables, then the resulting Jacobian matrix is sparse and one can scan the equation in order to determine the nonzero elements without explicitly performing the differentiation of the zero elements.

For instance, consider the equation $f_1(x_1, x_2) = x_1 + \sin(x_2)$. Differentiating this equation with respect to its independent variables yields the Jacobian elements j_{11} equals 1 and j_{12} equals $\cos(x_2)$. Now consider the indexed system of equations $f_i(x_1, x_2, x_3, \ldots, x_n) = x_i + \sin(x_2)$ for $i = 1, 2, 3, \ldots, n$. Since this system can compactly be represented as a single indexed expression, it would not be desirable to enumerate this expression (that is, explicitly write the expression for $i = 1, i = 2$, etc)., differentiate each resulting expression with respect to every independent variable and store the results. Alternatively, one could scan the expression to form a list of the expression's independent variables and only differentiate the expression with respect to variables in the list. Scanning the indexed expression above would, if done properly, produce a list containing the independent variables x_i and x_2 (as opposed to $x_1, x_2, x_3, \ldots, x_n$ and x_2). Differentiating this equation with respect to independent variables x_i and x_2 yields the Jacobian elements j_{ii} equals 1 and j_{i2} equals $\cos(x_2)$. The correct value for j_{22} is equal to $1 + \cos(x_2)$, which is the sum of j_{ii} and j_{i2} when $i = 2$. The additive property of the elements of the Jacobian matrix follows from the normal rules of differentiation.

Given a general indexed equation, one can use the method outlined above to generate a list of contributions to the Jacobian matrix elements. To construct the Jacobian matrix, one can take each contribution in the list, determine where it applies and add it to all of the appropriate elements in the Jacobian matrix. One can then move on the next contribution in the list. For example, reconsider the indexed expression above. The two contributions are j_{ii} equals 1 and j_{i2} equals $\cos(x_2)$. The first contribution j_{ii} implies that as the first step we add 1 on the entire diagonal of the Jacobian matrix. The second contribution implies that we should next add $\cos(x_2)$ on the entire second column of the Jacobian matrix.

If the Jacobian matrix is sparse, each row (or column) is usually stored as a linked list. Since one cannot efficiently locate an arbitrary element in a matrix represented by linked lists, one must construct the Jacobian matrix row by row (or column by column). An efficient way to build a sparse Jacobian matrix is to loop through each row and transfer its linked list to a vector (a one-dimensional array). The vector's dimension should be the same as that of a row of the Jacobian matrix. One can construct the row by considering only those contributions which affect that row. Finally one can transfer the vector back to the linked list.

In this section a brief description of code buffer objects was given. This was followed by general methods of calculating the elements of the Jacobian matrix. Finally a method of constructing the Jacobian matrix from a system of indexed equations was outlined.

Summary

In this section the SGFramework translator has been discussed. The SGFramework translator parses the equation specification file and generates C++ code, which, when linked to a NAM generates an executable simulation. The simulation can be invoked to solve the equations listed in the equation specification file for one or more sets of boundary conditions. The SGFramework translator consists of eight modules, which have been discussed in separate subsections.

6.4   The SGFramework Minimum-Degree Ordering Program

In order to reduce the memory requirement or to increase the computational efficiency of a linear solution algorithm, it is often desirable to rearrange the order in which the Jacobian matrix is   constructed. This is particularly true of direct methods, where matrix fill-in may be excessive. The SGFramework minimum-degree ordering program generates a row and column permutation based on the minimum-degree algorithm of Tinney [92]. In addition, the SGFramework ordering program generates an ordering which permutes the Jacobian matrix to block triangular form. This feature may be exploited in order to implement hybrids of direct and indirect linear solution methods such as the   Gummel method [108]. The SGFramework also provides filters which convert the SGFramework topology format to that of the sparse matrix manipulation system (SMMS) [109, 110] and vice versa. The SMMS framework contains several ordering routines as well as a host of other sparse-matrix operations and visualization tools. (These filters are not included on the distribution CD ROM, however.)

6.5   The SGFramework Mesh Generation Program

In this section we describe the specification of irregular meshes and introduce the mesh naming (labeling) paradigm, which is used in the equation specification file, and we describe the mesh-related data structures. The section begins with the layout of the mesh specification file, which itself involves the declaration of points and edges leading up to the naming of regions. The file also handles mesh refinement. The data structures associated with the mesh and their use in setting up and refining the mesh are presented, starting in Subsection 6.5.4. While these structures are in some sense straightforward, they are presented for completeness and clarity since the details of data structures used in generating irregular meshes are rarely enumerated.

The material is divided among several subsections. Subsection 6.5.1 provides a brief description of the mesh specification file. Subsection 6.5.2 describes the specification of the mesh skeleton (the points, edges and regions which specify the geometry of the mesh). Subsection 6.5.3 discusses the method by which an initial mesh is generated from the mesh skeleton. Finally, Subsection 6.5.7 summarizes the section on the mesh generation program. Subsection 6.5.4 describes the data structures which are used to generate the initial mesh. Subsections 6.5.5 and 6.5.6 outline the algorithms which divide the regions named in the mesh specification file into rectangular and triangular elements, respectively. A condensed version of much of this material is also given in the user's manual.

6.5.1   Mesh Specification File

SGFramework can automatically generate an arbitrary two-dimensional mesh consisting of triangular and rectangular elements based on the specification contained in an input file. The input file is divided into two parts: the mesh skeleton and the mesh refinement criteria.   The mesh skeleton contains the minimum amount of information necessary to accurately describe the mesh. The mesh skeleton consists of a hierarchical declaration of points, edges and regions. SGFramework performs several data-consistency checks and is designed so that additional checks may be implemented easily. The mesh refinement criteria specify the minimum and maximum spacing between nodes, the minimum and maximum numbers of refinement divisions and the conditions under which to refine (divide) triangles and/or rectangles. The naming of all of the mesh quantities and the data structures associated with the mesh are the principal issues discussed throughout this section. The data structures are presented for the sake of completeness and clarity and because they do not employ the more conventional tree structure [111].

6.5.2   Mesh Skeleton

As an example, consider the MOSFET mesh specification file listed in Section 10.4. The first part of the mesh specification file defines the mesh skeleton. The statements in this section are used by the grid-generation program to construct an initial mesh. In this example, the mesh skeleton consists of twelve points, fourteen edges and three regions. For clarity the mesh skeleton specification has been divided into five sections.

In the first section geometrical quantities such as the device width and depth are defined as constants. These constants are used by the following statements. The second section consists of a series of comments which provide a schematic of the mesh skeleton. This section is ignored by the mesh input-file parser. In the third section the skeleton points are declared. A point is defined by two coordinates. The coordinates are enclosed by parentheses. Each point is given a name. In the example given here, point names consist of the letter 'p' followed by other letters. For instance, the first point is named pA and the last point is named pL. In general, the names of objects must begin with a letter and may be followed by one or more letters and/or digits.

In the fourth section the skeleton edges are declared. Edges may be either line segments or circular arcs. A line segment is defined by its two endpoints. A circular arc is defined by its two endpoints and a third point on the arc. In each case the points defining the edge are enclosed by brackets. The edges in the example MOSFET mesh specification file are all line-segment edges. Each edge is given a name. In this example, edge names consist of the letter 'e' followed by two letters which indicate the points at the ends of the edges. In addition, an edge may be given a label. For example, edge eAB is labeled GATE and edge eCD is labeled DRAIN. Edges without labels are considered internal edges. Labels provide links between the mesh specification and problem specification files. A detailed explanation of labels will be presented in the following sections. The distinction between names and labels is that one label (such as 'GATE') can be used for several edges (or for several regions) which have different names (such as 'eAB'). The equation specification file refers to edges and regions by their labels (such as 'GATE' and 'DRAIN') while the mesh specification refers to points, edges and regions by their names (such as 'eAB' and 'eCD').

The edge definition may also include an initial node spacing and a grade. The grid-generation program will use these parameters to control the insertion of nodes on the edge. If these parameters are omitted, the grid-generation program will compute appropriate values for these parameters. For example, edge eAE is a line segment that connects points pE and pA. Edge eAE has an initial node spacing of 10 angstroms (10^{-7} cm) and a grade of 0.50. This declaration causes the grid-generation program to insert a node on the edge that is 10 angstroms from point pE. The next node will be spaced 15 angstroms from the previous node or 25 angstroms from point pE, so that the distance between the first and second nodes is 50 percent larger than the distance between the first node and point pE.

Finally, the regions are declared in the last section of the mesh skeleton. Regions are defined by a list of three or more edges. The list is enclosed between braces. The list of edges must form a simple closed curve. Regions include both the curve and its interior. The interiors of regions cannot overlap; however, regions may share boundaries. Each region has a name. However, unlike edges, where labels are optional, regions must be given a label. In this example, region names consist of the letter 'r' followed by a number. By default, regions are divided into triangular elements. Rectangular regions may be divided into rectangular elements if the opposite sides of the region have the same initial spacing and grade. The user may specify rectangular elements by inserting the keyword 'RECTANGLES' at the end of the region statement.

The second part of the MOSFET mesh specification file gives the refinement criteria. The statements in this section are used by the mesh refinement program to refine the initial mesh into a solution mesh. A discussion of these statements will be presented in Section 6.6. We will now discuss the method by which initial meshes are generated and the data structures that facilitate this procedure.

6.5.3   Generating the Initial Mesh

In the previous section the mesh skeleton was introduced. The mesh skeleton consists of a hierarchy of regions, edges and points. Two programs are used to construct the initial mesh: the mesh parser and the mesh generator. The mesh parser program parses the mesh specification file, validates the point, edge and region declarations and generates a mesh skeleton file. The mesh generator program reads the mesh skeleton file and generates the initial mesh. An outline of this procedure is as follows.

After the mesh generator program reads the mesh skeleton file, it adds nodes to each edge specified in the mesh skeleton. Nodes are mesh points in the simulation domain. Nodes are added to edges according to the specified initial node spacing and grade. As mentioned before, if the initial node spacing and grade are not specified, the mesh generator program computes appropriate values for these parameters. After nodes have been added to each edge, each straight and curved edge in the mesh specification file consists of one or more piecewise continuous segments. Segments are straight-line segments that connect two nodes. Segment objects will be referred to as edges. In order to avoid confusion, the word 'edge' will always refer to mesh edges or segments from this point forward. The edges specified in the mesh specification file will be referred to as 'skeleton edges'.

The next step in generating the initial mesh is to create a polygon object for each region. Polygons are defined by the vertices which consist of all the nodes that were added to the skeleton edges which define the boundary of the polygon. Nodes and vertices are similar. Vertices may be viewed as nodes that contain information that is useful to the mesh generation algorithm. One important distinction is that nodes are unique and vertices are not. In other words, each node corresponds to a unique point in the simulation domain, whereas two or more vertices may refer to the same node, hence the same point in the simulation domain. For example, consider the vertices on edges which are common to two polygons. Although the nodes to which the vertices refer on these edges are unique, the different polygons will contain distinct vertices that correspond to the same nodes.

Once the polygons are created, they are recursively divided into elements. Elements are the smallest objects into which the mesh is divided. Elements may either be triangles or rectangles. The algorithms by which polygons are divided into elements are discussed in Subsections 6.5.5 and 6.5.6. Once again, the material which makes up these subsections comprises details of the inner workings of the mesh generation routine and can be skipped by those not interested in writing their own mesh generation code.

6.5.4   Data Structures for Mesh Generation

Previously the mesh generation algorithm was outlined. In addition, node, edge, element, polygon and vertex objects were defined. In this subsection, the data structures used to represent these objects will be discussed.

Each node, each edge and each triangular or rectangular element is represented

Node Structure	
type	*variable*
real	coordinate 1
real	coordinate 2

Edge Structure	
type	*variable*
index	node 1
index	node 2
index	edge label
boolean	circular arc
index	adj. element

Element Structure	
type	*variable*
index	edge 1
index	edge 2
index	edge 3
index	edge 4
index	adj. elem 1
index	adj. elem 2
index	adj. elem 3
index	adj. elem 4
index	element label

Polygon Structure	
type	*variable*
integer	number of vertices
index	head of vertex list
index	vertex w/ max. angle
index	vertex w/ min. angle
index	element label
boolean	rectangular elements

Vertex Structure	
type	*variable*
index	node
index	edge
index	left neighbor
index	right neighbor
real	angle
real	density

Figure 6.7. The data structures of the initial mesh generation program.

by a data structure (see Figure 6.7). These structures are stored in dynamic arrays (arrays whose size increases as elements are added) and are referred to by indices within the arrays. A node structure consists of the node's two coordinates. An edge structure contains the indices of the edge's two endpoint nodes, an edge label index, a Boolean variable which indicates whether the edge is a straight-line segment or a circular arc and an index to an adjacent triangle or rectangle. The information needed to specify a circular arc (the center of the circle and its radius) is stored in a separate array. The purpose of the adjacent-triangle index will be described in the following paragraphs. An element structure contains the indices of the element's edges, the indices of its adjacent elements, and an element label index. If the element is a triangle, then the 'edge 4' and 'adjacent-element 4' fields are not used and they are assigned a null value. Boundary elements are not required to have elements adjacent to all of their edges. Such elements would have null values in the appropriate adjacent-element fields.

The data structures for regions are more complicated, since regions may contain an arbitrary number of nodes and edges. Regions geometrically form polygons whose vertices correspond to mesh nodes. Thus a polygon can be described by a list of vertices on its boundary. Each vertex on the polygon's boundary has two neighbors and there is an interior and exterior angle associated with the vertex (see Figure 6.10(a). The angle of a vertex is the interior angle between the line segments that connect the vertex to its neighbors. The mesh generation program represents regions by polygon structures. Polygon structures contain six fields. The

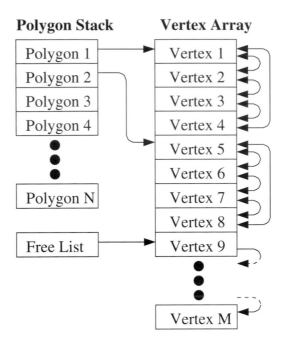

Figure 6.8. A schematic diagram of the polygon stack and the dynamic vertex array. The vertex structures are linked together and form doubly-linked lists.

first field is the number of vertices in the polygon. Vertices are represented by vertex structures, which will be discussed in the next paragraph. The second field of the polygon structure is the head of the vertex linked list. The third and fourth fields of the polygon structure reference the vertices with the minimum and maximum interior angles. The fifth field is a region label index and the final field is a flag that specifies whether the region should be divided into triangular or rectangular regions. Polygon structures are stored in a stack as they are created.

Vertices are represented by vertex structures and stored in a dynamic array. In addition, vertices form doubly-linked lists. In other words, vertices have pointers to their right and left neighbors, as shown in Figure 6.8. The vertex structure contains six fields. The first field is an index to the node which corresponds to the vertex. The second field is an index to the edge which is formed by the nodes which correspond to the vertex and its right neighbor. The third and fourth fields are the indices which link the vertices into a doubly-linked list. The fifth field is the interior angle which is formed by the vertex and its neighbors. The final field is a node density, which will be discussed in the following sections. Vertex structures are stored in a dynamic array as they are created.

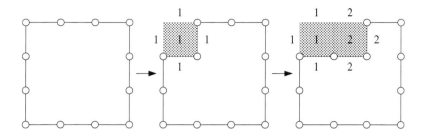

Figure 6.9. The division of a region into rectangular elements. The circles represent the polygon's vertices, the numbers inside the rectangular elements are the elements' indices and the numbers next to the elements' edges are the values of adjacent-element fields.

6.5.5   Polygon Division into Rectangular Elements

After the mesh skeleton file has been read by the mesh generator program and the initial data structures have been built (after the insertion of nodes and the construction of polygons), the polygons are then divided into triangular or rectangular elements according to their position on the polygon stack. Dividing a polygon into rectangular elements is simple, since the polygon has a rectangular geometry. The mesh generation program chops the vertex with the smallest angle from the polygon and forms a rectangular element (see Figure 6.9). This process continues until the entire region is divided into rectangles. Consider the left polygon in Figure 6.9. The circles represent the polygon's vertices. In the center polygon in Figure 6.9 the vertex in the upper left-hand corner has been chopped and a rectangular element has been created. This process creates a new node and two new edges.

In order to initialize the adjacent-element fields (AEFs) of the element structure, the AEF of the edge structures is used. The value of this field is shown by each edge. If an edge does not have a number by it, then the AEF of that edge contains a null value. The following steps are performed in order to properly initialize the AEFs.

1. When a new edge is created, its AEF is initialized to a null value.

2. When a new element is created, its AEFs are initialized to the value of the AEFs of the corresponding edges.

3. After the creation of a new element, the values of the AEFs of the element's edges are checked.

To determine if the edge is already referenced by another element, its AEF is checked. If the edge's AEF is null, the edge is not referenced by another element; otherwise the edge is referenced by another element. If an edge is not referenced by another element, its AEF is initialized to the index of the newly created element, as shown in the center polygon of Figure 6.9. If an edge is referenced by another element, then a link must be created between the element which the AEF references and the newly created element.

To illustrate the procedure described in the previous paragraph, consider the edge which is shared between rectangular elements 1 and 2, as shown in the right polygon of Figure 6.9. The value of this edge's AEF is 1. This means that element 1 is adjacent to the new element, element 2. When element 2 is created, its AEFs are initialized to the values of the AEFs of the corresponding edges. Since the value of the adjacent-element field of the edge which is shared by elements 1 and 2 has the value of 1, a link is created during this initialization. In other words, element 2 knows that element 1 is adjacent to it. To complete the link between these elements, the appropriate AEF of element 1 must be initialized to 2.

6.5.6   Polygon Division into Triangular Elements

Dividing a region into triangular elements is not as straightforward as dividing a region into rectangular elements. SGFramework's triangulation algorithm is based on the heuristics of Bank [112] and can summarized as follows. Every node in a region is assigned a 'density' which is a function of the distances to all of the neighboring nodes. Regions are recursively subdivided by splitting them along internal boundaries formed by connecting two nodes. New boundary nodes are created by grading the densities between edge endpoints. This assures a smooth transition of grid density which improves the conditioning of the Jacobian matrices of the discretized system of PDEs [113]. The boundary nodes are heuristically chosen to produce 'nice' triangles. For instance, splitting two straight edges such that the resulting internal angles are $\pi/3$ and $2\pi/3$ may produce equilateral triangles. Triangle 'niceness' is determined by a quality rating. Equilateral triangles are given a rating of one, and triangles whose vertices are nearly collinear are given a rating approaching zero.

In order to implement this algorithm, two operations must be performed on the polygons: splitting and chopping. Splitting is the process of dividing a polygon into two polygons as shown in Figure 6.10(a). Splitting introduces new nodes and edges in the simulation domain. It also creates a new polygon and modifies the original. Chopping is the process of removing one of the polygon's vertices and forming a triangular element, as shown in Figure 6.10(b). The procedure which is used to chop a triangle is identical to the procedure which is used to chop a rectangle. In addition to splitting and chopping, a 4-, 5-, 6-, 7- or 8-sided polygon may be divided by introducing a node at the polygon's centroid and forming triangular elements, as shown in Figure 6.10(c).

After the mesh is generated, the mesh generation program attempts to improve the quality of the triangular elements. Interior nodes are perturbed in attempts to make the triangles equilateral. The perturbing algorithm is identical to that of Bank [112]. In order to correct for problematic obtuse angles which may introduce nonphysical spikes in the solution or cause divergence [114], edges that subtend

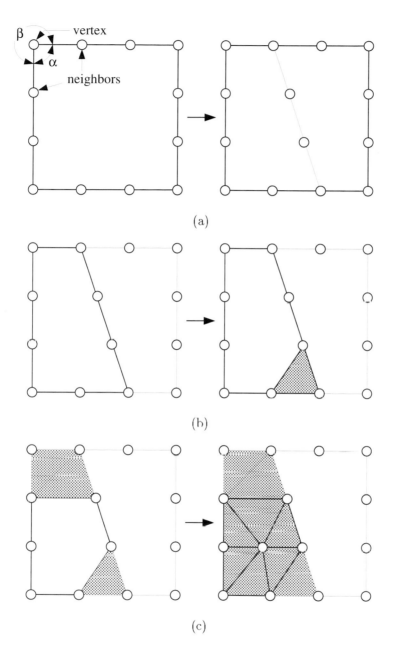

Figure 6.10. The division of a region into triangular elements by (a) splitting, (b) chopping and (c) inserting a centroid. Each vertex has an interior and an exterior angle. In this figure they are α and β, respectively.

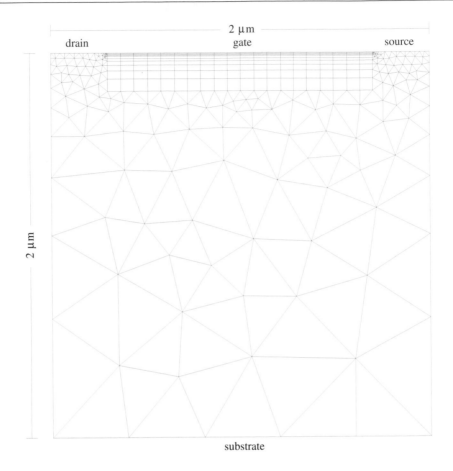

Figure 6.11. The initial MOSFET mesh generated from the specification file listed in the chapter on MOSFETs.

angles which sum to more than π in the two triangles on either side of the edge are removed and the quadrilateral made up from the two triangles is split along the line between the other pair of nodes. The initial mesh for the MOSFET is shown in Figure 6.11.

6.5.7   Summary

In this section we have discussed the specification of arbitrary 2D triangular- and rectangular-element meshes. We have also introduced the mesh naming (labeling) paradigm which provides an interface between the mesh and the equation specification files. Finally the mesh generation algorithm used by this algorithm has been presented. The data structures used are discussed at the end of this chapter. In the next section we will discuss the techniques which are used to refine an irregular mesh.

6.6 The SGFramework Mesh Refinement Program

In this section we discuss the refinement of arbitrary 2D triangular and rectangular element meshes. Typically the initial mesh must be refined in order to increase the accuracy of the solution. This is especially true if the system of PDEs is singularly perturbed (see Section 5.3.1) and has domains of smooth behavior and other domains of rapid variation. The mesh refinement program uses the refinement criteria specified in the mesh specification file as described in the first two parts of this section. As in the previous section, we will consider the MOSFET mesh specification file listed in Section 10.4 as an example.

This material is presented in a number of subsections. Subsection 6.6.1 outlines the mesh refinement algorithm. Subsection 6.6.4 summarizes the presentation of the refinement of 2D rectangular and triangular meshes, but more details are given at the end of the chapter. Subsection 6.6.2 discusses the refinement and splitting of triangular and rectangular elements, and Subsection 6.6.3 describes the data structures used to refine the mesh.

6.6.1 Mesh Refinement

The refinement criteria are located in the last half of the input file following the region statements. For clarity, this part of the file is further divided into four sections. As noted before, it is not necessary to structure an input file in this manner. The first section declares constants which are used by the following sections. The second section contains one statement which defines the coordinate labels. In the MOSFET example, the coordinates are labeled x and y, since the MOSFET simulation will use Cartesian coordinates.

The third section also contains one statement. This statement defines a refinement function named C. Refinement functions consist of a name, two refinement parameters and a body. The function name is declared after the keyword refine. The refinement parameters are enclosed by parentheses and follow the function's name. The first refinement parameter is the refinement measure and the second is the refinement distance. The body is the expression which follows the equal sign and may be a function of position. In this example, the body of the refinement statement is the MOSFET's doping profile. The refinement measure may be linear, log or signed log, as defined by equations (6.1), (6.2) and (6.3), respectively.

$$M_{\text{lin}}(x) = x \tag{6.1}$$

$$M_{\text{log}}(x) = \log(x) \tag{6.2}$$

$$M_{\text{slog}}(x) = \begin{cases} +\log(1+x) & \text{if } x \geq 0 \\ -\log(1-x) & \text{if } x < 0 \end{cases} \tag{6.3}$$

In order to determine whether a triangular or rectangular element should be refined, the mesh refinement program will evaluate the refinement functions at each of the element's vertices. If the measure between any two vertices exceeds the refinement distance, the element is refined. For example, consider the MOSFET example. The refinement measure is signed log and the refinement distance is 4. In order to determine if a triangular element with vertices (x_1, y_1), (x_2, y_2) and (x_3, y_3) should be refined, the mesh refinement program would first evaluate the mesh refinement function C at each vertex ($C_i = C(x_i, y_i)$ for $i = 1, 2, 3$). It would then check the measure between the vertices ($M_{12} = |M_{slog}(C_1) - M_{slog}(C_2)|$, $M_{23} = |M_{slog}(C_2) - M_{slog}(C_3)|$ and $M_{13} = |M_{slog}(C_1) - M_{slog}(C_3)|$). If $M_{12} > 4$, $M_{23} > 4$ or $M_{13} > 4$, then the triangle would be refined. Multiple mesh refinement functions may be declared. The aforementioned procedure is performed using each refinement function. The results of these tests are logically OR'ed together. Thus an element is refined if any one of the tests signifies that it should be refined.

The last section of the refinement criteria defines the minimum and maximum number of divisions and the minimum and maximum lengths of edges. Each triangular or rectangular element in the initial grid is assigned a division level of zero. If an element is refined, then the resulting elements are assigned the division level of their parent plus one. Thus elements formed from the division of a level-zero element would have division level one and elements formed from the division of a level-one element would have division level two.

The minimum and maximum edge lengths may be functions of position. Furthermore, the maximum edge length may be specified independently in both coordinates. The minimum edge length in the MOSFET example is set to the Debye length. This local specification of the minimum edge length is advantageous for devices with multiple junctions and would distribute the amount of net charge in each element roughly equally and would refine each junction appropriately. The maximum edge length is set to 1 micron in this example.

The following rules are used to determine whether an element should be refined. Once a decision has been reached, the remaining rules are not involved on that element. An element can be refined only once on each pass over the mesh.

1. If two or more of the element's edges have been divided by the refinement of adjacent elements, then the element is divided.

2. If an element's division level is less than the minimum division level specified in the mesh input file, then the element is divided.

3. If the element's division level is greater than the maximum division level specified in the mesh input file, then the element is not divided.

4. If any of the lengths of the element's edges are larger than the maximum edge length specified in the mesh input file, then the element is divided.

5. If all of the lengths of the element's edges are smaller than the minimum edge length specified in the mesh input file, then the element is not divided.

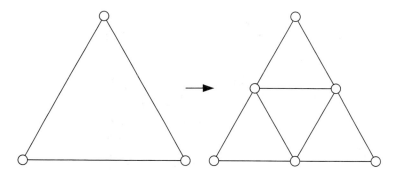

Figure 6.12. The refinement of a triangular element. The circles represent nodes while the lines represent edges.

6. If the criteria based on the refinement functions are not satisfied as discussed above, then the element is divided.

6.6.2   Refinement of Triangular and Rectangular Elements

Triangular elements are refined by inserting nodes at the midpoints of each edge as shown in Figure 6.12. Such a refinement perserves the quality of the original triangle. It also allows smooth transitions between coarse and fine domains. Rectangular elements are refined in one of three ways. They may be divided vertically, horizontally or both vertically and horizontally, as shown in Figure 6.13. In each case the rectangle is divided by inserting nodes at the midpoints of horizontal and/or vertical edges. The ability to refine rectangles in only one direction is useful when the problem has steep gradients in a direction that is parallel to one of the axes of the rectangle and small gradients in the other direction. This is the situation which arises in the vertical direction beneath the MOSFET's oxide/semiconductor boundary.

The mesh refinement program loops through each element in the mesh and tests whether the element should be refined according to the six refinement rules. The rules are tested in order; thus if rule one indicates an element should be refined and rule three indicates this element should not be refined, the element is refined. Before an element is refined, a check is performed on its neighbors. If a neighboring element's division level is less than the division level of the element considered, then the neighboring element is first refined. For example, consider Figure 6.14. The division level of each triangle is shown in the middle of the triangle. Before the gray triangle can be refined, the adjacent triangle with division level zero must first be refined as shown.

This process of looping through the elements and refining them is repeated until no elements are further refined. At this point, there may be triangular and rectangular elements with one side divided. These elements are split as shown in Figures 6.15 and 6.16, provided the resulting triangles are 'nice'. If the resulting tri-

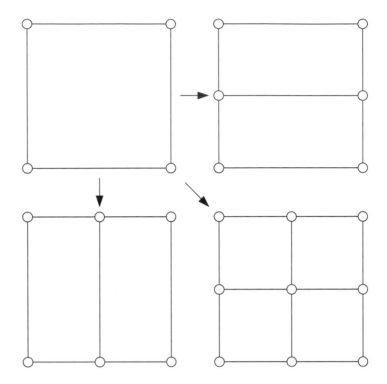

Figure 6.13. The refinement of a rectangular element. The circles represent nodes while the lines represent edges.

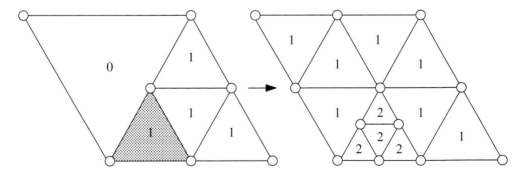

Figure 6.14. The refinement of the shaded element causes the neighboring element to be refined. The circles represent nodes and the lines represent edges. The numbers inside the triangles are the elements' division level.

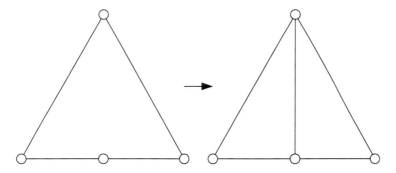

Figure 6.15. The splitting of a triangular element into two triangular elements. The circles represent nodes and the lines represent edges.

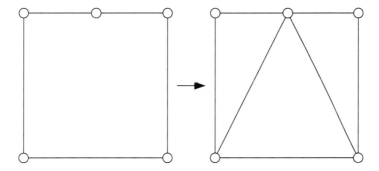

Figure 6.16. The splitting of a rectangular element into three triangular elements. The circles represent nodes and the lines represent edges.

angles are not 'nice', then the original elements are refined as shown in Figures 6.12 and 6.13 rather than split. This process of splitting and/or refining elements is repeated until no element has a divided edge.

6.6.3 Data Structures for Mesh Refinement

We now turn our attention to the data structures which are used by the mesh refinement program (see Figure 6.17). Each node, each edge and each triangular or rectangular element is represented by a data structure. Since the mesh refinement data structures are very similar to the mesh generation data structures, only the differences will be discussed. The mesh refinement node structure is identical to the mesh generation node structure. The same is true for the edge structure, except that the adjacent-element field is not needed by the mesh refinement program. The mesh refinement element structure is similar to the mesh generation element structure in that it contains the indices of the element's edges and the indices of the elements that are adjacent to its edges. The difference between these structures is that the mesh

Node Structure		Edge Structure		Element Structure	
type	*variable*	*type*	*variable*	*type*	*variable*
real	coordinate 1	index	node 1	index	edge 1a
real	coordinate 2	index	node 2	index	edge 1b
		index	edge label	index	edge 2a
		boolean	circular arc	index	edge 2b
				index	edge 3a
				index	edge 3b
				index	edge 4a
				index	edge 4b
				index	adj. elem. 1a
				index	adj. elem. 1b
				index	adj. elem. 2a
				index	adj. elem. 2b
				index	adj. elem. 3a
				index	adj. elem. 3b
				index	adj. elem. 4a
				index	adj. elem. 4b
				integer	level
				integer	status info.

Figure 6.17. The data structures of the mesh refinement program. These structures are identical to data structures of the mesh generation program except the edge structure does not have an adjacent-element field and the element structure has two edge and adjacent-element indices for each side of the element.

refinement element structure contains two sets of edge and adjacent-element index fields instead of one. The need for two sets of indices will soon become apparent. The mesh refinement element structure also contains an element-level field and some status information. The status information contains flags that specify whether the mesh refinement program first needs to divide any of the element's neighboring elements prior to dividing the element itself.

Consider Figure 6.14. When the shaded triangle is divided, nodes are created at the midpoints of the triangle's edges. The triangles which are adjacent to the shaded triangle now contain divided edges. One approach is to split the adjacent triangles, as shown in Figure 6.15. Splitting reduces the triangle's quality factor. Hence, if these split triangles need to be divided, they must first be 'unsplit' and the original triangle should be divided. SGFramework uses a slightly different approach to deal with this problem. Instead of splitting the adjacent triangles, both segments of its divided edge are stored in the element's data structure, and the segments' indices are then available for appropriate refinement of the elements. With this approach, it is not necessary to 'unsplit' triangles. Although we have focused on triangular elements, the same holds true for rectangular elements. Once all elements have been divided according to the refinement criteria, elements that contain divided edges are either divided or split. After the mesh is generated, it is not necessary to store the second set of edge and adjacent-element index fields. This eliminates the need to

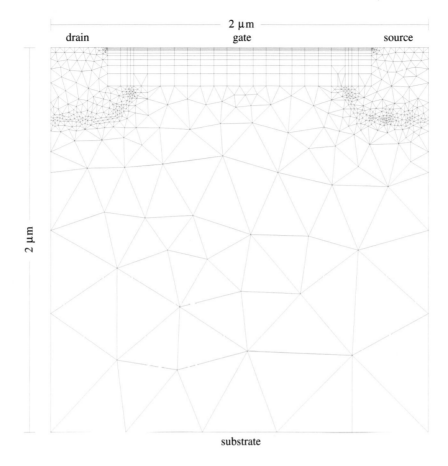

Figure 6.18. The refined MOSFET mesh generated from the specification file listed in Chapter 10, on MOSFETs.

store the mesh elements in the traditional tree structure.   However, since the tree structure is useful for adaptive meshes, the mesh refinement program can easily be modified to generate a refinement tree.

The refined MOSFET mesh is shown in Figure 6.18. As can be seen, the mesh refinement program refined the triangles and rectangles in the depletion regions (the regions where the doping profile C changes sign). An alternative MOSFET mesh that is generated from an initial mesh comprised solely of rectangular elements is shown in Figure 6.19. The mesh specification file for this mesh is not listed. In this mesh one can see the splitting of rectangular elements into triangular elements to match domains of fine resolution to domains of coarse resolution.

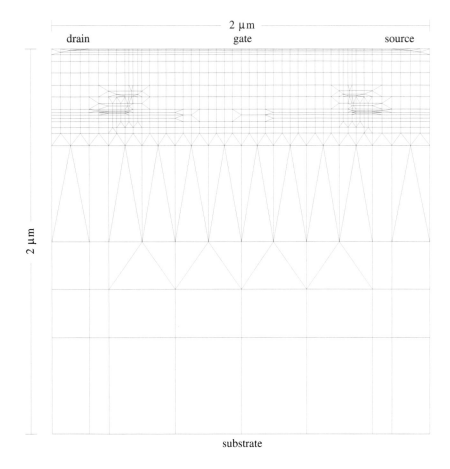

Figure 6.19. A MOSFET mesh generated from an initial mesh comprised solely of rectangular elements.

6.6.4 Summary

In this section we have discussed techniques which can be used to refine 2D triangular- and rectangular-element meshes. The data structures which are used by the mesh refinement program are outlined at the end of the chapter. Along with them we shall show two sample MOSFET meshes. In the next section we will describe the specification of a system of PDEs on irregular 2D meshes.

6.7 Interfacing the Mesh and Equation Specifications

In this section we will discuss how the naming (labeling) paradigm is used to specify a system of PDEs on irregular 2D meshes. The layout of this section is as follows. We will first discuss the issue of mesh connectivity. We will then present an overview of the box integration method and how it is implemented in SGFramework. Following this presentation, we will describe how complicated discretization schemes may easily be implemented as user-defined functions. Finally, we will bring all of these topics together and discuss the implementation of the PDEs themselves and how this is simplified using the naming paradigm.

6.7.1 Mesh Connectivity

In order to implement a finite-difference scheme, one needs to know how the mesh nodes are connected. In other words, each node must know who are its neighbors. This problem is trivial with rectangular meshes, since the simulation variables may be stored in multidimensional arrays which reflect the mesh geometry. In general, it is not convenient with irregular meshes to store the simulation variables in multidimensional arrays; thus they are usually stored in one dimensional arrays. Since the nodes are not connected in any predictable manner, one must explicitly store a list of neighboring nodes for each node on the mesh.

We will now discuss how SGFramework handles the issue of mesh connectivity. SGFramework stores the values of the simulation variables at each node in one-dimensional arrays and generates a list of neighboring nodes for each node on the mesh. SGFramework uses a set of data structures to represent the lists: node connectivity structures and neighbor structures. Figure 6.20 shows these data structures and Figure 6.21 labels the fields of these data structures. Each node has a node connectivity structure. This structure has three fields. The first field specifies the number of neighbor nodes that are connected to the current node. The second field specifies the area of the integration 'box' which surrounds the node. The 'box' is formed by connecting the perpendicular bisecting segments (see Section 5.2.3) of the edges which connect the node to its neighbors. The perpendicular bisecting segments are referred to as integration edges. The necessity of this field, as well as that of the terminology, will become apparent in the presentation of the box integration method. Each of the node's neighbors has a corresponding neighbor structure. These structures are stored in a list, and the last field of the node connectivity structure points to this list.

Node Connectivity Structure	
type	*variable*
integer	number of neighbors
real	box integration area
array	neighbor structures

Neighbor Structure	
type	*variable*
index	neighbor node
index	connecting edge
real	edge length
real	integration path length
real	partial box integration area

Figure 6.20. The node connectivity and neighbor data structures.

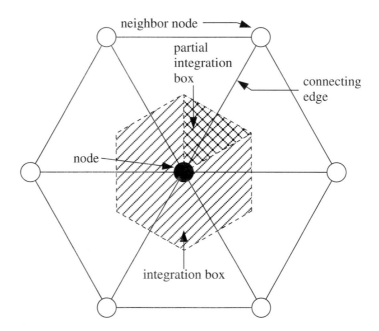

Figure 6.21. A schematic diagram illustrating the fields of the node connectivity and neighbor data structures.

The neighbor structure has five fields. The first field is the index of the neighboring node. The second field is the index of the edge that connects the node to this neighbor. The third field is the length of the edge that connects the node to this neighbor. The fourth field is the length of the integration edge which perpendicularly bisects the edge which connects the node to this neighbor. The last field is the partial area of the integration box. This is the area of the triangle whose vertices are the node and the endpoints of the integration edge between the node and this neighbor. The significance of the last two fields will become clear in the presentation of the box integration method.

It is not necessary to explicitly store the values of every field in the neighbor structure. For instance, one may store a list of the mesh's edge and node structures. These structures are discussed in the mesh refinement section. The index of the neighbor node, as well as the edge length, can be computed at runtime from the information in the edge structure. The advantage of not explicitly storing the values of each field is a reduction in memory requirements. The disadvantage is an increase in execution time. Thus a trade-off can be made between the simulation size and execution speed. The reduction in memory requirements and the increase in execution time are both typically small for 2D simulations.

Figure 6.22 shows a triangular mesh and its connectivity lists. Each box in Figure 6.22(b) represents either a node connectivity structure or a neighbor structure. The numbers in the node connectivity structures are the nodes' indices. The numbers in the neighbor structures are the indices of the neighboring nodes. Notice that there are thirteen node connectivity structures and neighbor structure lists although there are only seven nodes. The last six sets of structures are for the nodes which are on the boundary between materials A and B. The presence of these extra sets of structures will be discussed in Section 6.8. This finishes the discussion on mesh connectivity. We will now present a brief overview of the box integration method.

6.7.2   Box-Integration Method

In order to implement the box integration method, which was also discussed in Section 5.2.3, it is necessary that each node knows its neighbors. In addition to the lengths of the mesh edges which connect the node to its neighbors, it is usually necessary to compute the flux along the mesh edges. Furthermore, the area of the integration box and the lengths of integration edges are needed. As discussed in the beginning of this section, SGFramework stores these values in the node connectivity and neighbor structures. The only field of these structures which has not been described so far in this presentation is the partial integration area. Partial integration areas are useful if the source term is computed at the midpoints of the edges rather than the node. If this is the case, the area integration in Equation (5.13) can be approximated by the sum of the products of the source terms at the midpoints of the mesh edges and the corresponding partial integration areas. This concludes the overview of the box integration method. The discretization of boundary conditions will be discussed in Section 6.8. We now describe how complicated finite-difference

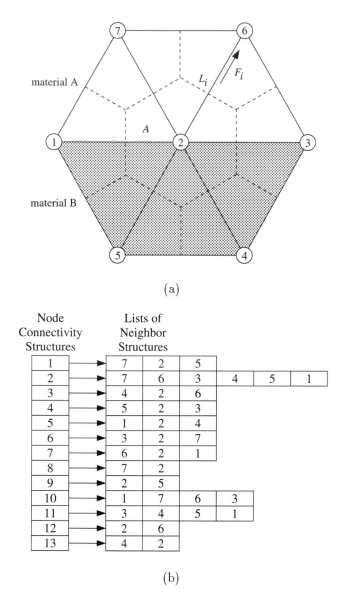

(a)

Node Connectivity Structures		Lists of Neighbor Structures					
1	→	7	2	5			
2	→	7	6	3	4	5	1
3	→	4	2	6			
4	→	5	2	3			
5	→	1	2	4			
6	→	3	2	7			
7	→	6	2	1			
8	→	7	2				
9	→	2	5				
10	→	1	7	6	3		
11	→	3	4	5	1		
12	→	2	6				
13	→	4	2				

(b)

Figure 6.22. (a) A simple triangular mesh. The solid lines are the element edges while the dashed lines are the integration edges. A is the area of the integration box surrounding node 2, F_i is the flux along the edge that connects node 2 to its ith neighbor and L_i is the length of the corresponding integration edge. (b) Node connectivity and neighbor structure lists corresponding to the mesh.

schemes may easily be implemented as user-defined functions, followed by the use of the labeling scheme to simplify the specification of the PDEs.

6.7.3 Finite-Difference Schemes and User-Defined Functions

In this discussion we will first consider the semiconductor equations which will serve as an example. Next we will discretize the semiconductor equations using central differences and discuss the problems that result from this discretization. An alternative discretization using the Scharfetter-Gummel method will then be discussed. Finally, we will show how the Scharfetter-Gummel method may be implemented in SGFramework using user-defined functions.

The semiconductor equations consist of three coupled, nonlinear PDEs: Poisson's equation, an electron continuity equation and a hole continuity equation. These are given by equations (6.4), (6.5) and (6.6) and have been scaled as in reference [9]. They are of the form given by Equation (5.12). The notation used is as follows. λ^2 is a small dimensionless quantity which results from the choice of scaling. V, n and p are the scaled electrostatic potential, electron concentration and hole concentration, respectively. R is the net scaled rate of recombination (recombination minus generation) per unit volume. Finally, $\mathbf{E}$, $\mathbf{J}_n$ and $\mathbf{J}_p$ are the scaled electric field, electron current density and hole current density, respectively and are given by Equations (6.7), (6.8) and (6.9). The Einstein relationship has been assumed in deriving Equations (6.8) and (6.9).

$$\lambda^2 \left(\nabla \cdot (\epsilon_r \mathbf{E}) \right) - (p - n + C) = 0 \tag{6.4}$$

$$+\nabla \cdot \mathbf{J}_n - R = \frac{\partial n}{\partial t} \tag{6.5}$$

$$-\nabla \cdot \mathbf{J}_p - R = \frac{\partial p}{\partial t} \tag{6.6}$$

$$\mathbf{E} = -\nabla V \tag{6.7}$$

$$\mathbf{J}_n = \mu_n \left(n\mathbf{E} + \nabla n \right) \tag{6.8}$$

$$\mathbf{J}_p = \mu_p \left(p\mathbf{E} - \nabla p \right) \tag{6.9}$$

To discretize the semiconductor equations using the box integration method, one must compute an approximate value for the fluxes F_i at the midpoints of the edges which connect the mesh nodes. One of the simplest approximations is central differencing. The central-differenced versions of Equations (6.7), (6.8) and (6.9) are given by equations (6.10), (6.11) and (6.12). The subscripts i and j are node indices. The notation $u|_{\text{mdpt}}$ represents the quantity u at the midpoint of the edge which connects nodes i and j. The quantity $h_{i,j}$ is the length of the edge which connects nodes i and j.

$$E|_{\text{mdpt}} = -\left(V_j - V_i \right) / h_{i,j} \tag{6.10}$$

$$J_n|_{\text{mdpt}} = \mu_n|_{\text{mdpt}} \left((n_i + n_j)/2 \, E|_{\text{mdpt}} + (n_j - n_i)/h_{i,j} \right) \qquad (6.11)$$

$$J_p|_{\text{mdpt}} = \mu_p|_{\text{mdpt}} \left((p_i + p_j)/2 \, E|_{\text{mdpt}} - (p_j - p_i)/h_{i,j} \right) \qquad (6.12)$$

Although these central-difference approximations are valid, in principle, on a fine enough mesh, in practice they are seldom used. If the mesh spacing is not sufficiently fine, the numerical solution of the electron and hole continuity equations will exhibit large nonphysical oscillations [84] since Poisson's equation is singularly perturbed (see Section 5.3.1). To overcome this problem, the electron and hole continuity equations can be discretized using the Scharfetter-Gummel method.

To discretize the electron current density (Equation (6.8)) using the Scharfetter-Gummel method, one assumes that both the current densities and electric field are constant along the edges that connect the mesh nodes. With these assumptions, the one-dimensional analog of equation (6.8) is a first-order ordinary differential equation which can be solved by introducing an integrating factor (see Subsection 5.3.3). The Scharfetter-Gummel discretization of the electron current density found using the solution obtained from the integrating-factor method is given by equations (6.13)–(6.17) and is useful because it more accurately represents the electron density between the nodes than does the simple discretization.

$$\text{aux1}(x) = \frac{x}{\sinh x} \qquad (6.13)$$

$$\text{aux2}(x) = \frac{1}{1 + \exp(x)} \qquad (6.14)$$

$$n|_{\text{mdpt}} = n_i \, \text{aux2}\left(\frac{V_i - V_j}{2U_t} \right) + n_j \, \text{aux2}\left(\frac{V_j - V_i}{2U_t} \right) \qquad (6.15)$$

$$\left. \frac{dn}{dx} \right|_{\text{mdpt}} = \text{aux1}\left(\frac{V_i - V_j}{2U_t} \right) \frac{n_j - n_i}{h_{i,j}} \qquad (6.16)$$

$$J_n|_{\text{mdpt}} = \mu_n|_{\text{mdpt}} \left(n|_{\text{mdpt}} \frac{V_i - V_j}{h_{i,j}} + \left. \frac{dn}{dx} \right|_{\text{mdpt}} \right) \qquad (6.17)$$

Equation (6.17) may elegantly be implemented as a user-defined function, as shown in the MOSFET equation specification file which is listed in Section 10.5. The function is given below.

```
// scaled electron current
func Jn(n1,n2,E,MU1,MU2,<h>)          // aux1(x) and aux2(x) are
   const  Ut   = (kb*T/e)/V0;         // functions that are internal
   assign MU   = ave(MU1,MU2);        // to the SimGen translator
   assign dV   = E*h/(2*Ut);
   assign n    = n1*aux2(dV)+n2*aux2(-dV);
   assign dndx = aux1(-dV)*(n2-n1)/h;
   return MU*(n*E+dndx);
```

A user-defined function consists of a function declaration and a body. A function declaration consists of the function name and an argument list. In general, SGFramework will symbolically differentiate a user-defined function with respect to each of its arguments. If, however, the user knows that it is not necessary to differentiate the function with respect to a particular argument, the user may enclose the argument in a pair of 'less than' and 'greater than' signs. This feature may also be used to control what information is incorporated into the elements of the Jacobian matrix. The function body consists of a list of three types of statements: local constants, local variables and a return statement. Local constants are declared by the 'const' keyword followed by the constant's name, an equal sign, its value and a semicolon. Local variables are declared by the 'assign' keyword followed by the variable's name, an equal sign, its value and a semicolon. The scope of local constants and variables is limited to the function body. The return statement marks the end of the function's statement list and the value of the return expression is returned to the expression that called the function.

The auxiliary functions given by Equations (6.13) and (6.14) must be carefully evaluated in order to avoid loss of precision and overflows. Consequently, these functions are incorporated in SGFramework as internal functions. (In general, users can declare functions that need special attention as external functions, as opposed to the user-defined functions included in the input file. They may then write code that evaluates the functions and their derivatives and link it with the SGFramework-generated code and a NAM.) The first statement in the electron current density user-defined function is a constant which computes the scaled thermal voltage. The next four statements declare local variables. The second statement computes the mobility at the midpoint of the edge, the third statement computes the argument of the auxiliary functions and the fourth and fifth statements solve equations (6.15) and (6.16). The last statement is a return statement which evaluates the electron current density (Equation (6.17)).

In this discussion we have shown how a complicated finite-difference scheme can easily be implemented as a user-defined function. We have used the Scharfetter-Gummel method to discretize the electron continuity equation as an example. We now address the use of the labeling scheme which simplifies the specification of the PDEs.

6.7.4 Naming Scheme and Specification of PDEs

Each simulation variable that is tagged as an unknown must have an equation associated with it, so that the system of the equations is not underspecified. In all but the simplest of simulations one typically partitions the simulation domain into distinct regions and solves a different system of PDEs in each region. Furthermore, different boundary conditions between regions may also need to be specified. SGFramework accomplishes this task through the use of the labels. As mentioned in the mesh specification section, each region must be given a label. Edges may also be given labels. The labels provide a link between the mesh specification file and the equation specification file. After the solution mesh has been generated, the mesh refinement program makes node lists for each edge and region label. The lists contain the indices of the nodes that are on the edges, or in the regions, that are associated with each label. The same node may be contained in the lists corresponding to several labels.

Equations are specified in SGFramework by equation statements. Equation statements have two parts: an equation header and an equation body. The equation header is specified between the keyword 'equ' and the arrow '→'. The header may refer to a variable or an array element. The body appears between the arrow '→' and the terminating semicolon. SGFramework solves the equation's bodies in order to determine the values of the variables and array elements referenced in the equation's headers. See the user's manual, Section 15.1.9, for details.

As an example, consider the MOSFET mesh specification and equation specification files. The MOSFET specification file declares two region labels ('SI' and 'SIO2') and six edge labels ('GATE', 'SOURCE', 'DRAIN', 'SUB', 'NOFLUX' and 'SISIO2'). For this simulation we wish to solve the semiconductor equations in the two regions labeled 'SI' and Laplace's equation in the one region labeled 'SIO2'. Furthermore, appropriate boundary conditions must be specified on the boundary between the silicon and silicon dioxide regions as well the external boundaries.

The first statement of the MOSFET equation specification file gives the name of the mesh file. This statement causes the SGFramework translator to read the MOSFET mesh file data. Consequently, the constants and labels that are declared in the mesh specification file are stored in the translator's symbol table. Furthermore, the translator automatically declares the constants 'NODES' and 'EDGES' which correspond to the number of mesh nodes and edges. The coordinates of the nodes and the values of the refinement statement expressions at each node are stored as one-dimensional arrays. Lastly, the translator stores the node connectivity and neighbors' structure lists.

Following the mesh statement, constants and user-defined functions are declared. The first several user-defined functions are models for the electron and hole lifetimes, the recombination rate, and the electron and hole mobilities. The discretized equations for the electric field, electron current density and hole current density are also implemented as user-defined functions. The next section of the specification file declares five one-dimensional arrays and specifies which elements of these arrays are unknown. These arrays store the electrostatic potential, the electron and hole concentrations and the electron and hole lifetimes at each node. The two statements, which begin with the keyword 'unknown', declare the array elements that follow to be unknowns. The values of unknown variables and array elements are determined when the simulation solves the equations. The four statements, which begin with the keyword 'known', declare the array elements that follow to be knowns. The values of known variables and array elements need not be solved for. In these statements the mesh labels are used to determine which nodes should be tagged as knowns and unknowns.

A straightforward evaluation of the elements of the Jacobian matrix would require the simulation to calculate the electron and hole mobilities several times at each node and the electric field and electron and hole current densities several times at each edge. In order to eliminate this inefficiency, the next set of statements instruct the translator to build a simulation which precomputes these user-defined functions and their derivatives prior to constructing the Jacobian matrix during each Newton iteration. This improves the performance of the simulation at the expense of additional memory requirements.

The next section of the MOSFET equation specification file declares the equations which the simulation will solve in order to determine the values of the variables tagged as unknowns. The header of the first equation, for example, indicates that the equation which follows the arrow '$\rightarrow$' should be associated with all of the elements of the 'V' (electrostatic potential) array which correspond to the nodes on the boundary between the 'SI' and 'SIO2' regions. Likewise, the header of the second equation indicates that the equation which follows the arrow '$\rightarrow$' should be associated with all of the elements of the 'V' array which correspond to the nodes in the 'SI' region.

The bodies of the equations are the finite-difference forms of the semiconductor PDEs. The box integration method is used to discretize these equations. In order to discretize the equations using the box integration method, one must loop over each neighboring node and sum the products $F_i L_i$ of the fluxes F_i along the mesh edges which connect the node to its neighbors and the corresponding integration edge lengths L_i. To accomplish this task, SGFramework provides a 'neighbor sum' ('nsum') function. The 'nsum' function has four arguments. The first argument is the index of the node whose neighbors should be looped over. The second argument is the summation's index. The third argument is a mesh label. If the label is an edge label, the 'nsum' function will only sum over the neighboring nodes whose connecting edge has the same edge label. If the label is a region label, the 'nsum' function will only sum over the neighboring nodes which are in that region. If the label is the keyword 'all', then every neighboring node will be included in the sum. The final argument of the 'nsum' function is an expression to sum. In the MOSFET simulation these expressions are products of the fluxes $\mathbf{E}$, $\mathbf{J}_n$ and $\mathbf{J}_p$ and the lengths of the integration edges. The '@' symbol indicates a precomputed function. As mentioned in the section on the box integration method, the index of the neighbor nodes, the area of the integration box, the lengths of the edges 'elen' and the lengths of the integration segments 'ilen' are stored in node connectivity and node structures. These quantities can be accessed by the 'node', 'area', 'elen' and 'ilen' functions, as shown in the MOSFET equation specification file.

In this section we have discussed the issues of mesh connectivity. We have also presented a brief overview of the box integration method and shown how to implement complicated discretization schemes as user-defined functions in SGFramework. Finally, we have put everything together and shown how the naming (labeling) paradigm simplifies the specification of PDEs on irregular meshes. In the next section we will discuss the implementation of boundary conditions on these meshes.

6.8   Specification of Boundary Conditions on Irregular 2D Meshes

In this section we will discuss how boundary conditions may be implemented on irregular 2D meshes. We will show how Dirichlet boundary conditions may be implemented with 'known', 'unknown' and 'assignment' statements and how Neumann boundary conditions may be implemented with the 'nsum' function. In addition, we will discuss how boundary conditions can be imposed at the interface between different materials.

To specify a Dirichlet boundary condition where the boundary value does not vary during the solution of the equations, one simply instructs the translator that the array elements which correspond to nodes on that boundary are known. The value of these array elements must also be initialized in the specification file prior to the instruction to solve the equations. The gate, source, drain and substrate contacts in the MOSFET provide examples of Dirichlet boundary conditions. The values of the electrostatic potential, electron concentration and hole concentration at the nodes on these boundaries are tagged as knowns and their values are initialized by procedures that are called from the simulation's main procedure. Other Dirichlet boundary conditions, where the value at the boundary must be calculated during the solution of the equations, are imposed using equation statements.

Zero flux Neumann (reflective) boundary conditions are also very easy to implement. In fact, these boundary conditions are implicit in the box integration method. Consider the diagram in Figure 6.23. Node i is on a zero-flux boundary and has four neighbors on one side of the boundary and none on the other side. To discretize a PDE which has the form of equation (5.12) at such a node, one constructs a box around the node and integrates both sides of the equation over the area of the box. The box is shaded in gray on the diagram.

As before, Green's theorem is used to convert the surface integral on the left side of the equation to a line integral. In order to compute this line integral, one must integrate the flux normal to the boundary along all six edges of the integration box. The integral of the normal flux along the segments numbered 1 to 4 (see Figure 6.23) is approximated by sum of the products of the fluxes along the respective edges that connect node i to its neighbors and the lengths of the corresponding integration segments. These products are then added together to approximate the line integral. Since the normal flux along segments 5 and 6 is zero, these segments do not contribute to the summation. Thus, the implementation of the finite-difference equation for nodes with zero-flux boundary conditions is identical in SGFramework to that for interior nodes. Zero-flux boundary conditions can be implemented with the 'nsum' function. The following SGFramework code would discretize $\nabla \cdot F\left(u_i, u_j, h_{i,j}\right) = S\left(v_i\right)$ on a zero-flux boundary.

```
nsum(i,j,all,F(u[i],u[node(i,j)],elen(i,j))*ilen(i,j)) =
S(v[i])*area(i);
```

To implement a non-zero-flux boundary condition on such a one-sided boundary, one cannot neglect the flux along the last two segments. Suppose the flux normal to

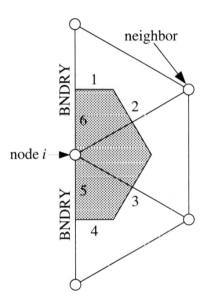

Figure 6.23. A typical boundary node with neighbors on only one side of the boundary.

these segments was F_{ext}. Assume further that these segments were labeled BNDRY, as shown in Figure 6.23. The following SGFramework code would discretize $\nabla \cdot F(u_i, u_j, h_{i,j}) = S(v_i)$ on this boundary.

```
nsum(i,j,all,F(u[i],u[node(i,j)],elen(i,j))*ilen(i,j) +
nsum(i,j,BNDRY,Fext*elen(i,j)/2) = S(v[i])*area(i);
```

The last type of boundary condition we will consider is at a physical boundary between different materials (one with nodes on both sides). As mentioned in the previous section, each node has its own connectivity structure and list of neighbor structures. Boundary nodes, however, have several connectivity structures and lists of neighbor structures. Consider node 2 of Figure 6.22a. It has three connectivity structures and lists of neighbor nodes. The first set of structures corresponds to an integration box that lies in both material A and B. It is represented by the second node connectivity structure and neighbor structure list. The second and third sets of structures correspond to the integration boxes that lie completely in material A and completely in material B, respectively. They are represented by the tenth and eleventh connectivity structures and neighbor structure lists. If node 2 were on the boundary of three or more regions, it would have additional sets of structures.

The user can choose which integration box to use by specifying its name in the label field of the 'nsum' function. For instance, nsum(2,j,all,*expr*) would sum *expr* over all of the second node's neighbors, nsum(2,j,MATERIALa,*expr*) would sum *expr* over all of the second node's neighbors which are in material A or on the boundary of material A, and nsum(2,j,MATERIALb,*expr*) would sum *expr* over all of the second node's neighbors which are in material B or on the boundary of material B. To efficiently access these structures, SGFramework orders the nodes and the labels such that the boundary nodes and region labels are indexed first. SGFramework can then use a simple formula to calculate which node connectivity structure and neighbor structure list to use.

We will now discuss how to discretize a PDE at a boundary between two materials. As an example we will consider Poisson's equation (Equation (4.28)) at the silicon/silicon dioxide boundary. The relative permittivities of silicon and silicon dioxide are $\epsilon_{Si} = 11.8$ and $\epsilon_{SiO2} = 3.9$, respectively. The permittivity of free space is incorporated in the constant L_0. We will assume that there is a surface charge σ on the boundary and that there is no free charge in the oxide. The following SGFramework code would implement equation 5 at this boundary.

```
equ V[i=SISIO2] ->
  L0*(nsum(i,j,SI,  eSi  *(V[i]-V[node(i,j)])/elen(i,j)*ilen(i,j)) +
      nsum(i,j,SIO2,eSiO2*(V[i]-V[node(i,j)])/elen(i,j)*ilen(i,j))) =
  (p[i]-n[i]+C[i])*area(i,SI) + nsum(i,j,SISIO2,sigma*elen(i,j)/2);
```

The first and second 'nsum' functions compute the integral of the electric displacement over the boundary of the integration box contained in the silicon and silicon dioxide regions, respectively. The source on the right hand side contains two parts. The first part is the free charge in the silicon and the second term is the surface charge on the boundary. The sum over SISIO2 picks up only nodes which on the boundary. The charge per unit length on the boundary is then multiplied by half the distance from each neighbor which is on the boundary, to give the total surface charge in the integration box. In the MOSFET simulation we have not modeled the surface charge, hence the source contains only the free charge in the silicon region.

In this section we have discussed the implementation of boundary conditions in SGFramework. Although we have focused on particular boundary conditions, it is possible to implement several other types. In the next section we will show the results of the MOSFET simulation.

6.9   SGFramework Build Script

As mentioned in the previous sections, the SGFramework generates C++ code which needs to be linked with an appropriate NAM. The SGFramework build script provides a simple method of building a simulation program. This script can construct a mesh refinement program from the code generated from the mesh program. (In addition, it provides an option to generate a simulation which may be run on Condor [115], a sophisticated multicomputer batch-processing system. However, this feature is not included in the build scripts generated by the setup program.)

6.10   SGFramework Visualization Tools

One of the most important components of modeling and simulation is the analysis and interpretation of the simulation results. SGFramework provides tools which allow users to visualize their results via surface and contour plots. Many of the surface plots in this book have been generated using these tools. If a simulation's results cannot adequately be displayed using the SGFramework visualization tools, a data extraction utility within the SGFramework can be used to generate data files which may be imported into more sophisticated data analysis and visualization programs.

6.11   Summary

In this chapter the entire SGFramework was described in considerable detail.

PROBLEMS

1. Set up a triangular mesh inside an elliptical region. Plot the mesh points and the connecting lines.

2. Using the mesh from the previous problem, solve Laplace's equation in the interior of the elliptical region, and plot the potential. Convert Laplace's equation to a form

where it is written in terms of the divergence of a gradient and use 'nsum' to implement the Laplacian. Let the potential on the exterior boundary of the ellipse be $10\,x^2$, where x is the distance along the major axis of the ellipse, measured from the left tip of the ellipse.

3. Divide the interior of the ellipse into two regions, and define labels for the regions, 'FREESPACE' and 'METAL'. Let the 'METAL' region be a line along the major axis of the ellipse, stretching from $x = 0.3L$ to $x = 0.7L$, where L is the length of the ellipse along that axis. Let the metal be at a potential of zero volts and the boundary of the ellipse be at the same potential as in the previous example. Find and plot the potential.

Part III

Semiconductor Devices

Chapter 7

PN Junction Diodes

The pn junction and diodes formed from it are discussed in this chapter. Junction electrostatics is the topic of Section 7.1. Steady-state (i, v) characteristics are handled in Section 7.2. Section 7.3 presents a review of the numerical model for a two-dimensional device. Section 7.4 is on small-signal analysis, and Section 7.5 is on transients. Finally, Section 7.6 is on photodiodes.

The goals of our presentation of pn junction diodes as well as other devices are

1. to establish as clearly as possible the basis for the mathematical description of the device;

2. to show how the device can be modeled numerically; and

3. to compare the analytic description of the device with the results of simulations and to critically evaluate both of them in the light of the comparison. (Provided that we show how comparisons can be done methodically, we can defer much of the work of comparison to the problems; this is necessary in order to save space.)

The combination of analytic and numerical approaches is very useful for understanding the device behavior. Both will be used side by side in this chapter. The analytic and numerical methods which are used here have been introduced already, but they will be reviewed as they are needed. This chapter is the first which uses the tools we have been developing in a real device analysis, so the review given here is also an introduction to what follows in the remaining chapters.

While we attempt to be very clear and to display all the steps in the analysis (as far as we are able) and to explain all the assumptions, this is not an introductory presentation of semiconductor behavior (although it may be considered a simple introduction to modeling of such devices). It is assumed that the structure of the device and its basic mode of operation are familiar. The basics of device operation are explained here, but a gentler introduction could be found elsewhere [116], [117].

The chapter begins with the electrostatic behavior of a one-dimensional semiconductor, with emphasis as usual on the proper boundary conditions, and reviews several simple models of pn junctions. The (i, v) characteristics of such a junction at low and high currents, under different conditions including steady state, sinusoidally varying and transient behavior, and breakdown are outlined. Finally photodiodes are introduced. Numerical models are brought in as illustrative examples throughout this discussion.

7.1   Electrostatic Description of a Simple Junction

The theory of semiconductor devices which is developed in introductory texts [116], [117], [118], [119], [120], makes extensive use of Poisson's equation. Poisson's equation describes the electrostatic potential V due to a charge density. The equilibrium expressions for the carrier number densities n and p are accurate enough close to equilibrium, when the electron and hole currents are small compared to the magnitude they are capable of having, to be used to specify n and p in terms of V. (Even though minority-carrier currents may be small in an absolute sense, the minority carriers are frequently not in equilibrium with the fields.)   When these expressions are used in the charge density, which in turn is used in Poisson's equation, a nonlinear equation for V is obtained. If this equation

$$
\begin{aligned}
\nabla^2 V &= -\frac{e}{\epsilon}(N_D - N_A + p - n) \\
&= -\frac{e}{\epsilon}\left(N_D - N_A + N_V \, \exp\left(\frac{E_V - E_F}{k_B T}\right)\right. \\
&\qquad \left. - N_C \, \exp\left(\frac{E_F - E_C}{k_B T}\right)\right)
\end{aligned}
\tag{7.1}
$$

is solved for V, then n and p can be found subsequently. (The energies at the band edges, E_C and E_V, vary linearly with V.) Since n and p were specified by equilibrium expressions, the net current carried by each species will be zero.

This equation is frequently solved approximately, analytically in textbook treatments, usually in one dimension and in many different circumstances, to show how different devices operate. It can be extended slightly to allow steady-state situations to be examined, but the analytic treatments are limited in their scope. The software employed here allows this equation to be solved in the same straightforward way for the analytically soluble cases and also for more general cases, which are not analytically soluble. (It also allows nonequilibrium densities to be calculated, using the continuity equation.) This does not mean that the analytic solutions are not useful. Numerical modeling is most successful when it provides enough insight into the system being modeled that it allows appropriate analytic models of the system to be developed. With this in mind, the usual one-dimensional examples will be modeled first, before modeling diodes in higher dimensions. However, before any modeling is attempted, the corresponding analytic treatments will be developed. The assumptions used in the analysis will be supported by the numerical results.

Two important one-dimensional cases which will be examined in detail are: the 1D pn junction; and the 1D metal-oxide-semiconductor (MOS) structure. The MOS structure will be studied in Chapter 10. The modeling of the pn junction and its analytic analysis will be handled in the next few sections.

7.1.1 One-Dimensional Semiconductor Device Electrostatics

In one dimension x the Poisson's equation inside a semiconductor in equilibrium is

$$\frac{d^2V}{dx^2} = -\frac{e}{\epsilon}\left(N_D - N_A + n_i\left[\exp\left(\frac{E_{Fi} - E_F}{k_BT}\right) - \exp\left(\frac{E_F - E_{Fi}}{k_BT}\right)\right]\right). \quad (7.2)$$

The Intrinsic Fermi level is $E_{Fi} = E_{Fi0} - eV$, where E_{Fi0} is the value of the Intrinsic Fermi level at the reference point where $V = 0$. (The carrier densities are the same but are written slightly differently here compared to eq. 7.1. There is a factor n_i outside the exponentials now, and $\pm(E_{Fi} - E_F)$ appears in the exponent.) This equation is written in finite-difference (FD) form by replacing $\frac{d^2V}{dx^2}$ with its FD equivalent.

$$\frac{d^2V}{dx^2} \simeq \frac{(V[i+1] - 2V[i] + V[i-1])}{(\Delta x)^2}. \quad (7.3)$$

The FD version is implemented in the file pn01.sg.

In an insulator the charge density is usually taken to be zero so the right-hand side of the above is zero and Poisson's equation becomes Laplace's equation. One of the electrostatic boundary conditions (see Section 2.10) which is needed between different materials states that the 'jump' in the normal component of the field $D = \epsilon E$ is equal to the free charge per unit area ρ_s on the surface.

$$D_{n2} - D_{n1} = \rho_s. \quad (7.4)$$

Using $D = \epsilon E = -\epsilon\frac{dV}{dx}$ in FD form, this gives an equation to be evaluated at the boundary between regions 1 and 2. The derivatives are found on the appropriate (opposite) sides of the boundary. If x is in the direction normal to the boundary,

$$-\epsilon_2\frac{dV}{dx}_{region2} + \epsilon_1\frac{dV}{dx}_{region1} = \rho_s. \quad (7.5)$$

This boundary condition is needed in the case of the MOS structure — see mos1.sg in Chapter 10, on MOS structures. For metallic regions the potential V is set to a constant value. Semiconductors resemble metals with a finite shielding length - the electrostatic field drops to zero with distance from a junction or contact. This behavior can be used as a boundary condition in the models of both the pn junction and the MOS structure.

The FD Poisson's equation cannot usually be solved by a straightforward method which employs integration from one side of the semiconductor to the other. This approach tends to lead to an exponential divergence of V as the integration proceeds. Instead the values of V at all the mesh points are found simultaneously for all values of x on the mesh. This 'implicit' solution of the discrete equations is found automatically when suitable input files are fed into SGFramework.

7.1.2   PN Junctions — Analytic Treatment

To solve Poisson's equation analytically and obtain an explicit expression for V the right-hand side of the equation must be simplified. In fact, it is often possible to divide the interior of the semiconductor into two types of region to obtain a major simplification in Poisson's equation. In one type of region (such as near a junction) we assume there are no carriers. In the other type of region, the net carrier density approximately equals the net dopant density, so that for some purposes the net charge density can be considered to be zero in this 'quasi-neutral' type of region. In the case of a pn junction the regions far from the junction are taken to be quasi-neutral, but immediately on either side of the junction the carriers are absent, so the charge density is that of the ionized impurities. It is important to realize that in reality 'quasi-neutral' regions do have a finite charge density. It is not safe to neglect that charge in general. It is permissible to do so in what follows, because we are focusing on the junction fields, not the weaker fields further away.

Since the field goes to (roughly) zero far from the junction on either side, there can be no net charge at the junction. All the electric-field lines which must start close to the junction on one side of the junction must end close to it on the other side. As a result the charges in the junction region on opposite sides of the junction are themselves equal and opposite.

In a pn junction there is a transition from the doping being mostly n-type with $N_d > N_a$ to the doping being mostly p-type with $N_d < N_a$. The point where $N_d = N_a$ is called the metallurgical junction. The simplest assumption about the doping profile (used in the next section) is that it is constant on either side of the metallurgical junction, with a sudden jump at the junction. A linear variation in the net dopant density, which passes through zero at the metallurgical junction, will be treated second.

7.1.3   Step Junction

A step junction and a linear junction are set up in the file pn01.sg. The charge density is not forced to take any particular values in the setting up of the problem but the doping profile has a step or a linear variation. The profiles found from this input are shown in Figures 7.1, 7.2 and 7.3. To understand these results the analytic solutions will be obtained and compared to them. We now derive the analytic formulae for the step junction.

The first quantity which should be obtained is the electric field E. Since $E = -\frac{dV}{dx}$, Poisson's equation can be written in the above form or as

$$\frac{dE}{dx} = \frac{\rho}{\epsilon} \tag{7.6}$$

where $\rho = 0$ in the quasi-neutral regions, $\rho = eN_D$ on the n-type side of the junction and $\rho = -eN_A$ on the p-type side. (It has been assumed that all the impurities are ionized, here.) Moving out from the metallurgical junction, the electric field decreases in magnitude. The field must go to zero when ρ goes to zero, because (in this approximation) there is no more charge for field lines to end on beyond this point. Let $x = 0$ be at the metallurgical junction. The width of the n-type side is x_n and let this region be from $x = -x_n$ to $x = 0$. The p-type region will be of width x_p from $x = 0$ to $x = x_p$. Then on the n-type side the equation for E is

$$\frac{dE}{dx} = \frac{eN_D}{\epsilon} \tag{7.7}$$

which can be integrated to give

$$E = \frac{eN_D}{\epsilon} \int\limits_{x=-x_n}^{x} dx = \frac{eN_D}{\epsilon}(x + x_n) \tag{7.8}$$

for $-x_n < x \leq 0$. This was made to be zero at $x = -x_n$, since E must be zero at the lower limit. On the p-type side

$$\frac{dE}{dx} = -\frac{eN_A}{\epsilon} \tag{7.9}$$

so that

$$E = E(x = 0) - \frac{eN_A}{\epsilon} \int\limits_{x=0}^{x} dx = \frac{e}{\epsilon}(N_D x_n - N_A x) \tag{7.10}$$

for $0 < x \leq x_p$. These expressions for E give the same value at $x = 0$, where the two regions meet. The field must also go to zero at the upper edge of the junction region, at $x = x_p$. For this to happen $N_D x_n = N_A x_p$ which confirms that the total charge in the junction is zero.

Now that E is known, V can be found from $E = -\frac{dV}{dx}$ and using $V(x = -x_n) = 0$. For $-x_n < x \leq 0$,

$$E = \frac{eN_D}{\epsilon}(x + x_n) \tag{7.11}$$

and

$$V = V(x = -x_n) - \int\limits_{x=-x_n}^{x} E \, dx \tag{7.12}$$

so

$$\begin{aligned}
V &= -\frac{eN_D}{\epsilon} \int\limits_{x=-x_n}^{x} (x + x_n) \, dx \\
&= -\frac{eN_D}{\epsilon} \left[\frac{x^2}{2} + x x_n \right]_{-x_n}^{x} \\
&= -\frac{eN_D}{\epsilon} \left(\frac{x^2}{2} + x x_n - \frac{x_n^2}{2} + x_n^2 \right) = -\frac{eN_D}{2\epsilon}(x + x_n)^2 \tag{7.13}
\end{aligned}$$

which goes to zero at $x = -x_n$.

At $x = 0$ this is $V(x = 0) = -\frac{eN_D x_n^2}{2\epsilon}$, so for $0 < x \le x_p$ the potential obeys

$$V = V(x = 0) - \int_{x=0}^{x} E\, dx \qquad (7.14)$$

or

$$V = -\frac{eN_D x_n^2}{2\epsilon} - \frac{e}{\epsilon} \int_{x=0}^{x} (N_D x_n - N_A x)\, dx = -\frac{e}{2\epsilon}\left(N_D x_n^2 + 2N_D x_n x - N_A x^2\right).$$

$$(7.15)$$

At the upper limit $x = x_p$ and

$$V(x = x_p) = -\frac{e}{2\epsilon}(N_D x_n^2 + 2N_D x_n x_p - N_A x_p^2). \qquad (7.16)$$

Since $N_D x_n = N_A x_p$ for overall charge neutrality, this can be written as

$$V(x = x_p) = -\frac{e}{2\epsilon}(N_D x_n^2 + N_A x_p^2) = -\frac{e}{2\epsilon} N_A x_p (x_n + x_p). \qquad (7.17)$$

Now $x_n = N_A x_p / N_D$ so $x_n + x_p = \frac{N_D + N_A}{N_D} x_p$, $x_p = (x_n + x_p)\frac{N_D}{N_D + N_A}$, and

$$V(x = x_p) = -\frac{e}{2\epsilon} \frac{N_A N_D}{N_D + N_A}(x_n + x_p)^2. \qquad (7.18)$$

7.1.4   The Built-In Potential

To find the value of the voltage across the junction it is useful to start by asking why there is a nonzero potential at a pn junction. In equilibrium the potential is responsible for ensuring that there is no net current flowing of either type of carrier. This means that the potential must repel the majority carriers on both sides of the junction, if it is going to stop their diffusion from pushing them across the junction. This in turn implies that the electrostatic-potential barrier multiplied by the electronic charge must be several times the mean carrier energy to stop most of the carriers, so the variation in V must be several times $\frac{k_B T_e}{e}$. The electric field which is needed to repel the majority carriers on both sides must push electrons back into the n-type region, and it must push holes back into the p-type region. The electric field can do this if its direction is from the n-side to the p-side, since the direction of the electric field is the direction a positive charge is pushed by the field. The ionized dopants on the n-side by themselves have a positive charge which field lines can start on. Similarly the ionized dopants on the p-side have a negative charge which field lines can end on. Consequently, when carriers are removed from the junction region, the dopants which are left behind have the right charge to set up the field which repels the carriers.

To find the magnitude of the potential in equilibrium, the electrons are assumed to obey the Boltzmann relation

$$n = n_0 \exp\left(\frac{e(V - V_0)}{k_B T_e}\right). \tag{7.19}$$

The electron density in the quasi-neutral n-type region is $n(x = x_n) \simeq N_D$. In the quasi-neutral p-type region the hole density is $p(x = x_p) \simeq N_A$, but $np = n_i^2$ so $n(x = x_p) = n_i^2/N_A$. The ratio of these two electron densities is

$$n(x = x_n)/n(x = x_p) = N_D N_A/n_i^2.$$

From the Boltzmann relation,

$$V - V_0 = \frac{k_B T_e}{e} \ln\left(\frac{n}{n_0}\right) \tag{7.20}$$

so the voltage between two points can be found from the ratio of the electron densities at those points. Across the junction the voltage is found from the ratio of the electron densities on either side of the junction, which is

$$V_{bi} = \frac{k_B T_e}{e} \ln\left(\frac{N_D N_A}{n_i^2}\right). \tag{7.21}$$

7.1.5 Linearly Graded Junction

We consider now a junction with a linear variation in the net doping, $N_D - N_A = -ax$, which will again have the n-type side on the left and the p-type side on the right. The magnitude of the dopant density is symmetric about $x = 0$. $x = 0$ is at the metallurgical junction where the dopant density is zero. This problem is set up in pn01.sg, and the results are plotted in the accompanying figures. Poisson's equation in the junction region is

$$\frac{dE}{dx} = \frac{\rho}{\epsilon} = \frac{e}{\epsilon}(N_D - N_A) = -\frac{eax}{\epsilon} \tag{7.22}$$

so

$$E = -\int_{x=-x_n}^{x} \frac{eax}{\epsilon} dx = -\left[\frac{eax^2}{2\epsilon}\right]_{-x_n}^{x} = \frac{ea}{2\epsilon}[x_n^2 - x^2]. \tag{7.23}$$

Symmetry means $x_n = x_p = W/2$, where W is the junction width, and E must go to zero at $x = \pm W/2$.

Integrating again with $V = 0$ at $x = -x_n$ gives

$$\begin{aligned}
V &= -\int_{x=-x_n}^{x} E \, dx = -\frac{ea}{2\epsilon}\left[x_n^2 x - \frac{x^3}{3}\right]_{-x_n}^{x} \\
&= -\frac{ea}{2\epsilon}\left[x_n^2 x - \frac{x^3}{3} + x_n^3 - \frac{x_n^3}{3}\right] = -\frac{ea}{2\epsilon}\left[x_n^2 x - \frac{x^3}{3} + \frac{2x_n^3}{3}\right]. \tag{7.24}
\end{aligned}$$

At $x = x_p$ (and since $x_p = x_n$)

$$V(x = x_p) = -\frac{ea}{2\epsilon}\frac{4x_p^3}{3} = -\frac{eaW^3}{12\epsilon}. \tag{7.25}$$

This is the junction's built-in potential.

7.1.6   Numerical Modeling of the PN Junction

The model set up in the file pn01.sg employs the nonlinear Poisson's equation. The other major points to notice about the problems which are set up in these files are related to the boundary conditions. The quantity being calculated is the potential V. One boundary condition that is usually used in solving Poisson's equation involves the zero of potential. The zero might be chosen to be on one side of the junction, so that for instance $V(x < -x_n) = 0$. The potential at the other side would then be set to the voltage dropped across the junction, $V(x > x_p) = V_{junct}$. We have another important condition which must also be considered, however. The electric field should go to zero a little way past $x = -x_n$ and $x = x_p$. The semiconductor shields itself from electric fields by allowing charge to move around. The shielding means that electric fields decay exponentially with distance into the semiconductor. Setting the field to zero on both sides provides two boundary conditions, which is the total number available to be imposed when solving a second-order differential equation. The fact that no electric-field lines can leave the junction region also means that the junction must be overall charge neutral.

On the other hand, saying that there is no field leaving implies only that the junction is overall charge neutral. The voltage across the junction would not be determined and the value would be left to float. We need to be able to impose the value of the voltage across the junction. In reality a low value of the electric field at the edges should emerge naturally from the simulation in order to keep the overall voltage reasonably small. If the field is nonzero only near the junction, then the potential dropped is moderate in size. If the field were substantial everywhere, the potential dropped would also tend to be very large. Other undesirable consequences would also result from having a large field everywhere, but having a very large potential would immediately violate the boundary condition which will be imposed on the potential.

The zero of potential is important in relation to the equilibrium expressions for the carrier densities, which are

$$n = N_C \exp\left(\frac{E_F - E_C}{k_B T}\right) \tag{7.26}$$

and

$$p = N_V \exp\left(\frac{E_V - E_F}{k_B T}\right). \tag{7.27}$$

The energies of the band edges, E_C and E_V, are measured relative to points far from the junction where that carrier's density is known. For the electrons it is useful to express all the quantities relative to the n-type bulk region at $x < -x_n$. The density in this region is

$$n(x < -x_n) = N_C \exp\left(\frac{E_F - E_C(x < -x_n)}{k_B T}\right) \qquad (7.28)$$

but

$$E_C(x) = E_C(x < -x_n) - e(V(x) - V(x < -x_n)) \qquad (7.29)$$

so the density of electrons at any point is

$$n(x) = N_C \exp\left(\frac{E_F - E_C(x < -x_n) + e(V - V(x < -x_n))}{k_B T}\right) \qquad (7.30)$$

which can be rearranged to give

$$n(x) = n(x < -x_n) \exp\left(\frac{e(V - V(x < -x_n))}{k_B T}\right). \qquad (7.31)$$

Similarly, the density of holes at any point is

$$p(x) = p(x > x_p) \exp\left(\frac{e(V(x > x_p) - V)}{k_B T}\right) \qquad (7.32)$$

The input file pn01.sg which follows uses expressions for the densities similar to these, and an imposed voltage across the junction.

Input File pn01.sg

```
// FILE:   sgdir\ch07\pn01\pn01.sg
// Demonstration of solution of 1D nonlinear Poisson's equation applied to
// an abrupt pn junction and a linearly graded junction.  To switch between
// the abrupt and linearly graded junctions, it will be necessary to comment
// and uncomment some of the lines in this file.  The lines ending with [A]
// should be uncommented for the abrupt junction and the lines ending with
// [L] should be uncommented for the linearly graded junction.


// mesh constants
const NX  = 101;                 // number of x-mesh points
const LX  = NX - 1;              // index of last x-mesh point
const DX  = 3.0e-6;              // mesh spacing (cm)

// physical constants
const T   = 300.0;               // operating temperature
const e   = 1.602e-19;           // electron charge (C)
const kb  = 1.381e-23;           // Boltzmann's constant (J/K)
const e0  = 8.854e-14;           // permittivity of vacuum (F/cm)
const eSi = 11.8;                // relative permittivity of Si
const eps = eSi*e0;              // permittivity of Si (F/cm)
```

```
const Ni  = 1.25e10;                 // intrinsic concentration of Si (cm^-3)
const Vt  = kb*T/e;                   // thermal voltage

// doping profile constants
// note:  the lines ended by [*] are only used for the linearly graded case
const Na  = 1.0e+15;                  // acceptor concentration on p-side (cm^-3)
const Nd  = 1.0e+15;                  // donor concentration on n-side (cm^-3)
const Vbi = Vt*ln(Na*Nd/sq(Ni));      // built-in potential
const WJ  = 1.0e-4;                   // linearly graded junction width (cm)     [*]
const XC  = LX/2*DX;                  // center of graded junction               [*]
const XL  = XC - WJ/2.0;              // start of graded junction                [*]
const XR  = XC + WJ/2.0;              // end of graded junction                  [*]


// set numerical algorithm parameters
set NEWTON DAMPING     = 3;
set NEWTON ACCURACY    = 1.0e-12;
set NEWTON ITERATIONS  = 100;
set LINSOL ALGORITHM   = GAUSSELIM;
set LINSOL FILL        = INFINITY;


// declare variables
var x[NX], C[NX], V[NX];


// Poisson's equation
equ V[i=1..LX-1] ->
  {V[i-1] - 2*V[i] + V[i+1]}/sq(DX) =
  -e/eps*{Na*exp((V[LX]-V[i])/Vt) - Nd*exp((V[i]-V[0])/Vt) + C[i]};


begin SetUpDopingProfile
  assign x[i=all] = i*DX;            // initialize mesh points
  assign C[i=0..LX/2]    = +Nd;      // ionized donor concentration             [A]
  assign C[i=LX/2+1..LX] = -Na;      // ionized acceptor concentration          [A]
//assign C[i=all] =                  // dopant profile                          [L]
//  (x[i]<=XL)           * {+Nd} +                                            // [L]
//  (x[i]>XL)*(x[i]<XR)  * {Nd - (Na+Nd)/(XR-XL)*(x[i]-XL)} +                 // [L]
//  (x[i]>=XR)           * {-Na};                                             // [L]
end

begin SetUpBoundaryConditions
  assign V[i=0..LX/2] = 0.0;
  assign V[i=LX/2+1..LX] = -Vbi;
end

begin main
```

```
   call SetUpDopingProfile;
   call SetUpBoundaryConditions;
   solve;
   write;
end
```

Interpretation of the Simulation Results

The expressions for the densities of electrons and holes should be examined closely to make sure the signs are correct and the reference potentials are right, so that the densities equal the doping density at the appropriate end of the junction.

The results for the step junction are plotted first, in Figures 7.1 and 7.2. The theoretical curve is shown as a dotted line and the numerical curve as a solid line. Figure 7.1 shows a symmetric junction. The equivalent results for a highly asymmetric abrupt junction follow in Figure 7.2. Then plots of the potential for several different doping profiles in a linear junction are shown in Figure 7.3.

The plot for the linear junction looks very similar to its equivalent for the abrupt junction. The lower doping density near the junction leads to a wider junction region, and the width of the junction in each of these plots similarly varies because of the different slopes of the doping in the linear junction.

See also the exercises at the end of this chapter.

It may be useful to see the matlab scripts used to make these plots, especially since the analytic results used for comparison are given in the matlab script.

Script File: Abrupt Junction Plots

```
% this matlab script plots the
% 1.   electrostatic-potential profile for the device modeled in the pn01.sg
%      file (solid line).
% 2.   depletion-approximation theoretical profile for an abrupt junction
%      (dashed line)
% 3.   depletion-approximation theoretical junction width

% load the data files, this will create vectors x, V and C which contain the
% position, electrostatic potential and doping concentration at each mesh point
load 'x.out';
load 'V.out';
load 'C.out';

% define variables needed to compute depletion-approximation profile ... the
% following variables should be kept in sync with the constants defined in the
% simulation file
Nd  = +max(C);
Na  = -min(C);
T   = 300.0;
e   = 1.602e-19;
```

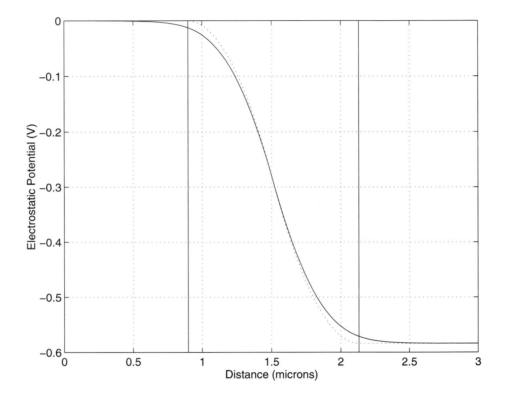

Figure 7.1. The profile of the electrostatic potential for a symmetric abrupt junction. The numerical result is the solid line, and the dotted line is the analytic result. The solid vertical lines are the theoretical positions of the junction edges. The metallurgical junction is at 1.5 μm.

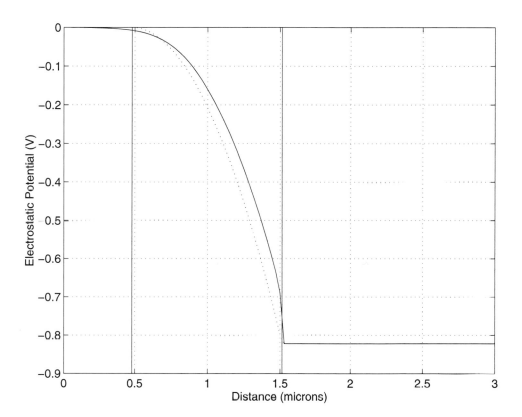

Figure 7.2. The profile of the electrostatic potential for an asymmetric abrupt junction. The solid vertical lines are the theoretical junction edges. The metallurgical junction is at 1.5 μm.

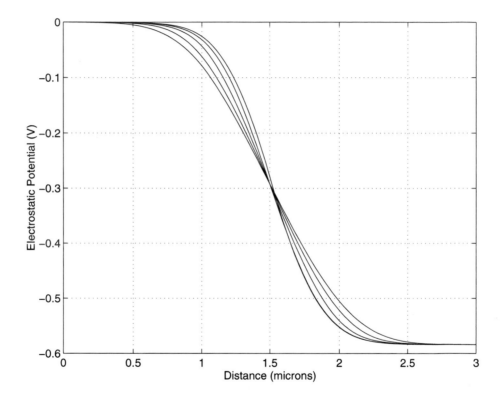

Figure 7.3. The electrostatic potential in linear junctions with different doping grade. The higher the doping level, the narrower the junction.

```
kb  = 1.381e-23;
e0  = 8.854e-14;
eSi = 11.8;
eps = eSi*e0;
Ni  = 1.25e10;
Vt  = kb*T/e;
Vbi = Vt*log(Na*Nd/Ni/Ni);
xn  = sqrt(2*eps*Vbi/e*Na/Nd/(Na+Nd));
xp  = Nd/Na*xn;

% find the location of the metal
xm = 0.0;
for i = 2:length(V),
  if xm == 0.0
    if C(i-1) * C(i) < 0
      xm = (x(i-1) + x(i)) / 2.0;
    end
  end
end

% compute the depletion-approximation profile
for i = 1:length(V),
  if x(i) <= xm - xn
    Vtheory(i) = 0.0;
  elseif x(i)  <= xm
    Vtheory(i) = -e*Nd/2/eps*(x(i)-xm+xn)^2;
  elseif x(i) <= xm + xp
    Vtheory(i) = -e/2/eps*(Nd*xn*xn - Na*((x(i)-xm)^2 - 2*(x(i)-xm)*xp));
  else
    Vtheory(i) = -Vbi;
  end
end

% plot the results ...
figure;
plot(x*1e4,V);
grid on;
xlabel('Distance (microns)');
ylabel('Electrostatic Potential (V)');
hold on;
plot(x*1e4, Vtheory, ':');
ymin = floor(10*min(V))/10;
plot([xm-xn,xm-xn]*1e4, [0.0,ymin]);
plot([xm+xp,xm+xp]*1e4, [0.0,ymin]);
hold off;
```

The file for the linear junction follows.

Script File: Linear Junction Plots

```
% this matlab script plots the electrostatic-potential profile for the device
% modeled in the pn01.sg file for an abrupt junction and four linearly graded
% junctions whose widths are 0.5, 1.0, 1.5 and 2 microns, respectively.

% load the data files and the position and electrostatic-potential profiles
load 'x.out';
load 'Vlg0.out';
load 'Vlg1.out';
load 'Vlg2.out';
load 'Vlg3.out';
load 'Vlg4.out';

% plot the results ...
figure;
plot(x*1e4,Vlg0);
hold on;
plot(x*1e4,Vlg1);
plot(x*1e4,Vlg2);
plot(x*1e4,Vlg3);
plot(x*1e4,Vlg4);
grid on;
xlabel('Distance (microns)');
ylabel('Electrostatic Potential (V)');
hold off;
```

Finally, we will briefly examine the role of n_i, and of the boundary conditions on the density, in determining the built-in potential. At $x = -x_n$ we impose charge neutrality,

$$n(x = -x_n) = N_D + p(x = -x_n). \tag{7.33}$$

In equilibrium, $np = n_i^2$ so $n(x = -x_n) = N_D + n_i^2/n(x = -x_n)$. This quadratic equation for $n(x = -x_n)$ has the solution

$$n(x = -x_n) = \left(N_D + \sqrt{N_D^2 + 4n_i^2} \right)/2. \tag{7.34}$$

Similarly

$$p(x = x_p) = \left(N_A + \sqrt{N_A^2 + 4n_i^2} \right)/2. \tag{7.35}$$

To find the equilibrium voltage across the junction we could try to impose these values for the densities at the junction edges. What would happen if these expressions were used to impose boundary values on the densities but n_i were neglected in these expressions? The expression for the built-in potential V_{bi} provides a clue. If n_i were zero, then charge neutrality would require, for instance, $n(x = -x_n) = N_D$, which in turn would mean $p(x = -x_n) = 0$. What potential is needed to make p go to zero? [If you change the input file to try this, make n_i small but not zero!]

7.2 (i, v) **Characteristics**

7.2.1 (i, v) **Characteristics for Low Currents and Voltages**

In this section we shall consider the currents, which are produced in a pn junction for low voltages. This case is relatively easy to describe analytically. A later section treats the more general case.

The simulation in the next section follows the analytic treatment to some extent, in that it uses the same ('near-equilibrium') method to find the electric field. That is, in the equation for the electric field it uses the exponential expressions for the densities that are valid close to equilibrium. However, once the electric field has been found, it uses that electric field and the equation of particle conservation to find the carrier densities and fluxes more accurately than the exponential epressions would allow.

This simulation is intended to emphasize the physics of the junction at low voltages, by conforming to the analytic method of analysis. It also provides a simple introduction to the Scharfetter-Gummel formulation of the equations for the carrier densities. In this simple case, we do not have to calculate the electric fields at the same time that the densities are found. The possibility that the electric field could take on any value, including zero, is a complication which we avoid in this simple simulation. The general form of the Scharfetter-Gummel discretization is used in Section 7.2.2 which treats 'arbitrary' voltages.

To find the current flowing in the pn junction we must consider nonequilibrium conditions. The Fermi level is not defined except at equilibrium. The behavior of the quasi-Fermi levels, which are the generalization of the Fermi level to nonequilibrium situations, is an important guide to understanding the carrier behavior. The quasi-Fermi levels are defined to have the same relationship to the densities as the Fermi level itself, with the relation still holding in nonequilibrium cases.

The quasi-Fermi levels of the majority carriers are pinned by the dopants at their usual positions with respect to the bands. Far from the junction the carriers are not much affected by the junction or the current flowing through it. The quasi-Fermi levels of the minority carriers far from the junction will also have their usual values relative to the bands, because in these regions the carriers are nearly in equilibrium.

In the junction, as elsewhere, we assume the currents are small. The current densities of electrons and holes vary as $-n\frac{\partial E_{Fn}}{\partial x}$ and $-p\frac{\partial E_{Fp}}{\partial x}$, where E_{Fn} is the electron quasi-Fermi level and E_{Fp} is the hole quasi-Fermi level. To keep the current small the quasi-Fermi levels of each species in the junction have to be flat between the junction and the side of the junction where that species is the majority carrier. In other words, the quasi-Fermi level for electrons in the junction is equal to the electron quasi-Fermi level in the n-type region. The hole quasi-Fermi level in the junction is equal to the hole quasi-Fermi level in the p-type region. If the quasi-Fermi levels were not flat in these regions, large currents would flow. This is not true when the carrier is a minority carrier, since even if the quasi-Fermi level varies, there will be too few carriers to give rise to a large current.

The quasi-Fermi levels are defined to obey

$$n = n_i \exp\left(\frac{E_{Fn} - E_i}{k_B T}\right) \tag{7.36}$$

and in the case of the holes

$$p = p_i \exp\left(\frac{E_i - E_{Fp}}{k_B T}\right) \tag{7.37}$$

so, multiplying these,

$$np = n_i^2 \exp\left(\frac{E_{Fn} - E_{Fp}}{k_B T}\right). \tag{7.38}$$

An applied voltage V_A causes the splitting of the levels, $eV_A = E_{Fn} - E_{Fp}$ and so

$$np = n_i^2 \exp\left(\frac{eV_A}{k_B T}\right) \tag{7.39}$$

in the junction although $np = n_i^2$ in the bulk. The hole density on the p-type side of the junction at $x = x_p$ is $p = N_A$, so

$$n(x_p) = \frac{n_i^2}{N_A} \exp\left(\frac{eV_A}{k_B T}\right). \tag{7.40}$$

Similarly on the n-type side at $x = -x_n$, the electron density is $n = N_D$ and

$$p(-x_n) = \frac{n_i^2}{N_D} \exp\left(\frac{eV_A}{k_B T}\right). \tag{7.41}$$

In equilibrium, each of these densities is given by the same expression but, with V_A set to zero. The excess density in each case is therefore given by the same expression but the exponential has one subtracted from it. The excess density of electrons is $\Delta n = n - n_{eq}$ and at $x = x_p$ is

$$\Delta n(x_p) = \frac{n_i^2}{N_A} \left[\exp\left(\frac{eV_A}{k_B T}\right) - 1\right]. \tag{7.42}$$

Similarly the excess hole density at $x = -x_n$ is

$$\Delta p(-x_n) = \frac{n_i^2}{N_D} \left[\exp\left(\frac{eV_A}{k_B T}\right) - 1\right]. \tag{7.43}$$

In the bulk regions where there is little electric field, the excess minority carriers diffuse into the junction and are not repelled by the electric field there. The diffusion-equation for the excess electrons in a p-type region is

$$-D_n \frac{\partial^2 \Delta n}{\partial x^2} = -\frac{\Delta n}{\tau_n}. \tag{7.44}$$

The negative source term is the excess density divided by τ_n, which is the time for excess electrons to recombine. The solution of this equation is

$$\Delta n = A \exp\left(-\frac{x}{L_n}\right) + B \exp\left(\frac{x}{L_n}\right) \tag{7.45}$$

where the diffusion length L_n is given by $L_n = \sqrt{D_n \tau_n}$. This diffusion-equation neglects mobility effects on the flux of excess carriers.

Suppose the excess electron density is to be calculated on the p-type side of the junction from the junction edge at $x = 0$ into the bulk at large x. The diffusion-equation solution given above needs boundary conditions. The boundary condition on Δn far into the bulk at large x is just that $\Delta n(bulk) = 0$, since the bulk is nearly in equilibrium. For this to happen we must have $B = 0$. The other boundary condition is that at the edge of the junction, which is where $x = 0$, the density should be $\frac{n_i^2}{N_A}(\exp(\frac{eV_A}{k_BT}) - 1)$ and, since the solution of the diffusion-equation at this point is just A now that B has been set to zero, A must be equal to this. The solution for $x \geq 0$ is

$$\Delta n(x \geq 0) = \frac{n_i^2}{N_A}\left[\exp\left(\frac{eV_A}{k_BT}\right) - 1\right]\exp\left(-\frac{x}{L_n}\right). \tag{7.46}$$

The electron current density is

$$J_n(x) = -e\Gamma_n \tag{7.47}$$

where Γ_n is the electron flux per unit area, and, since we are neglecting the contribution of mobility to this flux, it is entirely due to diffusion and

$$\Gamma_n = -D_n\frac{\partial \Delta n}{\partial x}. \tag{7.48}$$

Differentiating Δn with respect to x gives

$$\frac{\partial \Delta n}{\partial x} = -\frac{\Delta n}{L_n} \tag{7.49}$$

so the electron current density is

$$J_n = -eD_n\frac{\Delta n}{L_n}. \tag{7.50}$$

When $V_A > 0$ and so $\Delta n > 0$, the flux of electrons is out of the junction, which means the flux is in the positive x-direction, so the current is in the negative x-direction. The equivalent expression for the hole flux also shows that the flux is out of the junction when $\Delta p > 0$. The holes are the minority carrier on the n-type side, which is located beyond the junction on the negative-x side of the junction. The hole flux and current are in the same direction as each other, going to negative x when $V_A > 0$, so the total current density is

$$J = J_n + J_p$$

$$= -eD_n \left[\frac{\Delta n}{L_n}\right]_{JE} - eD_p \left[\frac{\Delta p}{L_p}\right]_{JE}. \tag{7.51}$$

The total current is $I = -JA$, with the minus sign to obtain the current in the negative x-direction, from the p-type region to the n-type region. The junction edge value of Δn is found by setting $x = 0$ as described above. Using this and the equivalent expression for Δp at the other edge of the junction, the current is found to be

$$I = eAn_i^2 \left(\frac{D_n}{L_n N_A} + \frac{D_p}{L_p N_D}\right) \left[\exp\left(\frac{eV_A}{k_B T}\right) - 1\right]. \tag{7.52}$$

Recombination in the junction is neglected, so the two components of the current should not change on their way across the junction.

(i, v) Characteristics for Low Currents and Voltages - Simplified Numerical Approach

An input file is given next, which uses a continuity equation based on a version of the Scharfetter-Gummel method. This is not the SG method used in most of this text, but it is a little easier to understand than the general SG method. It was derived in the examples at the end of Chapter 5 on numerical solution of PDEs. The densities in the Poisson's equation used in this file are given by (exponential) Boltzmann-type expressions. The electric field found from this equation is then used to calculate a more accurate density, from our simplified form of the Scharfetter-Gummel treatment of the continuity equation. [Compare the physical assumptions which went into this file with those used above. Some simplifications have been made, for instance in the generation rate.]

Input File pniv2.sg

```
// FILE:  sgdir\ch07\pniv2\pniv2.sg
// demonstration of solution of 1D nonlinear Poisson's equation
// plus SG continuity eqn. plus particle generation - roughly correct
// constants!
// - applied to a linear pn junction.

const NX = 101;                 // number of x-mesh points
const LX = NX - 1;              // index of last x-mesh point
const DX = 1.0e-7;              // mesh spacing
const VR = 0.7;                 // voltage at the rhs
const VT = 0.03;                // thermal voltage
const N0 = 1.0e+20;            // doping density
const ec = 1.6e-19;            // electronic charge
```

```
const eps0=8.8e-12;             // permittivity of vacuum;
const epsi=11.8;                // relative permittivity of Si
const eoeps = ec/eps0/epsi;     // e/epsilon
const mu  = 0.1 ;               // magnitude of electron mobility
const D   = VT * mu ;           // diffusion coefficient
const ni = 1.0e16;              // intrinsic density in MKS
const taun = 1.0e-6 ;           // electron lifetime
const taup = taun   ;           // hole lifetime

var V[NX],NA[NX],n[NX],E[NX],EP[NX],gamma[NX],x[NX];

set NEWTON ACCURACY = 1.0e-8 ;

// Poisson's equation
equ V[i=1..LX-1]  ->
  {V[i-1] - 2*V[i] + V[i+1]}/sq(DX) =
  eoeps*(NA[i] + N0*(exp((V[i]-VR)/VT) - exp(-V[i]/VT)));

// the following equations could be implemented as functions
equ E[i=0..LX-1]  -> E[i] = -{V[i+1] - V[i]}/DX;
equ EP[i=0..LX-1] -> EP[i] = exp(-E[i]*DX/VT);

// particle conservation equation
equ n[i=1..LX-1]  ->
 -mu/DX * [{EP[i]   *n[i]   - n[i+1]}*E[i]  /(EP[i]  -1.0) -
          {EP[i-1] *n[i-1] - n[i]   }*E[i-1]/(EP[i-1]-1.0)] =
 (sq(ni) - n[i]* N0*exp(-V[i]/VT))/(taup*(n[i]+ni) +
          taun*(N0*exp(-V[i]/VT) + ni));

begin main
  assign x[i=all] = i*DX ;                  // coordinate array
  assign NA[i=0..LX] = N0*(1.0 - 2.0*i/LX); // acceptors on left.
  assign V[i=0] = 0.0;
  assign V[i=LX] = VR;
  assign n[i=LX/2..LX] = N0;
  assign n[i=0] = ni*ni/N0;                 // eqbm. density
  assign E[i=all] = 1.0;
  assign EP[i=all]=exp(-E[i]*DX/VT);
  solve;
  assign gamma[i=0..LX-1]=-mu*E[i]*(EP[i]*n[i]-n[i+1])/(EP[i]-1.0);
  assign gamma[i=LX] = gamma[i-1];
end
```

Interpretation of the Simulation Results

The results of this run are shown in Figure 7.4, for the electron flux. The flux changes rapidly in the junction region. Since the flux is the integral of the particle production rate, this shows that the production is mainly in the junction. In equilibrium the net production rate (the production rate minus the recombination rate) is zero. The applied voltage in this run was close to the equilibrium potential across the junction, so the regions far from the junction should be close to having their equilibrium densities and hence zero net production.

When V_R is varied from about 0.5 to 1.0 (all of which are reverse-bias or only slightly forward-bias voltages for this junction), the electron flux from the p-side to the n-side shows the expected behavior. Positive values of this flux correspond to a reverse current. The flux changes sign at V_R equal to about 0.55 volts, which is low due to the low doping density.

The built-in voltage is given by $V_{bi} = V^{th} \ln\left(N_A N_D / n_i^2\right)$, where V^{th} is the 'thermal voltage', $V^{th} = k_B T_e / e$. In this case this is roughly $0.03 \times \ln\left(\frac{10^{40}}{10^{32}}\right) \simeq 0.5$ volts.

When N_0 is increased by a factor of a hundred (and Δx decreased by a factor of ten), the change from positive to negative currents takes place at V_R slightly greater than 0.8 volts. In the expression for V_{bi} the argument of the logarithm increases by 10^4 so V_{bi} increases by about 0.24. The set of results obtained by running the input file, for this case, for various values of V_R and entering the results, correct to the first two significant figures, into Matlab is shown in Figure 7.5. More elegantly plotted (I, V) curves will be demonstrated in the next section.

The generation of carriers in this example is predominantly in the junction region, where the product $np \ll n_i^2$. The rest of the device modeled here is not long enough for much effect to be caused by the relatively small generation rate per unit volume which is found outside the junction region.

To estimate the peak generation rate in the junction we return to the set of parameters given in the original input file and set
$$G_{max} \simeq n_i^2 / [\tau_n(n + n_1) + \tau_p(p + p_1)].$$

Since we set $\tau_p = \tau_n$ in the numerical example, this rate will be about $G_{max} \simeq n_i/2\tau_n$ in the 'center' of the junction. The rate will drop rapidly on either side, however. Using $n_i \simeq 10^{16}$ m^{-3} and $\tau_n \simeq 10^{-6}$ s we have $G_{max} \simeq 10^{22}$ m^{-3} s^{-1}. The width L_G of the part of the junction where the generation rate is as large as G_{max} will be a fraction of the total junction width. Suppose $L_G \simeq 10^{-6}$ m. Then $\Gamma = \int G\, dx$ will be roughly 10^{16} m^{-2}s^{-1}. This is consistent with the flux shown in the plot.

The expression for the electron flux per unit area, $n_i^2 \frac{D_n}{L_n N_A} (\exp[eV_A/k_B T_e] - 1)$, is not appropriate in this example. We shall see this, when we try to evaluate it. We would need:
$$n_i^2 = 10^{32} \text{ m}^{-6} \; ; \; N_A = 10^{20} \text{ m}^{-3}.$$
$$D_n = V^{th}\mu = 0.03 \times 0.1 = 3 \times 10^{-3} \text{ m}^2\text{s}^{-1}.$$
$$L_n = \sqrt{D_n \tau_n}. \; \tau_n = 10^{-6} \text{ so } L_n \simeq \sqrt{3 \times 10^{-3} \times 10^{-6}} \simeq 5.5 \times 10^{-5}\text{m}.$$
Since this is larger than the entire device, this is a clue that something is wrong. This expression was derived to describe currents produced by recombination/generation outside the junction, when the diffusion length is small. This is not the case here.

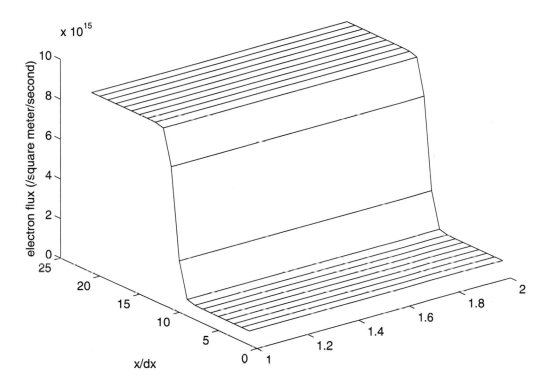

x 10^{15}

electron flux (/square meter/second)

x/dx

Figure 7.4. Electron flux at different points in a pn junction, calculated using a simplified Scharfetter-Gummel method. Since the flux changes dramatically as the junction region is crossed, the carrier production must be mainly in the junction region in this device.

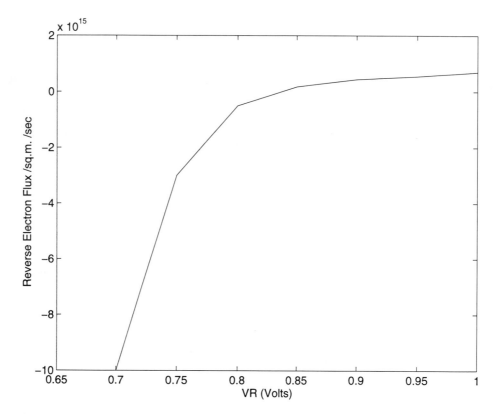

Figure 7.5. The maximum electron flux in a pn junction plotted for various values of the reverse voltage V_R.

7.2.2 Numerical Evaluation of PN Junction (i, v) Characteristics

The last section concentrated on small applied voltages, and, in Poisson's equation, it used expressions for the densities which are valid close to equilibrium (in some sense), so that the physical behavior of the device could be brought out more clearly by the model set up in the input file. In this section we turn to an input file which uses the continuity equations to find the carrier densities, and we use it to examine the (i, v) characteristics more thoroughly. The first file given is pn03a.sg, which is set up to compute (i, v) characteristics. The *grad* and *div* functions used in this file are included to allow the equations to be written in a more recognizable form. They also provide a simple example of the use of functions in SGFramework.

Nearly all of the input files in the remainder of the book are very similar to this file. New elements are added later. These include time dependence and the use of a mesh file that defines regions, in which different equations may hold. The main parts of the new input files are the same as this file, however. In many cases the whole files are given, for clarity, despite the duplication. (The input files are only a few pages long, at most.) In some cases, where a file which was given immediately before has been modified slightly, we show only the modifications to the file. In each case we aim for the clearest presentation.

Input File pn03a.sg

```
// FILE:   sgdir\pn03\pn03a.sg
// mesh constants
const NX  = 101;                    // number of x-mesh points
const LX  = NX - 1;                 // index of last x-mesh point
const DX  = 5.0e-6;                 // mesh spacing (cm)

// physical constants
const T   = 300.0;                  // ambient temperature (K)
const e   = 1.602e-19;              // electron charge (C)
const kb  = 1.381e-23;              // Boltzmann's constant (J/K)
const eo  = 8.854e-14;              // permittivity of vacuum (F/cm)
const eSi = 11.8;                   // relative permittivity of Si
const eps = eSi*eo;                 // permittivity of Si (F/cm)
const Ni  = 1.25e10;                // intrinsic concentration of Si (cm^-3)
const Vt  = kb*T/e;                 // thermal voltage (V)

// doping profile constants
const Na  = 1.0e+15;                // acceptor concentration on p-side (cm^-3)
const Nd  = 1.0e+15;                // donor concentration on n-side (cm^-3)
const Vbi = Vt*ln(Na*Nd/sq(Ni));    // built-in potential (V)

// scaling constants
const x0  = LX*DX;                  // distance scaling (cm)
const T0  = T;                      // temperature scaling (K)
const V0  = kb*T0/e;                // electrostatic-potential scaling (V)
```

```
const C0   = max(Na,Nd);                // concentration scaling (cm^-3);
const D0   = 35.0;                       // diffusion-coefficient scaling (cm^2/s)
const u0   = D0/V0;                       // mobility-coefficient scaling (cm^2/V/s)
const R0   = D0*C0/sq(x0);                // recombination rate scaling (cm^-3/s)
const t0   = sq(x0)/D0;                   // time scaling (s)
const E0   = V0/x0;                       // electric-field scaling (V/s)
const J0   = e*D0*C0/x0;                  // current-density scaling (A/cm^2)
const L0   = V0*eo/(e*sq(x0)*C0);         // Laplacian scaling constant

// 1D Cartesian discretized gradient operator
func grad(f1,f2,<h>)
   return (f2-f1)/h;


// 1D Cartesian discretized divergence operator
func div(f1,f2,<h>)
   return (f2-f1)/h;


// equilibrium electron concentration of uniformly doped silicon
func Neq(<N>)
   assign temp = [abs(N)+sqrt(sq(N)+4*sq(Ni))]/2.0;
   return (N>=0)*temp + (N<0)*sq(Ni)/temp;


// equilibrium hole concentration of uniformly doped silicon
func Peq(<N>)
   assign temp = [abs(N)+sqrt(sq(N)+4*sq(Ni))]/2.0;
   return (N<=0)*temp + (N>0)*sq(Ni)/temp;


// unscaled electron mobility
func MUn(<N>,<T>)
   const Al = 1430.0,  Bl = -2.20;
   const Ai = 4.61e17, Bi = 1.52e15;
   assign MUl = Al*pow((T/300.0),Bl);
   assign MUi = {Ai*T*sqrt(T)/N}/{ln(1+Bi*sq(T)/N)-Bi*T*T/(N+Bi*sq(T))};
   assign X   = sqrt(6.0*MUl/MUi);
   return MUl*{1.025/[1+pow(X/1.68,1.43)]-0.025};


// unscaled hole mobility
func MUp(<N>,<T>)
   const Al = 495.0,   Bl = -2.20;
   const Ai = 1.00e17, Bi = 6.25e14;
   assign MUl = Al*pow((T/300.0),Bl);
   assign MUi = {Ai*T*sqrt(T)/N}/{ln(1+Bi*sq(T)/N)-Bi*T*T/(N+Bi*sq(T))};
   assign X   = sqrt(6.0*MUl/MUi);
   return MUl*{1.025/[1+pow(X/1.68,1.43)]-0.025};


// scaled recombination
func R(n,p)
   const tn = 1.0e-6/t0;                  // scaled electron lifetime
```

```
  const tp = 1.0e-6/t0;            // scaled hole lifetime
  const ni = Ni/C0;                // scaled intrinsic concentration
  return (n*p-sq(ni))/[tp*(n+ni)+tn*(p+ni)];

// scaled electric field
func E(V1,V2,<h>)
  return -grad(V1,V2,h);

// scaled electron current density
func Jn(n1,n2,E,<un1>,<un2>,<h>)
  const Ut   = Vt/V0;              // scaled thermal voltage
  assign un   = ave(un1,un2);
  assign dV   = E*h/(2*Ut);
  assign n    = n1*aux2(dV)+n2*aux2(-dV);
  assign dndx = aux1(dV)*grad(n1,n2,h);
  return un*(n*E+dndx);

// scaled hole current density
func Jp(p1,p2,E,<up1>,<up2>,<h>)
  const Ut   = Vt/V0;              // scaled thermal voltage
  assign up   = ave(up1,up2);
  assign dV   = E*h/(2*Ut);
  assign p    = p1*aux2(-dV)+p2*aux2(dV);
  assign dpdx = aux1(dV)*grad(p1,p2,h);
  return up*(p*E-dpdx);


// set numerical algorithm parameters
set NEWTON DAMPING     = 3;
set NEWTON ACCURACY    = 1.0e-12;
set NEWTON ITERATIONS  = 100;
set LINSOL ALGORITHM   = GAUSSELIM;
set LINSOL FILL        = INFINITY;


// declare variables
var x[NX], C[NX], V[NX], n[NX], p[NX], un[NX], up[NX], h[LX], Vd, Jd;

// scaled Poisson's equation
equ V[i=1..LX-1] ->
  L0*div(eSi*E(V[i-1],V[i],h[i-1]),eSi*E(V[i],V[i+1],h[i]),ave(h[i],h[i-1])) -
  (p[i]-n[i]+C[i]) = 0.0;

// scaled electron continuity equation
equ n[i=1..LX-1] ->
  +div(Jn(n[i-1],n[i],E(V[i-1],V[i],h[i-1]),un[i-1],un[i],h[i-1]),
       Jn(n[i],n[i+1],E(V[i],V[i+1],h[i ]),un[i],un[i+1],h[i ]),
       ave(h[i],h[i-1])) -
```

```
   R(n[i],p[i]) = 0.0;

// scaled electron continuity equation
equ p[i=1..LX-1] ->
  -div(Jp(p[i-1],p[i],E(V[i-1],V[i],h[i-1]),up[i-1],up[i],h[i-1]),
       Jp(p[i],p[i+1],E(V[i],V[i+1],h[i  ]),up[i],up[i+1],h[i  ]),
       ave(h[i],h[i-1])) -
  R(n[i],p[i]) = 0.0;


begin SetUpDopingProfile
  assign x[i=all] = i*DX;        // initialize mesh points
  assign C[i=0..LX/2]    = +Nd;  // ionized donor concentration
  assign C[i=LX/2+1..LX] = -Na;  // ionized acceptor concentration
end

begin InitializeVariables
  assign  h[i=all] = x[i+1]-x[i];
  assign  V[i=all] = (i>LX/2)*-Vbi;
  assign  n[i=all] = Neq(C[i]);
  assign  p[i=all] = Peq(C[i]);
  assign un[i=all] = MUn(abs(C[i]),T);
  assign up[i=all] = MUp(abs(C[i]),T);
end

begin ScaleVariables
  assign  x[i=all] =  x[i]/x0;
  assign  h[i=all] =  h[i]/x0;
  assign  V[i=all] =  V[i]/V0;
  assign  C[i=all] =  C[i]/C0;
  assign  n[i=all] =  n[i]/C0;
  assign  p[i=all] =  p[i]/C0;
  assign un[i=all] = un[i]/u0;
  assign up[i=all] = up[i]/u0;
end

begin main
  call SetUpDopingProfile;
  call InitializeVariables;
  call ScaleVariables;

  // null the output file
  open "pn03a.out" write;
  close;

  assign Vd = 0.0;
  while (Vd < 0.9*Vbi) begin
    assign V[i=LX] = (Vd-Vbi)/V0;
```

```
      solve;
      assign Jd = -J0*{Jn(n[0],n[1],E(V[0],V[1],h[0]),un[0],un[1],h[0]) +
                       Jp(p[0],p[1],E(V[0],V[1],h[0]),up[0],up[1],h[0])};
      open "pn03a.out" append;
      file write Vd,Jd;
      close;
      assign Vd = Vd+0.01;
   end
end
```

The negative sign in the expression for J_d makes the forward-bias current, which in this case is from right to left, a positive current.

A similar file, pn03b.sg, is set up to compute electron and hole diffusion and drift-current densities. The parts of the file which differ from the above are given, as indicated.

Input File pn03b.sg

```
// (After the scaled electric field)

// unscaled electron drift-current density
func Jndr(<n1>,<n2>,<V1>,<V2>,<un1>,<un2>,<h>)
  const Ut    = Vt/V0;              // scaled thermal voltage
  assign un   = ave(un1,un2);
  assign dV   = (V1-V2)/(2*Ut);
  assign n    = n1*aux2(dV)+n2*aux2(-dV);
  return +un*n*E(V1,V2,h);

// unscaled electron diffusion-current density
func Jndf(<n1>,<n2>,<V1>,<V2>,<un1>,<un2>,<h>)
  const Ut    = Vt/V0;              // scaled thermal voltage
  assign un   = ave(un1,un2);
  assign dV   = (V1-V2)/(2*Ut);
  assign dndx = aux1(dV)*grad(n1,n2,h);
  return +un*dndx;

// unscaled hole drift-current density
func Jpdr(<p1>,<p2>,<V1>,<V2>,<up1>,<up2>,<h>)
  const Ut    = Vt/V0;              // scaled thermal voltage
  assign up   = ave(up1,up2);
  assign dV   = (V1-V2)/(2*Ut);
  assign p    = p1*aux2(-dV)+p2*aux2(dV);
  return +up*p*E(V1,V2,h);

// unscaled hole diffusion-current density
func Jpdf(<p1>,<p2>,<V1>,<V2>,<up1>,<up2>,<h>)
  const Ut    = Vt/V0;              // scaled thermal voltage
```

```
    assign up    = ave(up1,up2);
    assign dV    = (V1-V2)/(2*Ut);
    assign dpdx = aux1(dV)*grad(p1,p2,h);
    return -up*dpdx;


// (After the numerical algorithm parameters are set...)


// declare variables
var x[NX], C[NX], V[NX], n[NX], p[NX], un[NX], up[NX], h[LX];
var Jndrift[LX], Jndiff[LX], Jpdrift[LX], Jpdiff[LX];


// The main section becomes...


begin main
  call SetUpDopingProfile;
  call InitializeVariables;
  call ScaleVariables;
  solve;
  assign Jndrift[i=all] = J0*Jndr(n[i],n[i+1],V[i],V[i+1],un[i],un[i+1],h[i]);
  assign Jpdrift[i=all] = J0*Jpdr(p[i],p[i+1],V[i],V[i+1],up[i],up[i+1],h[i]);
  assign Jndiff [i=all] = J0*Jndf(n[i],n[i+1],V[i],V[i+1],un[i],un[i+1],h[i]);
  assign Jpdiff [i=all] = J0*Jpdf(p[i],p[i+1],V[i],V[i+1],up[i],up[i+1],h[i]);
  call UnscaleVariables;
  write;
end
```

Interpretation of the Simulation Results

Sample output from these files follows, in Figures 7.6 to 7.17.

When examining the plots, it is useful to compare them and try to understand why they have changed from one case to the next. The vertical scale is different in each case and should be taken note of. The shape of the electric field, and the fact that is 'widest' in reverse bias, can be explained using the analytic theory given above. The locations where the contributions to the current become large should be examined and compared in each case to the width of the junction. The electric field in the forward-bias case changes sign outside the junction. This is because of the Ohmic voltage drop, which follows Ohm's law. If the bulk is near enough to quasi-neutral that we can say that there is little charge density, and so $\nabla^2 V = 0$, then V must vary linearly in the bulk. This is the case, for large enough voltages that the applied voltage dominates the voltage set up by the charge in the bulk region. It is perhaps more evident from the plot of the electrostatic potential.

A few simple estimates will be given here. First, the built-in voltage for this diode is $V_{bi} = V^{th} \ln\left((10^{15})^2/(1.25 \times 10^{10})^2\right) \simeq 0.03 \ln\left(10^{30}/1.5 \times 10^{20}\right) \simeq 0.6$ volts. This is very close to the equilibrium potential which was computed.

The junction in question is abrupt, so its width can be estimated easily. The potential V dropped on one side of the junction, in a distance x, is $V = \frac{1}{2}\frac{\rho}{\epsilon}x^2$ (since the field is zero at the outside of the junction). Now in this device $N_A = N_D = 10^{15}$ cm$^{-3} = 10^{21}$ m^{-3} so if the carriers are all outside the junction region, the charge density in the junction is $\rho = 1.6 \times 10^{-19} \times 10^{21} = 1.6 \times 10^2$ C/m^3 and $\rho/\epsilon = \frac{1.6\times10^2}{11.8\times8.8\times10^{-12}} \simeq 1.6 \times 10^{12}$ V/m^2.

This means that the voltage drop on one side is $8 \times 10^{11}x^2$, and, if we set this equal to $\frac{1}{2}V_{bi}$, we have $0.3 = 8 \times 10^{11}x^2$, so $x = \sqrt{\frac{3}{8}} \times 10^{-6} \simeq 0.6\mu$ m. This is half the width of the junction. Again, this is in reasonable agreement.

Finally, we will estimate the size of the contributions to the current; the diffusive part of the current and the mobility part of the current.

The diffusive flux is $-D\nabla n$ which is in magnitude roughly $D\Delta n/L$ where L is the length over which the density is dropped. $D \simeq 3.5 \times 10^{-3}$ m^2 s^{-1}, $\Delta n \simeq 10^{21}$ m^{-3} and $L \simeq 10^{-6}$ m.

The flux is then about $\frac{3.5\times10^{-3}\times10^{21}}{10^{-6}} \simeq 3.5 \times 10^{24}$m^{-2} s^{-1}. The current associated with this is the flux multiplied by e, or about 5.6×10^5A m^{-2} s^{-1} which is 56 A cm^{-2} s^{-1}.

The mobility part almost exactly cancels this, so there is no need to check it also. If we did want to, we would get the same answer, since the electric field is roughly V^{th}/L and the mobility is D/V^{th}.

The diffusion flux is into the junction, in each case, while the mobility flux is out from the junction. The difference in the magnitude of the two parts of the electron and hole fluxes (the hole mobility flux being smaller than the electron mobility flux, and similarly for the diffusion fluxes) in this symmetric junction, is because of the difference in their mobilities. Their diffusion coefficients are proportional to their mobilities, so both diffusion and mobility fluxes are larger for electrons.

7.2.3   Reverse-Bias Breakdown and Avalanching

Diodes all eventually break down (that is, they fail and allow large undesired currents to flow) if a large enough reverse-bias is applied. Depending on the profile of the doping, three types of breakdown are possible. First, punch-through may occur. (See also the section on MOS devices in VLSI circuits, which discusses punch-through between source and drain in a short-channel MOSFET.) The depletion layer may extend at high voltage from the junction to one of the electrodes. This can happen if one side of the junction has (or both sides have) a low enough level of doping and the electrode is close enough to the junction. Any extra voltage beyond that needed to reach this state of punch-through cannot be dropped in the semiconducting bulk, which thus acts as a short with respect to the extra voltage. The extra current is only limited by the external circuit. Whether damage is done by punch through depends on the amount of heating done by the current, which depends on the external circuit's ability to limit the current.

By contrast with punch-through, Zener breakdown can occur if the doping level is high on both sides of the junction. The doping level has to be high to prevent punch-through, by keeping the junction narrow. The high doping level and narrow

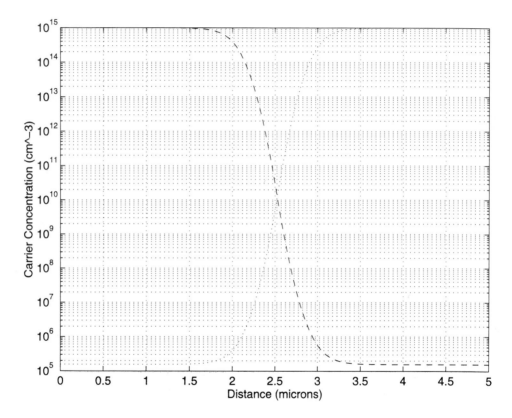

Figure 7.6. Equilibrium densities in the pn junction. The *n*-type region is on the left; *p*-type on the right. The dashed line denotes electrons and the dotted line denotes holes.

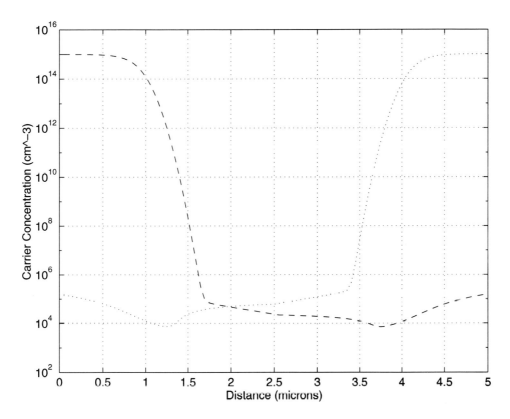

Figure 7.7. Reverse-bias densities in the pn junction. The n-type region is on the left; p-type on the right. The dashed line denotes electrons and the dotted line denotes holes.

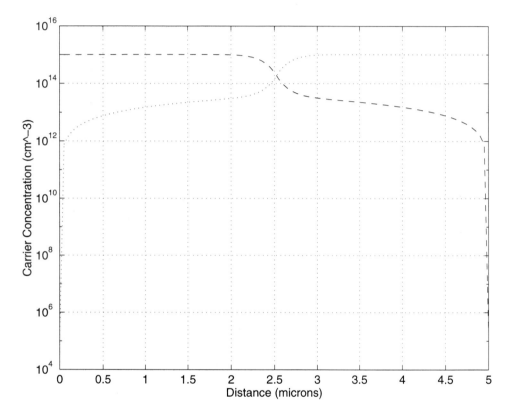

Figure 7.8. Forward-bias densities in the pn junction. The n-type region is on the left; p-type on the right. The dashed line denotes electrons and the dotted line denotes holes. The sudden drop in the density, at the edge where each species is the minority carrier, is determined by the boundary condition at that edge; see the text.

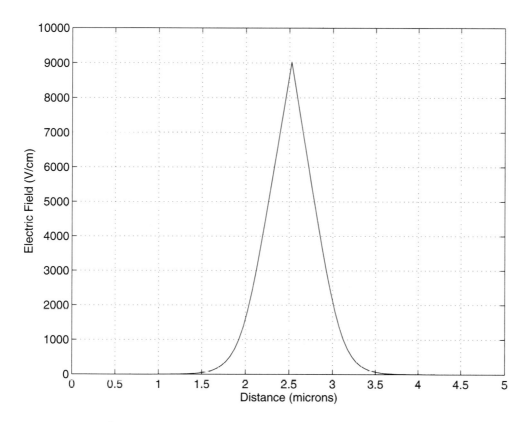

Figure 7.9. Equilibrium electric field in the pn junction.

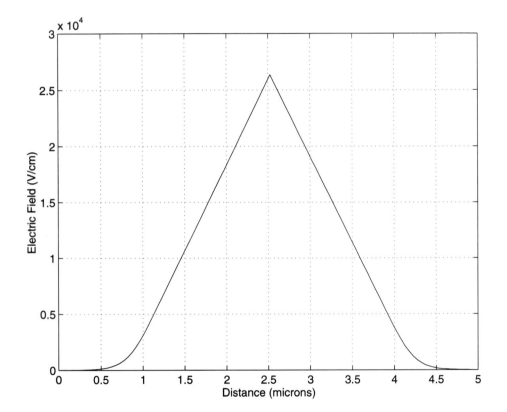

Figure 7.10. Reverse-bias electric field in the pn junction. The width of the junction is larger than in equilibrium because of the larger voltage drop.

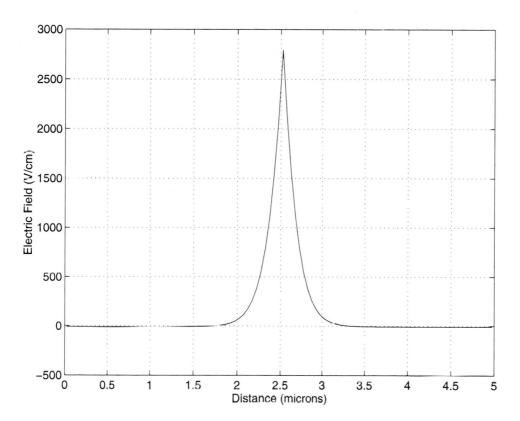

Figure 7.11. Forward-bias electric field in the pn junction. The width of the junction is less than in equilibrium because of the lower voltage drop.

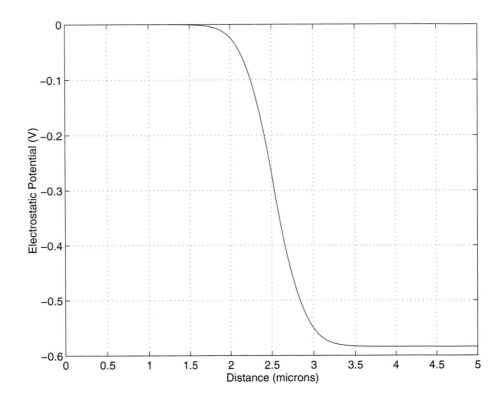

Figure 7.12. Equilibrium potential in the pn junction.

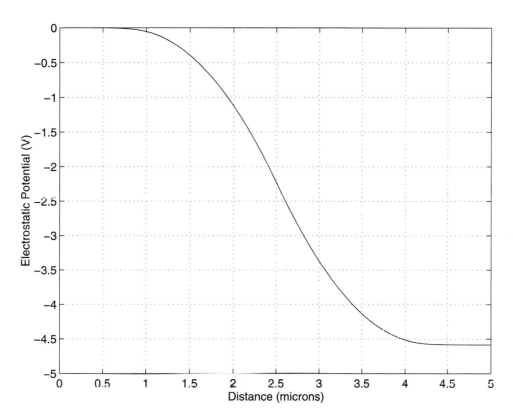

Figure 7.13. Reverse-bias potential in the pn junction.

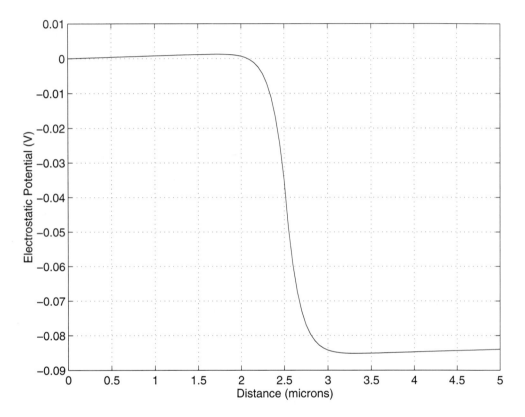

Figure 7.14. Forward-bias potential in the pn junction. Note the linear variation of V in the bulk region due to the Ohmic drop.

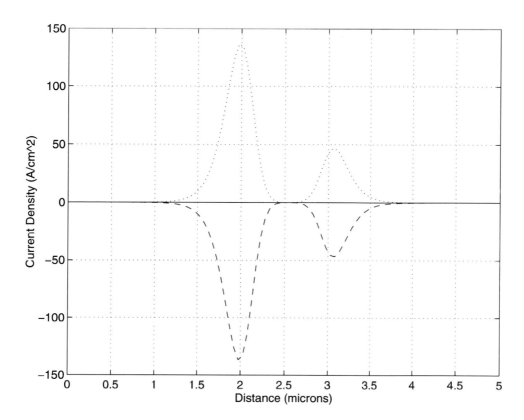

Figure 7.15. Equilibrium contributions to the current density in the pn junction. The n-type region is on the left; p-type on the right. The diffusion (dashed line) and mobility (dotted line) current densities are shown. These contributions to the fluxes are dominated by the majority carriers, but they cancel each other to very high accuracy. The diffusive part of the flux is away from the high-density region for each species, that is, into the junction. Since the n-type region is on the left, the diffusive electron flux is positive, which is a negative current. The diffusive hole flux is toward the left, which is also a negative current. The mobility fluxes for each species are opposite to the diffusive fluxes for that species, and cancel each other in equilibrium. The built-in electric field repels the majority carriers on each side of the junction, so the mobility fluxes are away from the junction. The asymmetry in the fluxes is due to the difference in electron and hole mobilities (and $D = \frac{k_B T}{e}\mu$ for both species).

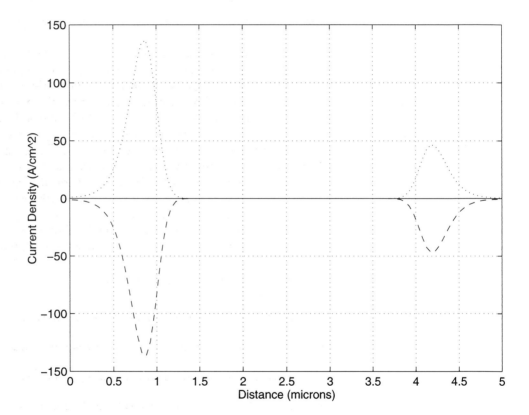

Figure 7.16. Reverse-bias contributions to the current density in the pn junction. The *n*-type region is on the left; *p*-type on the right.

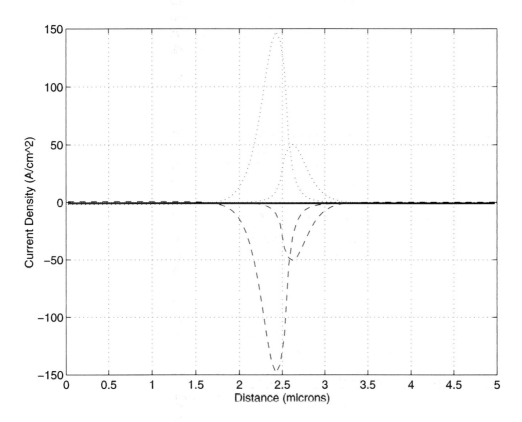

Figure 7.17. Forward-bias contributions to the current density in the pn junction. The n-type region is on the left; p-type on the right.

junction will also leave too little room for an avalanche to build up. (Avalanching is the third and most common cause of breakdown, in properly designed devices, except for Zener diodes.) For a high enough field and a narrow enough junction, electrons can tunnel quantum-mechanically from the p-side valence band to the n-side conduction band which is at a lower energy than the band they just left.

If the reverse field reaches values of around 2×10^5 V/m or more, then some of the minority carriers leaking through the junction will acquire enough kinetic energy to ionize the atoms in the crystal. The ionization results in pair generation, since it consists of taking an electron from the valence band to the conduction band, leaving a hole behind. These carriers then move in opposite directions, causing more ionization as they travel through the junction, which may lead to an avalanche.

The ionization coefficient α is the number of ionizations a carrier causes per unit length. For now the values of α for the two types of carrier will be set equal. To find the total current density J^M, where the M indicates the current density with multiplication taking place, two observations are used:

1. J^M is constant across the junction in steady state, and $J^M = J_n + J_p$.

2. In going a distance dx the electrons each produce αdx ionizations.

The flux of electrons, which is the number crossing a unit area per second, is $-J_n/e$, so the number produced in dx per unit area per second by the electrons is $(J_n/e)\alpha dx$. The number produced is positive, irrespective of the sign of the current, and both J_n and J_p will be positive in what follows. This must be added to the corresponding number produced by the holes to find the total number of electrons produced in dx per unit area per second, which is also the change in the electron flux in dx, which is $\pm dJ_n/e$. The sign of the change in the electron flux depends on which way the electron flux is going. In the situation we have analyzed previously the electrons are the minority carriers on the p-side, which is on the right. The electrons leak to the left, so the increase in their flux occurs as they go to the left. In other words, further to the right their flux must decrease in magnitude. For holes, on the other hand, the flux will increase as we go to the right. The same number of holes are produced in dx as electrons, so the equation for the hole current is the same as for electrons except for the sign to allow for the fluxes increasing in magnitude in opposite directions.

In going a distance dx to the right the electron flux must decrease in magnitude and so must the electron current, so

$$\frac{1}{e}dJ_n = -\left[\frac{1}{e}J_n\alpha dx + \frac{1}{e}J_p\alpha dx\right] \tag{7.53}$$

so the change in electron current is

$$dJ_n = -(J_n + J_p)\alpha dx = -J^M \alpha dx \tag{7.54}$$

which also means

$$\frac{\partial J_n}{\partial x} = -\alpha J^M. \tag{7.55}$$

The equation for the holes is the same except for the sign, so

$$\frac{\partial J_p}{\partial x} = \alpha J^M. \tag{7.56}$$

The hole flux is injected at $x = -x_n$ and moves in the positive x-direction. The hole current behaves the same way as the hole flux, and the solution of the equation for the hole current is

$$J_p = J_p(x = -x_n) + \alpha(x + x_n)J^M. \tag{7.57}$$

Integrating gave $J_p = A + \alpha x J^M$, and the constant of integration A was chosen to be $A = J_p(x = -x_n) + \alpha x_n J^M$ to provide the correct value at $x = -x_n$. Similarly, if the electron flux is injected at $x = x_p$, then the current, which flows towards $x = x_p$, as the flux is away from $x = x_p$, is

$$J_n = J_n(x = x_p) - \alpha(x - x_p)J^M. \tag{7.58}$$

Now the injected currents are unaffected by the multiplication, so $J_n(x = x_p) = J_{no}$ and $J_p(x = -x_p) = J_{po}$, where the subscripts indicate values in the absence of multiplication. Adding the two currents gives

$$J_p + J_n = J^M = J_{po} + J_{no} - \alpha(x - x_p)J^M + \alpha(x + x_p)J^M = J_{po} + J_{no} + \alpha W J^M \tag{7.59}$$

where $W = x_n + x_p$ is the junction width. The current J^M is thus

$$J^M = \frac{J_{po} + J_{no}}{1 - \alpha W}. \tag{7.60}$$

The multiplication factor M is defined to be the increase in the current caused by the multiplication, so

$$M = \frac{J^M}{J_{po} + J_{no}} = \frac{1}{1 - \alpha W} \tag{7.61}$$

The breakdown voltage BV is the voltage that makes α large enough that $\alpha W = 1$, which in turn gives infinite multiplication. But αW is commonly assumed to vary as a constant times V^r. For αW to be one when $V = BV$ we must have $\alpha W = (V/BV)^r$, in which case

$$M = \left[1 - \left(\frac{V}{BV}\right)^r\right]^{-1}. \tag{7.62}$$

Calculation of BV is discussed in considerable detail in Chapter 11, for real diode geometries and for α having a realistic dependence on the electric-field strength.

7.2.4   High-Current and High-Level Injection

When a device such as a pin diode, which has a region of low doping next to a highly doped region, is forward-biased, a large density of carriers is swept into the lightly doped region. The conductivity of the lightly doped region increases greatly when this happens, an effect known as conductivity modulation. The condition of having a carrier density injected which greatly exceeds the dopant density is known as high-level injection.

At high injection levels the injected carriers attract carriers of the opposite sign into the region where they are injected, to satisfy the requirement of charge neutrality. This second group of carriers usually has to come from the opposite side of the low-doped region. The densities of these carriers in the low-doped region will be flat if their diffusion lengths are large compared to the drift-region width W, which means they have time to diffuse around before they recombine. Since the densities are nearly equal due to the charge neutrality requirement, with value n_{inj}, the conductivity of the drift region is roughly given by $e(\mu_n + \mu_p)n_{inj}$.

If a stored-charge approach is applied, the current flowing into the drift region is equal to the stored charge of one species divided by the recombination time τ. The charge per unit area $Q = en_{inj}W$ (where W is the width of the drift region) divided by τ is the current density, which is also given by $J = en_{inj}(\mu_n + \mu_p)E$. The electric field E is the voltage across the drift region V_{dr} divided by W. Combining these,

$$J = \frac{en_{inj}W}{\tau} = e(\mu_n + \mu_p)n_{inj}\frac{V_{dr}}{W} \tag{7.63}$$

which can be rearranged to give an expression for the voltage needed across the drift region,

$$V_{dr} = \frac{W^2}{(\mu_n + \mu_p)\tau}. \tag{7.64}$$

These points are illustrated in two simulations, pin01.sg and pin02.sg, and Figures 7.18-7.22.

Input File pin01.sg

```
// FILE:  sgdir\ch07\pin01\pin01.sg
// mesh constants
const NX  = 401;                      // number of x-mesh points
const LX  = NX - 1;                   // index of last x-mesh point
const Wd  = 5.00e-4;                  // width of p+ diffusion (cm)
const We  = 30.0e-4;                  // width of n- epitaxial layer (cm)
const Ws  = 10.0e-4;                  // width of simulated n+ substrate (cm)
const DX  = (We+Ws)/LX;               // mesh spacing (cm)

// physical constants
const T   = 300.0;                    // ambient temperature (K)
const e   = 1.602e-19;                // electron charge (C)
const kb  = 1.381e-23;                // Boltzmann's constant (J/K)
```

```
const eo  = 8.854e-14;          // permittivity of vacuum (F/cm)
const eSi = 11.8;               // relative permittivity of Si
const eps = eSi*eo;             // permittivity of Si (F/cm)
const Ni  = 1.25e10;            // intrinsic concentration of Si (cm^-3)
const Vt  = kb*T/e;             // thermal voltage (V)

// doping profile constants
const Nd  = 5.0e+19;            // peak acceptor concentration (cm^-3)
const Ne  = 7.0e+14;            // epitaxial donor concentration (cm^-3)
const Ns  = 1.0e+19;            // substrate donor concentration (cm^-3)
const Vbi = Vt*ln(Nd*Ns/sq(Ni));  // built-in potential (V)

// scaling constants
const x0  = LX*DX;              // distance scaling (cm)
const T0  = T;                  // temperature scaling (K)
const V0  = kb*T0/e;            // electrostatic-potential scaling (V)
const C0  = max(Nd,Ns);         // concentration scaling (cm^-3);
const D0  = 35.0;               // diffusion-coefficient scaling (cm^2/s)
const u0  = D0/V0;              // mobility-coefficient scaling (cm^2/V/s)
const R0  = D0*C0/sq(x0);       // recombination-rate scaling (cm^-3/s)
const t0  = sq(x0)/D0;          // time scaling (s)
const E0  = V0/x0;              // electric-field scaling (V/s)
const J0  = e*D0*C0/x0;         // current-density scaling (A/cm^2)
const L0  = V0*eo/(e*sq(x0)*C0);  // Laplacian scaling constant


// dopant profile of a diffused epitaxial layer of concentration Ne and width
// d depositted on a substrate with a background concentration of Nb driven
// with a diffusion coefficient-time product of Dt
func Nepi(<x>,<Dt>,<Nb>,<Ne>,<d>)
  assign sqrt4Dt = sqrt(4.0*Dt);
  return Nb+(Ne-Nb)*{erf([d+x]/sqrt4Dt)+erf([d-x]/sqrt4Dt)}/2.0;

// dopant profile of a Gaussian profile of peak concentration Nmax and width
// d deposited on a substrate with a background concentration of Nb
func Ngauss(<x>,<Nmax>,<Nb>,<d>)
  assign alpha = ln(abs(Nmax/Nb))/sq(d);
  return (Nmax-Nb)*exp(-alpha*sq(x));

// pin dopant profile
func Npin(<x>)
  const Dt = 1.0e-9;
  return Ngauss(x,-Nd,Ne,Wd)+Nepi(x,Dt,Ns,Ne,We);

// 1D Cartesian discretized gradient operator
func grad(f1,f2,<h>)
  return (f2-f1)/h;
```

```
// 1D Cartesian discretized divergence operator
func div(f1,f2,<h>)
  return (f2-f1)/h;

// equilibrium electron concentration of uniformly doped silicon
func Neq(<N>)
  assign temp = [abs(N)+sqrt(sq(N)+4*sq(Ni))]/2.0;
  return (N>=0)*temp + (N<0)*sq(Ni)/temp;

// equilibrium hole concentration of uniformly doped silicon
func Peq(<N>)
  assign temp = [abs(N)+sqrt(sq(N)+4*sq(Ni))]/2.0;
  return (N<=0)*temp + (N>0)*sq(Ni)/temp;

// electron lifetime
func LTn(<N>)
  const LT0  = 3.95e-4;
  const Nref = 7.1e15;
  return LT0/(1.0+N/Nref);

// hole lifetime
func LTp(<N>)
  const LT0  = 3.52e-5;
  const Nref = 7.1e15;
  return LT0/(1.0+N/Nref);

// scaled electron mobility
func MUn(<N>,<T>,pn)
  const Al = 1430.0,   Bl = -2.20;
  const Ai = 4.61e17, Bi = 1.52e15;
  assign NU  = abs(N)*C0;
  assign pnU = (pn>0.0)*pn*sq(C0)+(pn<0.0);
  assign MUl = Al*pow((T/300.0),Bl);
  assign MUi = {Ai*T*sqrt(T)/NU}/{ln(1+Bi*sq(T)/NU)-Bi*T*T/(NU+Bi*sq(T))};
  assign MUc = {2e17*T*sqrt(T/pnU)}/ln(1+8.28e8*sq(T)*pow(pnU,-1/3));
  assign X   = sqrt(6*MUl*(MUi+MUc)/(MUi*MUc));
  return MUl*{1.025/[1+pow(X/1.68,1.43)]-0.025}/u0;

// scaled hole mobility
func MUp(<N>,<T>,pn)
  const Al = 495.0,    Bl = -2.20;
  const Ai = 1.00e17, Bi = 6.25e14;
  assign NU  = abs(N)*C0;
  assign pnU = (pn>0.0)*pn*sq(C0)+(pn<0.0);
  assign MUl = Al*pow((T/300.0),Bl);
  assign MUi = {Ai*T*sqrt(T)/NU}/{ln(1+Bi*sq(T)/NU)-Bi*T*T/(NU+Bi*sq(T))};
  assign MUc = {2e17*T*sqrt(T/pnU)}/ln(1+8.28e8*sq(T)*pow(pnU,-1/3));
  assign X   = sqrt(6*MUl*(MUi+MUc)/(MUi*MUc));
```

```
    return MU1*{1.025/[1+pow(X/1.68,1.43)]-0.025}/u0;

// scaled recombination
func R(n,p,<tn>,<tp>)
  const Cn = 2.8e-31/[D0/sq(C0*x0)];// scaled electron Auger coefficient
  const Cp = 1.2e-31/[D0/sq(C0*x0)];// scaled hole Auger coefficient
  const ni = Ni/C0;                 // scaled intrinsic concentration
  const pn = sq(ni);                // scaled intrinsic concentration squared
  assign Rsrh = (n*p-pn)/[tp*(n+ni)+tn*(p+ni)];
  assign Raug = (n*p-pn)*(Cn*n+Cp*p);
  return Rsrh+Raug;

// scaled electric field
func E(V1,V2,<h>)
  return -grad(V1,V2,h);

// scaled electron current density
func Jn(n1,n2,p1,p2,E,<C1>,<C2>,<h>)
  const Ut  = Vt/V0;                // scaled thermal voltage
  assign un1  = MUn(C1,T,n1*p1);
  assign un2  = MUn(C2,T,n2*p2);
  assign un   = ave(un1,un2);
  assign dV   = E*h/(2*Ut);
  assign n    = n1*aux2(dV)+n2*aux2(-dV);
  assign dndx = aux1(dV)*grad(n1,n2,h);
  return un*(n*E+dndx);

// scaled hole current density
func Jp(n1,n2,p1,p2,E,<C1>,<C2>,<h>)
  const Ut  = Vt/V0;                // scaled thermal voltage
  assign up1  = MUp(C1,T,n1*p1);
  assign up2  = MUp(C2,T,n2*p2);
  assign up   = ave(up1,up2);
  assign dV   = E*h/(2*Ut);
  assign p    = p1*aux2(-dV)+p2*aux2(dV);
  assign dpdx = aux1(dV)*grad(p1,p2,h);
  return up*(p*E-dpdx);


// set numerical algorithm parameters
set NEWTON DAMPING    = 3;
set NEWTON ACCURACY   = 1.0e-12;
set NEWTON ITERATIONS = 100;
set LINSOL ALGORITHM  = GAUSSELIM;
set LINSOL FILL       = INFINITY;


// declare variables
```

```
var x[NX], C[NX], V[NX], n[NX], p[NX], tn[NX], tp[NX], h[LX], Vd, Jd, dVd;

// scaled Poisson's equation
equ V[i=1..LX-1] ->
  L0*div(eSi*E(V[i-1],V[i],h[i-1]),eSi*E(V[i],V[i+1],h[i]),ave(h[i],h[i-1])) -
  (p[i]-n[i]+C[i]) = 0.0;

// scaled electron continuity equation
equ n[i=1..LX-1] ->
  +div(Jn(n[i-1],n[i],p[i-1],p[i],E(V[i-1],V[i],h[i-1]),C[i-1],C[i],h[i-1]),
       Jn(n[i],n[i+1],p[i],p[i+1],E(V[i],V[i+1],h[i  ]),C[i],C[i+1],h[i  ]),
       ave(h[i],h[i-1])) -
  R(n[i],p[i],tn[i],tp[i]) = 0.0;

// scaled hole continuity equation
equ p[i=1..LX-1] ->
  -div(Jp(n[i-1],n[i],p[i-1],p[i],E(V[i-1],V[i],h[i-1]),C[i-1],C[i],h[i-1]),
       Jp(n[i],n[i+1],p[i],p[i+1],E(V[i],V[i+1],h[i  ]),C[i],C[i+1],h[i  ]),
       ave(h[i],h[i-1])) -
  R(n[i],p[i],tn[i],tp[i]) = 0.0;


begin SetUpDopingProfile
  assign x[i=all] = i*DX;          // initialize mesh points
  assign C[i=all] = Npin(x[i]);    // initialize dopant profile
end

begin InitializeVariables
  assign h[i=all] = x[i+1]-x[i];
  assign n[i=all] = Neq(C[i]);
  assign p[i=all] = Peq(C[i]);
  assign V[i=all] = Vt*ln(n[i]/Ns);
  assign tn[i=all] = LTn(abs(C[i]));
  assign tp[i=all] = LTp(abs(C[i]));
end

begin ScaleVariables
  assign x[i=all] = x[i]/x0;
  assign h[i=all] = h[i]/x0;
  assign V[i=all] = V[i]/V0;
  assign C[i=all] = C[i]/C0;
  assign n[i=all] = n[i]/C0;
  assign p[i=all] = p[i]/C0;
  assign tn[i=all] = tn[i]/t0;
  assign tp[i=all] = tp[i]/t0;
end

begin UnscaleVariables
```

```
    assign x[i=all] = x[i]*x0;
    assign h[i=all] = h[i]*x0;
    assign V[i=all] = V[i]*V0;
    assign C[i=all] = C[i]*C0;
    assign n[i=all] = n[i]*C0;
    assign p[i=all] = p[i]*C0;
    assign tn[i=all] = tn[i]*t0;
    assign tp[i=all] = tp[i]*t0;
end

begin main
  call SetUpDopingProfile;
  call InitializeVariables;
  call ScaleVariables;

  // solve for equilibrium
  assign Vd = 0.0;
  assign V[i=0] = (Vd-Vbi)/V0;
  solve;

  // write all of the simulation variables to result file
  call UnscaleVariables;
  write;
  call ScaleVariables;

  // compute the current and write it to a file
  open "pin01.out" write;
  assign Jd = J0*{Jn(n[0],n[1],p[0],p[1],E(V[0],V[1],h[0]),C[0],C[1],h[0]) +
                  Jp(n[0],n[1],p[0],p[1],E(V[0],V[1],h[0]),C[0],C[1],h[0])};
  file write Vd,Jd;
  close;

  // forward-bias and solve
  assign dVd = 0.05;
  while (Vd+dVd < Vbi) begin
    assign Vd = Vd+dVd;
    assign V[i=0] = (Vd-Vbi)/V0;
    solve;
    open "pin01.out" append;
    assign Jd = J0*{Jn(n[0],n[1],p[0],p[1],E(V[0],V[1],h[0]),C[0],C[1],h[0]) +
                    Jp(n[0],n[1],p[0],p[1],E(V[0],V[1],h[0]),C[0],C[1],h[0])};
    file write Vd,Jd;
    close;
  end

  // write the results to result file
  call UnscaleVariables;
  write;
```

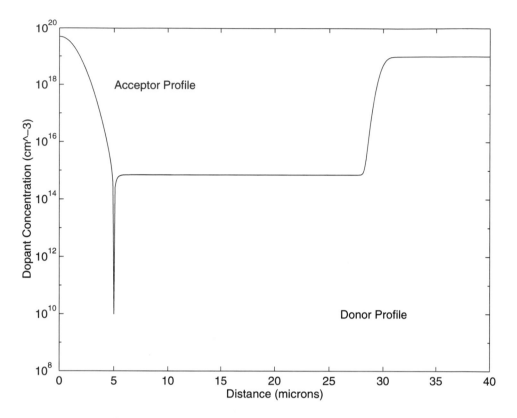

Figure 7.18. Dopant concentration in the pin diode.

```
    call ScaleVariables;
end
```

The file which follows, pin02.sg, performs a time-dependent calculation. The mesh data is read from a file mesh.dat.

Input File pin02.sg

```
// FILE:  sgdir\pin02\pin02.sg
// mesh constants ... these correspond to the data in mesh.dat
const NX  = 176;                  // number of x-mesh points
const LX  = NX-1;                 // index of last mesh point
const Wd  = 160.0e-4;             // device width
const Nd  = 5.0e+19;              // peak acceptor concentration (cm^-3)
const Ne  = 1.0e+14;              // epitaxial donor concentration (cm^-3)
const Ns  = 1.0e+19;              // substrate donor concentration (cm^-3)
```

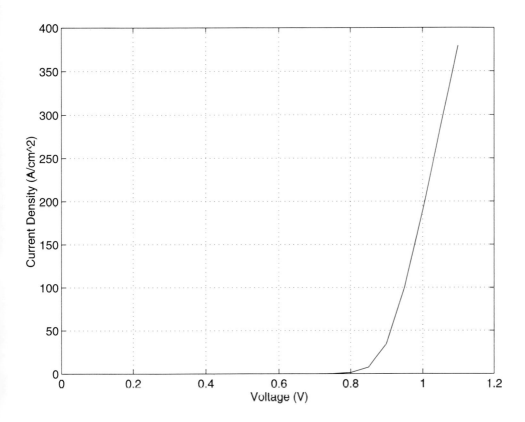

Figure 7.19. Current density as a function of voltage in the pin diode. The current varies linearly with voltage instead of exponentially, for large voltages; see the text.

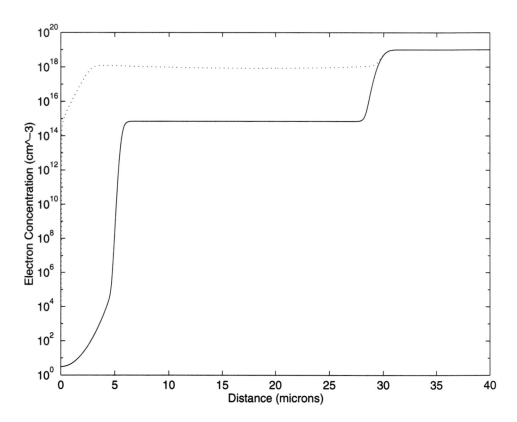

Figure 7.20. Electron concentration in equilibrium (solid line) and at a high level of injection (dotted line).

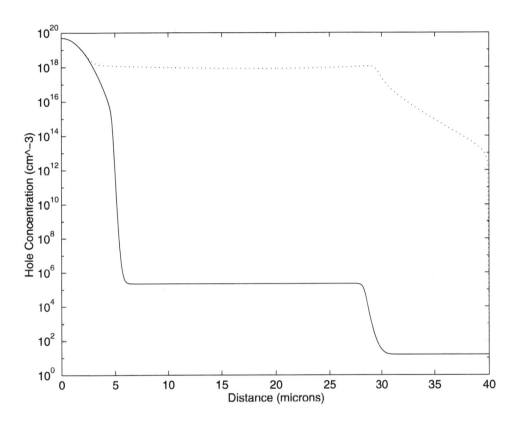

Figure 7.21. Hole concentration in equilibrium (solid line) and at a high level of injection (dotted line).

```
// physical constants and built-in potential
const T    = 300.0;                    // ambient temperature (K)
const e    = 1.602e-19;                // electron charge (C)
const kb   = 1.381e-23;                // Boltzmann's constant (J/K)
const eo   = 8.854e-14;                // permittivity of vacuum (F/cm)
const eSi  = 11.8;                     // relative permittivity of Si
const eps  = eSi*eo;                   // permittivity of Si (F/cm)
const Ni   = 1.25e10;                  // intrinsic concentration of Si (cm^-3)
const Vt   = kb*T/e;                   // thermal voltage (V)
const Vbi  = Vt*ln(Nd*Ns/sq(Ni));      // built-in potential (V)


// scaling constants
const x0   = Wd;                       // distance scaling (cm)
const T0   = T;                        // temperature scaling (K)
const V0   = kb*T0/e;                  // electrostatic-potential scaling (V)
const C0   = max(Nd,Ns);               // concentration scaling (cm^-3)
const D0   = 35.0;                     // diffusion-coefficient scaling (cm^2/s)
const u0   = D0/V0;                    // mobility-coefficient scaling (cm^2/V/s)
const R0   = D0*C0/sq(x0);             // recombination-rate scaling (cm^-3/s)
const t0   = sq(x0)/D0;                // time scaling (s)
const E0   = V0/x0;                    // electric-field scaling (V/s)
const J0   = e*D0*C0/x0;               // current-density scaling (A/cm^2)
const L0   = V0*eo/(e*sq(x0)*C0);      // Laplacian scaling constant



// 1D Cartesian discretized gradient operator
func grad(f1,f2,<h>)
  return (f2-f1)/h;

// 1D Cartesian discretized divergence operator
func div(f1,f2,<h>)
  return (f2-f1)/h;

// equilibrium electron concentration of uniformly doped silicon
func Neq(<N>)
  assign temp = [abs(N)+sqrt(sq(N)+4*sq(Ni))]/2.0;
  return (N>=0)*temp + (N<0)*sq(Ni)/temp;

// equilibrium hole concentration of uniformly doped silicon
func Peq(<N>)
  assign temp = [abs(N)+sqrt(sq(N)+4*sq(Ni))]/2.0;
  return (N<=0)*temp + (N>0)*sq(Ni)/temp;

// electron lifetime
func LTn(<N>)
  const LT0  = 3.95e-4;
  const Nref = 7.1e15;
  return LT0/(1.0+N/Nref);
```

```
// hole lifetime
func LTp(<N>)
  const LT0  = 3.52e-5;
  const Nref = 7.1e15;
  return LT0/(1.0+N/Nref);

// scaled electron mobility
func MUn(<N>,<T>,pn)
  const Al = 1430.0,  Bl = -2.20;
  const Ai = 4.61e17, Bi = 1.52e15;
  assign NU  = abs(N)*C0;
  assign pnU = (pn>0.0)*pn*sq(C0)+(pn<0.0);
  assign MUl = Al*pow((T/300.0),Bl);
  assign MUi = {Ai*T*sqrt(T)/NU}/{ln(1+Bi*sq(T)/NU)-Bi*T*T/(NU+Bi*sq(T))};
  assign MUc = {2e17*T*sqrt(T/pnU)}/ln(1+8.28e8*sq(T)*pow(pnU,-1/3));
  assign X   = sqrt(6*MUl*(MUi+MUc)/(MUi*MUc));
  return MUl*{1.025/[1+pow(X/1.68,1.43)]-0.025}/u0;

// scaled hole mobility
func MUp(<N>,<T>,pn)
  const Al = 495.0,    Bl = -2.20;
  const Ai - 1.00e17, Bi = 6.25e14;
  assign NU  = abs(N)*C0;
  assign pnU = (pn>0.0)*pn*sq(C0)+(pn<0.0);
  assign MUl = Al*pow((T/300.0),Bl);
  assign MUi = {Ai*T*sqrt(T)/NU}/{ln(1+Bi*sq(T)/NU)-Bi*T*T/(NU+Bi*sq(T))};
  assign MUc = {2e17*T*sqrt(T/pnU)}/ln(1+8.28e8*sq(T)*pow(pnU,-1/3));
  assign X   = sqrt(6*MUl*(MUi+MUc)/(MUi*MUc));
  return MUl*{1.025/[1+pow(X/1.68,1.43)]-0.025}/u0;

// scaled recombination
func R(n,p,<tn1>,<tp1>)
  const tn = 4.9e-5/t0;
  const tp = 4.2e-5/t0;
  const Cn = 2.8e-31/[D0/sq(C0*x0)];// scaled electron Auger coefficient
  const Cp = 1.2e-31/[D0/sq(C0*x0)];// scaled hole Auger coefficient
  const ni = Ni/C0;                 // scaled intrinsic concentration
  const pn = sq(ni);                // scaled intrinsic concentration squared
  assign Rsrh = (n*p-pn)/[tp*(n+ni)+tn*(p+ni)];
  assign Raug = (n*p-pn)*(Cn*n+Cp*p);
  return Rsrh+Raug;

// scaled electric field
func E(V1,V2,<h>)
  return -grad(V1,V2,h);

// scaled electron current density
```

```
func Jn(n1,n2,p1,p2,E,<C1>,<C2>,<h>)
  const  Ut   = Vt/V0;                    // scaled thermal voltage
  assign un1  = MUn(C1,T,n1*p1);
  assign un2  = MUn(C2,T,n2*p2);
  assign un   = ave(un1,un2);
  assign dV   = E*h/(2*Ut);
  assign n    = n1*aux2(dV)+n2*aux2(-dV);
  assign dndx = aux1(dV)*grad(n1,n2,h);
  return un*(n*E+dndx);

// scaled hole current density
func Jp(n1,n2,p1,p2,E,<C1>,<C2>,<h>)
  const  Ut   = Vt/V0;                    // scaled thermal voltage
  assign up1  = MUp(C1,T,n1*p1);
  assign up2  = MUp(C2,T,n2*p2);
  assign up   = ave(up1,up2);
  assign dV   = E*h/(2*Ut);
  assign p    = p1*aux2(-dV)+p2*aux2(dV);
  assign dpdx = aux1(dV)*grad(p1,p2,h);
  return up*(p*E-dpdx);


// set numerical algorithm parameters
set NEWTON DAMPING     = 3;
set NEWTON ACCURACY    = 1.0e-12;
set NEWTON ITERATIONS  = 100;
set LINSOL ALGORITHM   = GAUSSELIM;
set LINSOL FILL        = INFINITY;


// declare variables
var x[NX], C[NX], V[NX], n[NX], p[NX], tn[NX], tp[NX], h[LX], Vd, Jd;
var nlast[NX], plast[NX], dJd, dt, t, dJdt, fEQ;

// scaled Poisson's equation
equ V[i=1..LX-1] ->
  L0*div(eSi*E(V[i-1],V[i],h[i-1]),eSi*E(V[i],V[i+1],h[i]),ave(h[i],h[i-1])) -
  (p[i]-n[i]+C[i]) = 0.0;

// scaled electron continuity equation
equ n[i=1..LX-1] ->
  +div(Jn(n[i-1],n[i],p[i-1],p[i],E(V[i-1],V[i],h[i-1]),C[i-1],C[i],h[i-1]),
       Jn(n[i],n[i+1],p[i],p[i+1],E(V[i],V[i+1],h[i ]),C[i],C[i+1],h[i ]),
       ave(h[i],h[i-1])) -
  R(n[i],p[i],tn[i],tp[i]) = (n[i]-nlast[i])/(dt/t0);

// scaled hole continuity equation
equ p[i=1..LX-1] ->
```

```
  -div(Jp(n[i-1],n[i],p[i-1],p[i],E(V[i-1],V[i],h[i-1]),C[i-1],C[i],h[i-1]),
       Jp(n[i],n[i+1],p[i],p[i+1],E(V[i],V[i+1],h[i  ]),C[i],C[i+1],h[i  ]),
       ave(h[i],h[i-1]))) -
   R(n[i],p[i],tn[i],tp[i]) = (p[i]-plast[i])/(dt/t0);

// current boundary condition
equ V[i=0] -> 0.0 =
     fEQ *{V[0]+Vbi/V0} +
   (1-fEQ)*{Jn(n[0],n[1],p[0],p[1],E(V[0],V[1],h[0]),C[0],C[1],h[0])+
            Jp(n[0],n[1],p[0],p[1],E(V[0],V[1],h[0]),C[0],C[1],h[0])-Jd/J0};

begin SetUpMesh
  open "mesh.dat" read;
  file goto "mesh points";
  file read x;
  file goto "dopant concentration";
  file read C;
  close;
end

begin InitializeVariables
  assign h[i=all] = x[i+1]-x[i];
  assign n[i=all] = Neq(C[i]);
  assign p[i=all] = Peq(C[i]);
  assign V[i=all] = Vt*ln(n[i]/Ns);
  assign tn[i=all] = LTn(abs(C[i]));
  assign tp[i=all] = LTp(abs(C[i]));
  assign nlast[i=all] = n[i];
  assign plast[i=all] = p[i];
end

begin ScaleVariables
  assign x[i=all] = x[i]/x0;
  assign h[i=all] = h[i]/x0;
  assign V[i=all] = V[i]/V0;
  assign C[i=all] = C[i]/C0;
  assign n[i=all] = n[i]/C0;
  assign p[i=all] = p[i]/C0;
  assign tn[i=all] = tn[i]/t0;
  assign tp[i=all] = tp[i]/t0;
  assign nlast[i=all] = nlast[i]/C0;
  assign plast[i=all] = plast[i]/C0;
end

begin UnscaleVariables
  assign x[i=all] = x[i]*x0;
  assign h[i=all] = h[i]*x0;
  assign V[i=all] = V[i]*V0;
```

```
    assign C[i=all] = C[i]*C0;
    assign n[i=all] = n[i]*C0;
    assign p[i=all] = p[i]*C0;
    assign tn[i=all] = tn[i]*t0;
    assign tp[i=all] = tp[i]*t0;
    assign nlast[i=all] = nlast[i]*C0;
    assign plast[i=all] = plast[i]*C0;
end

begin UpdateLastVariables
    assign nlast[i=all] = n[i];
    assign plast[i=all] = p[i];
end

begin main
    call SetUpMesh;
    call InitializeVariables;
    call ScaleVariables;

    // solve for equilibrium and write the results to the output file
    assign fEQ = 1;
    assign t  = 0.0;
    assign dt = 1.0e100;
    assign Jd = 0.0;
    solve;
    assign Vd = Vbi+(V[0]-V[LX])*V0;
    open "pin02.out" write;
    file write t,Vd,Jd;
    close;
    call UpdateLastVariables;

    // ramp the current up quickly
    assign fEQ = 0;
    assign dJdt = 1.0e9;
    do begin
      assign dt = max(t/4.0,1.0e-9);
      assign dJd = dJdt*dt;
      assign t = t+dt;
      assign Jd = Jd+dJd;
      solve;
      assign Vd = Vbi+(V[0]-V[LX])*V0;
      open "pin02.out" append;
      file write t,Vd,Jd;
      close;
      call UpdateLastVariables;
    end while (Jd < 300.0);
end
```

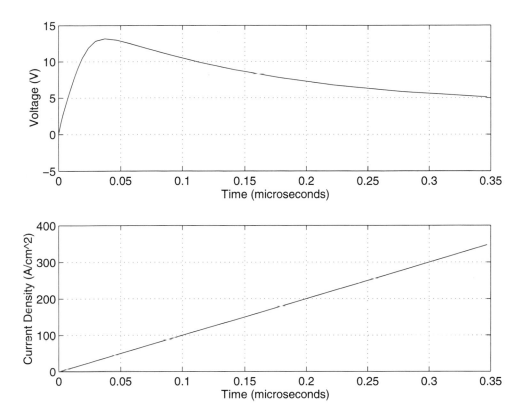

Figure 7.22. Voltage and current density as functions of time in the pin diode. The current is being ramped up linearly in this simulation. The voltage exhibits a characteristic overshoot, which is eventually reduced, in part by the high-level injection, since the injected carriers decrease the resistance of the drift region of the diode.

7.3   Two-Dimensional Modeling of a Device

The behavior of a pn diode will now be illustrated with further numerical studies, in two dimensions. First, we need to examine carefully the input files we shall use. This will be the first time in this text where we have used a 2D input file in an example where the emphasis was on the physical system it represents. A detailed review of the purpose of the file's contents is given below, although the file is remarkable for how similar it is to the 1D files which have gone before. The main thing which is new here is the use of region labels (such as 'ANODE') in the setting up of the equations. These regions are defined in the mesh specification file. The mesh specification file defines the mesh that is used, and is described in the second section below. The equation specification file is, to a large extent, independent of the details of the geometry, because of the use of the labels.

7.3.1   2D Input File

The file which follows is pn05.sg, which uses the mesh specification file pn05.sk.

Input File pn05.sg

```
// FILE:  sgdir\pn05\pn05.sg
// specify irregular mesh
mesh "pn05.msh";

// physical constants and built-in potential
const T   = 300.0;              // ambient temperature (K)
const e   = 1.602e-19;          // electron charge (C)
const kb  = 1.381e-23;          // Boltzmann's constant (J/K)
const eo  = 8.854e-14;          // permittivity of vacuum (F/cm)
const eSi = 11.8;               // relative permittivity of Si
const eps = eSi*eo;             // permittivity of Si (F/cm)
const Ni  = 1.25e10;            // intrinsic concentration of Si (cm^-3)
const Vt  = kb*T/e;             // thermal voltage (V)

// scaling constants
const x0  = max(Wdiode,Ddiode); // distance scaling (cm)
const T0  = T;                  // temperature scaling (K)
const V0  = kb*T0/e;            // electrostatic-potential scaling (V)
const C0  = max(Na,Nd);         // concentration scaling (cm^-3);
const D0  = 35.0;               // diffusion-coefficient scaling (cm^2/s)
const u0  = D0/V0;              // mobility-coefficient scaling (cm^2/V/s)
const R0  = D0*C0/sq(x0);       // recombination-rate scaling (cm^-3/s)
const t0  = sq(x0)/D0;          // time scaling (s)
const E0  = V0/x0;              // electric-field scaling (V/s)
const J0  = e*D0*C0/x0;         // current-density scaling (A/cm^2)
const L0  = V0*eo/(e*sq(x0)*C0);// Laplacian scaling constant


// 1D Cartesian discretized gradient operator
func grad(f1,f2,<h>)
  return (f2-f1)/h;

// equilibrium electron concentration of uniformly doped silicon
func Neq(<N>)
  assign temp = [abs(N)+sqrt(sq(N)+4*sq(Ni))]/2.0;
  return (N>=0)*temp + (N<0)*sq(Ni)/temp;

// equilibrium hole concentration of uniformly doped silicon
func Peq(<N>)
  assign temp = [abs(N)+sqrt(sq(N)+4*sq(Ni))]/2.0;
```

```
  return (N<=0)*temp + (N>0)*sq(Ni)/temp;

// electron lifetime
func LTn(<N>)
  const LT0  = 3.95e-4;
  const Nref = 7.1e15;
  return LT0/(1.0+N/Nref);

// hole lifetime
func LTp(<N>)
  const LT0  = 3.52e-5;
  const Nref = 7.1e15;
  return LT0/(1.0+N/Nref);

// scaled electron mobility
func MUn(<N>,<T>,pn)
  const Al = 1430.0,  Bl = -2.20;
  const Ai = 4.61e17, Bi = 1.52e15;
  assign NU  = abs(N)*C0;
  assign pnU = (pn>0.0)*pn*sq(C0)+(pn<0.0);
  assign MUl = Al*pow((T/300.0),Bl);
  assign MUi = {Ai*T*sqrt(T)/NU}/{ln(1+Bi*sq(T)/NU)-Bi*T*T/(NU+Bi*sq(T))};
  assign MUc = {2e17*T*sqrt(T/pnU)}/ln(1+8.28e8*sq(T)*pow(pnU,-1/3));
  assign X   = sqrt(6*MUl*(MUi+MUc)/(MUi*MUc));
  return MUl*{1.025/[1+pow(X/1.68,1.43)]-0.025}/u0;

// scaled hole mobility
func MUp(<N>,<T>,pn)
  const Al = 495.0,   Bl = -2.20;
  const Ai = 1.00e17, Bi = 6.25e14;
  assign NU  = abs(N)*C0;
  assign pnU = (pn>0.0)*pn*sq(C0)+(pn<0.0);
  assign MUl = Al*pow((T/300.0),Bl);
  assign MUi = {Ai*T*sqrt(T)/NU}/{ln(1+Bi*sq(T)/NU)-Bi*T*T/(NU+Bi*sq(T))};
  assign MUc = {2e17*T*sqrt(T/pnU)}/ln(1+8.28e8*sq(T)*pow(pnU,-1/3));
  assign X   = sqrt(6*MUl*(MUi+MUc)/(MUi*MUc));
  return MUl*{1.025/[1+pow(X/1.68,1.43)]-0.025}/u0;

// scaled recombination
func R(n,p,<tn1>,<tp1>)
  const tn = 4.9e-5/t0;
  const tp = 4.2e-5/t0;
  const Cn = 2.8e-31/[D0/sq(C0*x0)];// scaled electron Auger coefficient
  const Cp = 1.2e-31/[D0/sq(C0*x0)];// scaled hole Auger coefficient
  const ni = Ni/C0;                 // scaled intrinsic concentration
  const pn = sq(ni);                // scaled intrinsic concentration squared
  assign Rsrh = (n*p-pn)/[tp*(n+ni)+tn*(p+ni)];
  assign Raug = (n*p-pn)*(Cn*n+Cp*p);
```

```
    return Rsrh+Raug;

// scaled electric field
func E(V1,V2,<h>)
  return -grad(V1,V2,h);

// scaled electron current density
func Jn(n1,n2,E,un1,un2,<h>)
  const Ut    = Vt/V0;              // scaled thermal voltage
  assign un   = ave(un1,un2);
  assign dV   = E*h/(2*Ut);
  assign n    = n1*aux2(dV)+n2*aux2(-dV);
  assign dndx = aux1(dV)*grad(n1,n2,h);
  return un*(n*E+dndx);

// scaled hole current density
func Jp(p1,p2,E,up1,up2,<h>)
  const Ut    = Vt/V0;              // scaled thermal voltage
  assign up   = ave(up1,up2);
  assign dV   = E*h/(2*Ut);
  assign p    = p1*aux2(-dV)+p2*aux2(dV);
  assign dpdx = aux1(dV)*grad(p1,p2,h);
  return up*(p*E-dpdx);


// set numerical algorithm parameters
set NEWTON DAMPING    = 0;          // try setting this to 3 ... why
set NEWTON ACCURACY   = 1.0e-12;    // does it take more iterations to
set NEWTON ITERATIONS = 100;        // converge with damping than without?
set LINSOL ALGORITHM  = GAUSSELIM;
set LINSOL FILL       = INFINITY;
set DISTANCE SCALE    = x0;


// declare variables and specify which ones are knowns and unknowns
var     V[NODES],   n[NODES],   p[NODES],   tn[NODES],   tp[NODES];
unknown V[SI],      n[SI],      p[SI];
known   V[ANODE],   n[ANODE],   p[ANODE];
known   V[CATHODE], n[CATHODE], p[CATHODE];


// scaled Poisson's equation
equ V[i=SI] ->
  L0*nsum(i,j,all,{eSi*E(V[i],V[node(i,j)],elen(i,j))}*ilen(i,j)) -
  (p[i]-n[i]+C[i])*area(i) = 0.0;

// scaled electron continuity equation
equ n[i=SI] ->
```

```
  +nsum(i,j,all,
    {Jn(n[i],n[node(i,j)],E(V[i],V[node(i,j)],elen(i,j)),MUn(C[i],T,n[i]*p[i]),
        MUn(C[node(i,j)],T,n[node(i,j)]*p[node(i,j)]),elen(i,j))}*ilen(i,j)) -
  R(n[i],p[i],tn[i],tp[i])*area(i) = 0.0;

// scaled hole continuity equation
equ p[i=SI] ->
  -nsum(i,j,all,
    {Jp(p[i],p[node(i,j)],E(V[i],V[node(i,j)],elen(i,j)),MUp(C[i],T,n[i]*p[i]),
        MUp(C[node(i,j)],T,n[node(i,j)]*p[node(i,j)]),elen(i,j))}*ilen(i,j)) -
  R(n[i],p[i],tn[i],tp[i])*area(i) = 0.0;


begin InitializeVariables
  assign n[i=all] = Neq(C[i]);
  assign p[i=all] = Peq(C[i]);
  assign V[i=all] = Vt*ln(n[i]/Nd);
  assign tn[i=all] = LTn(abs(C[i]));
  assign tp[i=all] = LTp(abs(C[i]));
end

begin ScaleVariables
  assign x[i=all] = x[i]/x0;
  assign V[i=all] = V[i]/V0;
  assign C[i=all] = C[i]/C0;
  assign n[i=all] = n[i]/C0;
  assign p[i=all] = p[i]/C0;
  assign tn[i=all] = tn[i]/t0;
  assign tp[i=all] = tp[i]/t0;
end

begin UnscaleVariables
  assign x[i=all] = x[i]*x0;
  assign V[i=all] = V[i]*V0;
  assign C[i=all] = C[i]*C0;
  assign n[i=all] = n[i]*C0;
  assign p[i=all] = p[i]*C0;
  assign tn[i=all] = tn[i]*t0;
  assign tp[i=all] = tp[i]*t0;
end


begin main
  call InitializeVariables;
  call ScaleVariables;
  solve;
  call UnscaleVariables;
  write;
```

end

The mesh specification file pn05.sk which is used to generate the mesh file pn05.msh is described in the next section.

The first part of the mesh specification file defines constants. Some of the constants are scaling constants. Some of the others are scaled to make them more convenient for computations. Functions are then defined for use in the equations which follow. Next a short section declares that certain variables are either 'known' or 'unknown'. The matrix solver does not attempt to solve for 'known' variables.

Use of Region Labels

The variables and array elements that we specify in a 'known' or 'unknown' statement or in an equation header is not simply (say) the voltage at all points on the mesh. When we say that the voltage is known or unknown, we also must specify that we mean the voltage at a given point or points. In other words, when specifying what is known or unknown, the points at which the quantity is known or unknown are part of the specification. The same is true for equation headers.

The list of points after the variable name is given in terms of the region. The use of the word ANODE in $V[ANODE]$, makes it clear precisely which voltages are involved. Finally, if we say $V[all]$ is unknown but later say that $V[ANODE]$ is known, there is a clear contradiction. A later statement, such as this, overrides an earlier one, when declaring known and unknown variables. In this case, the voltages at all points except the anode and cathode are unknown. In contrast, if two equation headers refer to the same variables, the one which came first is used.

Next in the input file comes a list of equations, which are used to determine the value of the 'unknown' quantity specified after the keyword 'equ'. These equations also use region labels such as $i = SI$, in this instance to determine where the equations apply. In the equation statements, as was explained in the last paragraph, earlier equations overrule later ones. This was done because at a boundary we deal with points specified on an edge such as $i = SISIO2$, but these edge points are also in the region $i = Si$. Using the earlier equation to overrule the later one simplifies the specification of the boundary condition. The boundary condition is given first, so that the points on the boundary will use the boundary condition, not the bulk equation. This is despite the fact that the boundary points are part of both the boundary region and the bulk region.

There should be as many equations as unknowns, and the unknown quantity must appear in the equation which is supposed to determine it. One function of these statements is to determine the topology of the Jacobian matrix, which is used in the solution of the set of coupled equations. The derivative of the equation with respect to the variable after the keyword 'equ' will be on the diagonal of the Jacobian matrix. These statements are all relatively clear, except that some of them make use of the functions 'lsum' and 'nsum' to implement finite-difference expressions for divergences on an irregular mesh. (See Section 15.3.4). The way these work is explained in Chapter 15. The meaning of the expressions appearing inside nsum will be reviewed next.

The Neighbor Sum

The nsum functions in the file above often contain a difference of two voltages, divided by 'elen' and then multiplied by 'ilen'. The quantity 'elen(i,j)' is the distance between the node i and its jth neighbor — j is not the index of the jth node. Thus the difference in voltages which appears in the file is $V[i] - V[node(i,j)]$, $node(i,j)$ being the index of the jth neighbor of node i, and this is divided by $elen(i,j)$. This is just $-\frac{\partial V}{\partial \ell}$, where the derivative with respect to ℓ means the derivative along the line from the node i to the jth neighbor. This, in turn, is equal to the electric-field component in that direction, along the line from node i to the jth neighbor. This electric field is multiplied by the length of the surface it passes through at right angles to find the flux of electric field leaving the node i toward the jth neighbor. (As in all the 2D simulations, we assume a unit height in the direction perpendicular to the mesh. The unit height converts $ilen$ into an area, and $area(i)$ into a volume.) That length is '$ilen$', the length of the perpendicular bisector of the line from node i to the jth neighbor. By summing over all the neighbouring nodes using nsum, the total flux leaving node i can then be approximated.

The summation functions are implemented in such a way that if the bulk equations are used all the way up to a boundary, there will be no flux through the boundary. For this reason the 'default' boundary condition in these simulations is a no-flux boundary condition. Whenever a boundary condition is not explicitly specified in the input file, there will be no flux through the boundary in question. See Section 6.8. In the case of Poisson's equation, this means there will be no flux of electric field, i.e. no normal electric field, through the boundary.

The next statements specify how the system of coupled, nonlinear equations will be solved. Again, these were described in Chapter 6. The Newton-method parameters are given; the linear matrix solution will be done by Gaussian elimination; and so on.

Three routines, which initialize variables (using equilibrium values of the densities, for instance), which scale the variables and which restore them (by undoing the scaling) are defined next. These are called later, in the 'main' routine.

The main routine first calls initializing and scaling routines. Next, the statement 'solve' causes the values of the unknowns to be computed, using the equations given by 'equ' statements. Then the scaling is undone; p and n are found from V; and the solution is written to data files.

Precomputed Functions

The next input file is pn06.sg. It is included to show the use of 'precomputed functions'. As was discussed in Section 6.3.7, precomputed functions are employed to speed up the solution. (See the user's manual, Section 15.3.6.) Those parts of the file which are identical to pn05.sg are left out, where possible, in which case the title of the section or a comment is given.

Input File pn06.sg

```
// FILE:  sgdir\ch07\pn06\pn06.sg
// specify irregular mesh
// physical constants and built-in potential
// scaling constants
// 1D Cartesian discretized gradient operator
// equilibrium electron concentration of isotropically doped silicon
// equilibrium hole concentration of isotropically doped silicon
// electron lifetime
// hole lifetime
// scaled electron mobility
// unscaled hole mobility
// scaled recombination
// scaled electric field
// scaled electron current density
// scaled hole current density
// set numerical algorithm parameters
// declare variables and specify which ones are knowns and unknowns
// for each Newton iteration, precompute the following functions prior to
// constructing the Jacobian matrix
precompute @ NODE i -> MUn(C[i],T,n[i]*p[i]);
precompute @ NODE i -> MUp(C[i],T,n[i]*p[i]);
precompute @ EDGE i (j,k) ODD -> E(V[j],V[k],elen);
precompute @ EDGE i (j,k) ODD -> Jn(n[j],n[k],@E(i,j,k),@MUn(j),@MUn(k),elen);
precompute @ EDGE i (j,k) ODD -> Jp(p[j],p[k],@E(i,j,k),@MUp(j),@MUp(k),elen);


// scaled Poisson's equation
equ V[i=SI] ->
  L0*nsum(i,j,all,{eSi*@E(edge(i,j),i,node(i,j))}*ilen(i,j)) -
  (p[i]-n[i]+C[i])*area(i) = 0.0;

// scaled electron continuity equation
equ n[i=SI] ->
  +nsum(i,j,all,{@Jn(edge(i,j),i,node(i,j))}*ilen(i,j)) -
  R(n[i],p[i],tn[i],tp[i])*area(i) = 0.0;

// scaled hole continuity equation
equ p[i=SI] ->
  -nsum(i,j,all,{@Jp(edge(i,j),i,node(i,j))}*ilen(i,j)) -
  R(n[i],p[i],tn[i],tp[i])*area(i) = 0.0;


// initialize variables; scale them; main program; unscale them and end
// all as in the previous file pn05.sg
```

7.3.2   Review of Mesh Construction

The first thing either of these (equation specification) '.sg' files does is look for another file called 'pn05.msh' or 'pn06.msh', which is a mesh file. (Reading the mesh file does not have to be done first, but it usually is.) This in turn is generated from a mesh specification file such as 'pn05.sk'. The next topic we shall address will be the layout and contents of the mesh specification file pn05.sk, which is given below.

Skeleton File pn05.sk

```
CONST Wdiode   = 9.0e-4;
CONST Ddiode   = 9.0e-4;
CONST Wanode   = 2.0e-4;
CONST Wcathode = 2.0e-4;
CONST Wspacing = 0.5e-4;

point pA = (0.0e-4,             0.0e-4);
point pB = (Wanode,            0.0e-4);
point pC = (Wdiode-Wcathode,  0.0e-4);
point pD = (Wdiode,            0.0e-4);
point pE = (0.0e-4,           -Ddiode);
point pF = (Wanode,           -Ddiode);
point pG = (Wdiode-Wcathode,  -Ddiode);
point pH = (Wdiode,           -Ddiode);

edge eAB = ANODE    [pA, pB] (Wspacing, 0.0);
edge eBC = NOFLUX   [pB, pC] (Wspacing, 0.0);
edge eCD = CATHODE  [pC, pD] (Wspacing, 0.0);
edge eEF = NOFLUX   [pE, pF] (Wspacing, 0.0);
edge eFG = NOFLUX   [pF, pG] (Wspacing, 0.0);
edge eGH = NOFLUX   [pG, pH] (Wspacing, 0.0);
edge eAE = NOFLUX   [pA, pE] (Wspacing, 0.0);
edge eBF =          [pB, pF] (Wspacing, 0.0);
edge eCG =          [pC, pG] (Wspacing, 0.0);
edge eDH = NOFLUX   [pD, pH] (Wspacing, 0.0);

region rAEFB = SI {eAE, eEF, eBF, eAB} RECTANGLES;
region rBFGC = SI {eBF, eFG, eCG, eBC} RECTANGLES;
region rCGHD = SI {eCG, eGH, eDH, eCD} RECTANGLES;


const Na  = 1.00e+19;
const Nd  = 1.00e+14;
const Rx  = 2.00e-04;
const Ry  = 2.50e-04;
const Ax  = ln(Na/Nd)/sq(Rx);
const Ay  = ln(Na/Nd)/sq(Ry);
```

```
coordinates x, y;

refine C (SignedLog, 1.0) = Nd-(Na+Nd)*ngdep(x,y,2.0*Wanode,Ax,Ay);

set minimum divisions = 0;
set maximum divisions = 3;
```

The first parts of the mesh specification file are largely self-explanatory. (A schematic of the device is often given first, with each line starting with a double slash to make that line into a comment.) Constants are defined which will be used to specify the mesh. Then points, edges and regions are built up. Finally, constants describing the doping, and the refinement function (which makes use of the doping level) are specified. These parts of the file are discussed, in order, in the sections which follow.

Points, Edges and Regions

After the constants we come to the heart of the mesh specification file, which defines the mesh skeleton. The mesh skeleton is the minimum amount of information needed to specify the entire mesh. The mesh is specified hierarchically.

First, points are defined, in terms of their coordinates (which, in this file, are x and y.) The points are identified, in this case, as 'pA', 'pB', etc. The names given to points, edges and regions are arbitrary, but a notation which makes names of points begin with a 'p' is useful.

From points, edges can be built up. The edge 'eAB' is the line from point 'pA' to 'pB', and is the anode, according to the line which defines it. The data at the end of the line specifies the mesh spacing along the edge, before any refinement of the mesh is done. Similarly other edges are constructed.

Next, regions are defined which are enclosed by groups of edges. For instance region $rAEFB$ is made up of the four edges joining points A, E, F and B. (Edges must be listed in counterclockwise order. Opposite edges must be identical, except for a shift in one coordinate, if the region is to be divided into rectangles. They must have the same orientation with the same spacing. For example, edge eAB has the same orientation as eEF but edge eBA does not.) The word 'rectangles' implies that the mesh inside the region is to be made up of rectangles. Three regions are set up here, which include the whole problem.

Mesh Refinement

After the initial mesh geometry has been given, the remaining data deals with refinement of that mesh. Constants which are needed to specify the refinement function are defined. (These constants describe the doping, in the cases we set up in this book.) Then a refinement function C is given which indicates to the refinement program how much the initial mesh needs to be refined as a function of position. After the refinement function is specified, the maximum and minimum allowable mesh spacings and numbers of divisions are stated. The doping profile, which is used to specify the refinement function, is set up using functions n_{sdep} and n_{gdep}, which will now be described.

The Doping Profile; n_{sdep} and n_{gdep}

The doping profile is modeled analytically, by means of two functions, n_{sdep} and n_{gdep}, which are defined using the following expressions:

```
-----------------------------------------------------------------------
 * nsdep:   This function returns an approximate deposition profile of a step
 *          implant driven in an inert environment.
 *
 *                      1        W/2 + x          W/2 - x
 *          nsdep(x,W,Dt) = - (erf(---------) + erf(----------))
 *                      2        2*sqrt(Dt)       2*sqrt(Dt)
 */
REAL nsdep(REAL x, REAL W, REAL Dt)
{
  REAL D  = 2.0 * sqrt(Dt);
  REAL Wh = W / 2.0;
  return 0.5 * (erf((Wh + x)/D) + erf((Wh - x)/D));
}



/*
 -----------------------------------------------------------------------
 * ngdep:   This function returns an approximate Gaussian deposition.
 */
REAL ngdep(REAL x, REAL y, REAL W, REAL ax, REAL ay)
{
  REAL xprime = fabs(x) - W/2.0;
  return ((xprime <= 0.0) ? 1.0 : exp(-ax*xprime*xprime))*exp(-ay*y*y);
}
```

The Error Function

To make sense of the function n_{sdep} we have to go back to the definition of the error function, which was given in Chapter 3, on transport.

$$erf(x) = \frac{2}{\sqrt{\pi}} \int_0^x exp(-s^2)\, ds. \tag{7.65}$$

The error function can be plotted using matlab. The set of commands given in the table produced Figure 7.23. (They were put in a file called errorf.m, to avoid a naming conflict with matlab's built-in function erf.) The figures suggest a potential source of confusion: if the argument of the error function is less than zero, should the error function be taken to be zero, since the range of integration is now from zero

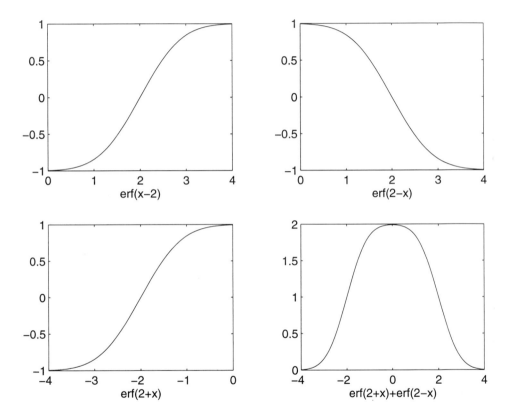

Figure 7.23. Plots of the error function, for different arguments of the function, and for combinations of error functions.

Script File to Plot Error Functions

```
subplot(2,2,1),fplot('erf(x-2)',[0 4]),xlabel('erf(x-2)')
subplot(2,2,2),fplot('erf(2-x)',[0 4]),xlabel('erf(2-x)')
subplot(2,2,3),fplot('erf(2+x)',[-4 0]),xlabel('erf(2+x)')
subplot(2,2,4),fplot('erf(2+x)+erf(2-x)',[-4 4]),xlabel('erf(2+x)+erf(2-x)')
```

to a negative number, or should we evaluate the integral out to a negative upper limit? If the argument is allowed to be negative and the integral evaluated out to the (negative) argument, the error function will be antisymmetric. This is actually the correct interpretation. The first two frames of Figure 7.23 show $erf(x-2)$ and $erf(2-x)$. erf is indeed antisymmetric and $erf(x-a) + erf(a-x)$ is zero. However, the function used in n_{sdep} actually uses $erf(a+x) + erf(a-x)$, as is shown in the last frame for $a = 2$. The third frame shows $erf(2+x)$, and this must be added to the second frame, which shows $erf(2-x)$, to get the final frame, which is equal to twice n_{sdep}. Notice the different bottom axis in each case.

All of the plots, except the first, show error functions which depend on the variable x, in the form $erf(2 \pm x)$. The fourth plot is centered on $x = 0$ and the curve shown extends about 2 units on either side, from about $x = -2$ to $x = 2$. The function n_{sdep} involves $erf\left(\frac{W/2 \pm x}{\sqrt{4Dt}}\right)$. We can thus expect n_{sdep} to look like the fourth of these figures. It will still be centered on $x = 0$. Its width will be roughly $W/2$ on either side of this center, from $x = -W/2$ to $x = W/2$. This is because the width of n_{sdep} is determined by the distance between the points where the two error functions 'switch'. One has a transition near where $W/2 + x$ is zero, and the other has a transition where $W/2 - x$ is zero.

If n_{sdep} is expressed in terms of a 'shifted' coordinate x, so that we use $n_{sdep}(x - \Delta, W, Dt)$, then the function is shifted over by a distance Δ. It then extends from about $x = \Delta - W/2$ to $x = \Delta + W/2$ and is centered at $x = \Delta$. At the edges of this region n_{sdep} drops from its maximum value to nearly zero over a distance of a few times $\sqrt{4Dt}$ (which is a measure of the width of the Gaussians which were integrated over to get the error functions).

A physical diffusion will have its peak at the edge where the dopant diffuses in. The shift Δ will be chosen so that the peak, which is the 'center' of the one-dimensional profile, is at the surface so at the surface $x = \Delta$. In two dimensions, with (say) y varying normal to the surface and x varying parallel to the surface, a product of functions is used which is of the form $n_{sdep}(x - \Delta_x, W_x, Dt) \times n_{sdep}(y - \Delta_y, W_y, Dt)$. If the surface is at $y = L_1$, then we must have $\Delta_y = L_1$.

Mesh Generation Commands

The commands given at the terminal to generate the mesh and simulation are now demonstrated. It is quite interesting to run this mesh specification file, to see the

Script to Run pn05.sg

```
mesh pn05.sk
sggrid -vc -aspect pn05.xsk
sgbuild ref pn05_ref
pn05_ref -vc -aspect
sgxlat -p16 pn05.sg
sgbuild sim pn05
order pn05.top pn05.prm
pn05 -vs2 -v12 -Lpn05.log
```

Windows which are opened to show the mesh must be closed before the simulation can be continued.

graphical output it produces; this is discussed in more detail in the next section. The first four lines the script given in the table generate the mesh. The last four build the simulation and run it. The meaning of all of these commands is described in the User's Manual, which is Chapter 15.

7.3.3   Graphical Output

This run produces two pictures of the mesh along the way. First, the mesh skeleton is plotted. (Click on the plot when ready to to move on. The run will not continue until the plot is removed.) Next it will plot the refined mesh. (Again, click on the plot to continue, or the run will remain paused.) It then runs SGFramework.

The results may be plotted using SGFramework's plotting routines. Because it is not possible in general to solve semiconductor problems on a regular mesh, and because most plotting software is only useful for regular meshes, SGFramework has its own plotting capabilities. The command

```
triplot pn05 V
```

will use the data in the file 'pn05.res' and plot the variable V as a surface plot. The other variables n and p can also be substituted for V. To make a postcript file of the results, the option '-ps' should be added at the end of the line, i.e., it should say

```
triplot pn05 V -ps
```

The postcript file will have the same name as the simulation, with a '.ps' on the end, so in this case it will be pn05.ps. If the variable V were omitted the user would be prompted for the variable to be plotted. If there is more than one set of data (say from multiple time steps) an option '-i2' after the simulation name will plot the second; '-i1' plots the first, and so on. If in doubt what to do, the plotter has a help feature; type

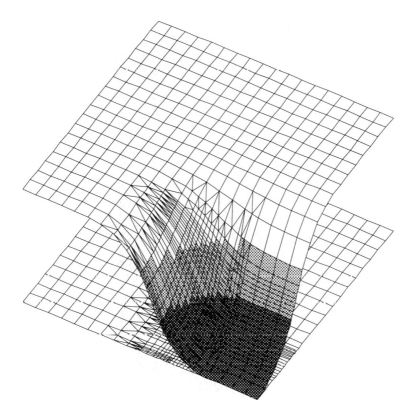

Figure 7.24. The potential V produced by the simulation file pn05.sg. The mesh is based on rectangles - see the mesh specification file pn05.sk.

```
triplot -h
```

These plotting routines make use of the mesh generated from the mesh specification file, so they will not work in cases where we did not generate such a file.

Using the command

```
triplot pn05 V -ps
```

a file such pn05.ps can be made, as shown. If the command is changed to (say)

```
triplot pn05 V -az-100 -ps
```

the azimuthal angle will be altered from its default of 30 degrees to -100 degrees. A similar option exists for the elevation angle.

7.4   Small-Signal Analysis

Small-signal analysis is conveniently divided into the two cases of forward and reverse bias. In reverse-bias the device is primarily capacitive, with some small conductance.

7.4.1   Reverse-Bias Characteristics

In reverse-bias the pn junction behaves like a capacitor. This capacitor is formed when the majority carriers are repelled from the junction. When the voltage across the junction is changed, the distance the majority carriers are repelled is also changed. This causes more or less of the ionized impurities to be exposed in the region vacated by the carriers. Equal and opposite charges $\pm Q$ are set up in the junction, on opposite sides, so that the junction remains overall charge neutral. Because the junction is neutral overall, field lines which start on one side of the junction can end on the other side without penetrating the bulk semiconductor. Suppose the junction is of width W, and a charge of Q is added on one side of the junction, and a charge $-Q$ is added on the other. (The charge Q may be the result of changing the amount of ionized impurities which are exposed when W changes slightly.) Then the extra voltage due to the charge $\pm Q$ is $V = EW$, where E is the extra electric field due to Q. The extra electric displacement D is equal to the extra charge per unit area, Q/A, where A is the junction area, and $D = \epsilon E$, so $V = \frac{Q}{\epsilon A} W$. The capacitance is

$$C = \frac{Q}{V} = \frac{\epsilon A}{W}. \tag{7.66}$$

The reverse-bias conductance can be found simply if we can assume conditions are quasi-static. This means that the relationship between I and V is the same as in steady state, but each varies only slightly about its mean value. The conductance is defined to be $G = \frac{dI}{dV}$ evaluated at the bias point. Since $I = I_O(\exp(\frac{eV_A}{k_BT}) - 1)$ then

$$G = \frac{eI_O}{k_BT} \exp\left(\frac{eV_A}{k_BT}\right) = \frac{e}{k_BT}(I + I_O). \tag{7.67}$$

7.4.2   Forward-Bias Characteristics

In addition to the majority carriers, whose movement gives rise to the capacitance that is found in reverse-bias, in forward-bias there may be significant numbers of minority carriers near the junction. The two sets of carriers give rise to two contributions to the capacitance which are combined in parallel with each other. The minority carriers respond less quickly than the majority carriers to fluctuations. The minority carriers are created and destroyed by generation and recombination. They are thus more likely to oscillate out of phase with the applied voltage. Their response, therefore, may not be purely capacitive.

The small oscillations in carrier density are added to the steady-state density at the bias point. The steady-state density satisfies the diffusion equation, and since the equation is linear, the small extra density must also satisfy the diffusion-equation. The oscillations depend on time, so the diffusion-equation is

$$\frac{\partial \tilde{p}}{\partial t} - D_p \frac{\partial^2 \tilde{p}}{\partial x^2} = -\frac{\tilde{p}}{\tau_p}. \tag{7.68}$$

If $\tilde{p} = Re[p_o \exp(j\omega t)]$, where $Re[A]$ is the real part of A, and p_o is the phasor density perturbation, then $\frac{\partial \tilde{p}}{\partial t}$ equals $Re[j\omega p_o \exp(j\omega t)]$. Omitting the real-part symbols and the factor of $\exp(j\omega t)$ in each term, the equation for the phasor p_o is

$$D_p \frac{\partial^2 p_o}{\partial x^2} = (j\omega + \frac{1}{\tau_p})p_o. \tag{7.69}$$

This is just the usual steady-state diffusion equation, but with the recombination time τ_p replaced by $\tau_p/(1 + j\omega\tau_p)$.

The boundary condition on the perturbed density far from the junction is still that it is zero. At the junction edge the usual boundary condition on the density is used, namely

$$\Delta p(x = -x_n) = \frac{n_i^2}{N_D} \left[\exp\left(\frac{e(V_A + v_a)}{k_B T} \right) - 1 \right] \tag{7.70}$$

where the voltage is a dc part V_A plus an alternating part v_a. If $v_a/k_B T \ll 1$, then $\exp(\frac{e(V_A + v_a)}{k_B T}) \simeq \exp(\frac{eV_A}{k_B T})(1 + \frac{v_a}{k_B T})$, so Δp is

$$\Delta p(x = -x_n) = \frac{n_i^2}{N_D} \left[\exp\left(\frac{eV_A}{k_B T} \right) \left(1 + \frac{v_a}{k_B T} \right) - 1 \right]. \tag{7.71}$$

The oscillating part of this is the term with the oscillating voltage v_a in it. The rest, which is left over when $v_a = 0$, is the usual dc excess density. The oscillating part is

$$p_o(x = -x_n) = \frac{n_i^2}{N_D} \exp\left(\frac{eV_A}{k_B T} \right) \frac{v_a}{k_B T}. \tag{7.72}$$

This boundary condition on the oscillating part of the density is the same as the dc boundary condition but with an extra factor of $\frac{v_a}{k_B T}$. The overall oscillating density amplitude will be the same as the dc density except

1. it will be multiplied by $\frac{v_a}{k_B T}$,

2. τ_p will be replaced by $\tau_p/(1 + j\omega\tau_p)$.

The dc current varies as the density divided by $\sqrt{\tau_p}$, so the current due to the oscillation in density has an extra factor of $\frac{v_a}{k_B T}$ from the density, and when $\tau_p^{-1/2}$ is replaced by $\sqrt{\frac{1+j\omega\tau_p}{\tau_p}}$ it has an extra factor of $\sqrt{1 + j\omega\tau_p}$.

$$i = I\frac{v_a}{k_B T}\sqrt{1 + j\omega\tau_p} = I_O \exp\left(\frac{eV_A}{k_B T} \right) \frac{v_a}{k_B T}\sqrt{1 + j\omega\tau_p}. \tag{7.73}$$

The ratio of i to v_a was found in exactly the same way as the low-frequency conductance, with one exception. The low-frequency conductance was found using a quasi-static approximation, so ω was set to zero. The voltage in the exponential was expanded, using a small v_a to find p_o, then dividing by v_a. This is equivalent to simply differentiating p with respect to V, which is effectively what was done when finding $\frac{dI}{dV}$.

The ac admittance is therefore just the dc conductivity multiplied by $\sqrt{1 + j\omega\tau_p}$.

$$Y = \sqrt{1 + j\omega\tau_p}\, G_{dc} = G_{rf} + j\omega C_{rf} \tag{7.74}$$

so the rf parameters are

$$G_{rf} = \left(\sqrt{1 + \omega^2 \tau_p^2} + 1\right)^{1/2} G_{dc}/\sqrt{2} \tag{7.75}$$

and the rf capacitance is

$$C_{rf} = \left(\sqrt{1 + \omega^2 \tau_p^2} - 1\right)^{1/2} G_{dc}/\sqrt{2}\omega. \tag{7.76}$$

7.4.3   Numerical Modeling of Small-Signal Behavior

In this section we present a simulation which is only slightly different from the dc simulations given in Section 7.2.2, and apply it to the small-signal analysis which has been handled analytically in this section. This input file could be used with only very minor changes to model transients, and other time-dependent phenomena, in the pn junction. For this reason the modeling of transients will be left to the exercises. The input file is pn04.sg. The differencing with respect to time in this file results in an implicit scheme.

Input File pn04.sg

```
// FILE:  sgdir\pn04\pn04.sg
// mesh constants
const NX  = 101;                      // number of x-mesh points
const LX  = NX - 1;                   // index of last x-mesh point
const DX  = 5.0e-6;                   // mesh spacing (cm)

// physical constants
const T   = 300.0;                    // ambient temperature (K)
const e   = 1.602e-19;                // electron charge (C)
const kb  = 1.381e-23;                // Boltzmann's constant (J/K)
const eo  = 8.854e-14;                // permittivity of vacuum (F/cm)
const eSi = 11.8;                     // relative permittivity of Si
const eps = eSi*eo;                   // permittivity of Si (F/cm)
const Ni  = 1.25e10;                  // intrinsic concentration of Si (cm^-3)
const Vt  = kb*T/e;                   // thermal voltage (V)
const PI  = 3.14159265358979323846; //
```

```
// simulation constants
const Vdc = -1.0;                   // dc bias (V)
const dVd = 0.25;                   // dc voltage step (V)
const Ass = 0.05;                   // small-signal amplitude (V)
const fss = 1.0e3;                  // small-signal frequency (Hz)
const wss = 2*PI*fss;               // small-signal angular frequency (radians)

// doping profile constants
const Na  = 1.0e+15;                // acceptor concentration on p-side (cm^-3)
const Nd  = 1.0e+15;                // donor concentration on n-side (cm^-3)
const Vbi = Vt*ln(Na*Nd/sq(Ni));    // built-in potential (V)

// scaling constants
const x0  = LX*DX;                  // distance scaling (cm)
const T0  = T;                      // temperature scaling (K)
const V0  = kb*T0/e;                // electrostatic-potential scaling (V)
const C0  = max(Na,Nd);             // concentration scaling (cm^-3);
const D0  = 35.0;                   // diffusion-coefficient scaling (cm^2/s)
const u0  = D0/V0;                  // mobility-coefficient scaling (cm^2/V/s)
const R0  = D0*C0/sq(x0);           // recombination-rate scaling (cm^-3/s)
const t0  = sq(x0)/D0;              // time scaling (s)
const E0  = V0/x0;                  // electric-field scaling (V/s)
const J0  = e*D0*C0/x0;             // current-density scaling (A/cm^2)
const L0  = V0*eo/(e*sq(x0)*C0);    // Laplacian scaling constant

// 1D Cartesian discretized gradient operator
func grad(f1,f2,<h>)
  return (f2-f1)/h;

// 1D Cartesian discretized divergence operator
func div(f1,f2,<h>)
  return (f2-f1)/h;

// equilibrium electron concentration of uniformly doped silicon
func Neq(<N>)
  assign temp = [abs(N)+sqrt(sq(N)+4*sq(Ni))]/2.0;
  return (N>=0)*temp + (N<0)*sq(Ni)/temp;

// equilibrium hole concentration of uniformly doped silicon
func Peq(<N>)
  assign temp = [abs(N)+sqrt(sq(N)+4*sq(Ni))]/2.0;
  return (N<=0)*temp + (N>0)*sq(Ni)/temp;

// unscaled electron mobility
func MUn(<N>,<T>)
  const Al = 1430.0,  Bl = -2.20;
  const Ai = 4.61e17, Bi = 1.52e15;
```

```
  assign MUl = Al*pow((T/300.0),Bl);
  assign MUi = {Ai*T*sqrt(T)/N}/{ln(1+Bi*sq(T)/N)-Bi*T*T/(N+Bi*sq(T))};
  assign X   = sqrt(6.0*MUl/MUi);
  return MUl*{1.025/[1+pow(X/1.68,1.43)]-0.025};

// unscaled hole mobility
func MUp(<N>,<T>)
  const Al = 495.0,   Bl = -2.20;
  const Ai = 1.00e17, Bi = 6.25e14;
  assign MUl = Al*pow((T/300.0),Bl);
  assign MUi = {Ai*T*sqrt(T)/N}/{ln(1+Bi*sq(T)/N)-Bi*T*T/(N+Bi*sq(T))};
  assign X   = sqrt(6.0*MUl/MUi);
  return MUl*{1.025/[1+pow(X/1.68,1.43)]-0.025};

// scaled recombination
func R(n,p)
  const tn = 1.0e-6/t0;            // scaled electron lifetime
  const tp = 1.0e-6/t0;            // scaled hole lifetime
  const ni = Ni/C0;               // scaled intrinsic concentration
  return (n*p-sq(ni))/[tp*(n+ni)+tn*(p+ni)];

// scaled electric field
func E(V1,V2,<h>)
  return -grad(V1,V2,h);

// scaled electron current density
func Jn(n1,n2,E,<un1>,<un2>,<h>)
  const Ut  = Vt/V0;              // scaled thermal voltage
  assign un   = ave(un1,un2);
  assign dV   = E*h/(2*Ut);
  assign n    = n1*aux2(dV)+n2*aux2(-dV);
  assign dndx = aux1(dV)*grad(n1,n2,h);
  return un*(n*E+dndx);

// scaled hole current density
func Jp(p1,p2,E,<up1>,<up2>,<h>)
  const Ut  = Vt/V0;              // scaled thermal voltage
  assign up   = ave(up1,up2);
  assign dV   = E*h/(2*Ut);
  assign p    = p1*aux2(-dV)+p2*aux2(dV);
  assign dpdx = aux1(dV)*grad(p1,p2,h);
  return up*(p*E-dpdx);


// set numerical algorithm parameters
set NEWTON DAMPING    = 3;
set NEWTON ACCURACY   = 1.0e-12;
set NEWTON ITERATIONS = 100;
```

```
set LINSOL ALGORITHM  = GAUSSELIM;
set LINSOL FILL       = INFINITY;


// declare variables
var x[NX], C[NX], V[NX], n[NX], p[NX], un[NX], up[NX], h[LX], Vd, Jd;
var nlast[NX], plast[NX], dt, t;

// scaled Poisson's equation
equ V[i=1..LX-1] ->
  L0*div(eSi*E(V[i-1],V[i],h[i-1]),eSi*E(V[i],V[i+1],h[i]),ave(h[i],h[i-1])) -
  (p[i]-n[i]+C[i]) = 0.0;

// scaled electron continuity equation
equ n[i=1..LX-1] ->
  +div(Jn(n[i-1],n[i],E(V[i-1],V[i],h[i-1]),un[i-1],un[i],h[i-1]),
       Jn(n[i],n[i+1],E(V[i],V[i+1],h[i  ]),un[i],un[i+1],h[i  ]),
       ave(h[i],h[i-1])) -
  R(n[i],p[i]) = (n[i]-nlast[i])/(dt/t0);

// scaled electron continuity equation
equ p[i=1..LX-1] ->
  -div(Jp(p[i-1],p[i],E(V[i-1],V[i],h[i-1]),up[i-1],up[i],h[i-1]),
       Jp(p[i],p[i+1],E(V[i],V[i+1],h[i  ]),up[i],up[i+1],h[i  ]),
       ave(h[i],h[i-1])) -
  R(n[i],p[i]) = (p[i]-plast[i])/(dt/t0);


begin SetUpDopingProfile
  assign x[i=all] = i*DX;           // initialize mesh points
  assign C[i=0..LX/2]    = +Nd;     // ionized donor concentration
  assign C[i=LX/2+1..LX] = -Na;     // ionized acceptor concentration
end

begin InitializeVariables
  assign h[i=all] = x[i+1]-x[i];
  assign V[i=all] = (i>LX/2)*-Vbi;
  assign n[i=all] = Neq(C[i]);
  assign p[i=all] = Peq(C[i]);
  assign un[i=all] = MUn(abs(C[i]),T);
  assign up[i=all] = MUp(abs(C[i]),T);
end

begin ScaleVariables
  assign x[i=all] =  x[i]/x0;
  assign h[i=all] =  h[i]/x0;
  assign V[i=all] =  V[i]/V0;
  assign C[i=all] =  C[i]/C0;
```

```
      assign  n[i=all] =  n[i]/C0;
      assign  p[i=all] =  p[i]/C0;
      assign un[i=all] = un[i]/u0;
      assign up[i=all] = up[i]/u0;
end

begin UpdateLastVariables
   assign nlast[i=all] = n[i];
   assign plast[i=all] = p[i];
end

begin main
   call SetUpDopingProfile;
   call InitializeVariables;
   call ScaleVariables;
   call UpdateLastVariables;

   // null the output file
   open "pn04.out" write;
   close;

   // increment the diode-bias from equilibrium to the desired voltage
   assign dt = 1.0e100;
   assign Vd = 0.0;
   while (sign(Vd) < sign(Vdc)) begin
     assign V[i=LX] = (Vd-Vbi)/V0;
     solve;
     assign Vd = Vd+sign(Vdc)*abs(dVd);
     call UpdateLastVariables;
   end;
   assign Vd = Vdc;
   assign V[i=LX] = (Vd-Vbi)/V0;
   solve;
   call UpdateLastVariables;

   // add the small signal to the bias voltage and run until the transients
   // decay and steady state is reached
   assign t  = 0.0;
   assign dt = 0.05/fss;
   while (t < 10.0/fss) begin
     assign Vd = Vdc + Ass*sin(wss*t);
     assign V[i=LX] = (Vd-Vbi)/V0;
     solve;
     assign t = t+dt;
     call UpdateLastVariables;
   end

   // simulate one period of sinusoidal excitation
```

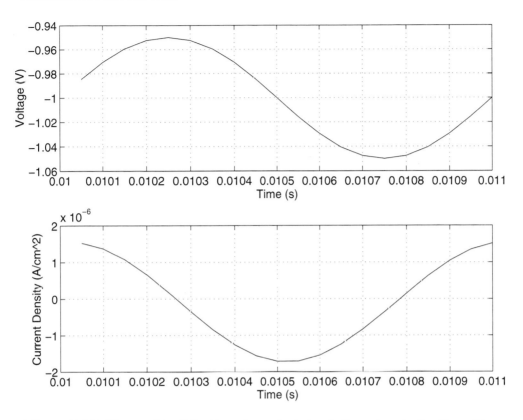

Figure 7.25. The results of the time-dependent simulation set up in the file pn04.sg for small-signal analysis of the pn junction.

```
while (t <= 11.0/fss) begin
   assign Vd = Vdc + Ass*sin(wss*t);
   assign V[i=LX] = (Vd-Vbi)/V0;
   solve;
   assign Jd = -J0*{Jn(n[0],n[1],E(V[0],V[1],h[0]),un[0],un[1],h[0]) +
                    Jp(p[0],p[1],E(V[0],V[1],h[0]),up[0],up[1],h[0])};
   open "pn04.out" append;
   file write t,Vd,Jd;
   close;
   assign t = t+dt;
   call UpdateLastVariables;
end
end
```

Interpretation of the Simulation Results

To interpret the small-signal behavior we need to recognize that the device is in reverse-bias and is expected to behave like a capacitor. The current should be $\simeq \pi/2$ ahead of the voltage, since $\omega C \gg R$, which it appears to be.

The capacitor has capacitance per unit area $C = \frac{\epsilon}{d}$. We know that $\epsilon = \epsilon_r \epsilon_0 \simeq 10^{-10}$ F/m. We need to know the 'plate separation' d. We find this, as in the dc abrupt junction case, by setting the voltage dropped across half the junction equal to $\frac{1}{2}\frac{\rho}{\epsilon}x^2$, where the junction width $d = 2x$.

$\rho \simeq 1.6 \times 10^{-19} \times 10^{21}$ C/m^3 since the doping level is 10^{21} m^{-3}.

The total voltage is the built-in voltage plus the one volt which is applied. Now $V_{bi} \simeq 0.03 \ln\left(\frac{10^{30}}{1.5 \times 10^{20}}\right) \simeq 0.6$ volts. Half of the total voltage is thus 0.8 V. This should be equal to $\frac{1}{2}\frac{1.6 \times 10^2}{10^{-10}}x^2 \simeq 0.8 \times 10^{12}x^2$. This means that $x^2 \simeq 10^{-12}$ m^2 so $x \simeq 10^{-6}$ m, and $C = \frac{\epsilon}{d} \simeq \frac{10^{-10}}{2.0 \times 10^{-6}} \simeq 5 \times 10^{-5}$ F/m^2.

Now from the plot the current has amplitude $\tilde{I} \simeq 1.75 \times 10^{-6}$ A/cm^2, and from the data file, $\omega = 2\pi 10^3$r/s while the voltage amplitude is $\tilde{V} = 0.05$ V. Since the magnitude of the current in a capacitor should equal $\omega C \tilde{V}$, we can estimate C from $C = \frac{\tilde{I}}{\omega \tilde{V}} \simeq \frac{1.75 \times 10^{-6}}{100\pi} \simeq 6 \times 10^{-9}$ F/cm$^2 = 6 \times 10^{-5}$ F/m^2.

The estimate is only to limited accuracy, but the two ways of finding C do agree quite well at this level.

7.5   Transients in PN Junctions

Two cases will be considered here, a turn-off transient and a turn-on transient.

7.5.1   Transient Behavior During Turn-Off

In switching a diode from an on-state to an off-state, switching is not instantaneous. There is a delay before the off-state is reached. The main reason for this is again that minority carriers take some time to respond. The minority carriers have to be removed before the device state can change. They can recombine, which will take several times τ_p, or they can flow across the junction to the side where they are the majority. The external circuit will not usually allow a very large current, except in power electronic circuits. In any case, the maximum current which can flow in the external circuit limits the carriers' ability to cross the junction rapidly. Otherwise the carriers' drift velocity in the junction could easily be large enough to get them across very rapidly, once they enter the junction.

The time response can be analyzed using an equation describing the conservation of charge, which states that the rate of change in the hole-charge Q on the n-type side is equal to the hole current i flowing in minus the rate at which the charge recombines.

$$\frac{dQ}{dt} = i - \frac{Q}{\tau_p}. \tag{7.77}$$

Since this is a conservation equation, it can be derived from the diffusion equation, or from the original conservation equation introduced in Chapter 3. However, its meaning should be clear by itself. The transition from forward to reverse-bias occurs when the minority-carrier density drops below the equilibrium value. In forward-bias the minority-carrier density is above the equilibrium value. Before the transition a nearly constant hole current flows out of the n-type side, denoted by I_h. Since it flows out, $i = -I_h$ and the conservation of charge becomes

$$\frac{dQ}{dt} = -I_h - \frac{Q}{\tau_p}. \tag{7.78}$$

Collecting the terms with Q in them,

$$\frac{dQ}{dt} + \frac{Q}{\tau_p} = -I_h. \tag{7.79}$$

The left side may be rewritten using an integrating factor method (which can be verified by differentiating the expression in square brackets below) so that

$$\exp\left(-\frac{t}{\tau_p}\right)\frac{d}{dt}\left[Q \exp\left(\frac{t}{\tau_p}\right)\right] = -I_h. \tag{7.80}$$

The equation can then be rearranged as

$$\frac{d}{dt}\left[Q \exp\left(\frac{t}{\tau_p}\right)\right] = -I_h \exp\left(\frac{t}{\tau_p}\right). \tag{7.81}$$

Integrating this gives

$$Q \exp\left(\frac{t}{\tau_p}\right) = A - \tau_p I_h \exp\left(\frac{t}{\tau_p}\right) \tag{7.82}$$

and so the charge is

$$Q = A \exp\left(-\frac{t}{\tau_p}\right) - \tau_p I_h. \tag{7.83}$$

Let t_s be the time taken by the switching process. If we know the stored charge $Q(t = 0)$ at $t = 0$ and the stored charge at $t = t_s$ which is denoted $Q(t = t_s)$, then we can solve for t_s. First we set

$$Q(t = 0) = A - \tau_p I_h \tag{7.84}$$

so that the constant A is

$$A = Q(t = 0) + \tau_p I_h. \tag{7.85}$$

The charge at time t is then

$$Q(t) = (Q(t = 0) + \tau_p I_h) \exp\left(-\frac{t}{\tau_p}\right) - \tau_p I_h. \tag{7.86}$$

Substituting in that $t = t_s$ and setting $Q = Q(t = t_s)$, we have

$$Q(t_s) + \tau_p I_h = (Q(t = 0) + \tau_p I_h) \exp\left(-\frac{t_s}{\tau_p}\right) \tag{7.87}$$

which has the solution

$$t_s = \tau_p \ln\left(\frac{Q(t = 0) + \tau_p I_h}{Q(t = t_s) + \tau_p I_h}\right). \tag{7.88}$$

The stored charge is not yet zero when the direction of bias changes, but t_s can be estimated by setting $Q(t = t_s) = 0$ to obtain an upper bound on t_s. The initial charge at $t = 0$ is the charge in steady state, when all the hole current flowing into the n-type region was recombining there. The hole current entering the n-type region is the hole current leaving the p-type region and is thus the total current in the forward-biased steady state, I_{ss}^F. In steady state $Q = \tau_p i$, so in this case $Q(t = 0) = \tau_p I_{ss}^F$ and

$$t_s = \tau_p \ln\left(\frac{I_{ss}^F + I_h}{I_h}\right). \tag{7.89}$$

7.5.2   Transient Behavior During Turn-On

Suppose the turn-on is achieved using a constant current I^F. The charge conservation equation is then

$$\frac{dQ}{dt} = I^F - \frac{Q}{\tau_p}. \tag{7.90}$$

This means that the current flows into the n-type region as hole-current. This equation is the same as that used in the previous section, except that I^F has replaced $-I_h$. The solution is therefore

$$Q(t) = (Q(t = 0) - \tau_p I^F) \exp\left(-\frac{t}{\tau_p}\right) + \tau_p I^F. \tag{7.91}$$

The initial charge will be approximated as being zero now, so

$$Q(t) = \tau_p I^F \left[1 - \exp\left(-\frac{t}{\tau_p}\right)\right]. \tag{7.92}$$

The charging may take place quasi-statically, in which case the steady-state charge is obtained which is equal to τ_p times the steady-state current, as can be seen from the charge conservation equation with the time derivative set to zero. The steady-state current is $I_O \exp(\frac{eV_A}{k_B T}) - 1)$. This gives an expression for the charge during a quasi-static charge-up,

$$Q = \tau_p I_O \left[\exp\left(\frac{eV_A}{k_B T}\right) - 1\right]. \tag{7.93}$$

If we set this equal to the expression for the charge found from the charge conservation equation

$$\tau_p I^F \left[1 - \exp \left(-\frac{t}{\tau_p} \right) \right] = \tau_p I_O \left[\exp \left(\frac{eV_A}{k_B T} \right) - 1 \right] \qquad (7.94)$$

we can find $V_A(t)$,

$$V_A(t) = \frac{k_B T}{e} \ln \left(1 + \frac{I^F}{I_O} \left[1 - \exp \left(-\frac{t}{\tau_p} \right) \right] \right). \qquad (7.95)$$

7.6 Photodiodes Based on PN and PIN Junctions

Optoelectronic or photonic devices are generally designed to absorb, emit or transmit light. This section will consider primarily devices which are intended to function by absorbing light. First we shall briefly describe the effects of light absorption in a semiconductor. Then we shall consider how this process may be used to detect light and give an example simulation of a semiconductor light detector.

7.6.1 Light Absorption in a Semiconductor

Most light absorption in a semiconductor is associated with an electron in the valence band absorbing the photon and being promoted to the conduction band. In this way, the absorption generates a pair of carriers: a hole in the valence band and an electron in the conduction band. The photon energy needed to do this must be at least the energy of the band gap, so since the photon energy is $hf = \hbar\omega$, then we need

$$\hbar\omega \geq E_g. \qquad (7.96)$$

The frequency of the photon is f, its angular frequency is ω, Planck's constant is h and $\hbar = h/2\pi$. The band-gap energy is E_g.

The photon flux Γ_{ph} per unit area per second at a depth z in the semiconductor decays exponentially

$$\frac{d\Gamma_{ph}}{dz} = -\alpha \Gamma_{ph} \qquad (7.97)$$

since the rate of absorption per unit depth (and per unit area) must be proportional to the number of photons passing through the unit area per second. The number of photons absorbed per second, per unit depth and area is the absorption rate per unit volume, and this is also the electron-hole generation rate per unit volume,

$$G_{ph} = -\frac{d\Gamma_{ph}}{dz}. \qquad (7.98)$$

The solution of the equation for Γ_{ph} is

$$\Gamma_{ph} = \eta \Gamma_{ph0} \exp(-\alpha z) \qquad (7.99)$$

where η is the fraction of the photons which are not reflected from the outer surface of the material, and Γ_{ph0} is the incident flux arriving at the surface of the material.

$$G_{ph} = \alpha\eta\Gamma_{ph0}\exp(-\alpha z). \qquad (7.100)$$

These expressions will allow us to estimate a generation rate of carriers in a semiconductor. Our model of the simple pn junction can be augmented with this extra generation rate and we can then examine how its (i, v) characteristics are affected.

7.6.2   Semiconductor Light Detectors

The most straightforward semiconductor light detector is based on a pn (or a pin) junction constructed so light can penetrate to the junction region. The importance of the pn junction as a photodiode starts from the fact that it is straightforward to make a simple light detector out of a reverse-biased pn junction. In reverse bias the current is small, so any change should be easy to detect. Photons of the correct frequency can create carrier pairs and the minority carriers will be swept across the junction. We can make a sensitive light detector to exploit this, for instance, by fabricating a thin p-type layer on top of an n-type substrate. The p-type layer is kept thin, so that if the light comes into the surface of the p-type layer, the carriers will be produced close to the junction.

The pin diode has an extra, nearly intrinsic, layer, hence the extra i in the name. The pin diode makes a somewhat better light detector. The entire intrinsic region can be kept reverse-biased with a modest reverse voltage and can be used to collect light and convert it to carriers. The response time of the pin structure is determined by the transit time of carriers across the intrinsic region. The time can be small enough at high fields for the pin diode to be useful in optical communications.

7.6.3   A Model of a PIN Diode Light Detector

The role of the carriers generated by the light can be illustrated with an example of a pin diode. The device is reverse-biased with the intention of causing all the minority carriers which are created by the light to flow across the junction and contribute to a reverse current. This extra reverse current is simply subtracted from the usual current, so the (i, v) curve is shifted to more negative currents by the effects of the light absorption.

A model of a pn junction used as a light detector is given below, in the file pniv3.sg. This file differs from pniv2.sg due to two simple changes. First, the doping density is a hundred times higher, so the spatial scale has been shrunk by a factor of ten. As the analytic results given above show, the junction width scales as the charge density to the power minus one-half, if the charge density is independent of position. Second, carrier generation has been added using the array Gopt. Only the parts of pniv3.sg which are different from pniv2.sg are included. This simulation does not use the models presented in Chapter 4. To do so would require a 2D simulation.

Input File pniv3.sg

```
// FILE:  sgdir\ch07\pniv3\pniv3.sg
// demonstration of solution of 1D nonlinear Poisson's equation
// plus SG continuity eqn. plus particle generation - roughly correct
// constants!
// - applied to a linear pn junction. Higher doping level than pniv2.sg .
// Also can have extra generation Gopt - due to photons, say.

const NX = 101;              // number of x-mesh points
const LX = NX - 1;           // index of last x-mesh point
const DX = 1.0e-7;           // mesh spacing
const VR = 0.7;              // voltage at the rhs
const VT = 0.03;             // thermal voltage
const N0 = 1.0e+20;          // doping density
const ec = 1.6e-19;          // electronic charge
const eps0=8.8e-12;          // permittivity of vacuum;
const epsi=11.8;             // relative permittivity of Si
const eoeps = ec/eps0/epsi;  // e/epsilon
const mu  = 0.1;             // magnitude of electron mobility
const D   = VT * mu;         // diffusion coefficient
const ni = 1.0e16;           // intrinsic density in MKS
const taun = 1.0e-6;         // electron lifetime
const taup = taun;           // hole lifetime

var V[NX],NA[NX],n[NX],E[NX],EP[NX],gamma[NX],Gopt[NX],x[NX];

set NEWTON ACCURACY = 1.0e-8 ;

// Poisson's equation
equ V[i=1..LX-1]  ->
  {V[i-1] - 2*V[i] + V[i+1]}/sq(DX) =
  eoeps*(NA[i] + N0*(exp((V[i]-VR)/VT) - exp(-V[i]/VT)));

// the following equations could be implemented as functions
equ E[i=0..LX-1]  -> E[i]  = -{V[i+1] - V[i]}/DX;
equ EP[i=0..LX-1] -> EP[i] = exp(-E[i]*DX/VT);

// particle conservation equation
equ n[i=1..LX-1]  ->
 -mu/DX * [{EP[i]   *n[i]   - n[i+1]}*E[i]  /(EP[i]  -1.0) -
          {EP[i-1] *n[i-1] - n[i]  }*E[i-1]/(EP[i-1]-1.0)] =
 (sq(ni) - n[i]* N0*exp(-V[i]/VT))/(taup*(n[i]+ni) +
          taun*(N0*exp(-V[i]/VT) + ni)) + Gopt[i];

begin main
  assign x[i=all] = i*DX ;               // coordinate array
  assign NA[i=0..LX] = N0*(1.0 - 2.0*i/LX);  // acceptors on left .
```

```
assign Gopt[i=LX/4..3*LX/4] = 2.0e+20 ;
assign V[i=0] = 0.0;
assign V[i=LX] = VR;
assign n[i=LX/2..LX] = NO;
assign n[i=0] = ni*ni/NO;                          // eqbm. density
assign E[i=all] = 1.0;
assign EP[i=all]=exp(-E[i]*DX/VT);
solve;
assign gamma[i=0..LX-1]=-mu*E[i]*(EP[i]*n[i]-n[i+1])/(EP[i]-1.0);
assign gamma[i=LX] = gamma[i-1];
end
```

Interpretation of the Simulation Results

The electron flux is shown in Figure 7.26. The change in the flux between points x_1 and x_2 due to a source of particles with source density $S(x)$ is $\Delta\Gamma = \int_{x_1}^{x_2} S(x)\,dx$. For a constant source $S(x) = S_0$ this increases linearly with distance, being equal to $S_0\,(x_2 - x_1)$. In our example $x_2 - x_1$ is 6×10^{-7} m, so the value of the source has to be chosen to be of the order of 10^{21} m^{-3}s^{-1} to give a substantial contribution to the flux. In fact G_{opt} is 2×10^{20} (the exponential decay is in the direction normal to the x axis) and the expected change in flux is $2 \times 10^{20} \times 6 \times 10^{-7} = 1.2 \times 10^{14}$m^{-2} s^{-1}, which is about equal to the change in flux seen on crossing the lighted region. The variation of the flux with the applied voltage obtained using the second set of parameters discussed earlier, with all dimensions ten times smaller, with a hundred times the doping level and with $G_{opt} = 10^{22}$ m^{-3}s^{-1}, is shown in Figure 7.27. To the accuracy of the data used (two figures) the results do indeed show that the curve is simply shifted by a uniform amount toward greater reverse flux, as all the extra carriers are swept across the junction in the 'reverse' direction.

7.7   Summary

This chapter began the discussion of semiconductor devices with an examination of the simplest such device, the pn junction. The treatment was primarily intended to provide the mathematical basis for the analytic theory of the pn junction, followed by the procedure for simulating a pn junction, and then to compare the two. Simulations, which go beyond what can be done analytically, can then be done and meaningful results obtained.

PROBLEMS

1. Refer to Figures 7.1, 7.2 and 7.3. Why are the theoretical and simulation results different? What criteria should be used to measure the junction width? Why are the differences between 'theory' and simulation greater in the asymmetric case? What assumption does the analytic solution make that the simulation does not? Why is the junction narrower when the grade is increased?

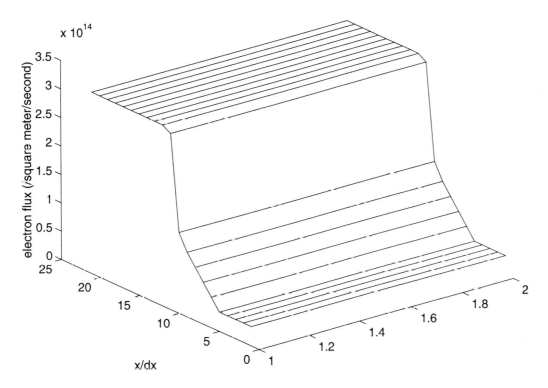

Figure 7.26. The electron flux in the photodiode, as a function of position.

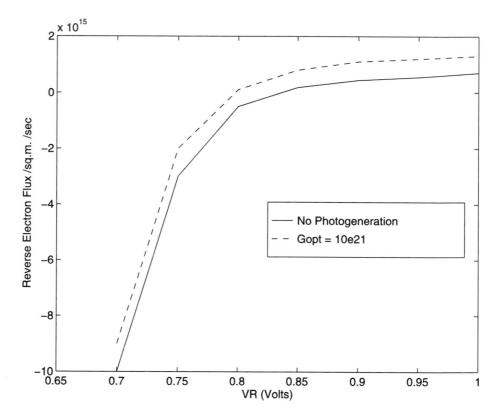

Figure 7.27. The (reverse) electron flux in the photodiode, plotted as a function of reverse-bias voltage.

2. The next few problems here are concerned with solving Poisson's equation in a semiconductor, and lead up to solving the nonlinear Poisson's equation in a pn junction. The nonlinear Poisson's equation is notoriously difficult to solve, even in equilibrium and in one dimension. If we assume $qV \ll k_B T_e$, then the exponential in the Boltzmann expression for the electron density can be expanded. The charge density is then $\rho = e/\epsilon_0 (n_i - n_0(1 + eV/k_B T_e))$. If the electron density at infinity n_0 is set equal to the ion density n_i (n_0 is really the electron density where V is zero), then Poisson's equation can be solved analytically. Find the solution for a small negative surface-charge density ρ_s on an infinite metal plate. Let $V = 0$ far from the plate. The plate is at $x = 0$.

   [For hints, look up the Debye length in the index.]

3. Try to solve the previous problem using the finite-difference form of the full nonlinear Poisson's equation, as follows. Choose $E(x = 0) = D(x = 0)/\epsilon_0 = \rho_s/\epsilon_0$. $\rho = e[n_i - n_0 \exp(eV/k_B T_e)]$ and, since $n_0 = n_i$, this becomes $\rho = e n_i \left[1 - \exp\left(\frac{eV}{k_B T_e}\right)\right]$. The potential is assumed to go to zero at infinity. Choose a value of $V(x = 0)$ and integrate the FD form of Poisson's equation outwards, and see whether V decreases in magnitude as x increases. Use the values of $V[1]$ found from $E[0] = -(V[1]-V[0])/\Delta x$ (say) and of $V[0]$ as the initial or boundary conditions on the integration. Each time V diverges by getting too large and positive (negative), this puts an upper (lower) limit on the guess for $V(x = 0)$. Repeat the integration for different guesses for $V(x = 0)$, trying to narrow down the initial guess. Plot the results for different initial guesses. Let $n_i = 10^{12}$ cm^{-3} and $V^{th} \equiv \frac{k_B T_e}{e} = 0.03$ V. A 'good' initial guess might allow a reasonable-looking solution to be found for positions close to the starting point, but V will probably diverge if followed far.

4. Repeat the previous problem, using an implicit scheme employing the 'equ' and 'solve' statements in SGFramework.

5. Find the electrostatic potential in a pn junction, using the nonlinear Poisson's equation with the equilibrium expressions for the carrier densities in one dimension. Solve the equation using an implicit scheme. Try several different doping profiles in the vicinity of the junction. Compare the potential profiles obtained. Impose the difference in voltage at the ends of the device. In other words, let $V(x = 0) = 0$ and set $V(x = L) = V_L$, where V_L is a value specified in the input file. The equilibrium densities must be expressed in terms of the potential so that if (say) the electrons are the majority carrier at $x = L$ and their equilibrium density there is $n_{eq}(x = L)$, then n must equal $n_{eq}(x = L)$ when V equals V_L. Vary V_L to obtain different applied voltages. Concentrate on reverse bias, since the assumption of equilibrium which was used to find the expressions for the density is not valid in forward-bias. (Is it valid for all the species everywhere in reverse-bias? If not, why do we still use the equilibrium densities?)

6. This problem is intended to address the issue of majority-carrier transport in a pn junction. The junction is at $x = 0$. On one side of the junction, where $0 \le x < L$, where electrons are the majority carriers, any electric field near the junction is expected to repel the electrons. Choose a suitable analytic approximation to the electric field. To find the electron density we need to know the production rate and

the boundary conditions, in addition to the electric field. Let $n(x = 0) = 0$ and let $n(x = L)$ equal the equilibrium value there. Choose a moderately large value of the donor density to use at $x = L$. Let $L = 4\,\mu m$. At $x = L$ the production rate must be zero if the point is in equilibrium. Let the carrier production rate per unit length be $S(x) = S_0 (L - x)$ with S_0 set to 10^{10} cm^{-2}/s. Which of these assumptions is least realistic? Find the carrier density that is produced in steady state under these conditions. The problem we have set up is linear, so an implicit solution is not needed. The diffusion-equation can be stepped forward in time using a simple explicit scheme. Plot the flux of electrons. What do you expect given the assumptions used? Is this realistic?

7. Using the file pn01.sg (which describes junctions with a step in the doping and junctions with linearly graded doping) check the analytic expressions for the potential. Plot the numerical results and the analytic results on the same figure.

8. Evaluate the expressions for the small-signal characteristics of a pn junction, (a) in reverse-bias (C and G) and (b) in forward-bias (C_{rf} and G_{rf}), for the parameters of one of the pn junctions modeled in the text. Compare your results to the results of simulations.

9. Compare the dc and time-dependent simulation files given in the text. What changes were made to allow for the time dependence? Use the file pn04.sg to estimate numerically the small-signal parameters of the device modeled, and compare the answers to the analytic results.

10. Calculate analytically the stored charge as a function of time in a pn diode during (a) turn-off and (b) turn-on, for the parameters of the pn diode modeled in the file pn04.sg. Evaluate the charge numerically by modifying pn04.sg to describe the transient, and compare the results to the analytic values.

11. Plot G versus x for the example set up in the file pniv2.sg. Estimate the integral $\Gamma = \int G\,dx$ and compare the answer to the flux shown in the plot obtained from pniv2.sg.

12. For a 'typical' pin diode, such as was modeled in the text, compute the resistance of the lightly doped region, both in equilibrium and in high-level injection conditions. What percentage change in the conductivity occurs during high-level injection?

13. Refer to Figure 7.22. Why does the voltage overshoot and then drop down again? Why is the final voltage $5V$? What will the power dissipation be after $0.35\mu s$? Estimate the device temperature for this rate of power dissipation. What is the major defect in this device, and how could it be redesigned for better performance? Repeat the simulation, with an improved design.

Chapter 8

Bipolar Junction Transistors

This chapter introduces basic ideas of bipolar junction transistors and compares simple models of their behavior to numerical results. First, Section 8.1, treats static characteristics. Section 8.2 derives the Ebers-Moll model of the bjt. Section 8.3 discusses time-dependent, that is, small-signal and transient behavior. 'Lumped' models of bjts, derived from the Ebers-Moll model, are used in modeling important bjt circuits, in Section 8.4. Section 8.5 handles 2D numerical modeling of the internal working of bjts.

Material from Chapter 7, on pn junctions will be drawn on repeatedly, because the BJT is composed of two back-to-back pn junctions.

8.1 Static Characteristics

Figure 8.1 shows a schematic of a low power bipolar transistor.

The BJT behaves like two diodes back to back, provided the base is much thicker than a diffusion length L_B for minority carriers in the base. The discussion below

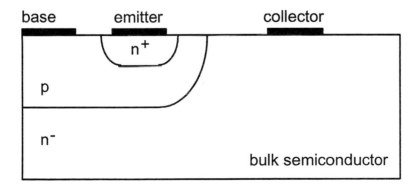

Figure 8.1. A schematic of a low power bipolar transistor.

assumes that holes are the minority carriers in the base. (The diffusion length is the typical net distance a minority carrier diffuses, in a 'sea' of majority carriers, before recombining. In the hole recombination time τ_p a hole diffuses on average a distance $L_B = \sqrt{D_p \tau_p}$.) If the base width $W \gg L_B$, the minority carriers will recombine before crossing the base, so the two diodes in a sense do not interfere with each other's operation. The base is more usually narrow, however, with $W \ll L_B$. The device still resembles two diodes back to back in this case, but the expressions for the currents are modified somewhat, as will now be shown.

The diffusion equation for minority carriers in the base, if we can assume that they are moving in a region where the electric fields are not affecting them strongly, is

$$-D_p \frac{\partial^2 \Delta p}{\partial x^2} = -\frac{\Delta p}{\tau_p}. \tag{8.1}$$

The equation was written in this form, with the minus signs left in, because this is the steady-state diffusion equation, when the particle production rate per unit volume is given by $S = -\Delta p/\tau_p$. However, using the fact that $L_B^2 = D_p \tau_p$, it can be written as

$$\frac{\partial^2 \Delta p}{\partial x^2} = \frac{\Delta p}{L_B^2}. \tag{8.2}$$

This has the solution

$$\Delta p = A \exp(x/L_B) + B \exp(-x/L_B). \tag{8.3}$$

This can be simplified, because when L_B is large compared to W, the exponentials can be approximated as being linear. Alternatively, we can say that when L_B is large compared to W then τ_p is also large compared to the time to diffuse across the base. The time to diffuse a distance W is about $\tau_W = W^2/2D_p$. In fact the left side of the diffusion equation is roughly equal in magnitude to $\Delta p/\tau_W$. To see this, we estimate the second derivative with respect to x as being about as big as the quantity being differentiated divided by W^2. This is appropriate because W is the length scale on which the density can be expected to vary in the base. The diffusion term is thus proportional to $\frac{D_p}{W^2}$, and this is being compared to a term proportional to $1/\tau_p$. It was already decided that $L_B \gg W$, however, and $L_B = \sqrt{D_p \tau_p}$. In other words, saying $L_B \gg W$ is equivalent to saying the diffusion term is much bigger than the recombination term in the base. There is too little time for carriers to recombine before they cross the base. Then the second derivative is equal to zero and the solution for Δp is linear (as can be seen from the exponential solution when $\frac{x}{L_p} \ll 1$). The approximate solution to the diffusion equation when recombination in the base can be neglected (when $W \ll L_B$) is

$$\Delta p(x) = \Delta p(x = 0) \left(1 - \frac{x}{W}\right) + \Delta p(x = W)\frac{x}{W}. \tag{8.4}$$

This matches the boundary conditions at $x = 0$ and $x = W$, once they are specified in terms of values of Δp at those locations. The hole current is

$$I_p = -eAD_p \frac{\partial \Delta p}{\partial x} = eAD_p \frac{\Delta p(x=0) - \Delta p(x=W)}{W}. \tag{8.5}$$

The boundary conditions which allow these expressions for the density and current to be evaluated are

$$\Delta p(x=0) = \frac{n_i^2}{N_D} \left[\exp\left(\frac{eV_{EB}}{k_B T}\right) - 1 \right] \tag{8.6}$$

and at $x = W$

$$\Delta p(x=W) = \frac{n_i^2}{N_D} \left[\exp\left(\frac{eV_{CB}}{k_B T}\right) - 1 \right]. \tag{8.7}$$

The base hole current can be split into two parts, $I_p = I_{p1} - I_{p2}$. The expressions for I_{p1} and I_{p2} make it clear that I_{p1} is associated with the diode made from one junction, and I_{p2} with the other junction.

$$I_{p1} = \frac{eAD_p}{W} \frac{n_i^2}{N_D} \left[\exp\left(\frac{eV_{EB}}{k_B T}\right) - 1 \right] = I_{pO} \left[\exp\left(\frac{eV_{EB}}{k_B T}\right) - 1 \right] \tag{8.8}$$

and, using I_{pO} again,

$$I_{p2} = I_{pO} \left[\exp\left(\frac{eV_{CB}}{k_B T}\right) - 1 \right]. \tag{8.9}$$

Then the overall hole current is

$$I_p = I_{pO} \left[\exp\left(\frac{eV_{EB}}{k_B T}\right) - \exp\left(\frac{eV_{CB}}{k_B T}\right) \right]. \tag{8.10}$$

The minority carriers in the emitter and the collector are not affected by the base width, so the electron currents between the base and these regions are the same as they would be if the junctions were simple diodes. The diode expressions for the electron currents give the electron current from emitter to base,

$$I_{En} = \frac{eAD_n}{L_n} \frac{n_i^2}{N_{AE}} \left[\exp\left(\frac{eV_{EB}}{k_B T}\right) - 1 \right] \tag{8.11}$$

and the electron current from base to collector (which is opposite to the direction for which the diode current is defined, so it has a negative sign) is

$$I_{Cn} = -\frac{eAD_n}{L_n} \frac{n_i^2}{N_{AC}} \left[\exp\left(\frac{eV_{CB}}{k_B T}\right) - 1 \right]. \tag{8.12}$$

The total emitter current is then

$$\begin{aligned} I_E \;=\; & I_p + I_{En} = eAn_i^2 \left(\left[\frac{D_n}{L_n N_A} + \frac{D_p}{W N_D} \right] \left[\exp\left(\frac{eV_{EB}}{k_B T}\right) - 1 \right] \right. \\ & \left. - \frac{D_p}{W N_D} \left[\exp\left(\frac{eV_{CB}}{k_B T}\right) - 1 \right] \right) \end{aligned} \tag{8.13}$$

while the total collector current is

$$
I_C = I_p + I_{Cn} = eAn_i^2 \left(-\left[\frac{D_n}{L_n N_A} + \frac{D_p}{W N_D} \right] \left[\exp\left(\frac{eV_{CB}}{k_B T} \right) - 1 \right] \right.
$$

$$
\left. + \frac{D_p}{W N_D} \left[\exp\left(\frac{eV_{EB}}{k_B T} \right) - 1 \right] \right). \tag{8.14}
$$

These are identical to the equivalent diode results, except that the base width appears instead of L_p in the term $\frac{D_p}{W N_D}$. For a wide base the device is exactly like two diodes back to back and this term is $\frac{D_p}{L_p N_D}$.

8.2   The Ebers-Moll Equations

The Ebers-Moll equations result from thinking of the BJT as an equivalent circuit, shown in Figure 8.2. The same equations can be derived from the currents found previously, and the values of the parameters appearing in the Ebers-Moll equations obtained. We shall begin with the equations for I_E and I_C.

If the voltage V_{CB} is set to zero in the expression for I_E, then

$$
I_E(V_{CB} = 0) = I_{FO} \left[\exp\left(\frac{eV_{EB}}{k_B T} \right) - 1 \right] \tag{8.15}
$$

where the factor I_{FO} is given by

$$
I_{FO} = eAn_i^2 \left(\frac{D_n}{L_n N_A} + \frac{D_p}{W N_D} \right). \tag{8.16}
$$

Similarly, if $V_{EB} = 0$, the expression for I_C is

$$
I_C(V_{EB} = 0) = -I_{RO} \left[\exp\left(\frac{eV_{CB}}{k_B T} \right) - 1 \right] \tag{8.17}
$$

with I_{RO} given by

$$
I_{RO} = eAn_i^2 \left(\frac{D_n}{L_n N_A} + \frac{D_p}{W N_D} \right). \tag{8.18}
$$

Strictly I_{FO} and I_{RO} can be different, because D_n, L_n and N_A appearing in each of them refer to the corresponding bulk regions, so a distinction should be made between them.

These expressions for I_E and I_C resemble diode equations with I_{FO} and I_{RO} being the saturation currents. The part of I_C which varies as $\exp(\frac{eV_{CB}}{k_B T}) - 1$ is introduced into the base at the collector. The part of I_E which varies in the same way can be thought of as being the part of the current introduced into the base at the collector which made it across the base to the emitter. The fraction making it is given by

$$
\alpha_F = \frac{D_p}{W N_D} \bigg/ \left(\frac{D_p}{W N_D} + \frac{D_n}{L_n N_A} \right). \tag{8.19}
$$

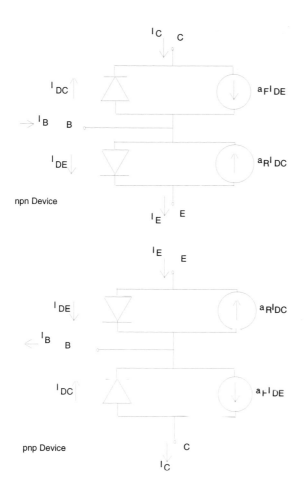

Figure 8.2. Schematic of the Ebers-Moll model of the BJT. In this figure, I_{DC} is the 'diode' part of the collector current, and similarly for the other currents.

Again, D_n, L_n and N_A in this expression refer to the collector region.

Similarly, I_E has a term which varies as $\exp(\frac{eV_{CB}}{k_B T}) - 1$. A fraction of this gets through the base and contributes to I_C. The coefficient of this term in I_E is

$$e A n_i^2 \left(\frac{D_n}{L_n N_A} + \frac{D_p}{W N_D} \right)$$

and the coefficient of this term in I_C is

$$e A n_i^2 \frac{D_p}{W N_D}.$$

The fraction of this current getting through the base is

$$\alpha_R = \frac{D_p}{W N_D} \Big/ \left(\frac{D_p}{W N_D} + \frac{D_n}{L_n N_A} \right). \tag{8.20}$$

In this case the emitter values are implied for D_n, L_n and N_A.

These expressions can be simplified if we use $\alpha_F I_{FO} = \alpha_R I_{RO} = e A n_i^2 \frac{D_p}{W N_D}$. The total currents can now be written in the form of the Ebers-Moll equations, which are

$$I_E = I_{FO} \left[\exp\left(\frac{eV_{EB}}{k_B T} \right) - 1 \right] - \alpha_R I_{RO} \left[\exp\left(\frac{eV_{CB}}{k_B T} \right) - 1 \right] \tag{8.21}$$

whereas for I_C

$$I_C = \alpha_F I_{FO} \left[\exp\left(\frac{eV_{EB}}{k_B T} \right) - 1 \right] - I_{RO} \left[\exp\left(\frac{eV_{CB}}{k_B T} \right) - 1 \right]. \tag{8.22}$$

The base current is given by $I_B = I_E - I_C$.

8.3 BJT Time-Dependent Behavior

The time-dependent behavior of the BJT will be handled in two parts. First, small oscillatory time variations will be treated. Next, transients during turn-on and turn-off will be described.

8.3.1 Small-Signal Analysis

To find the response of the BJT to small ac signals, it is generally assumed that the response is quasi-static. The above equations for I_E and I_C, from which I_B can also be found, should still apply in this case. Suppose that I_B and I_C are expressed in terms of V_{BE} and V_{CE}. This reflects a change from measuring voltages relative to the base, which is the common connection in the Ebers-Moll model, to measuring voltages relative to the emitter, which is likely to be the common connection in an amplifier circuit.

It can be seen that small changes in these voltages, V_{BE} and V_{CE} (the small changes being denoted by v_{be} and v_{ce}) give rise to small changes in the currents (which will be denoted i_b and i_c.) The currents are given by

$$i_b = \left[\frac{\partial I_B}{\partial V_{BE}}\right]_{V_{CE}} v_{be} + \left[\frac{\partial I_B}{\partial V_{CE}}\right]_{V_{BE}} v_{ce} \tag{8.23}$$

and the oscillating part of the collector current is

$$i_c = \left[\frac{\partial I_C}{\partial V_{BE}}\right]_{V_{CE}} v_{be} + \left[\frac{\partial I_C}{\partial V_{CE}}\right]_{V_{BE}} v_{ce}. \tag{8.24}$$

The derivatives

$$g_{11} \equiv \left[\frac{\partial I_B}{\partial V_{BE}}\right]_{V_{CE}} \tag{8.25}$$

and

$$g_{22} \equiv \left[\frac{\partial I_C}{\partial V_{CE}}\right]_{V_{BE}} \tag{8.26}$$

can be interpreted as conductances, since they each relate the current between a pair of points to the voltage between those points. The pairs of points are (1) the base and the emitter and (2) the collector and the emitter. They will thus be represented as resistors in the equivalent-circuit model.

The other two derivatives,

$$g_{12} \equiv \left[\frac{\partial I_B}{\partial V_{CE}}\right]_{V_{BE}} \tag{8.27}$$

and

$$g_{21} = \left[\frac{\partial I_C}{\partial V_{BE}}\right]_{V_{CE}} \tag{8.28}$$

have to be represented as voltage-controlled current sources. This leads to an equivalent circuit with two combinations of a resistor and a current source, connected in parallel with each other. One of these combinations is between the base and the emitter; the other is between the collector and the emitter.

The hybrid-pi model is essentially the above circuit with g_{12} and g_{22} omitted. These two terms provide the dependence on v_{ce}, so leaving them out amounts to saying that v_{ce} has no effect. In active mode the Ebers-Moll equations are approximated as

$$I_E \simeq I_{FO} \exp\left(\frac{eV_{EB}}{k_B T}\right) \tag{8.29}$$

and the collector current is

$$I_C \simeq \alpha_F I_{FO} \exp\left(\frac{eV_{EB}}{k_B T}\right) \tag{8.30}$$

which do not involve v_{ce}. More complicated hybrid-pi models also exist, which will not be discussed here.

8.3.2   Turn-On Transients in the BJT

If the base current is constant during turn-on and equal to I_{BB}, then the base charge obeys the charge conservation equation

$$\frac{dQ_B}{dt} = I_{BB} - \frac{Q_B}{\tau_B}. \tag{8.31}$$

Since the initial charge is roughly zero, $Q_B(t = 0) = 0$, the equation can be solved exactly as was done for the equivalent equation on one side of a pn junction, to give

$$\exp\left(-\frac{t}{\tau_B}\right) \frac{d}{dt}\left[\exp\left(\frac{t}{\tau_B}\right) Q_B\right] = I_{BB} \tag{8.32}$$

so that

$$\frac{d}{dt}\left[\exp\left(\frac{t}{\tau_B}\right) Q_B\right] = I_{BB}\exp\left(\frac{t}{\tau_B}\right) \tag{8.33}$$

and then

$$\exp\left(\frac{t}{\tau_B}\right) Q_B = A + \tau_B I_{BB} \exp\left(\frac{t}{\tau_B}\right). \tag{8.34}$$

Finally,

$$Q = A\exp\left(-\frac{t}{\tau_B}\right) + \tau_B I_{BB}. \tag{8.35}$$

The initial condition is $Q(t = 0) = 0 = A + \tau_B I_{BB}$, so

$$Q_B = \tau_B I_{BB}\left[1 - \exp\left(-\frac{t}{\tau_B}\right)\right]. \tag{8.36}$$

Now the time to diffuse a distance L is $L^2/2D$ if the diffusion coefficient is D, so each carrier in the base will stay there for a time $\tau = W^2/2D_B$ in active mode. After this time it leaks to the collector, so $i_c = \frac{Q_B}{\tau}$. Once the BJT reaches saturation biasing i_c reaches a steady value I_{CC}. The time to reach this current is found from

$$Q_B(t_{sat}) = \tau I_{CC} = \tau_B I_{BB}\left[1 - \exp\left(-\frac{t_{sat}}{\tau_B}\right)\right]. \tag{8.37}$$

From this, t_{sat} is

$$t_{sat} = -\tau_B \ln\left(1 - \frac{\tau I_{CC}}{\tau_B I_{BB}}\right). \tag{8.38}$$

8.3.3   Turn-Off Transients in the BJT

During turn-off, the base current is $i_B = -x I_{BB}$ and the initial stored charge is $Q(t = 0) = I_{BB}\tau_B$, as follows from the above discussion. Charge conservation means

$$\frac{dQ_B}{dt} = -x I_{BB} - \frac{Q_B}{\tau_B} \tag{8.39}$$

which is the same as during turn-on but with $-xI_{BB}$ instead of I_{BB}. The solution is therefore

$$Q_B = A \exp\left(-\frac{t}{\tau_B}\right) - x\tau_B I_{BB} \qquad (8.40)$$

and at time $t = 0$ this should equal $I_{BB}\tau_B$, so $Q(t = 0) = A - x\tau_B I_{BB} = \tau_B I_{BB}$ or A is $A = (1+x)\tau_B I_{BB}$, which makes $Q(t)$ be

$$Q(t) = \tau_B I_{BB}\left[(1+x)\exp\left(-\frac{t}{\tau_B}\right) - x\right]. \qquad (8.41)$$

At turn-off Q_B is again equal to τI_{CC}, so at $t = t_{off}$

$$\tau I_{CC} = \tau_B I_{BB}\left[(1+x)\exp\left(-\frac{t_{off}}{\tau_B}\right) - x\right] \qquad (8.42)$$

or

$$t_{off} = -\tau_B \ln\left[\left(\frac{\tau I_{CC}}{\tau_B I_{BB}} + x\right)/(1+x)\right]. \qquad (8.43)$$

8.4   BJT Circuits

In this section the Ebers-Moll equations will be used to demonstrate the use of the BJT in some of its more important applications. In what follows we assume an npn device.

A BJT Invertor

The BJT invertor for use in digital circuits will be considered first. An input voltage supplied to the base causes the BJT to carry a current. A current I_C enters the collector and a current I_E leaves the emitter. The current I_C enters the collector through a load resistor R_C. One end of this load resistor is connected to the power supply at a voltage denoted by V_{CC}, and the other end, which is connected to the collector, is therefore at a voltage $V_C = V_{CC} - I_C R_C$. V_C is the output voltage, so when no current flows, the output voltage is high, being equal to V_{CC}, and if I_C is large enough V_C is low. The circuit is an invertor, because a low input voltage means the device does not conduct and the output is high. A high input voltage makes the device conduct, so the output is low.

An example of such an invertor is modeled using the file embjt01.sg which follows. The response of the circuit is shown in the accompanying Figure 8.3. The plot of V_{out} (which is V_{CE}) as a function of V_{in} (which is V_{BB}) illustrates some of the important parameters of the invertor. The idealized invertor output voltage consists of three straight lines with two breakpoints. The first breakpoint is when the output voltage starts to drop and the second breakpoint is when it has fallen to its low value and flattens out again. The plot shown resembles this idealization reasonably well. The output high voltage V_{OH} is the initial high output voltage on this plot and the output low voltage V_{OL} is the final low output voltage shown.

The input voltage at the first breakpoint is called V_{IL}, which is the lowest input voltage which turns the BJT on. It is the input voltage at which the invertor output starts to leave its high state. The input voltage at the second breakpoint is called V_{IH}. This is the input voltage which is just high enough to drive the output to its low value, by just saturating the transistor.

The change in input voltage between the two breakpoints is called the transition width; $TW = V_{IH} - V_{IL}$. The output voltages are designed to be such that the low output is lower than the low input, and the high output is higher than the high input. Frequently this output is fed into the input of another identical invertor or logic component. When its output is high, it will be higher than the high input of the next device, so the second device will turn on. Similarly, the low output of the first stage is lower than the low input of the second stage, so the second device will turn off when the first device's output is low. If either of the output levels fell between the two input levels, the second device would be between its high and low states. The margin by which the critical outputs exceed the critical inputs is called the noise margin, and there is a noise margin for high and low voltages. $NM_H = V_{OH} - V_{IH}$ and $NM_L = V_{IL} - V_{OL}$.

In the example shown, the values are roughly $V_{OH} = 10\text{V}$, $V_{OL} = 0.2\text{V}$, $V_{IL} = 1\text{V}$ and $V_{IH} = 2\text{V}$. As a result, $NM_H \simeq 8\text{V}$ and $NM_L \simeq 0.8\text{V}$.

Input File embjt01.sg

```
// FILE:   sgdir\ch08\embjt01\embjt01.sg
// This file contains a demonstration of how to
// model a BJT using the Ebers-Moll equations.
// The collector is connected to VCC through a 1.1-kilohm resistor.
// The base is connected to VBB through an 11-kilohm resistor,
// and the emitter is grounded.
// NOTE: AF*IE0 = AR*IR0.

// Circuit Parameters
const TMAX = 100;
const VCC  = 10.0;
const VT = 0.025;
const AF = 0.99;
const AR = 0.495;
const IC0= 4.0e-16;
const IE0= 2.0e-16;
const RB = 1.1e+04;
const RC = 1.1e+03;
const ICM = VCC/RC;
const VBCMAX = 0.75 ;
const VBCMIN=VT;
const VBEMAX = 0.79;

var IC[TMAX],IB[TMAX];
```

```
var VBC[TMAX],VBE[TMAX],VBE0[TMAX];
var VBB[TMAX],VBEM[TMAX];

set NEWTON ACCURACY  = 1.0e-15;

equ IB[i=all] ->
  (IB[i] - (IE0*(exp(VBE[i]/VT)-1.0) -
        AR*IC0*(exp(VBC[i]/VT)-1.0) -
  (VCC - (VBE[i]-VBC[i]))/RC)) = 0.0 ;

equ VBC[i=all] ->
  ((IC[i] - (AF*IE0*(exp(VBE[i]/VT)-1.0) -
              IC0*(exp(VBC[i]/VT)-1.0))) * (VBC[i] >= VBCMIN) -
  (VCC - RC*IC[i] - (VBE[i] - VBC[i]))       * (VBC[i] <  VBCMIN)) *
  (VBC[i] <= VBCMAX) = (VBC[i]>VBCMAX)*(VBC[i]-VBCMAX);

equ IC[i=all] ->
  (IC[i]>=0.0) * (IC[i] <= ICM) *
  ((IC[i] - (AF*IE0*(exp(VBE[i]/VT)-1.0) -
              IC0*(exp(VBC[i]/VT)-1.0))) * (VBC[i] < VBCMIN) -
  (VCC - RC*IC[i] - (VBE[i] - VBC[i]))       * (VBC[i]>=VBCMIN)) =
  (IC[i]<0.0)*IC[i] + (IC[i]>ICM)*(IC[i]-ICM);

equ VBB[i=all]->
  VBB[i] - RB*IB[i] = VBE[i] ;

equ VBE[i=1..TMAX-1]->
  (VBB[i] < VCC)*(VBE[i]-VBEM[i]) + (VBB[i] > VCC)*(VBE[i]-VBE[i-1]) = 0.0;

begin main
  assign VBB[i=all]  = VCC*(1.01*i/TMAX);
  assign VBE0[i=all] = VBEMAX*(VBB[i]>VBEMAX) + (VBB[i]<=VBEMAX)*VBB[i];
  assign VBEM[i=all] = VBE0[i]*(1.0 - 0.01*(1.01 - 1.01*i/TMAX));
  solve;
  write;
end
```

Fan-Out in a BJT Invertor

The connection of another invertor to the output lowers the output high level because a BJT invertor draws a current at the input. If enough such invertors are connected in parallel, the noise margin will be reduced to zero. The maximum number of invertors which can be connected is called the fan-out of the invertor. The effect of connecting $N = 50$ invertors is shown in the example file embjt02.sg and Figure 8.4. Fifty invertors have reduced NM_H to almost zero, but not quite, so a fan-out of fifty is possible in this case.

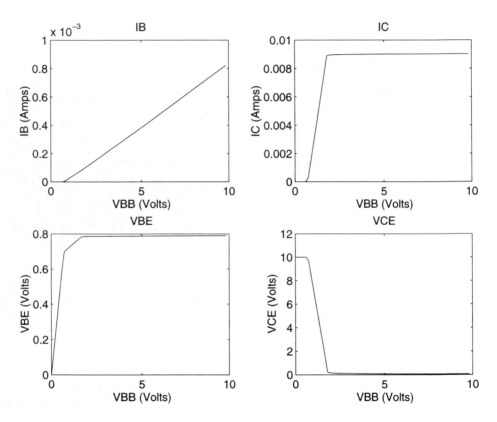

Figure 8.3. BJT modeled using Ebers-Moll equations.

The file embjt02.sg which follows is almost identical to embjt01.sg, but the current I_{load} has been introduced and appears in several of the equations.

Input File embjt02.sg

```
// File: sgdir\ch08\embjt02\embjt02.sg
// This file contains a demonstration of how to
// model a BJT using the Ebers-Moll equations.
// The collector is connected to VCC through a 1.1-kilohm resistor.
// The base is connected to VBB through an 11-kilohm resistor,
// and the emitter is grounded.
// NOTE: AF*IE0 = AR*IR0.
// To show the effects of fan-out, N=50 load devices,
// each identical to the driver, are connected through their own
// base resistance RB to the collector of the driver.

// Circuit Parameters
const TMAX = 100;
const VCC  = 10.0;
const VT = 0.025;
const AF = 0.99;
const AR = 0.495;
const IC0= 4.0e-16;
const IE0= 2.0e-16;
const RB = 1.1e+04;
const RC = 1.1e+03;
const ICM = VCC/RC;
const VBCMAX = 0.75 ;
const VBCMIN=VT;
const VBEMAX = 0.79;
const N = 50;

var IC[TMAX],IB[TMAX];
var VBC[TMAX],VBE[TMAX],VBEO[TMAX];
var VBB[TMAX],VLOAD[TMAX],ILOAD[TMAX];

set NEWTON ACCURACY  = 1.0e-15;

equ IB[i=all]->
  (IB[i] - (IE0*(exp(VBE[i]/VT)-1.0) - AR*IC0*(exp(VBC[i]/VT)-1.0) -
  ((VCC - (VBE[i] - VBC[i]))/RC - ILOAD[i]))) = 0.0 ;

equ VBC[i=all]->
  ((IC[i] - (AF*IE0*(exp(VBE[i]/VT)-1.0) -
            IC0*(exp(VBC[i]/VT)-1.0))) * (VBC[i] >= VBCMIN) -
  (VCC - RC*(IC[i]+ILOAD[i]) - (VBE[i] - VBC[i])) *
  (VBC[i] < VBCMIN)) * (VBC[i] <= VBCMAX) =
  (VBC[i] > VBCMAX) * (VBC[i] - VBCMAX);
```

```
equ IC[i=all]->
  (IC[i] >= 0.0) * (IC[i] <= ICM) *
  ((IC[i] - (AF*IE0*(exp(VBE[i]/VT)-1.0) -
              ICO*(exp(VBC[i]/VT)-1.0))) * (VBC[i] < VBCMIN) -
  (VCC - RC*(IC[i]+ILOAD[i]) - (VBE[i] - VBC[i])) * (VBC[i] >= VBCMIN)) =
  (IC[i] < 0.0) * IC[i] + (IC[i] > ICM) * (IC[i] - ICM);

equ VBB[i=all]->
  (VBB[i] - RB*IB[i] - VBE[i]) = 0.0 ;

equ VLOAD[i=all] -> VLOAD[i] =
  (VBE[i]-VBC[i])*(VBE[i]-VBC[i] <= VBEMAX) + VBEMAX*(VBE[i]-VBC[i] > VBEMAX) ;

equ ILOAD[i=all] -> ILOAD[i] =
  N*(VBE[i]-VBC[i]-VLOAD[i])/RB ;

begin main
  assign VBB[i=all]  =   VCC*(1.01*i/TMAX);
  assign VBE0[i=all] = VBEMAX*(VBB[i]>VBEMAX) + (VBB[i]<=VBEMAX)*VBB[i];
  assign VBE[i=all]  = VBE0[i]*(1.0 - 0.01*(1.01 - 1.01*i/TMAX));
  solve;
  write;
end
```

8.5   Numerical Model of the BJT

We now turn to the numerical modeling of the carrier behavior and electrostatics of the BJT. The 2D simulations which follow require considerable amounts of memory; in some cases 16MB or more.

To model the BJT at a microscopic level, the files bjt01.sg and bjt01.sk have been set up. The file bjt01.sg is almost the same as other device model input files, the physical constants, physical models and equations being essentially identical. The only differences in the source files are really the call to the mesh specification file (where the geometry is set up) and the boundary conditions applied to the various regions. The input file bjt01.sg will be given next, despite its similarity to previous input files. The mesh specification file bjt01.sk follows after bjt01.sg. It is similar in format to the other mesh specification files, but this file contains the bjt-specific part of the problem specification. The ability to keep the changes to a minimum on going from one device to another makes the modeling task much easier and less error prone than it would otherwise be.

One feature of this file, which is somewhat remarkable, is the apparent absence of boundary conditions. As described previously, if no boundary condition is specified explicitly at a given boundary, the bulk equation used at that boundary will lead

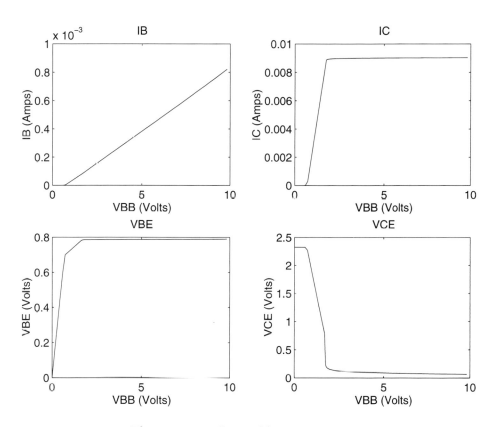

Figure 8.4. Effects of fan-out on a BJT.

to a 'no-flux' boundary condition.

Input File bjt01.sg

```
// FILE:  sgdir\ch08\bjt01\bjt01.sg
// specify irregular mesh
mesh "bjt01.msh";

// physical constants and built-in potential
const T   = 300.0;              // ambient temperature (K)
const e   = 1.602e-19;          // electron charge (C)
const kb  = 1.381e-23;          // Boltzmann's constant (J/K)
const eo  = 8.854e-14;          // permittivity of vacuum (F/cm)
const eSi = 11.8;               // relative permittivity of Si
const eps = eSi*eo;             // permittivity of Si (F/cm)
const Ni  = 1.25e10;            // intrinsic concentration of Si (cm^-3)
const Vt  = kb*T/e;             // thermal voltage (V)

// scaling constants
const x0  = max(Wbjt,Dbjt);     // distance scaling (cm)
const T0  = T;                  // temperature scaling (K)
const V0  = kb*T0/e;            // electrostatic-potential scaling (V)
const C0  = max(Ne,max(Nb,Nc)); // concentration scaling (cm^-3);
const D0  = 35.0;               // diffusion-coefficient scaling (cm^2/s)
const u0  = D0/V0;              // mobility-coefficient scaling (cm^2/V/s)
const R0  = D0*C0/sq(x0);       // recombination rate scaling (cm^-3/s)
const t0  = sq(x0)/D0;          // time scaling (s)
const E0  = V0/x0;              // electric-field scaling (V/s)
const J0  = e*D0*C0/x0;         // current-density scaling (A/cm^2)
const L0  = V0*eo/(e*sq(x0)*C0); // Laplacian scaling constant


// 1D Cartesian discretized gradient operator
func grad(f1,f2,<h>)
  return (f2-f1)/h;

// equilibrium electron concentration of uniformly doped silicon
func Neq(<N>)
  assign temp = [abs(N)+sqrt(sq(N)+4*sq(Ni))]/2.0;
  return (N>=0)*temp + (N<0)*sq(Ni)/temp;

// equilibrium hole concentration of uniformly doped silicon
func Peq(<N>)
  assign temp = [abs(N)+sqrt(sq(N)+4*sq(Ni))]/2.0;
  return (N<=0)*temp + (N>0)*sq(Ni)/temp;

// electron lifetime
func LTn(<N>)
```

```
  const LT0  = 3.95e-4;
  const Nref = 7.1e15;
  return LT0/(1.0+N/Nref);

// hole lifetime
func LTp(<N>)
  const LT0  = 3.52e-5;
  const Nref = 7.1e15;
  return LT0/(1.0+N/Nref);

// scaled electron mobility
func MUn(<N>,<T>,pn)
  const Al = 1430.0,  Bl = -2.20;
  const Ai = 4.61e17, Bi = 1.52e15;
  assign NU  = abs(N)*C0;
  assign pnU = (pn>0.0)*pn*sq(C0)+(pn<0.0);
  assign MUl = Al*pow((T/300.0),Bl);
  assign MUi = {Ai*T*sqrt(T)/NU}/{ln(1+Bi*sq(T)/NU)-Bi*T*T/(NU+Bi*sq(T))};
  assign MUc = {2e17*T*sqrt(T/pnU)}/ln(1+8.28e8*sq(T)*pow(pnU,-1/3));
  assign X   = sqrt(6*MUl*(MUi+MUc)/(MUi*MUc));
  return MUl*{1.025/[1+pow(X/1.68,1.43)]-0.025}/u0;

// scaled hole mobility
func MUp(<N>,<T>,pn)
  const Al = 495.0,   Bl = -2.20;
  const Ai = 1.00e17, Bi = 6.25e14;
  assign NU  = abs(N)*C0;
  assign pnU = (pn>0.0)*pn*sq(C0)+(pn<0.0);
  assign MUl = Al*pow((T/300.0),Bl);
  assign MUi = {Ai*T*sqrt(T)/NU}/{ln(1+Bi*sq(T)/NU)-Bi*T*T/(NU+Bi*sq(T))};
  assign MUc = {2e17*T*sqrt(T/pnU)}/ln(1+8.28e8*sq(T)*pow(pnU,-1/3));
  assign X   = sqrt(6*MUl*(MUi+MUc)/(MUi*MUc));
  return MUl*{1.025/[1+pow(X/1.68,1.43)]-0.025}/u0;

// scaled recombination
func R(n,p,<tn1>,<tp1>)
  const tn = 4.9e-5/t0;
  const tp = 4.2e-5/t0;
  const Cn = 2.8e-31/[D0/sq(C0*x0)];// scaled electron Auger coefficient
  const Cp = 1.2e-31/[D0/sq(C0*x0)];// scaled hole Auger coefficient
  const ni = Ni/C0;                 // scaled intrinsic concentration
  const pn = sq(ni);                // scaled intrinsic concentration squared
  assign Rsrh = (n*p-pn)/[tp*(n+ni)+tn*(p+ni)];
  assign Raug = (n*p-pn)*(Cn*n+Cp*p);
  return Rsrh+Raug;

// scaled electric field
func E(V1,V2,<h>)
```

```
    return -grad(V1,V2,h);

// scaled electron current density
func Jn(n1,n2,E,un1,un2,<h>)
  const Ut   = Vt/V0;                // scaled thermal voltage
  assign un  = ave(un1,un2);
  assign dV  = E*h/(2*Ut);
  assign n   = n1*aux2(dV)+n2*aux2(-dV);
  assign dndx = aux1(dV)*grad(n1,n2,h);
  return un*(n*E+dndx);

// scaled hole current density
func Jp(p1,p2,E,up1,up2,<h>)
  const Ut   = Vt/V0;                // scaled thermal voltage
  assign up  = ave(up1,up2);
  assign dV  = E*h/(2*Ut);
  assign p   = p1*aux2(-dV)+p2*aux2(dV);
  assign dpdx = aux1(dV)*grad(p1,p2,h);
  return up*(p*E-dpdx);


// set numerical algorithm parameters
set NEWTON DAMPING     = 3;
set NEWTON ACCURACY    = 1.0e-12;
set NEWTON ITERATIONS = 100;
set LINSOL ALGORITHM  = GAUSSELIM;
set LINSOL FILL       = INFINITY;
set DISTANCE SCALE    = x0;


// declare variables and specify which ones are knowns and unknowns
var     V[NODES], n[NODES], p[NODES], tn[NODES], tp[NODES];
unknown V[SI], n[SI], p[SI];
known   V[EMITTER], V[BASE], V[COLLECTOR];
known   n[EMITTER], n[BASE], n[COLLECTOR];
known   p[EMITTER], p[BASE], p[COLLECTOR];

// for each Newton iteration, precompute the following functions prior to
// constructing the Jacobian matrix
precompute @ NODE i -> MUn(C[i],T,n[i]*p[i]);
precompute @ NODE i -> MUp(C[i],T,n[i]*p[i]);
precompute @ NODE i -> R(n[i],p[i],tn[i],tp[i]);
precompute @ EDGE i (j,k) ODD -> E(V[j],V[k],elen);
precompute @ EDGE i (j,k) ODD -> Jn(n[j],n[k],@E(i,j,k),@MUn(j),@MUn(k),elen);
precompute @ EDGE i (j,k) ODD -> Jp(p[j],p[k],@E(i,j,k),@MUp(j),@MUp(k),elen);


// scaled Poisson's equation
```

```
equ V[i=SI] ->
  L0*nsum(i,j,all,{eSi*@E(edge(i,j),i,node(i,j))}*ilen(i,j)) -
  (p[i]-n[i]+C[i])*area(i) = 0.0;

// scaled electron continuity equation
equ n[i=SI] ->
  +nsum(i,j,all,{@Jn(edge(i,j),i,node(i,j))}*ilen(i,j)) -
  @R(i)*area(i) = 0.0;

// scaled hole continuity equation
equ p[i=SI] ->
  -nsum(i,j,all,{@Jp(edge(i,j),i,node(i,j))}*ilen(i,j)) -
  @R(i)*area(i) = 0.0;


begin InitializeVariables
  open "bjt01eq.dat" read;
  file read V;
  close;
  assign n[i=all] = Nc*exp(V[i]/Vt);
  assign p[i=all] = sq(Ni)/n[i];
  assign tn[i=all] = LTn(abs(C[i]));
  assign tp[i=all] = LTp(abs(C[i]));
end

begin ScaleVariables
  assign x[i=all] = x[i]/x0;
  assign V[i=all] = V[i]/V0;
  assign C[i=all] = C[i]/C0;
  assign n[i=all] = n[i]/C0;
  assign p[i=all] = p[i]/C0;
  assign tn[i=all] = tn[i]/t0;
  assign tp[i=all] = tp[i]/t0;
end

begin UnscaleVariables
  assign x[i=all] = x[i]*x0;
  assign V[i=all] = V[i]*V0;
  assign C[i=all] = C[i]*C0;
  assign n[i=all] = n[i]*C0;
  assign p[i=all] = p[i]*C0;
  assign tn[i=all] = tn[i]*t0;
  assign tp[i=all] = tp[i]*t0;
end

var Ie, Ib, Ic;
begin ComputeCurrent
  assign Ie = J0*lsum(i,EMITTER,nsum(i,j,SI,ilen(i,j)*{
```

```
    Jn(n[i],n[node(i,j)],E(V[i],V[node(i,j)],elen(i,j)),
MUn(C[i],T,n[i]*p[i]),
MUn(C[node(i,j)],T,n[node(i,j)]*p[node(i,j)]),elen(i,j))+
    Jp(p[i],p[node(i,j)],E(V[i],V[node(i,j)],elen(i,j)),
MUp(C[i],T,n[i]*p[i]),
MUp(C[node(i,j)],T,n[node(i,j)]*p[node(i,j)]),elen(i,j))}));

    assign Ib = J0*lsum(i,BASE,nsum(i,j,SI,ilen(i,j)*{
    Jn(n[i],n[node(i,j)],E(V[i],V[node(i,j)],elen(i,j)),
MUn(C[i],T,n[i]*p[i]),
MUn(C[node(i,j)],T,n[node(i,j)]*p[node(i,j)]),elen(i,j))+
    Jp(p[i],p[node(i,j)],E(V[i],V[node(i,j)],elen(i,j)),
MUp(C[i],T,n[i]*p[i]),
MUp(C[node(i,j)],T,n[node(i,j)]*p[node(i,j)]),elen(i,j))}));

    assign Ic = J0*lsum(i,COLLECTOR,nsum(i,j,SI,ilen(i,j)*{
    Jn(n[i],n[node(i,j)],E(V[i],V[node(i,j)],elen(i,j)),
MUn(C[i],T,n[i]*p[i]),
MUn(C[node(i,j)],T,n[node(i,j)]*p[node(i,j)]),elen(i,j))+
    Jp(p[i],p[node(i,j)],E(V[i],V[node(i,j)],elen(i,j)),
MUp(C[i],T,n[i]*p[i]),
MUp(C[node(i,j)],T,n[node(i,j)]*p[node(i,j)]),elen(i,j))}));
end

begin main
  call InitializeVariables;
  call ScaleVariables;
  solve;
  Call ComputeCurrent;
  open "bjt01.out" write;
  file write Ie, Ib, Ic;
  close;
  call UnscaleVariables;
  write;
end
```

Some of the last few lines of this file calculate the currents through each electrode. They make use of the 'lsum' function, (see Section 15.3.4) which permits a sum to be done over a list of points. The calculation is done in the same way as the current is calculated in the main file. Since the transport equations conserve current, and they are numerically implemented in conservative form, the steady-state currents to the electrodes should, when summed over all the electrodes, give 'exactly' zero. There should be no current through the other boundaries, since 'NOFLUX' boundary conditions were used. Because the numerical expressions only have a finite accuracy, the currents will actually fail to add up to zero at some level. It is also important to be careful about the sign conventions used, both with regard to the flow being in or out of the device, and with regard to the relative sign of electron and hole currents.

The mesh specification file bjt01.sk follows.

Skeleton File bjt01.sk

```
//    emitter          base              collector
//    A-----B--------C-----D-------------E-----F---------G
//    |     | /      |     |     |       |     |         |
//    |-----|-       |     |   /         |     |         |
//    |     |        |     |  /          |     |         |
//    |-----|--------|-----|-            |     |         |
//    |     |        |     |             |     |         |
//    |     |        |     |             |     |         |
//    |     |        |     |             |     |         |
//    H-----I--------J-----K-------------L-----M---------N

CONST WCe   = 2.0e-4;                     // emitter contact width
CONST WCb   = 2.0e-4;                     // base contact width
CONST WCc   = 2.0e-4;                     // collector contact width
CONST WSeb  = 3.0e-4;                     // emitter-base spacing width
CONST WSbc  = 7.0e-4;                     // base emitter spacing width
CONST Wbjt  = WCe+WCb+WCc+WSeb+WSbc+5.0e-4;  // bjt width
CONST Dbjt  = 10.0e-4;                    // bjt depth
CONST Wspacing = 0.5e-4;                  // initial grid-point spacing

point pA = (0.0e-4,                 0.0e-4);
point pB = (WCe,                    0.0e-4);
point pC = (WCe+WSeb,               0.0e-4);
point pD = (WCe+WSeb+WCb,           0.0e-4);
point pE = (WCe+WSeb+WCb+WSbc,      0.0e-4);
point pF = (WCe+WSeb+WCb+WSbc+WCc,  0.0e-4);
point pG = (Wbjt,                   0.0e-4);
point pH = (0.0e-4,                 -Dbjt);
point pI = (WCe,                    -Dbjt);
point pJ = (WCe+WSeb,               -Dbjt);
point pK = (WCe+WSeb+WCb,           -Dbjt);
point pL = (WCe+WSeb+WCb+WSbc,      -Dbjt);
point pM = (WCe+WSeb+WCb+WSbc+WCc,  -Dbjt);
point pN = (Wbjt,                   -Dbjt);

edge eAB = EMITTER   [pA, pB] (Wspacing, 0.0);
edge eBC = NOFLUX    [pB, pC] (Wspacing, 0.0);
edge eCD = BASE      [pC, pD] (Wspacing, 0.0);
edge eDE = NOFLUX    [pD, pE] (Wspacing, 0.0);
edge eEF = COLLECTOR [pE, pF] (Wspacing, 0.0);
edge eFG = NOFLUX    [pF, pG] (Wspacing, 0.0);
edge eHI = NOFLUX    [pH, pI] (Wspacing, 0.0);
```

```
edge eIJ = NOFLUX     [pI, pJ] (Wspacing, 0.0);
edge eJK = NOFLUX     [pJ, pK] (Wspacing, 0.0);
edge eKL = NOFLUX     [pK, pL] (Wspacing, 0.0);
edge eLM = NOFLUX     [pL, pM] (Wspacing, 0.0);
edge eMN = NOFLUX     [pM, pN] (Wspacing, 0.0);
edge eAH = NOFLUX     [pA, pH] (Wspacing, 0.0);
edge eBI =            [pB, pI] (Wspacing, 0.0);
edge eCJ =            [pC, pJ] (Wspacing, 0.0);
edge eDK =            [pD, pK] (Wspacing, 0.0);
edge eEL =            [pE, pL] (Wspacing, 0.0);
edge eFM =            [pF, pM] (Wspacing, 0.0);
edge eGN = NOFLUX     [pG, pN] (Wspacing, 0.0);

region rAHIB = SI {eAH, eHI, eBI, eAB} RECTANGLES;
region rBIJC = SI {eBI, eIJ, eCJ, eBC} RECTANGLES;
region rCJKD = SI {eCJ, eJK, eDK, eCD} RECTANGLES;
region rDKLE = SI {eDK, eKL, eEL, eDE} RECTANGLES;
region rELMF = SI {eEL, eLM, eFM, eEF} RECTANGLES;
region rFMNG = SI {eFM, eMN, eGN, eFG} RECTANGLES;


const Ne  = 1.00e+19;
const Nb  = 1.00e+16;
const Nc  = 1.00e+14;
const Re  = 1.00e-04;
const Rb  = 3.00e-04;
const Ae  = ln(Ne/Nb)/sq(Re);
const Ab  = ln(Nb/Nc)/sq(Rb);
const dC  = 1.0e-4;

coordinates x, y;

refine C (SignedLog, 1.0) =
  Nc -
  (Nb+Nc)*ngdep(x,y,2.0*(WCe+WSeb+WCb),Ab,Ab) +
  (Ne+Nb)*ngdep(x,y,2.0*WCe,Ae,Ae);

set maximum length = 1.0 - (1.0-0.75*Wspacing)*{
  [(x<WCe+WSeb+WCb+Rb) and (y>-Rb)] or [(x>WCe+WSeb+WCb+WSbc-dC) and
   (x<WCe+WSeb+WCb+WSbc+WCc+dC) and (y>-dC)]};

set minimum divisions = 0;
set maximum divisions = 3;
```

The simulation may be run under Win32 or UNIX. The DOS script runbjt01.bat is given in a table, as is the UNIX script runbjt01. A successful device simulation is often helped by starting from an equilibrium solution. This is true for the simulation bjt01.sg, which uses the output from bjt01eq.sg (given next).

Input File bjt01eq.sg

```
// FILE:  sgdir\ch08\bjt01\bjt01eq.sg
// specify irregular mesh
mesh "bjt01.msh";

// physical constants and built-in potential
const T   = 300.0;              // ambient temperature (K)
const e   = 1.602e-19;          // electron charge (C)
const kb  = 1.381e-23;          // Boltzmann's constant (J/K)
const eo  = 8.854e-14;          // permittivity of vacuum (F/cm)
const eSi = 11.8;               // relative permittivity of Si
const eps = eSi*eo;             // permittivity of Si (F/cm)
const Ni  = 1.25e10;            // intrinsic concentration of Si (cm^-3)
const Vt  = kb*T/e;             // thermal voltage (V)

// scaling constants
const x0  = max(Wbjt,Dbjt);     // distance scaling (cm)
const T0  = T;                  // temperature scaling (K)
const V0  = kb*T0/e;            // electrostatic-potential scaling (V)
const C0  = max(Ne,max(Nb,Nc)); // concentration scaling (cm^-3);
const D0  = 35.0;               // diffusion-coefficient scaling (cm^2/s)
const u0  = D0/V0;              // mobility-coefficient scaling (cm^2/V/s)
const R0  = D0*C0/sq(x0);       // recombination-rate scaling (cm^-3/s)
const t0  = sq(x0)/D0;          // time scaling (s)
const E0  = V0/x0;              // electric-field scaling (V/s)
const J0  = e*D0*C0/x0;         // current-density scaling (A/cm^2)
const L0  = V0*eo/(e*sq(x0)*C0); // Laplacian scaling constant


// 1D Cartesian discretized gradient operator
func grad(f1,f2,<h>)
  return (f2-f1)/h;

// equilibrium electron concentration of uniformly doped silicon
func Neq(<N>)
  assign temp = [abs(N)+sqrt(sq(N)+4*Ni)]/2.0;
  return (N>=0)*temp + (N<0)*sq(Ni)/temp;

// scaled electric field
func E(V1,V2,<h>)
  return -grad(V1,V2,h);

// set numerical algorithm parameters
set NEWTON DAMPING    = 3;
set NEWTON ACCURACY   = 1.0e-12;
set NEWTON ITERATIONS = 100;
set LINSOL ALGORITHM  = GAUSSELIM;
```

```
set LINSOL FILL       = INFINITY;
set DISTANCE SCALE    = x0;


// declare variables and specify which ones are knowns and unknowns
var     V[NODES];
unknown V[SI];
known   V[EMITTER], V[BASE], V[COLLECTOR];

// for each Newton iteration, precompute the following functions prior to
// constructing the Jacobian matrix
precompute @ EDGE i (j,k) ODD -> E(V[j],V[k],elen);


// scaled Poisson's equation
equ V[i=SI] ->
  L0*nsum(i,j,all,{eSi*@E(edge(i,j),i,node(i,j))}*ilen(i,j)) =
  (sq(Ni)/Nc/C0*exp(-V[i]/(Vt/V0))-Nc/C0*exp(V[i]/(Vt/V0))+C[i])*area(i);

begin InitializeVariables
  assign V[i=all]       = Vt*ln(Neq(C[i])/Nc);
  assign V[i=EMITTER]   = Vt*ln(Ne/Nc);
  assign V[i=BASE]      = Vt*ln(sq(Ni)/Nb/Nc);
  assign V[i=COLLECTOR] = Vt*ln(Nc/Nc);
end

begin ScaleVariables
  assign x[i=all] = x[i]/x0;
  assign V[i=all] = V[i]/V0;
  assign C[i=all] = C[i]/C0;
end

begin UnscaleVariables
  assign x[i=all] = x[i]*x0;
  assign V[i=all] = V[i]*V0;
  assign C[i=all] = C[i]*C0;
end

begin main
  call InitializeVariables;
  call ScaleVariables;
  solve;
  call UnscaleVariables;
  open "bjt01eq.dat" write;
  file write V;
  close;
  write;
end
```

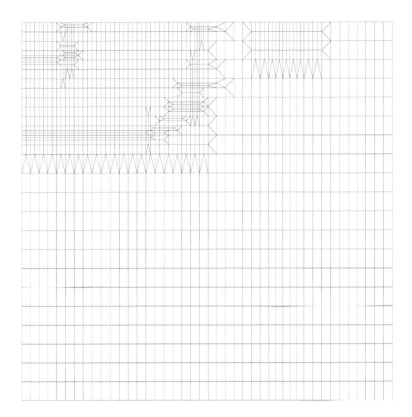

Figure 8.5. The mesh produced by the mesh specification file bjt01.sk

This run will plot the unrefined mesh, then the refined mesh. In both cases it is necessary to click on the plot in order to move on. The run will be paused until each plot is removed (although the pause can be removed by a command line option).

The maximum number of divisions has been set to 3 in bjt01.sk.   When set to 2, for instance, relatively little refinement is done, and such a simulation may ultimately diverge. A maximum of three divisions gives a better-looking mesh.

The Newton accuracy has been set to a relatively small value of 10^{-12}, which provides high accuracy. This number may need relaxing back to a larger value, in other runs with different applied voltages.

The mesh which is produced from bjt01.sk is shown in Figure 8.5.

Figure 8.6 shows the dopant concentration; Figure 8.7 the electron density; Figure 8.8 the hole density and Figure 8.9 the voltage obtained from this calculation.

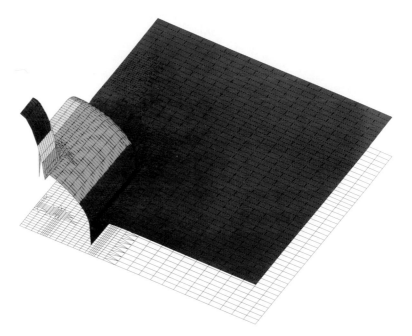

Figure 8.6. The log of the absolute value of the dopant concentration C in the bjt. The various regions of the device are clearly visible. Starting at the top left is the emitter, surrounded by the base, which in turn is surrounded by the collector. The mesh has also been refined near the collector contact.

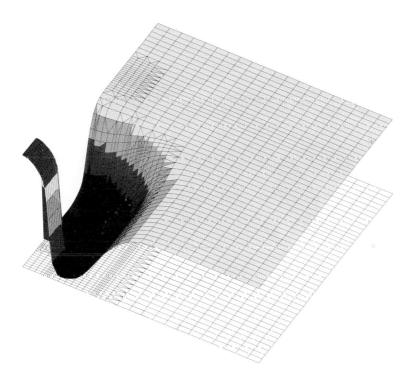

Figure 8.7. The log of the electron concentration n found in the bjt calculation.

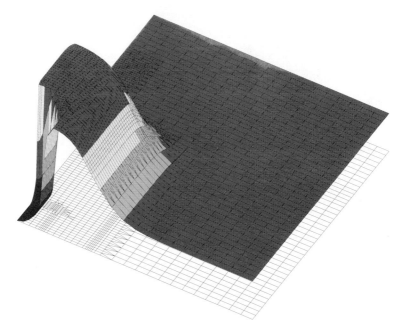

Figure 8.8. The log of the hole concentration p found in the bjt calculation.

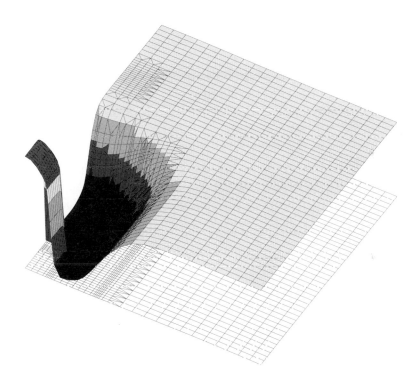

Figure 8.9. The potential V found in the bjt calculation. In equilibrium, the potential resembles the logarithm of the electron density. See Equation (4.12).

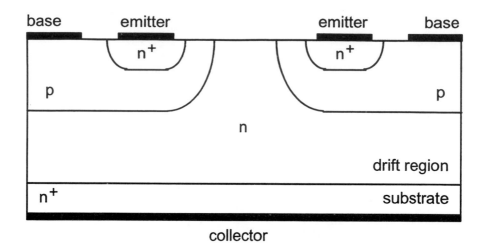

Figure 8.10. A schematic of a high power bipolar transistor.

Equation Specification File bjt02.sg

A second BJT simulation, bjt02.sg, is identical to bjt01.sg, except that it imports a mesh for a power bjt. A schematic of a power bipolar transistor is given in Figure 8.10. bjt02.sk follows, after the first lines of bjt02.sg.

```
// FILE:  sgdir\ch08\bjt02\bjt02.sg
// specify irregular mesh
mesh "bjt02.msh";
(then same as bjt01.sg....)
```

Mesh Specification File bjt02.sk

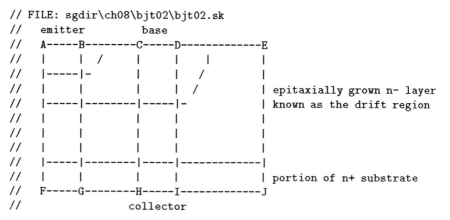

```
CONST WCe  = 2.0e-4;                      // emitter contact width
CONST WCb  = 2.0e-4;                      // base contact width
CONST WSeb = 3.0e-4;                      // emitter-base spacing width
CONST Wbjt = WCe+WCb+WSeb+6.0e-4;         // bjt width
CONST Depi = 30.0e-4;                     // epitaxial depth
cONST Dsub = 10.0e-4;                     // simulated substrate depth
CONST Dbjt = Depi+Dsub;                   // depth of bjt
CONST Wspacing = 0.5e-4;                  // initial grid point spacing
CONST Dspacing = 2.5e-4;                  // initial grid point spacing

point pA = (0.0e-4,         0.0e-4);
point pB = (WCe,            0.0e-4);
point pC = (WCe+WSeb,       0.0e-4);
point pD = (WCe+WSeb+WCb, 0.0e-4);
point pE = (Wbjt,           0.0e-4);
point pF = (0.0e-4,        -Dbjt);
point pG = (WCe,           -Dbjt);
point pH = (WCe+WSeb,      -Dbjt);
point pI = (WCe+WSeb+WCb,  -Dbjt);
point pJ = (Wbjt,          -Dbjt);

edge eAB = EMITTER    [pA, pB] (Wspacing, 0.0);
edge eBC = NOFLUX     [pB, pC] (Wspacing, 0.0);
edge eCD = BASE       [pC, pD] (Wspacing, 0.0);
edge eDE = NOFLUX     [pD, pE] (Wspacing, 0.0);
edge eFG = COLLECTOR  [pF, pG] (Wspacing, 0.0);
edge eGH = COLLECTOR  [pG, pH] (Wspacing, 0.0);
edge eHI = COLLECTOR  [pH, pI] (Wspacing, 0.0);
edge eIJ = COLLECTOR  [pI, pJ] (Wspacing, 0.0);
edge eAF = NOFLUX     [pA, pF] (Dspacing, 0.0);
edge eBG =            [pB, pG] (Dspacing, 0.0);
edge eCH =            [pC, pH] (Dspacing, 0.0);
edge eDI =            [pD, pI] (Dspacing, 0.0);
edge eEJ = NOFLUX     [pE, pJ] (Dspacing, 0.0);

region rAFGB = SI {eAF, eFG, eBG, eAB} RECTANGLES;
region rBGHC = SI {eBG, eGH, eCH, eBC} RECTANGLES;
region rCHID = SI {eCH, eHI, eDI, eCD} RECTANGLES;
region rDIJE = SI {eDI, eIJ, eEJ, eDE} RECTANGLES;


const Ne = 1.00e+19;                      // peak emitter donor conc.
const Nb = 1.00e+16;                      // peak base acceptor conc.
const Nd = 1.00e+15;                      // drift region donor conc.
const Nc = 1.00e+19;                      // substrate donor conc.
const Re = 1.00e-04;                      // emitter diffusion radius
```

```
const Rb   = 3.00e-04;                              // base diffusion radius
const Ae   = ln(Ne/Nb)/sq(Re);
const Ab   = ln(Nb/Nd)/sq(Rb);

coordinates x, y;

refine C (SignedLog, 1.0) =
   Nd -
   (Nb+Nd)*ngdep(x,y,2.0*(WCe+WSeb+WCb),Ab,Ab) +
   (Ne+Nb)*ngdep(x,y,2.0*WCe,Ae,Ae) +
   (Nc-Nd)*nsdep(y+Dbjt,Dsub,1.0e-9);

set maximum length = 1.0 - (1.0-0.75*Wspacing)*{
   [(x<WCe+WSeb+WCb+1.5*Rb) and (y>-1.5*Rb)]};
set minimum length = Wspacing/2.0;
set minimum divisions = 0;
set maximum divisions = 5;
```

The voltage it finds using this mesh is shown in Figure 8.11 which follows.

This mesh specification file is for a power transistor. The device has a 'drift region', as explained in Chapter 11, on power devices. The doping is controlled by the quantity C which is defined in the mesh specification file, and which is the dopant concentration. The functions n_{sdep} and n_{gdep} are used to specify the doping, which is shown in Figure 8.12.

8.6   Summary

This chapter extended the theme of Chapter 7, on pn junctions, by taking the same approach to the bipolar junction transistor. The analytic theory was again introduced and compared to simulation results. Extensive use was made of the analytic models to set up circuit simulations. Detailed microscopic models of devices were then set up, beginning with a low power device and followed by a power bjt. The difference in the models of the two devices was only in the mesh specification file which defined the geometry and the doping.

The comparison of the Ebers-Moll model to microscopic simulations has been postponed to the problems, which follow.

PROBLEMS

1. In each of the simulations of bjt circuits given in the text, (a) sketch the circuit; (b) verify the equations used in the simulation, and (c) suggest more appropriate parameter values for use in the simulations. Rerun the simulations using the better parameters and plot the results.

2. Estimate the electric field inside a BJT under the main normal operating conditions.

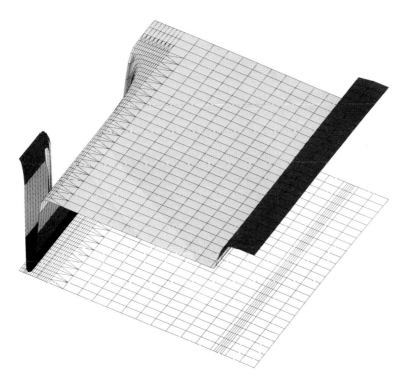

Figure 8.11. The voltage profile in a power transistor.

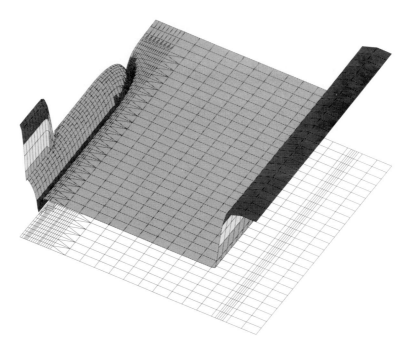

Figure 8.12. Doping profile in the power transistor specified by bjt02.sk

3. Using the electric field from the previous question, use a Monte Carlo method to follow carriers injected into a BJT through each terminal, under different operating conditions. Use the mean free path λ and collision frequency ν which can be estimated from data given in Chapter 4. How many collisions do you expect a typical particle to have before it leaves each junction where it is injected? How long will this take, in each case? Plot several trajectories, to confirm your estimates.

4. Using the mesh specification file bjt01.sk, draw a diagram showing the bjt modeled by means of bjt01.sk. Label the regions of the device and indicate the doping levels in each region. How is the refinement function C related to the layout of the device?

5. Choose a set of parameters suitable for a BJT for use in VLSI applications. Set up a suitable mesh specification file, by modifying bjt01.sk, and calculate the equilibrium voltage and carrier densities using bjt01.sg. Plot the results, including the mesh you generate.

6. For the device which was set up in the previous problem, or using the device specified using the mesh specification file bjt01.sk given in the text, plot the usual static (i, v) curves for the bjt. Make use of the functions I_e, I_b and I_c which are defined at the end of the file bjt01.sg. (The command 'extract bjt01.res' will cause all the results from the run to be written to files, one for each array with names such as n.000, and one file called vars.000 containing the variables which are not arrays. The currents will be in this file.) Compare the results to analytic theory.

7. Evaluate the small signal parameters g_{11}, g_{12}, g_{21} and g_{22}, for the device specified in the mesh specification file bjt01.sk. Use these results to plot i_c as a function of time when the bjt is suitably biased and v_{be} is oscillating sinusoidally with an amplitude of 0.1 volts. Find i_c numerically and compare the results.

8. Evaluate Q_B both analytically and numerically, during (a) turn-off and (b) turn-on transients. Compare the results.

DOS Script File runbjt01.bat

```
echo off
cls

rem generate the mesh
mesh bjt01.sk
sggrid bjt01.xsk
call sgbuild ref bjt0_ref
bjt0_ref

rem build and run the equilibrium simulation
sgxlat -p15 bjt01eq.sg
call sgbuild sim bjt01eq
order bjt01eq.top bjt01eq.prm
bjt01eq -vs2 -vl2 -Lbjt01eq.log

rem build and run the full simulation
sgxlat -p15 bjt01.sg
call sgbuild sim bjt01
order bjt01.top bjt01.prm
bjt01 -vs2 -vl2 -Lbjt01.log
group Ie Ib Ic <bjt01.out >bjt01.tab

rem plot the results
echo plotting electostatic potential
triplot bjt01 V -az300 -ps
if exist V.ps del V.ps
ren bjt01.ps V.ps
echo plotting electron concentration
triplot bjt01 n -az300 -log -ps
if exist n.ps del n.ps
ren bjt01.ps n.ps
echo plotting hole concentration
triplot bjt01 p -az300 -log -ps
if exist p.ps del p.ps
ren bjt01.ps p.ps
echo plotting dopant concentration
triplot bjt01 C -az300 -log -ps
if exist C.ps del C.ps
ren bjt01.ps C.ps

rem clean up
del bjt01.xsk
    (etc.)
del bjt01.out
```

UNIX Script File runbjt01

```
#!/bin/sh
# generate the mesh
mesh bjt01.sk
sggrid bjt01.xsk
sgbuild ref bjt0_ref
bjt0_ref
# build and run the equilibrium simulation
sgxlat -p15 bjt01eq.sg
sgbuild sim bjt01eq
order bjt01eq.top bjt01eq.prm
bjt01eq -vs2 -vl2 -Lbjt01eq.log
# build and run the full simulation
sgxlat -p15 bjt01.sg
sgbuild sim bjt01
order bjt01.top bjt01.prm
bjt01 -vs2 -vl2 -Lbjt01.log
group Ie Ib Ic <bjt01.out >bjt01.tab
# plot the results
echo plotting electrostatic potential
triplot bjt01 V -az150
echo plotting electron concentration
triplot bjt01 n -az150 -log
echo plotting hole concentration
triplot bjt01 p -az150 -log
echo plotting dopant concentration
triplot bjt01 C -az150 -log
# clean up
rm bjt01.xsk
   (etc.)
rm bjt01.out
```

Chapter 9

Junction Field-Effect Transistors

This very short chapter gives a brief overview of the derivation of the analytic models used to describe the junction field-effect transistor, and presents a simulation of a JFET. Section 9.1 describes the static characteristics. Section 9.2 outlines the small signal behavior. Section 9.3 contains the numerical model of the JFET.

9.1 Static Characteristics

An idealized JFET is shown in Figure 9.1.

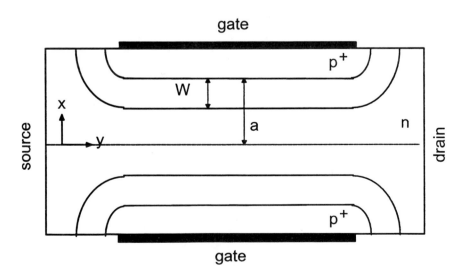

Figure 9.1. A schematic of an idealized junction field effect transistor.

The junction field-effect transistor or JFET is taken in this chapter to be arranged with source on the left and drain on the right and gate at top and bottom. The x-direction runs from the middle of the channel vertically, towards the gate. The device is of width $2a$ in the vertical direction. The gate depletion layer is of width W, however, so the channel runs from $W - a$ to $a - W$.

The y-direction is horizontal, from source to drain. The height h of the device is out of the plane of the diagram, in the z-direction.

We now find I_D, the drain current, in terms of V_D the drain voltage and V_G the gate voltage (defined relative to a source voltage of zero) for V_D and V_G below pinch-off. At pinch-off the depletion region of the gate fills the channel. Below pinch-off, $0 \leq V_D \leq V_{Dsat}$ and $0 \geq V_G \geq V_P$. The steps taken (which in some regards closely resemble the equivalent derivation for the MOSFET, Chapter 10) are:

1. The current density J_N in the channel is mostly due to the electric field so $J_N \simeq e\mu_n N_D E_y = -e\mu_n N_D \frac{\partial V}{\partial y}$, where the electron density n was replaced by the donor density N_D. To find the total channel current in the y direction, which is equal to $-I_D$, this is integrated over x and z to the edges of the channel:

$$I_D = - \int \int J_N \, dx \, dy = -h \int_W^{2a-W} J_N \, dx. \tag{9.1}$$

W is the width of the depletion layer.

$$I_D = 2eh\mu_n N_D \frac{\partial V}{\partial y}(a - W). \tag{9.2}$$

2. To get rid of the derivative with respect to y, this can be integrated over the length of the channel from $y = 0$ to $y = L$. Since I_D is independent of y, the integral of I_D over the length is $I_D L$, so

$$I_D L = 2eh\mu_n N_D \int (a - W)\frac{\partial V}{\partial y} \, dy. \tag{9.3}$$

The integral becomes an integral over V,

$$I_D = \frac{2eh\mu_n N_D}{L} \int_0^{V_D} (a - W) \, dV. \tag{9.4}$$

3. To find W as a function of y the channel is assumed to be much narrower than it is long. The channel electrostatics can then be treated as one-dimensional, with all the important variation being in the (shorter) x-direction. In one dimension the width of a depletion layer which has a constant charge density eN_D on one side of the junction is given by

$$\frac{\partial^2 V}{\partial x^2} = -\frac{\rho}{\epsilon} = -\frac{eN_D}{\epsilon_{Si}\epsilon_0}. \tag{9.5}$$

Integrating over the x coordinate,

$$E = -\frac{\partial V}{\partial x} = \frac{eN_D}{\epsilon_{Si}\epsilon_0}(x - W) \tag{9.6}$$

the constant being chosen to give $E(x = W) = 0$. Integrating a second time, from a point x to the quasi-neutral bulk at $x = W$, we find the difference in V:

$$\Delta V = -\frac{eN_D}{2\epsilon_{Si}\epsilon_0}(x - W)^2. \tag{9.7}$$

The magnitude of the voltage between $x = 0$ and $x = W$ is

$$\Delta V = \frac{eN_D W^2}{2\epsilon_{Si}\epsilon_0}.$$

The actual applied voltage in this case is $V_{bi} + V - V_G$, so W is found to be

$$W = \sqrt{\frac{2\epsilon_{Si}\epsilon_0}{eN_D}(V_{bi} + V - V_G)}. \tag{9.8}$$

Now when $V = V_D = 0$ and $V_G = V_P$, we need $W = a$, that is, the device is pinched off by the depletion layer of the gate filling the channel. $2a$ is the thickness of the device. Using this expression for W to find a when $V = 0$ and $V_G = V_P$, it can be shown that

$$\frac{W}{a} = \sqrt{\frac{V_{bi} + V - V_G}{V_{bi} - V_P}}. \tag{9.9}$$

Substituting W into I_D and integrating gives

$$I_D = \frac{2eh\mu_n N_D a}{L}\left(V_D - \frac{2}{3}(V_{bi} - V_P)^{-1/2}\left[(V_D + V_{bi} - V_G)^{3/2} - (V_{bi} - V_G)^{3/2}\right]\right) \tag{9.10}$$

for $0 \leq V_D \leq V_{Dsat}$ and $V_P \leq V_G \leq 0$.

4.) In pinch-off I_D is assumed to be roughly equal to $I_D(V_{Dsat}) = I_{Dsat}$, which is given by the above expression with V_D set equal to V_{Dsat}. But V_{Dsat} is the value of $V_D = V(y = L)$ which causes pinch-off, so that $W = a$. When $V_D = 0$ and $V_G = V_P$ we also had pinch-off. For $V_G \neq V_P$ a nonzero V can cause pinch-off provided the combination $V - V_G$ appearing in the expression for W has the same value of $-V_P$ as in the other case when pinch-off occurred. This implies that the value of V needed is $V = V_G - V_P$. Replacing V_{Dsat} by $V_G - V_P$ in the expression for I_{Dsat} gives

$$I_{Dsat} = \frac{2eh\mu_n N_D a}{L}\left(V_G - V_P - \frac{2}{3}(V_{bi} - V_P)^{-1/2}\left[(V_{bi} - V_P)^{3/2} - (V_{bi} - V_G)^{3/2}\right]\right). \tag{9.11}$$

9.2   Small-Signal Analysis

The junction next to the gate in the actual device structure is usually reverse biassed, so it behaves primarily like a capacitor. For low enough frequencies a capacitor resembles an open circuit, so the gate is shown disconnected in the equivalent circuit. At higher frequencies capacitors must be included between gate and source and between gate and drain. This is to be expected since the junction around the gate resembles a capacitor. In reality, there is always a small leakage current due to the reverse biased gate junction.

The ac part of the drain current i_d is found by assuming quasi-static conditions, so the static expressions for the drain current I_D as a function of V_G and V_D hold. Taylor-expanding the drain current gives

$$i_d = \left[\frac{\partial I_D}{\partial V_D}\right]_{V_G} v_d + \left[\frac{\partial I_D}{\partial V_G}\right]_{V_D} v_g = g_d v_d + g_m v_g. \tag{9.12}$$

g_d, the coefficient of v_d, is interpreted as a straightforward conductance from the drain to the source, which is the point to which all the voltages are referenced and to which the drain current flows. g_m must be represented as a voltage-controlled current source, in parallel with the conductance g_d. This completes the equivalent circuit.

The partial derivatives are found using the static equations given previously. The results below pinch-off, $V_D \leq V_{Dsat}$, are

$$g_d = G\left[1 - \sqrt{(V_D + V_{biG})/V_{biP}}\right] \tag{9.13}$$

and

$$g_m = G\left[\sqrt{(V_D + V_{biG})/V_{biP}} - \sqrt{V_{biG}/V_{biP}}\right]. \tag{9.14}$$

In these equations $G = 2eh\mu_n N_D a/L$, and the voltages were represented as $V_{biG} = V_{bi} - V_G$ and $V_{biP} = V_{bi} - V_P$. Above pinch-off, $V_D > V_{Dsat}$,

$$g_d = 0, \quad g_m = G\left[1 - \sqrt{V_{biG}/V_{biP}}\right]. \tag{9.15}$$

9.3   Numerical Model of a JFET

This section presents a model of a JFET, which uses the standard equation specification file and a mesh specification file specialized to correspond to a JFET. Some results of the simulation are presented.

JFET Equation Specification File

We now give the equation specification file for a JFET simulation. Its mesh specification file follows.

```
// FILE:  sgdir\ch09\jfet\jfet.sg
// specify irregular mesh
mesh "jfet.msh";

// physical constants and built-in potential
const T   = 300.0;              // ambient temperature (K)
const e   = 1.602e-19;          // electron charge (C)
const kb  = 1.381e-23;          // Boltzmann's constant (J/K)
const eo  = 8.854e-14;          // permittivity of vacuum (F/cm)
const eSi = 11.8;               // relative permittivity of Si
const eps = eSi*eo;             // permittivity of Si (F/cm)
const Ni  = 1.25e10;            // intrinsic concentration of Si (cm^-3)
const Vt  = kb*T/e;             // thermal voltage (V)

// scaling constants
const x0  = max(Wjfet,Djfet);   // distance scaling (cm)
const T0  = T;                  // temperature scaling (K)
const V0  = kb*T0/e;            // electrostatic-potential scaling (V)
const C0  = max(Na,Nd);         // concentration scaling (cm^-3);
const D0  = 35.0;               // diffusion-coefficient scaling (cm^2/s)
const u0  = D0/V0;              // mobility-coefficient scaling (cm^2/V/s)
const R0  = D0*C0/sq(x0);       // recombination-rate scaling (cm^-3/s)
const t0  = sq(x0)/D0;          // time scaling (s)
const E0  = V0/x0;              // electric-field scaling (V/s)
const J0  = e*D0*C0/x0;         // current-density scaling (A/cm^2)
const L0  = V0*eo/(e*sq(x0)*C0); // Laplacian scaling constant


// 1D Cartesian discretized gradient operator
func grad(f1,f2,<h>)
  return (f2-f1)/h;

// equilibrium electron concentration of uniformly doped silicon
func Neq(<N>)
  assign temp = [abs(N)+sqrt(sq(N)+4*sq(Ni))]/2.0;
  return (N>=0)*temp + (N<0)*sq(Ni)/temp;

// equilibrium hole concentration of uniformly doped silicon
func Peq(<N>)
  assign temp = [abs(N)+sqrt(sq(N)+4*sq(Ni))]/2.0;
  return (N<=0)*temp + (N>0)*sq(Ni)/temp;

// electron lifetime
func LTn(<N>)
  const LT0  = 3.95e-4;
  const Nref = 7.1e15;
  return LT0/(1.0+N/Nref);
```

```
// hole lifetime
func LTp(<N>)
  const LT0  = 3.52e-5;
  const Nref = 7.1e15;
  return LT0/(1.0+N/Nref);

// scaled electron mobility
func MUn(<N>,<T>,pn)
  const Al = 1430.0,  Bl = -2.20;
  const Ai = 4.61e17, Bi = 1.52e15;
  assign NU  = abs(N)*C0;
  assign pnU = (pn>0.0)*pn*sq(C0)+(pn<0.0);
  assign MUl = Al*pow((T/300.0),Bl);
  assign MUi = {Ai*T*sqrt(T)/NU}/{ln(1+Bi*sq(T)/NU)-Bi*T*T/(NU+Bi*sq(T))};
  assign MUc = {2e17*T*sqrt(T/pnU)}/ln(1+8.28e8*sq(T)*pow(pnU,-1/3));
  assign X   = sqrt(6*MUl*(MUi+MUc)/(MUi*MUc));
  return MUl*{1.025/[1+pow(X/1.68,1.43)]-0.025}/u0;

// scaled hole mobility
func MUp(<N>,<T>,pn)
  const Al = 495.0,   Bl = -2.20;
  const Ai = 1.00e17, Bi = 6.25e14;
  assign NU  = abs(N)*C0;
  assign pnU = (pn>0.0)*pn*sq(C0)+(pn<0.0);
  assign MUl = Al*pow((T/300.0),Bl);
  assign MUi = {Ai*T*sqrt(T)/NU}/{ln(1+Bi*sq(T)/NU)-Bi*T*T/(NU+Bi*sq(T))};
  assign MUc = {2e17*T*sqrt(T/pnU)}/ln(1+8.28e8*sq(T)*pow(pnU,-1/3));
  assign X   = sqrt(6*MUl*(MUi+MUc)/(MUi*MUc));
  return MUl*{1.025/[1+pow(X/1.68,1.43)]-0.025}/u0;

// scaled recombination
func R(n,p,<tn1>,<tp1>)
  const tn = 4.9e-5/t0;
  const tp = 4.2e-5/t0;
  const Cn = 2.8e-31/[D0/sq(C0*x0)];// scaled electron Auger coefficient
  const Cp = 1.2e-31/[D0/sq(C0*x0)];// scaled hole Auger coefficient
  const ni = Ni/C0;                 // scaled intrinsic concentration
  const pn = sq(ni);                // scaled intrinsic conc. squared
  assign Rsrh = (n*p-pn)/[tp*(n+ni)+tn*(p+ni)];
  assign Raug = (n*p-pn)*(Cn*n+Cp*p);
  return Rsrh+Raug;

// scaled electric field
func E(V1,V2,<h>)
  return -grad(V1,V2,h);

// scaled electron current density
func Jn(n1,n2,E,un1,un2,<h>)
```

```
  const Ut     = Vt/V0;                // scaled thermal voltage
  assign un    = ave(un1,un2);
  assign dV    = E*h/(2*Ut);
  assign n     = n1*aux2(dV)+n2*aux2(-dV);
  assign dndx  = aux1(dV)*grad(n1,n2,h);
  return un*(n*E+dndx);

// scaled hole current density
func Jp(p1,p2,E,up1,up2,<h>)
  const Ut     = Vt/V0;                // scaled thermal voltage
  assign up    = ave(up1,up2);
  assign dV    = E*h/(2*Ut);
  assign p     = p1*aux2(-dV)+p2*aux2(dV);
  assign dpdx  = aux1(dV)*grad(p1,p2,h);
  return up*(p*E-dpdx);


// set numerical algorithm parameters
set NEWTON DAMPING    = 3;
set NEWTON ACCURACY   = 1.0e-12;
set NEWTON ITERATIONS = 100;
set LINSOL ALGORITHM  = GAUSSELIM;
set LINSOL FILL       = INFINITY;
set DISTANCE SCALE    = x0;


// declare variables and specify which ones are knowns and unknowns
var     V[NODES], n[NODES], p[NODES], tn[NODES], tp[NODES];
unknown V[SI], n[SI], p[SI];
known   V[SOURCE], V[DRAIN], V[GATE];
known   n[SOURCE], n[DRAIN], n[GATE];
known   p[SOURCE], p[DRAIN], p[GATE];

// for each Newton iteration, precompute the following functions prior to
// constructing the Jacobian matrix
precompute @ NODE i -> MUn(C[i],T,n[i]*p[i]);
precompute @ NODE i -> MUp(C[i],T,n[i]*p[i]);
precompute @ NODE i -> R(n[i],p[i],tn[i],tp[i]);
precompute @ EDGE i (j,k) ODD -> E(V[j],V[k],elen);
precompute @ EDGE i (j,k) ODD -> Jn(n[j],n[k],@E(i,j,k),
@MUn(j),@MUn(k),elen);
precompute @ EDGE i (j,k) ODD -> Jp(p[j],p[k],@E(i,j,k),
@MUp(j),@MUp(k),elen);


// scaled Poisson's equation
equ V[i=SI] ->
  L0*nsum(i,j,all,{eSi*@E(edge(i,j),i,node(i,j))}*ilen(i,j)) -
```

```
  (p[i]-n[i]+C[i])*area(i) = 0.0;

// scaled electron continuity equation
equ n[i=SI] ->
  +nsum(i,j,all,{@Jn(edge(i,j),i,node(i,j))}*ilen(i,j)) -
@R(i)*area(i) = 0.0;

// scaled hole continuity equation
equ p[i=SI] ->
  -nsum(i,j,all,{@Jp(edge(i,j),i,node(i,j))}*ilen(i,j)) -
@R(i)*area(i) = 0.0;


begin InitializeVariables
  open "jfeteq.dat" read;
  file read V;
  close;
  assign n[i=all] = Nd*exp(V[i]/Vt);
  assign p[i=all] = sq(Ni)/n[i];
  assign tn[i=all] = LTn(abs(C[i]));
  assign tp[i=all] = LTp(abs(C[i]));
end

begin ScaleVariables
  assign x[i=all] = x[i]/x0;
  assign V[i=all] = V[i]/V0;
  assign C[i=all] = C[i]/C0;
  assign n[i=all] = n[i]/C0;
  assign p[i=all] = p[i]/C0;
  assign tn[i=all] = tn[i]/t0;
  assign tp[i=all] = tp[i]/t0;
end

begin UnscaleVariables
  assign x[i=all] = x[i]*x0;
  assign V[i=all] = V[i]*V0;
  assign C[i=all] = C[i]*C0;
  assign n[i=all] = n[i]*C0;
  assign p[i=all] = p[i]*C0;
  assign tn[i=all] = tn[i]*t0;
  assign tp[i=all] = tp[i]*t0;
end

var Is, Id, Ig;
begin ComputeCurrent
  assign Is = J0*lsum(i,SOURCE,nsum(i,j,SI,ilen(i,j)*{
    Jn(n[i],n[node(i,j)],E(V[i],V[node(i,j)],elen(i,j)),
  MUn(C[i],T,n[i]*p[i]),
```

```
  MUn(C[node(i,j)],T,n[node(i,j)]*p[node(i,j)]),elen(i,j))+
    Jp(p[i],p[node(i,j)],E(V[i],V[node(i,j)],elen(i,j)),
  MUp(C[i],T,n[i]*p[i]),
  MUp(C[node(i,j)],T,n[node(i,j)]*p[node(i,j)]),elen(i,j))}));

  assign Id = J0*lsum(i,DRAIN,nsum(i,j,SI,ilen(i,j)*{
    Jn(n[i],n[node(i,j)],E(V[i],V[node(i,j)],elen(i,j)),
  MUn(C[i],T,n[i]*p[i]),
  MUn(C[node(i,j)],T,n[node(i,j)]*p[node(i,j)]),elen(i,j))+
    Jp(p[i],p[node(i,j)],E(V[i],V[node(i,j)],elen(i,j)),
  MUp(C[i],T,n[i]*p[i]),
  MUp(C[node(i,j)],T,n[node(i,j)]*p[node(i,j)]),elen(i,j))}));

  assign Ig = J0*lsum(i,GATE,nsum(i,j,SI,ilen(i,j)*{
    Jn(n[i],n[node(i,j)],E(V[i],V[node(i,j)],elen(i,j)),
  MUn(C[i],T,n[i]*p[i]),
  MUn(C[node(i,j)],T,n[node(i,j)]*p[node(i,j)]),elen(i,j))+
    Jp(p[i],p[node(i,j)],E(V[i],V[node(i,j)],elen(i,j)),
  MUp(C[i],T,n[i]*p[i]),
  MUp(C[node(i,j)],T,n[node(i,j)]*p[node(i,j)]),elen(i,j))}));
end

begin main
  call InitializeVariables;
  call ScaleVariables;
  solve;
  Call ComputeCurrent;
  open "jfet.out" write;
  file write Is, Id, Ig;
  close;
  call UnscaleVariables;
  write;
end
```

JFET Mesh Specification File

```
// FILE:  sgdir\ch09\jfet\jfet.sk
//
//                            GATE
//     A-------B---------------------------------C-------D
//     |   \   |                p+               |  /    |
//  S  |       --|-------------------------------|--     | D
//  O  |         |                               |       | R
//  U  |         |                               |       | A
//  R  |         |               n               |       | I
//  C  |         |                               |       | N
//  E  |         |                               |       |
//     H-------I---------------------------------J-------K -- symmetry
//                                                          line of
```

```
CONST Wjfet = 14.0e-4;                    // jfet width
CONST Wgate = 10.0e-4;                    // gate width
CONST Djfet =  6.0e-4;                    // jfet depth
CONST Wspacing = 0.33334e-4;              // initial grid point spacing

point pA = (0.0e-4,            0.0e-4);
point pB = ((Wjfet-Wgate)/2.0, 0.0e-4);
point pC = ((Wjfet+Wgate)/2.0, 0.0e-4);
point pD = (Wjfet,            0.0e-4);
point pH = (0.0e-4,           -Djfet);
point pI = ((Wjfet-Wgate)/2.0, -Djfet);
point pJ = ((Wjfet+Wgate)/2.0, -Djfet);
point pK = (Wjfet,            -Djfet);

edge eAB = NOFLUX    [pA, pB] (Wspacing, 0.0);
edge eBC = GATE      [pB, pC] (Wspacing, 0.0);
edge eCD = NOFLUX    [pC, pD] (Wspacing, 0.0);
edge eHI = NOFLUX    [pH, pI] (Wspacing, 0.0);
edge eIJ = NOFLUX    [pI, pJ] (Wspacing, 0.0);
edge eJK = NOFLUX    [pJ, pK] (Wspacing, 0.0);
edge eAH = SOURCE    [pA, pH] (Wspacing, 0.0);
edge eBI =           [pB, pI] (Wspacing, 0.0);
edge eCJ =           [pC, pJ] (Wspacing, 0.0);
edge eDK = DRAIN     [pD, pK] (Wspacing, 0.0);

region rAHIB = SI {eAH, eHI, eBI, eAB} RECTANGLES;
region rBIJC = SI {eBI, eIJ, eCJ, eBC} RECTANGLES;
region rCJKD = SI {eCJ, eJK, eDK, eCD} RECTANGLES;


const Nd  = 1.00e+14;
const Na  = 1.00e+19;
const Rj  = 1.0e-4;
const Aj  = ln(Na/Nd)/sq(Rj);

coordinates x, y;

refine C (SignedLog, 1.0) =
  Nd - (Na+Nd)*ngdep(x-Wjfet/2.0,y,Wgate,Aj,Aj);

set minimum divisions = 0;
set maximum divisions = 3;
```

The doping profile and the potential in the JFET are shown in Figures 9.2 and 9.3.

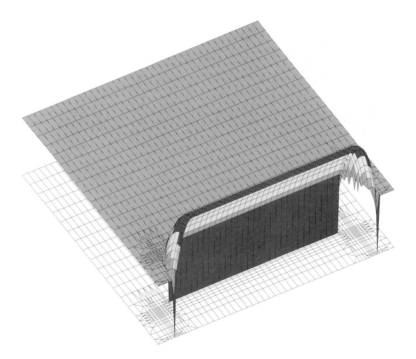

Figure 9.2. The doping concentration in the JFET

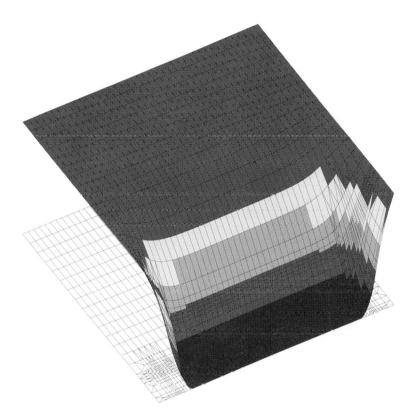

Figure 9.3. The voltage in the JFET.

9.4   Summary

A short discussion of the analytic basis for describing JFETs was given; the presentation was brief because of the similarity of much of the material to the equivalent for MOSFETs. MOSFETs are the topic of Chapter 10 and are treated in considerably more detail.

PROBLEMS

1. Make analytic estimates of the electric fields inside a JFET at (or close to) equilibrium.

2. Using the analytic estimates from the previous question, discuss where carriers which are injected through each terminal of the device are likely to go. Use a Monte Carlo method to test your answer. Use the mean free path λ and collision frequency ν which can be estimated from data given in Chapter 4. How many collisions do you expect a typical particle to have before it leaves each junction where it is injected? How long will this take, in each case? Plot several trajectories to confirm your estimates.

3. Evaluate the small-signal parameters g_d and g_m for a typical JFET geometry. Obtain the values from a numerical simulation and compare the results.

Chapter 10

Metal-Oxide-Semiconductor Structures

Metal-oxide-semiconductor (MOS) devices are described analytically and by means of simulations in this chapter. MOS capacitors and MOSFETs are some of the most important semiconductor devices and are emphasized throughout this book.

We begin with the analytic theory of one-dimensional MOS structures, which are MOS capacitors, in Section 10.1. The CV characteristics of the MOS capacitor are found. We then turn to the MOSFET and derive its (i, v) characteristics, including its small-signal analysis, in Section 10.2. This leads on, in Section 10.3, to a discussion of VLSI uses of MOSFETs and the effects which take place in very small MOSFETs. Numerical simulation of MOSFETs is begun with Section 10.4, which is devoted to the MOSFET mesh specification file. Section 10.5 is concerned with the MOSFET equation specification file, and Section 10.6 describes the results of simulations. Next, in Section 10.7, the analytic models we obtained of the MOSFET are used to simulate MOSFET circuits, using 'lumped' circuit-elements.

Microscopic modeling of the MOSFET was begun in chapter 5, as an example of semiconductor modeling. It is continued here, and in the exercises the results are compared to the analytic models which are derived in this chapter.

10.1 Electrostatics of the One-Dimensional MOS Structure

The metal-oxide-semiconductor structure by itself functions as a capacitor. A voltage applied to the metal gate attracts a charge onto the surface of the metal nearest to the semiconductor. Electric-field lines run through the oxide layer, between the charge on the surface of the metal and an equal and opposite charge in a layer near the surface of the semiconductor. The capacitance of this device is equal to the magnitude of the charge on the metal plate, divided by the voltage of the metal relative to the back of the semiconductor. To calculate the voltage, it is usual to use Poisson's equation in one dimension, x, which in the semiconductor is

$$\frac{\partial^2 V}{\partial x^2} = \frac{e}{\epsilon}(N_A - N_D + n - p). \qquad (10.1)$$

441

The semiconductor is in equilibrium since the oxide does not allow a flux of either species (assuming they do not recombine on the surface), so the equilibrium expressions for the densities are appropriate. In the oxide layer the charge density is (nearly) zero so the right-hand side of this equation (nearly) vanishes. The solution of $\frac{d^2V}{dx^2} = 0$ is just $V = Ax + B$, so the potential in the oxide varies linearly. The potential in the metal is constant. The boundary condition between regions is that the change in the normal component of D on crossing a boundary equals the charge per unit area on the surface. V is continuous across boundaries. It is possible to analytically solve Poisson's equation in the semiconductor. The trick which is employed involves multiplying both sides of the equation by $\frac{dV}{dx}$. The left-hand side is then $\frac{dV}{dx}\frac{d^2V}{dx^2} = \frac{1}{2}\frac{d}{dx}\left(\frac{dV}{dx}\right)^2$. The right-hand side, once multiplied by $\frac{dV}{dx}$, can also be written as an exact derivative with respect to x. Integrating both sides with respect to x yields an expression for the square of the electric field. Unfortunately this expression is quite complicated. In some cases the analytic expression needs to be evaluated numerically. In any case, it is convenient to solve the whole problem numerically.

To describe the electrostatics of the MOS structure some materials properties of the components of the device need to be defined. The Fermi level E_F is the energy at which the occupation probability of an energy level is one-half. The lowest energy at which an electron is completely free from the material is the vacuum level E_0. The workfunction Φ is the energy needed for an electron at the Fermi level E_F to reach the vacuum level. $\Phi = E_0 - E_F$. Now, in a metal the Fermi level is usually in the middle of the conduction band, so there are many levels below E_F which are likely to be occupied. This is why there are so many conduction electrons. Because of the abundance of conduction electrons, it is not likely that enough of them can be removed to alter E_F. The workfunction Φ is a useful way to characterize the metal because it measures the the energy required to remove the most energetic electrons and because $E_0 - E_F$ is a constant for any given metal.

In a semiconductor the Fermi level is usually between bands. This difference in the locations of the Fermi levels differentiates between metals and semiconductors. Because there are fewer conduction electrons in a semiconductor, a change in the doping or electrical contact with another material can change E_F. There may be few electron energy levels at E_F. For these reasons it is more useful in a semiconductor to make use of the electron affinity $\chi = E_0 - E_C$, the energy needed to to take an electron at the surface of the crystal which is at the bottom of the conduction band, at energy E_C, to the vacuum level E_0. χ depends only on the material.

If a metal and a semiconductor are put into contact, electrons will flow between the two until the Fermi level of the semiconductor changes to match that of the metal. The number of electrons exchanged will be much too small to change E_F in the metal. The density of states near to E_F is much smaller in the semiconductor, so the semiconductor E_F can be altered by adding or removing a relatively small number of electrons. After the Fermi levels line up, the workfunctions are also equal.

In the MOS structure the metal and semiconductor are not in contact but are separated by an insulator. If they are in equilibrium, their Fermi levels still line up and their workfunctions become equal. The equilibrium state provides a reference condition which would occur if the ends of the MOS structure were connected by a short circuit. If the metal which forms the gate has a potential V_G applied to it relative to the back of the semiconductor, then the Fermi levels are split apart. The energy of electrons is lowered by a positive potential, so if V_G is positive, the Fermi level at the gate is lower than that of the semiconductor.

$$E_F(metal) = E_F(semiconductor) - eV_G. \tag{10.2}$$

The file moscap1.sg sets up the equation given above along with the appropriate boundary conditions for all the regions. By solving this equation for several values of V_G, the gate voltage, it is possible to develop simple analytic approximations for $V(x)$ based on the solutions obtained numerically. The applied voltage V_G could

1. attract majority carriers (which is called accumulation), it could

2. repel the majority carriers to expose the 'bare charge' of the ionized impurities (called depletion) and

3. if it repels the majority carriers and is increased to a large enough magnitude it could attract a significant number of minority carriers. When the minority concentration at the oxide surface exceeds the majority concentration in the bulk, we have inversion.

Both the accumulated majority carriers and the inversion minority carriers that are attracted to the gate will form a thin layer at the semiconductor-oxide interface. The width of this layer is of the order of the Debye length λ_D. When the majority carriers are repelled to expose the ionized impurities, it will be assumed, for the moment, that all the majority carriers are repelled, out to some maximum distance W. Hence, the region from $x = 0$ to $x = W$ is depleted of carriers (except possibly at the semiconductor-oxide interface), and is called the depletion region. Beyond $x = W$, in the 'bulk' semiconductor, quasi-neutrality will be assumed to hold. Within the depletion region (for $x \leq W$) the charge consists of the charge on the ionized impurities. The charge density is consequently roughly constant in the depletion region. Typically W, the size of the depletion region, will be much larger than λ_D.

10.1.1   CV Characteristics of a MOS Capacitor

The charges set up in the semiconductor by attracting carriers to the surface of the oxide lie in a very narrow layer. The depletion charge consists of the charge of the impurities exposed after the majority carriers have been repelled out to a distance W. Once inversion begins, the thickness W of the depletion layer varies relatively slowly, which means that the extra charge attracted is due to minority carriers at the oxide surface.

In accumulation, this approximate picture indicates that the oxide behaves much like a conventional capacitor. The oxide electric field E_{ox} is responsible for the entire voltage drop in the MOS structure, since the field lines start on the charge on one side of the oxide and end on the equal and opposite charge on the other side of the oxide. If there is no charge in the oxide, the oxide field, E_{ox}, is constant and the voltage drop is just $V_G = V_{ox} = E_{ox}t_{ox}$, where t_{ox} is the oxide thickness. The potential in the semiconductor is approximately zero, because according to this model there is no electric field in the semiconductor in accumulation.

In depletion, the field penetrates the semiconductor out to a distance W, and since there is a nonzero charge density ρ in the depleted semiconductor, E is not constant. Suppose the charge on the gate is positive and the exposed impurities are negative, implying that this is a p-type semiconductor. Field lines start on the positive gate charge and end on the negative charges on the impurities. The impurity charge density is roughly constant with value $\rho = -eN_A$, so field lines end at a constant rate with increasing x, given by

$$\frac{dE}{dx} = \frac{\rho}{\epsilon} = -\frac{eN_A}{\epsilon} \tag{10.3}$$

so the electric field drops linearly

$$E = E(x=0) - \frac{eN_A x}{\epsilon}. \tag{10.4}$$

$x = 0$ is at the oxide-to-semiconductor interface and x increases with distance from the gate. The field is zero at $x = W$ because there are no fields outside the depleted region, so $E(x=0) = \frac{eN_A W}{\epsilon}$ and

$$E = \frac{eN_A(W-x)}{\epsilon}. \tag{10.5}$$

The potential in the semiconductor is found using $E = -\frac{dV}{dx}$. Integrating the equation

$$\frac{dV}{dx} = -\frac{eN_A(W-x)}{\epsilon} \tag{10.6}$$

gives the electrostatic potential in depletion, in terms of $V(x=0)$,

$$V(x) = V(x=0) - \frac{eN_A}{\epsilon}\left(Wx - \frac{x^2}{2}\right). \tag{10.7}$$

However, since the potential in the bulk semiconductor $V(x=W)$ is conventionally set to zero, we can set $V(x=0) = \frac{eN_A}{\epsilon}\frac{W^2}{2}$ and the potential $V(x)$ in depletion is

$$V(x) = \frac{eN_A}{2\epsilon}(W-x)^2. \tag{10.8}$$

The width W is found from this, if we know $V(x=0)$, by setting $x=0$ in this expression. If the potential $V(inv)$ at the oxide surface at the onset of inversion is used as the value of $V(x = 0)$, then the width of the depletion region W_m, at the onset of inversion, can be found by setting

$$V(inv) = \frac{eN_AW_m^2}{2\epsilon}.$$

(10.9)

In Section 4.2 it was shown that $V(inv) = 2\phi_F$.

The boundary condition at the oxide-to-semiconductor interface is a condition on the normal components of the electric displacement D, and the charge per unit area on the surface ρ_s. See Section 2.10.

$$D_{n,Si} - D_{n,ox} = \rho_s.$$

(10.10)

Even if there is no surface charge the electric fields differ, because $D_{ox} = \epsilon_{ox}\epsilon_0 E_{ox}$ is equal to $D_{Si} = \epsilon_{Si}\epsilon_0 E_{Si}$ and so $E_{ox} = \frac{\epsilon_{Si}}{\epsilon_{ox}}E(x = 0)$. Here $E(x = 0)$ is the field in the Si at $x = 0$. Now $E(x = 0) = \frac{eN_AW}{\epsilon_{Si}\epsilon_0}$ in the silicon. With the expression for the electric field in the oxide this is equivalent to saying that the D field in the oxide is equal in magnitude to the charge per unit area which the field lines end on. Both ways we find $D_{ox} = eN_AW$, since looking from the oxide into the semiconductor there is a charge $-eN_AW$ per unit area inside the semiconductor.

The voltage dropped in the oxide is $V_{ox} = E_{ox}t_{ox} = \frac{eN_AWt_{ox}}{\epsilon_{ox}\epsilon_0}$ When this is added to the voltage dropped in the semiconductor we get the gate voltage V_G in depletion.

$$V_G = V_{ox} + V(x = 0) = \frac{eN_AW}{\epsilon_0}\left(\frac{t_{ox}}{\epsilon_{ox}} + \frac{W}{2\epsilon_{Si}}\right).$$

(10.11)

Expressing W in terms of $V(x = 0)$ we find $W = \sqrt{2\epsilon_{Si}\epsilon_0 V(x = 0)/eN_A}$, so the gate voltage is

$$V_G = \frac{eN_At_{ox}}{\epsilon_{ox}\epsilon_0}\sqrt{2\epsilon_{Si}\epsilon_0 V(x = 0)/eN_A} + V(x = 0)$$

$$= \frac{t_{ox}}{\epsilon_{ox}}\sqrt{\frac{2e\epsilon_{Si}N_AV(x = 0)}{c_0}} + V(x = 0).$$

(10.12)

If the voltage is increased further, and inversion begins, the change of the gate voltage V_G leads to a pile-up of charge at the oxide surface. We assume, here, that it does not lead to changes in ρ, E or V inside the semiconductor. $V(x = 0)$ stays at its maximum depletion value as the inversion builds up. Similarly in accumulation $V(x = 0) = 0$ according to this model. The actual variation of V_G with $V(x = 0)$ can be obtained using the input file moscap1.sg, which is given below.

In the numerical formulation the narrowness of the layers of charge created by attracting carriers toward the gate can cause problems because the spatial structure of the layer may be difficult to resolve on the mesh. The densities of electrons and holes are both given in terms of their values at the back of the semiconductor where the potential is set to zero. (Indeed, we may assume that the potential is

zero in much of the quasi-neutral bulk, so that we do not have to simulate the
entire bulk. In general, however, quasi-neutrality does not guarantee that V is
constant, or even that the charge density is negligible.) There is no surface charge
by assumption. The carrier charge that builds up at the oxide surface is treated as
being spread over a finite thickness, as was just mentioned, so it does not form a
surface charge. The boundary condition at the oxide-to-semiconductor interface is
just $D_{ox} = \epsilon_{ox}\epsilon_0 E_{ox} = D_{Si} = \epsilon_{Si}\epsilon_0 E_{Si}$, as in the above analysis, while $E = -\frac{dV}{dx}$
in both regions. Finally, in the metal, $V = V_G$.

Input File moscap1.sg

```
// FILE:  sgdir\ch10\moscap1\moscap1.sg
// demonstration of solution of 1D nonlinear Poisson's equation
// - applied to a MOS capacitor structure.

const NX = 39;                    // number of x-mesh points
const LX = NX - 1;                // index of last x-mesh point
const LOX= 5;                     // mesh point on oxide layer/semi boundary
const DX = 1.0e-8;                // mesh spacing
const VL = 1.0;                   // voltage at the lhs
const VT = 0.03;                  // thermal voltage
const N0 = 1.0e+20;               // doping density
const ec = 1.6e-19;               // electronic charge
const eps0=8.8e-12;               // permittivity of vacuum
const epsi=11.8;                  // relative permittivity of Si
const eoeps = ec/eps0/epsi;       // e/epsilon
const epsio= 3.9;                 // eps-relative of SiO2
const VBI = 0.7 ;                 // built-in voltage

var V[NX],NA[NX],x[NX];

set NEWTON ACCURACY = 1.0e-8;

// Laplace equation in oxide
equ V[i=1..LOX-1] -> {V[i-1] - 2*V[i] + V[i+1]}/sq(DX) = 0.0;

// boundary condition, oxide - semiconductor interface
equ V[i=LOX] -> -epsio*{V[i] - V[i-1]}/DX = -epsi*{V[i+1] - V[i]}/DX;

// Poisson's equation in Si
equ V[i=LOX+1..LX-1] -> {V[i-1] - 2*V[i] + V[i+1]}/sq(DX) =
  eoeps * (NA[i] + N0 * (exp((V[i])/VT) - exp((-VBI-V[i])/VT)));

begin main
  assign x[i=all]    =  i*DX;    // Coordinate Array
  assign NA[i=0..LX] = -N0;      // donors on right .
  assign V[i=0]      =  VL;      // BC on potential
```

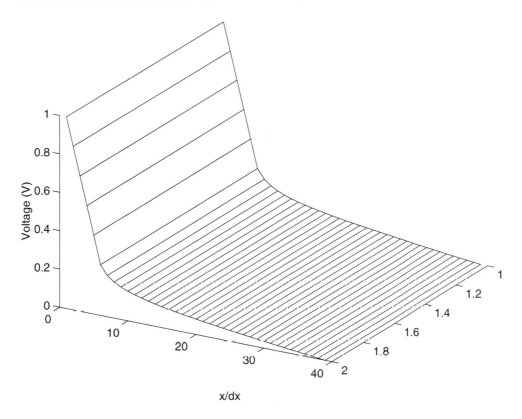

x/dx

Figure 10.1. The electrostatic potential in the MOS structure. The rapid, linear drop takes place in the oxide. The slower, exponential drop in voltage takes place inside the semiconductor. There is a jump in slope at the oxide-semiconductor interface, where the electric field changes. To find the capacitance of this device we need to know the charge on either of its plates. This is relatively easy to do for the metal, since $D = \epsilon E$ in the oxide equals ρ_s, the charge per unit area on the metal surface.

```
   assign V[i=LX]    = 0.0;    // BC on potential
   solve;
end
```

The result of the simulation is shown in Figure 10.1.

Interpretation of the Simulation Results

To interpret the results shown, we will first estimate the Debye length, λ_D.

$$\lambda_D = \sqrt{\frac{k_B T_e \epsilon}{N_D e^2}} = \sqrt{\frac{V^{th}\epsilon}{N_D e}} \text{ since } V^{th} = k_B T_e/e.$$

$\epsilon = \epsilon_r \epsilon_0 = 11.8 \times 8.8 \times 10^{-12} \simeq 10^{-10}$ F/m. $V^{th} \simeq 0.03$V. $N_D = 10^{20}$m^{-3}.
$\frac{V^{th}\epsilon}{N_D e} \simeq \frac{10^{-12} \times 0.03}{10^{20} \times 1.6 \times 10^{-19}} \simeq \frac{3 \times 10^{-12}}{16} \simeq 0.2 \times 10^{-12}$.
The square root of this is $\lambda_D \simeq 5 \times 10^{-7}m= 0.5\mu$m.

The mesh spacing in this simulation is $\Delta x = 10^{-8}$m, and there are 40 cells, so the entire simulation region is of length $L = 0.4\mu$m. This is probably too small to fully resolve the exponential decay, although it is plausible, from the plot, that λ_D is about the size predicted.

The electric field in the oxide accounts for nearly all of the potential drop, so we can estimate it as $E_{ox} \simeq V/t_{ox} = 1/(5 \times 10^{-8}) = 2 \times 10^7$ V/m .

The boundary condition at the interface between oxide and silicon requires continuity of normal D, so $\epsilon_{Si} E_{Si} = \epsilon_{ox} E_{ox}$ or $11.8 E_{Si} = 3.9 \times 2 \times 10^7$, so the electric field at the top of the silicon should be $E_{Si} \simeq \frac{2}{3} \times 10^7 \simeq 7 \times 10^6$ V/m.

From the plot, the voltage drop in the first cell in the silicon is (very roughly) 0.05 volts and the cell width is 10^{-8} m, so the field shown in the plot is $\Delta V/\Delta x \simeq \frac{0.05}{10^{-8}} = 5 \times 10^6$ V/m , which is good agreement, given the very rough estimates employed.

The electric field in the oxide of $E_{ox} \simeq 2 \times 10^7$ V/m corresponds to an electric displacement of $D = \epsilon E \simeq 4 \times 10^{-11} \times 2 \times 10^7 \simeq 8 \times 10^{-4}$ C/m^2. (Both D and E are constant in the oxide.) But D is equal to the charge per unit area on the surface of the metal, so $\rho_s \simeq 8 \times 10^{-4}$C/m^2. The capacitance per unit area is $C = \rho_s/V$, and since V is one volt, we have $C \simeq 8 \times 10^{-4}$F/m^2.

This estimate of C is based on finding the electric field above the plate. From the field we find the charge on the plate and divide it by the voltage. This is the same method we would use to find C more accurately, numerically. The value of C we found here should, in fact, be equal to the oxide capacitance, to the accuracy we are working. ($C_{ox} = \frac{\epsilon_{ox}}{t_{ox}} \simeq 4 \times 10^{-11}/5 \times 10^{-8} = 8 \times 10^{-4}$F/m^2. This agrees because it is essentially the same calculation. The value of E used in the first calculation actually meant that all the voltage was dropped across the oxide.) A numerical calculation would be needed to find the effect of the semiconductor on the capacitance.

10.2 Metal-Oxide-Semiconductor Field Effect Transistors

A schematic of a low power MOSFET is given in Figure 10.2.

We now derive the main results needed for describing the MOSFET's performance. For more details, see [117],[118]. The structure of an MOS transistor is shown in Figure 10.2. The horizontal direction along the channel is the y-direction. Distance down into the substrate and away from the gate is denoted by x.

10.2.1 (i, v) Characteristics of MOSFETs

We now derive the relationship between the drain current I_D, and the gate and drain voltages V_G and V_D (with the source voltage set to zero) in the regime where V_G is high enough for turn-on, $V_G \geq V_T$ and V_D is too low for pinch-off (when the channel width drops to zero), $0 \leq V_D \leq V_{Dsat}$. V_T is called the threshold (or turn-

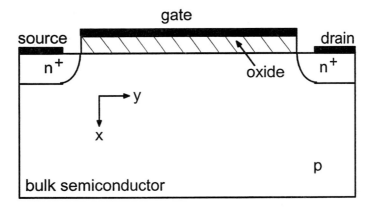

Figure 10.2. A schematic of a low-power metal-oxide-semiconductor field effect transistor.

on) voltage. It is the voltage necessary to create a channel, by attracting minority carriers to the oxide-semiconductor interface and causing inversion.

The steps followed are:

1. the current in the channel is assumed to be primarily due to the applied electric field, so the current density $J_N \simeq e\mu_n n E_y = -e\mu_n n \frac{\partial V}{\partial y}$ and the total current is the integral of this over the channel. The integral is

$$I_D = -h\frac{\partial V}{\partial y}\left[e\int \mu_n n\, dx\right] \tag{10.13}$$

where h is the channel height, which is in the z-direction. The quantity in the square bracket is Q, the total charge of the minority carriers (which we are assuming to be electrons) in the channel, multiplied by the average mobility $\bar{\mu}_n$.

2. Because the current I_D is roughly the same all the way along the channel, it can be integrated over y to give $\int_0^L I_D\, dy = I_D L$ and at the same time to get rid of the y-derivative in the expression for I_D

$$I_D L = -h\bar{\mu}_n \int Q\frac{\partial V}{\partial y}\, dy = -h\bar{\mu}_n \int_0^{V_D} Q\, dV. \tag{10.14}$$

3. The charge Q of the electrons in the channel is sometimes approximated as the charge on one plate of the capacitor, which is made up of the oxide layer and has capacitance per unit area $C_{ox} = \frac{\epsilon_{ox}\epsilon_0}{t_{ox}}$. t_{ox} is the thickness of the oxide layer. The voltage applied to this capacitor is $V_G - V$ between the gate at V_G and some point in the channel where the voltage is V. However, only the voltage in excess of V_T attracts minority carriers, so the charge is

$$Q = -C_{ox}((V_G - V) - V_T) \tag{10.15}$$

where V_T is the threshold voltage. The charge Q can be seen to have the correct sign since a large positive V_G attracts a negative charge.

If the depletion width increases, from W_m, the width at the onset of inversion, to $W > W_m$, then extra charge is exposed on the impurities. The amount of extra charge is $Q_{dep} = -eN_A[W - W_m]$, in the layer of width $W - W_m$. This amount of extra charge is available for field lines to terminate on, so the minority-carrier charge can decrease by this amount and Q becomes

$$Q = -C_{ox}((V_G - V) - V_T) - Q_{dep}. \tag{10.16}$$

Using the expressions for W found above,

$$W = \sqrt{\frac{2\epsilon_{Si}\epsilon_0}{eN_A}(2\phi_F + V)} \tag{10.17}$$

and W_m is obtained by setting $V = 0$ in eq. 10.17. Then

$$Q_{dep} = -eN_AW_m \left(\sqrt{1 + \frac{V}{2\phi_F}} - 1 \right). \qquad (10.18)$$

4. Integrating Q over V gives

$$I_D = \frac{h\bar{\mu}_n}{L} \left[C_{ox} \left((V_G - V_T)V_D - \frac{V_D^2}{2} \right) - \frac{4}{3}eN_AW_m\phi_F \left(\left(1 + \frac{V_D}{2\phi_F} \right)^{3/2} - \left(1 + \frac{3V_D}{4\phi_F} \right) \right) \right].$$
$$(10.19)$$

5. To find I_D for $V_D > V_{Dsat}$, (V_{Dsat} is the value of V which makes Q go to zero) assume it is equal to $I_D(V_D = V_{Dsat})$. We now set $Q = 0$ and solve for V, which will be equal to V_{Dsat}.

$$V_G - V - V_T = \frac{eN_AW_m}{C_{ox}} \left(\sqrt{1 + \frac{V}{2\phi_F}} - 1 \right)$$

so in terms of $V_M = eN_AW_m/C_{ox}$,

$$(V_G - V_T + V_M) - V = V_M \sqrt{1 + \frac{V}{2\phi_F}}.$$

Squaring this expression,

$$(V_G - V_T)^2 + V_M^2 + 2V_M(V_G - V_T) - 2V(V_G - V_T + V_M) + V^2 = V_M^2 + \frac{V_M^2 V}{2\phi_F}.$$

Rearranging this gives a quadratic in V,

$$V^2 - V \left[\frac{V_M^2}{2\phi_f} + 2(V_G - V_T + V_M) \right] + (V_G - V_T)(V_G - V_T + 2V_M) = 0.$$

In saturation, this V replaces V_D in equation 10.19.

$$
\begin{aligned}
V &= V_G - V_T + V_M + \frac{V_M^2}{4\phi_F} - \\
&\qquad \sqrt{\left(\frac{V_M^2}{4\phi_F} + V_G - V_T + V_M \right)^2 - (V_G - V_T)(V_G - V_T + 2V_M)} \\
&= V_G - V_T + V_M \left(1 + \frac{V_M}{4\phi_F} - \sqrt{\left(\frac{V_M}{4\phi_F} + 1 \right)^2 + \frac{V_G - V_T}{2\phi_F}} \right) (10.20)
\end{aligned}
$$

10.2.2   Small-Signal Analysis of MOSFETs

The MOSFET small-signal response is assumed to be given by its dc characteristics. The frequencies we are usually interested in are low enough compared to the response times of the MOSFET, which are short, that the response of the MOSFET to them is given by its static response. No separate numerical simulation of time-dependent behavior will be given in this chapter, since dc simulations with time-varying electrode voltages are sufficient in this case.

The MOSFET drain current I_D depends on V_D and V_G. When we examine small-signal response, we find the oscillatory part of the drain current i_d to vary with the oscillatory part of the voltages v_d and v_g.

$$i_d = \left[\frac{\partial I_D}{\partial V_D}\right]_{V_G} v_d + \left[\frac{\partial I_D}{\partial V_G}\right]_{V_D} v_g = g_d v_d + g_m v_g. \tag{10.21}$$

This expression implies a small-signal model where i_d is supplied by two circuit elements between source and drain. Since part of the current i_d is $g_d v_d$ (and v_d is equal to v_{ds} because v_s is our zero of voltage), this current is the current through a resistor $R = 1/g_d$. The second component of the current is supplied by a current source, with current $g_m v_g$.

10.3   Submicron Devices and VLSI

Modern very large scale integrated (VLSI) circuits primarily use MOS devices. The very small size of the devices and their closeness to each other makes parasitic and other nonideal elements into a major problem, even though some parasitics are reduced in magnitude at small sizes. The benefits of the size reduction are that carrier transit times are also reduced and most important that the density of elements is increased. In this section we shall discuss the parasitics and other problems associated with the scaling down of MOSFETs including failure modes, and techniques to deal with these effects. (A good discussion of this material is given in [118].)

10.3.1   Scaling MOS Structures

As device dimensions are made smaller, the ratios of the characteristic lengths are kept fixed, so that device parameters, such as conductances, are, as far as possible, left unchanged. It is also natural to choose the electric field in the device as a parameter which needs to be kept the same. (It is probably only the maximum field which needs to be kept the same, but if the device is to function in the same way as it is scaled, it seems reasonable to try to keep the field everywhere roughly the same.) Since the electric field E is a voltage divided by a length, keeping E fixed while decreasing the device dimensions means the voltages must also decrease. But supply voltages tend to be standardized. That is, they have to conform to a standard value which the designer of a device has no control over. A 3.3 V standard

supply voltage applied to a gate oxide of 20-nm width leads to an electric field
of $1.65\,10^6$ V/cm. Beyond this value of the field, too large a fraction of the oxide
layers produced in a large-scale process, which produces devices of varying quality,
will break down.

To decrease depletion-layer widths, the doping must be increased. The channel
length L is a critical dimension because it is one of the smallest dimensions. When
the channel length L approaches the size of the source and drain junctions' depletion width, 'short channel' effects can become significant. These effects introduce
nonidealities into the MOSFET's (i, v) characteristics.

To prevent the doping of source and drain going too far into the channel, and
so decreasing its length, the doping cannot be allowed to diffuse far sideways. This
will also prevent it from going very deep, since in the diffusion process the dopants
go sideways as well as down. (Even the oxide layer thicknesses are decreased, so
that the thin metal connections can be laid across the surface without having them
break. If they have to go down deep holes, at the corners of the oxide layers, there
will be a tendency for the metal to break at the corner.)

When the channel length $L \leq 1\mu$m, various short-channel effects appear. We
shall begin by discussing velocity saturation. The electric field along the channel can
be very large, because of the short channel length, if the drain and source voltages
are not decreased. The velocity of charged particles in a reasonably small electric
field, and in a material where the particles undergo collisions, will increase linearly
with the strength of the field. For high enough fields, however, the velocity saturates. (When the energy which the particle picks up from the electric field between
collisions is bigger than the thermal energy of the particle, the low-field mobility no
longer applies. After going a mean free path λ, the particle's directed velocity v_E
along E is given by $\frac{1}{2}mv_E^2 = eE\lambda$, so $v_E = \sqrt{2eE\lambda/m}$. In semiconductors the relationship between v_E and E is more complicated, since other mechanisms prevent
the linear scaling from occurring, apart from the effects of inertia.) In the simplest
model, the carrier velocity varies linearly with E for small E, so $v_{dr} - \mu E$. At high
electric fields, the velocity is constant and equal to v_{Esat}. (The different notation
for the velocity, v_E, is used to distinguish the high-electric-field case from the low
field. In the high field case, the velocity the carriers acquire from the field, which is
directed along the field, is comparable to the thermal velocity. The velocity in the
low electric field case is typically small compared to the thermal velocity.)

The derivation which follows is more involved than most of the material in this
book, and can be skipped.

To allow for velocity saturation, we can go back to the usual derivation of the
expression for the drain current,

$$I_D = \frac{h\bar{\mu}_n C_{ox}}{L}\left[(V_G - V_T)V_D - \frac{V_D^2}{2}\right] \qquad (10.22)$$

and use the same analysis, but instead of going straight to I_D we find the channel
voltage as a function of the distance along the channel y. This allows us to calculate
the electric field, in order to determine when it reaches the level where the velocity

saturates. It reaches this level first, at the drain end of the gate. Treating I_D as a known parameter, we set it equal to $hen_s v_{dr}$, where n_s is the carrier density per unit area below the gate. As discussed previously, $n_s = \frac{C_{ox}}{e}[V_G - V_T - V]$.

To find the electric field below the saturation voltage we substitute this value of n_s into I_D to obtain

$$I_D = -hC_{ox}[V_G - V_T - V]\mu E = \mu h C_{ox}[V_G - V_T - V]\frac{\partial V}{\partial y} \tag{10.23}$$

so the magnitude of the electric field, which is equal to $\frac{\partial V}{\partial y}$, is

$$\frac{\partial V}{\partial y} = \frac{I_D}{\mu h C_{ox}[V_G - V_T - V]}. \tag{10.24}$$

Rearranging this, and using $\hat{V} = V_G - V_T - V$, so that $\frac{\partial \hat{V}}{\partial y} = -\frac{\partial V}{\partial y}$, gives

$$-\hat{V}\frac{\partial \hat{V}}{\partial y} = \frac{I_D}{\mu h C_{ox}}. \tag{10.25}$$

The left-hand side can be rewritten so that

$$\frac{1}{2}\frac{\partial(\hat{V})^2}{\partial y} = -\frac{I_D}{\mu h C_{ox}} \tag{10.26}$$

and this can be integrated to give

$$\hat{V}^2 = A - \frac{2I_D y}{\mu h C_{ox}} \tag{10.27}$$

or, taking the square root,

$$V_G - V_T - V = \sqrt{A - \frac{2I_D y}{\mu h C_{ox}}}. \tag{10.28}$$

If we expect $V = 0$ at $y = 0$ we must choose $A = (V_G - V_T)^2$, so finally

$$V = V_G - V_T - \sqrt{(V_G - V_T)^2 - \frac{2I_D y}{\mu h C_{ox}}}. \tag{10.29}$$

If we further set $y = L$ and $V = V_D$, we find

$$[V_D - (V_G - V_T)]^2 = (V_G - V_T)^2 - \frac{2I_D L}{\mu h C_{ox}} \tag{10.30}$$

from which

$$V_D^2 - 2V_D(V_G - V_T) = -\frac{2I_D L}{\mu h C_{ox}} \tag{10.31}$$

confirming the previous result for I_D.

The saturation voltage can be defined to be the voltage that makes the electric field at the drain end of the gate reach the saturation value. This definition replaces the previous definition, which had saturation happen when the drain end of the channel was pinched-off. The saturation velocity is v_{Esat}, so the critical electric field is $E_{sat} = v_{Esat}/\mu$.

From the expression for $V_G - V_T - V$ and $\frac{\partial V}{\partial y} = \frac{I_D}{\mu h C_{ox}(V_G-V_T-V)}$ we can show

$$\frac{\partial V}{\partial y} = \frac{I_D}{\mu h C_{ox}\sqrt{(V_G - V_T)^2 - \frac{2I_D y}{\mu h C_{ox}}}} \tag{10.32}$$

which is biggest when $y = L$.

$$E(y = L) = \frac{I_D}{\mu h C_{ox}\sqrt{(V_G - V_T)^2 - \frac{2I_D L}{\mu h C_{ox}}}}. \tag{10.33}$$

If this field equals the saturation electric field E_{sat}, then I_D equals the saturation current, so

$$I_{Dsat} = \mu h C_{ox} E_{sat}\sqrt{(V_G - V_T)^2 - \frac{2I_{Dsat}L}{\mu h C_{ox}}}. \tag{10.34}$$

Rearranging this, using $\alpha = I_{Dsat}/\mu h C_{ox}$, gives

$$\left(\frac{\alpha}{E_{sat}}\right)^2 = (V_G - V_T)^2 - 2\alpha L \tag{10.35}$$

so

$$\alpha = \frac{I_{Dsat}}{\mu h C_{ox}} = \left[-2L + \sqrt{4L^2 + (V_G - V_T)^2/E_{sat}^2}\right] E_{sat}^2/2. \tag{10.36}$$

Finally I_{Dsat} is

$$I_{Dsat} = \mu h C_{ox} E_{sat}^2 L \left[\sqrt{1 + (V_G - V_T)^2/E_{sat}^2 L^2} - 1\right]. \tag{10.37}$$

To find V_{sat}, Equation 10.32 for $\frac{\partial V}{\partial y}$ in terms of I_D can be used:

$$\frac{\partial V}{\partial y} = \frac{I_D}{\mu h C_{ox}[V_G - V_T - V]} \tag{10.38}$$

so at saturation $\frac{\partial V}{\partial y} = E_{sat}$, $V = V_{sat}$ and

$$V_{sat} = V_G - V_T - \frac{I_{Dsat}}{\mu h C_{ox} E_{sat}} \tag{10.39}$$

with I_{Dsat} given above.

Other short-channel effects include charge sharing, which occurs when the depletion layer of the drain expands far enough under the gate to include a significant portion of the channel, an effect which is more important in a short channel. If the drain depletion layer encroaches on the channel, it reduces the effective impurity charge beneath the gate and this reduces V_T, which in turn increases I_D. In other words, increasing V_D increases I_D faster than it otherwise would, which leads to a lower output resistance for the device. If V_D is increased enough, the depletion layers of the source and of the drain can eventually meet, leading to a phenomenon known as punch-through. It might seem that this meeting of junctions would decrease the carrier density and hence the drain current, but increasing V_D causes another effect called drain induced barrier lowering (DIBL). The barrier is the potential the carriers must overcome to get from source to drain. As the drain voltage is increased, the potential barrier is decreased. Hence the carriers can easily 'travel' from the source to the drain. These effects are decreased by increasing the doping levels under the gate. Increasing the doping will decrease the junction widths, but will also increase the threshold voltage V_T.

The last major short-channel effect is subthreshold conduction, which simply means that some carriers are present at the oxide-semiconductor interface at gate voltages less than the threshold. In other words, a channel exists before $V_G = V_T$, as long as the Fermi level for the minority carrier has 'passed through' the intrinsic level E_i. However, the density of minority carriers is less than the dopant concentration under the gate.

The main failure modes of MOS devices in integrated circuits are breakdown of oxide layers due to charging, which is known as electrostatic discharge (ESD), damage due to hot electrons, and latch-up. ESD is the breakdown of the gate oxide due to inadvertent charging of the gate, through contact with the outside environment. It is also a major source of damage during processing, for instance in plasma etching. ESD may in some circumstances be prevented by means of a diode built in parallel with the gate or a similar device which will allow a current to flow before ESD happens.

Hot electrons are highly energetic electrons which are generated in high-electric-field regions such as at the drain during pinch-off operation. The hot electrons may avalanche in or near the region where they are formed. If the hot electrons enter the oxide layer, they may charge the oxide (or its interfaces). Alternatively, hot electrons may pass through it to the gate.

Latch-up happens because n and p regions are close on the chip and may form undesired bipolar transistors. So long as the pn junctions are reverse biased, which they normally are, there is no real problem with these transistors being present. If one of the junctions becomes forward biased, a large current will flow which can damage the device.

10.4   MOSFET Mesh Specification File

In this section we present the mesh specification file for a MOSFET. The equation specification file is given in Section 10.4, and the simulation is discussed in Section 10.6.

```
// mos.sk
// mesh constants
const WDEV  = 2.0e-4;      // device width           | (cm) | segment 1
const DDEV  = 2.0e-4;      // device depth           | (cm) | segment 2
const WOX   = 1.4e-4;      // oxide width            | (cm) | segment 3
const DOX   = 0.2e-6;      // oxide depth            | (cm) | segment 4
const WCONT = 0.2e-4;      // contact width          | (cm) | segment 5
const DRECT = 0.2e-4;      // depth of rect. region  | (cm) | segment 6


//        |--5--|   |------------3-----------|   |--5--|
//
//                       A-----------------------B        -
//                       |\\\\\\\\\\\\\\\\\\\\\\\|         4
//     -   C-----D---E-----------------------F---G-----H  -
//     6   |         |                        |         | |
//     -   |         I------------------------J         | |
//         |                                             | |
//         |                                             | 2
//         |                                             | |
//         |                                             | |
//         |                                             | |
//         K---------------------------------------------L  -
//
//         |----------------------1----------------------|


// define points
point pA = ((WDEV-WOX)/2, DOX), pG = (WDEV-WCONT, 0.0);
point pB = ((WDEV+WOX)/2, DOX), pH = (WDEV, 0.0);
point pC = (0.0, 0.0),         pI = ((WDEV-WOX)/2, -DRECT);
point pD = (WCONT, 0.0),       pJ = ((WDEV+WOX)/2, -DRECT);
point pE = ((WDEV-WOX)/2, 0.0), pK = (0.0, -DDEV);
point pF = ((WDEV+WOX)/2, 0.0), pL = (WDEV, -DDEV);


// define edges
edge eAB = GATE   [pA, pB] (WOX/20, 0.0);
edge eDE = NOFLUX [pE, pD] (1.0e-7, 0.5);
edge eEF = SISIO2 [pE, pF] (WOX/20, 0.0);
edge eFG = NOFLUX [pF, pG] (1.0e-7, 0.5);
edge eIJ =        [pI, pJ] (WOX/20, 0.0);
edge eCD = DRAIN  [pC, pD] (WCONT/8, 0.0);
edge eAE = NOFLUX [pE, pA] (1.0e-7, 0.5);
edge eGH = SOURCE [pG, pH] (WCONT/8, 0.0);
```

```
edge eBF = NOFLUX [pF, pB] (1.0e-7, 0.5);
edge eCK = NOFLUX [pC, pK] (WCONT/8, 0.5);
edge eEI =        [pE, pI] (1.0e-7, 0.5);
edge eHL = NOFLUX [pH, pL] (WCONT/8, 0.5);
edge eFJ =        [pF, pJ] (1.0e-7, 0.5);
edge eKL = SUB    [pK, pL] (WDEV/5, 0.0);

//define regions
region r1 = SIO2 {eAE, eEF, eBF, eAB} RECTANGLES;
region r2 = SI   {eEI, eIJ, eFJ, eEF} RECTANGLES;
region r3 = SI   {eCK, eKL, eHL, eGH, eFG, eFJ, eIJ, eEI, eDE, eCD};

// define coordinate labels
coordinates x, y;

// physical constants and properties of Si and SiO2
const T     = 300.0;              // operating temperature
const e     = 1.602e-19;          // electron charge            (C)
const kb    = 1.381e-23;          // Boltzmann's constant      (J/K)
const e0    = 8.854e-14;          // permittivity of vacuum   (F/cm)
const eSi   = 11.8;               // dielectric constant of Si
const eSiO2 = 3.9;                // dielectric constant of SiO2

// doping constants
const NS = 1.0e16;                // substrate doping        (cm^-3)
const NC = 1.0e19;                // contact doping          (cm^-3)
const WDIFF = (WDEV-WOX)/2;       // diffusion width          (cm)
const DDIFF = 0.25e-4;            // diffusion depth          (cm)
const DT    = 1.0e-11;            // diffusion coef. * time   (cm^2)

// doping profile
refine C (SignedLog, 3.0) = (y <= 0.0) * { -NS              +
  (NC+NS) * nsdep(x,      2*WDIFF,DT) * nsdep(y,2*DDIFF,DT) +
  (NC+NS) * nsdep(WDEV-x,2*WDIFF,DT) * nsdep(y,2*DDIFF,DT) };

// set min/max edge spacing and min/max refinement levels
set minimum length = sqrt(e0*eSi*(kb*T/e)/e/abs(C));
set maximum length = 1.0;
set minimum divisions = 0;
set maximum divisions = 20;
```

10.5   MOSFET Equation Specification File

We now turn to the equation specification file used in the MOSFET simulations, which are discussed in Section 10.6. As usual, the file depends little on the device being modeled, which is nearly all specified in the mesh skeleton file.

```
// FILE:  sgdir\ch10\mos\mos.sg
```

```
mesh "mos.msh";                    // read the mesh

// physical and scaling constants
const Ni  = 1.25e10;               // intrinsic conc. of Si    (cm^-3)
const T0  = 300.0;                 // temperature scaling         (K)
const V0  = kb*T0/e;               // potential scaling           (V)
const C0  = 1.0e19;                // concentration scaling    (cm^-3)
const D0  = 35.0;                  // diff. coef. scaling     (cm^2/s)
const u0  = D0/V0;                 // mob. coef. scaling    (cm^2/V s)
const X0  = 2.88e-4;               // distance scaling           (cm)
const J0  = e*D0*C0/X0;            // current scaling         (A/cm^2)
const tm0 = sq(X0)/D0;             // time scaling                (s)
const L0  = (V0*e0)/(sq(X0)*e*C0); // Laplacian scaling
const Pref = NS;                   // hole concentration reference
const Nref = sq(Ni)/Pref;          // electron concentration reference

// electron lifetime
func LTn(<N>)
  const LT0  = 3.95e-4;
  const Nref = 7.1e15;
  return LT0 / (1.0 + N / Nref);

// hole lifetime
func LTp(<N>)
  const LT0  = 3.52e-5;
  const Nref = 7.1e15;
  return LT0 / (1.0 + N / Nref);

// scaled recombination
func R(n,p,<tn>,<tp>)
  const Cn = 2.8e-31/[D0/sq(C0*X0)];
  const Cp = 1.2e-31/[D0/sq(C0*X0)];
  const ni = Ni/C0;
  const pn = ni*ni;
  assign Rsrh = (n*p-pn)/[tp*(n+ni)+tn*(p+ni)];
  assign Raug = (n*p-pn)*(Cn*n+Cp*p);
  return Rsrh + Raug;

// electron mobility
func MUn(<N>,<T>)
  const Al = 1430.0,  Bl = -2.20;
  const Ai = 4.61e17, Bi = 1.52e15;
  assign MUl = Al*pow((T/300.0),Bl);
  assign MUi = {Ai*T*sqrt(T)/N}/{ln(1+Bi*T*T/N)-Bi*T*T/(N+Bi*T*T)};
  assign X   = sqrt(6*MUl/MUi);
  return MUl*{1.025/[1+pow(X/1.68,1.43)]-0.025};

// hole mobility
```

```
func MUp(<N>,<T>)
  const Al = 495.0,    Bl = -2.20;
  const Ai = 1.00e17, Bi = 6.25e14;
  assign MU1 = Al*pow((T/300.0),Bl);
  assign MUi = {Ai*T*sqrt(T)/N}/{ln(1+Bi*T*T/N)-Bi*T*T/(N+Bi*T*T)};
  assign X   = sqrt(6*MU1/MUi);
  return MU1*{1.025/[1+pow(X/1.68,1.43)]-0.025};

// scaled electric field
func E(V1,V2,<h>)
  return (V1-V2)/h;

// scaled electron current
func Jn(n1,n2,E,MU1,MU2,<h>)          // aux1(x) and aux2(x) are
  const Ut   = (kb*T/e)/V0;          // functions that are internal
  assign MU   = ave(MU1,MU2);        // to the translator
  assign dV   = E*h/(2*Ut);
  assign n    = n1*aux2(dV)+n2*aux2(-dV);
  assign dndx = aux1(-dV)*(n2-n1)/h;
  return MU*(n*E+dndx);

// scaled hole current
func Jp(p1,p2,E,MU1,MU2,<h>)
  const  Ut   = (kb*T/e)/V0;
  assign MU   = ave(MU1,MU2);
  assign dV   = E*h/(2*Ut);
  assign p    = p1*aux2(-dV)+p2*aux2(dV);
  assign dpdx = aux1(-dV)*(p2-p1)/h;
  return MU*(p*E-dpdx);

// declare variables and specify which one are unknowns
var  V[NODES],  n[NODES],  p[NODES];
var tn[NODES], tp[NODES];
var un[NODES], up[NODES];
var switch;

unknown V[SIO2];
unknown V[SI], n[SI], p[SI];
known   V[SUB], n[SUB], p[SUB];
known   V[DRAIN], n[DRAIN], p[DRAIN];
known   V[SOURCE], n[SOURCE], p[SOURCE];
known   V[GATE];

// precompute these functions prior to each Newton iteration
precompute@EDGE i (j,k) ODD->E(V[j],V[k],elen);
precompute@EDGE i (j,k) ODD->Jn(n[j],n[k],@E(i,j,k),un[j],un[k],elen);
precompute@EDGE i (j,k) ODD->Jp(p[j],p[k],@E(i,j,k),up[j],up[k],elen);
```

```
// Poisson's equation (Si and SiO2 interface)
equ V[i=SISIO2] ->
  L0*(nsum(i,j,SI,  {eSi  *@E(edge(i,j),i,node(i,j))}*ilen(i,j)) +
      nsum(i,j,SIO2,{eSiO2*@E(edge(i,j),i,node(i,j))}*ilen(i,j))) -
  (p[i]-n[i]+C[i])*area(i,SI) = 0.0;

// Poisson's equation (Si bulk)
equ V[i=SI] ->
  L0*nsum(i,j,ALL,{eSi*@E(edge(i,j),i,node(i,j))}*ilen(i,j)) -
  (p[i]-n[i]+C[i])*area(i) = 0.0;

// Poisson's equation (SiO2 bulk)
equ V[i=SIO2] ->
  L0*nsum(i,j,ALL,{eSiO2*@E(edge(i,j),i,node(i,j))}*ilen(i,j)) = 0.0;

// electron continuity equation (Si and SiO2 interface)
equ n[i=SISIO2] ->
  switch*(
  nsum(i,j,SI,@Jn(edge(i,j),i,node(i,j))*ilen(i,j)))
  +
  (1.0-switch)*n[i]
   =
  switch*(
  R(n[i],p[i],tn[i],tp[i])*area(i,SI))
  +
  (1.0-switch)*(Nref/C0)*exp(V[i])
;

// electron continuity equation
equ n[i=SI] ->
  switch*(
  nsum(i,j,ALL,@Jn(edge(i,j),i,node(i,j))*ilen(i,j)))
  +
  (1.0-switch)*n[i]
   =
  switch*(
  R(n[i],p[i],tn[i],tp[i])*area(i))
  +
  (1.0-switch)*(Nref/C0)*exp(V[i])
;

// hole continuity equation (Si and SiO2 interface)
equ p[i=SISIO2] ->
  switch*(
  -nsum(i,j,SI,@Jp(edge(i,j),i,node(i,j))*ilen(i,j)))
  +
  (1.0-switch)*p[i]
   =
```

```
switch*(R(n[i],p[i],tn[i],tp[i])*area(i,SI))
+
(1.0-switch)*(Pref/C0)*exp(-V[i])
;
// hole continuity equation
equ p[i=SI] ->
switch*(
  -nsum(i,j,ALL,@Jp(edge(i,j),i,node(i,j))*ilen(i,j)) )
+
(1.0-switch)*p[i]
=
switch*(
  R(n[i],p[i],tn[i],tp[i])*area(i))
+
(1.0-switch)*(Pref/C0)*exp(-V[i])
;

// set the numerical algorithm parameters
set NEWTON DAMPING    = 3;
set NEWTON ACCURACY   = 1.0e-12;
set NEWTON ITERATIONS = 100;
set LINSOL ALGORITHM  = GAUSSELIM;
set LINSOL FILL       = INFINITY;
set DISTANCE SCALE    = X0;


// This procedure computes the electron and hole lifetimes.  It should
// be called prior to scaling the variables.
begin ComputeLifetimes
  assign tn[i=SI] = LTn(abs(C[i]));
  assign tp[i=SI] = LTp(abs(C[i]));
end

// This procedure computes the electron and hole lifetimes.  It should
// be called prior to scaling the variables.
begin ComputeMobilities
  assign un[i=SI] = MUn(abs(C[i]),T);
  assign up[i=SI] = MUp(abs(C[i]),T);
end

// This procedure scales the variables.
begin ScaleVars
  assign V[i=all] = V[i]/V0;
  assign n[i=all] = n[i]/C0;
  assign p[i=all] = p[i]/C0;
  assign C[i=all] = C[i]/C0;
  assign un[i=SI] = un[i]/u0;
  assign up[i=SI] = up[i]/u0;
```

```
    assign tn[i=SI] = tn[i]/tm0;
    assign tp[i=SI] = tp[i]/tm0;
end

begin RestoreVars
    assign V[i=all] = V[i]*V0;
    assign n[i=all] = n[i]*C0;
    assign p[i=all] = p[i]*C0;
    assign C[i=all] = C[i]*C0;
    assign un[i=SI] = un[i]*u0;
    assign up[i=SI] = up[i]*u0;
    assign tn[i=SI] = tn[i]*tm0;
    assign tp[i=SI] = tp[i]*tm0;
end


// This procedure increments the drain-to-source voltage.
var Ids, Vgs, Vds, deltaVds;
begin IncrementVds
    assign V[i=DRAIN] = V[i] + deltaVds / V0;
    assign Vds = Vds + deltaVds;
end

// initialize the variables
begin InitVars
    assign  n[i=SI] = (C[i] > 0) * +0.5*(C[i]+sqrt(sq(C[i])+4*Ni)) + (C[i] < 0);
    assign  p[i=SI] = (C[i] < 0) * -0.5*(C[i]-sqrt(sq(C[i])+4*Ni)) + (C[i] > 0);
    assign  n[i=SI] = (C[i] > 0) * n[i] + (C[i] < 0) * sq(Ni) / p[i];
    assign  p[i=SI] = (C[i] < 0) * p[i] + (C[i] > 0) * sq(Ni) / n[i];
    assign  V[i=SI] = (kb * T / e) * ln(n[i] / Nref);
end
var WriteCount;
begin main

    // initialize variables
    call InitVars;
    call ComputeLifetimes;
    call ComputeMobilities;
    call ScaleVars;
    assign switch = 0.0;
    solve;
    assign switch = 1.0;
    call RestoreVars;
    call ComputeLifetimes;
    call ComputeMobilities;
    call ScaleVars;
    write;
```

```
// compute IV data
// assign Vds = 0.0;
// assign deltaVds = 0.10;
// assign WriteCount = 0;
// while (Vds < 5.05) begin
   solve;
   assign Ids = -J0*X0*lsum(i,SOURCE,nsum(i,j,all,ilen(i,j)*
     {@Jn(edge(i,j),i,node(i,j))+@Jp(edge(i,j),i,node(i,j))}));
//   if (WriteCount >= 4.5) begin
   write;
//     assign WriteCount = 0;
//   end
//   call IncrementVds;
//   assign WriteCount = WriteCount + 1;
// end

end
```

(Note the use of the 'switch' in this file to go from an equilibrium simulation to a nonequilibrium simulation.)

10.6   Results of the MOSFET Simulations

This section displays the results of the sample MOSFET simulation that was used as an example throughout Chapter 6, on SGFramework. Figure 10.3 shows the drain-to-source current I_{ds} as a function of the drain-to-source voltage V_{ds} for various values of the gate-to-source voltage V_{gs}. Figure 10.4 depicts the current I_{ds} as a function of the voltage V_{gs} for $V_{ds} = 0.5$ V. This plot can be used to extract the MOSFET's threshold voltage, which is approximately one volt. Figures 10.5–10.7 show surface plots of the logarithm of the electron density first in equilibrium, then with an inversion channel and finally when the device is conducting between the drain and source and the channel is pinched off. [1]

Interpretation of the Simulation Results

We shall now attempt to estimate the quantities needed to calculate the drain current I_D, to see if the plots contain reasonable results. The expression for I_D contains the quantity $\frac{h\bar{\mu}_n}{L}C_{ox}$. This is the expression we need to evaluate.

The height of the device will be taken to be one centimeter, or $h = 10^{-2}$ m. The average mobility will be set to $\bar{\mu}_n \simeq D/V^{th} = 3.5 \times 10^{-3}/0.03 \simeq 0.1$, where the value for the diffusion coefficient was taken from the input file. Also from the input file, the length of the gate is $L = 1.4 \times 10^{-6}$ m while the thickness of the gate oxide is $t_{ox} = 2 \times 10^{-9}$ m.

---

[1] Several of these figures are reprinted from *Computer Physics Communications*, Vol. 93, 'Strategies for mesh-handling and model specification within a highly flexible simulation framework', 179-211, 1995, with permission from Elsevier Science — NL, Amsterdam, the Netherlands.

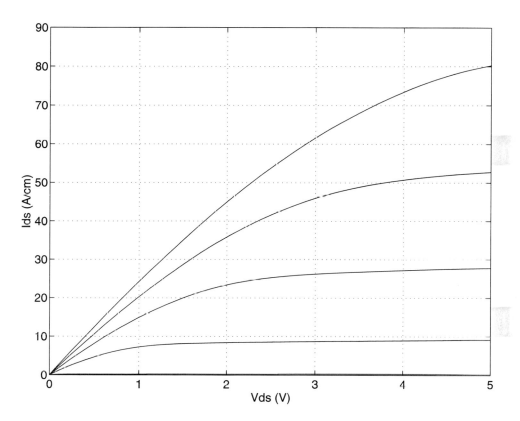

Figure 10.3. The MOSFET current-voltage curves for $V_{gs} = 1, 2, 3, 4$ and 5 V. The lower curve shows a slight 'short-channel' effect.

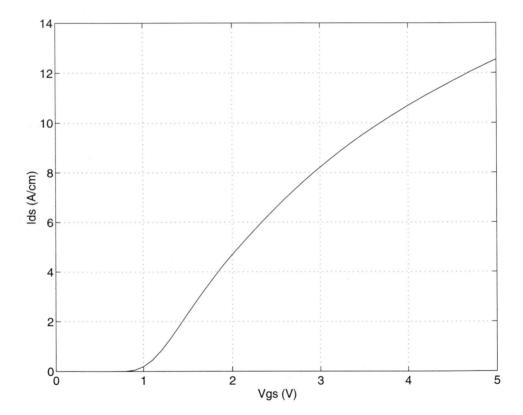

Figure 10.4. The MOSFET threshold curve for $V_{ds} = 0.5$ V.

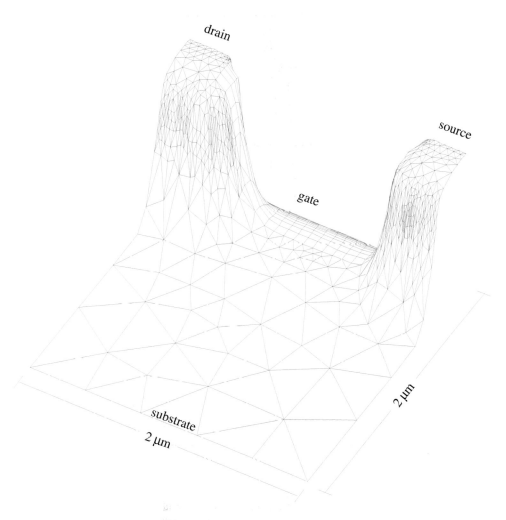

Figure 10.5. The log of the electron concentration profile for $V_{gs} = 0\text{V}$, $V_{ds} = 0\text{V}$.

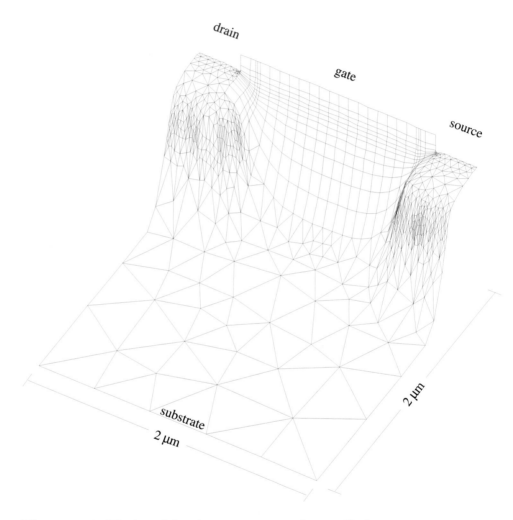

Figure 10.6. The log of the electron concentration profile for $V_{gs} = 5\text{V}$, $V_{ds} = 0\text{V}$.

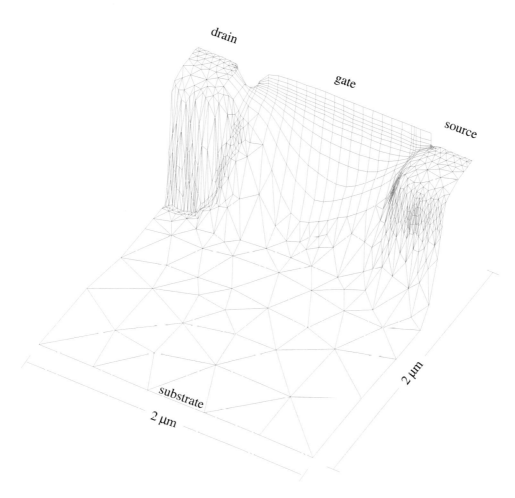

Figure 10.7. The log of the electron concentration profile for $V_{gs} = 3\text{V}$, $V_{ds} = 5\text{V}$.

The oxide capacitance $C_{ox} = \epsilon/t_{ox} \simeq \frac{3.9 \times 10^{-11}}{2 \times 10^{-9}} \simeq 2 \times 10^{-2}$ F/m^2.

Combining all these, $\frac{h\bar{\mu}_n}{L} C_{ox} \simeq \frac{10^{-2} \times 0.1}{1.4 \times 10^{-6}} \times 2 \times 10^{-2} \simeq 14$.

This factor appears in the expression for I_D, eq. 10.19, which is approximated in eq. 10.22 as $I_D = \frac{h\bar{\mu}_n C_{ox}}{L}[(V_G - V_T)V_D - \frac{1}{2}V_D^2]$. Since the current shown in the figure is of the order of tens of amps, this is the right order of magnitude. The derivative of I_D with respect to V_D can be estimated from the plot to be about 4 A/V at low values of V_G. According to our theoretical estimate, the value of $\frac{\partial I_D}{\partial V_D}$ is about $14V_D$, and since V_D is 0.5 in the plot, we would estimate $\frac{\partial I_D}{\partial V_G} \simeq 7$ A/V. This is not very good agreement, but it is within the range of errors we would expect for this sort of comparison.

10.7 MOSFET Circuits

In this section we shall illustrate several of the most important MOSFET circuits. The analytic formulae given in the preceding sections will be used to model MOS-FETs in external circuits.

Resistive Load

A simple digital inverter is demonstrated first, using a MOSFET with a resistor acting as a load, connected between the MOSFET's drain and the power supply at VDD volts. The source is grounded, the gate voltage is the input and the drain voltage is the output.

Setting up the problem is made complicated by the fact that the MOSFET switches between modes of operation. This is handled using conditional statements. An inequality enclosed in parentheses evaluates to one if it is true and to zero if it is false. In the file conditional statements test to see if there is a current at all: $V_D > 0$ and $V_G > V_T$ must hold for a current to flow. A second set of tests will distinguish between the Ohmic regime, with $V_D < V_G - V_T$, and saturation, when $V_D > V_G - V_T$. The appropriate expression for the current is multiplied by a conditional, which is one when that expression for the current should be used. The conditional is zero when that expression for the current should not be used.

This file is relatively easy to understand. As such it is a useful basis of comparison for the simulations which follow. This comparison should make it straightforward to draw the corresponding circuits. The results are shown in Figure 10.8.

Input File mos1.sg

```
// FILE:  sgdir\ch10\mos1\mos1.sg
// This file contains a demonstration of how to
// model a MOSFET with a load resistor.

// Circuit Parameters
const TMAX = 100;
const VDD  = 5.0;
```

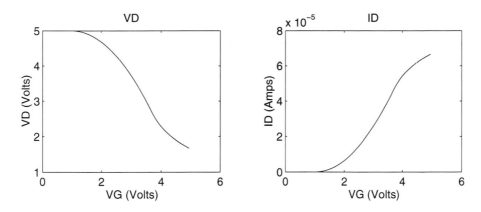

Figure 10.8. Performance of a MOSFET with a resistive load, modeled using lumped-circuit elements and the analytic model of the MOSFET terminal characteristics. The plot of V_D versus V_G shows the inverter characteristic shape, but does not have as sharp a transition as is desirable.

```
const RL  = 5.0e+04;
const VT = 1.0;
const k = 1.28e-5;

var ID[TMAX];
var VG[TMAX];
var VD[TMAX];

set NEWTON ACCURACY  = 1.0e-10;


equ VD[i=1..TMAX-1]->
  VD[i] = (VDD - RL*ID[i]);

equ ID[i=1..TMAX-1]->
  ID[i] =  0.0 + (VD[i] > 0.0) * (VG[i] > VT) *
          ((VD[i] > (VG[i]-VT)) * (k/2.0)*sq(VG[i]-VT) +
          (VD[i] <= (VG[i]-VT)) * k*((VG[i]-VT)*VD[i]-sq(VD[i])/2.0));

begin main
  assign VG[i=all] = VDD*(1.0*i/TMAX);
  solve;
  write;
end
```

Enhancement Load

The next example uses another MOSFET as the load, the MOSFET in this case being an enhancement device. An enhancement MOSFET requires an applied gate voltage to enhance the channel.There is no channel otherwise. A depletion MOSFET (which is used in the next simulation) has a channel when the gate voltage is zero. It is called a depletion device, since a gate voltage is used to deplete the channel of carriers. The load MOSFET replaces the resistor in the previous example, the current flowing its between source and drain, and its gate connected to its source so $V_G = 0$ relative to its source. The current through the load MOSFET requires another conditional, since it is either zero or equal to the saturation value. However, the load resistor is connected between V_{DD} and the output at V_D, so the conditional involves the relative values of these voltages. The results from this simulation are shown in Figure 10.9.

Input File mos2.sg

```
// FILE:  sgdir\ch10\mos2\mos2.sg
// This file contains a demonstration of how to
// model a MOSFET with an enhancement load.

// Circuit Parameters
const TMAX = 100;
const VDD  = 5.0;
const RL   = 5.0e+04;
const VT   = 1.0;
const k    = 1.28e-5;
const VT2  = 1.0;
const k2   = 1.0e-5;

var ID[TMAX];
var VG[TMAX];
var VD[TMAX];

set NEWTON ACCURACY = 1.0e-10;


equ VD[i=1..TMAX-1]->
  ID[i] = 0.0 + (VDD >= (VT2 + VD[i])) * (k2/2.0) * sq(VDD - VT2 - VD[i]);

equ ID[i=1..TMAX-1]->
  ID[i] = 0.0 + (VD[i] > 0.0) * (VG[i] > VT) *
          ((VD[i] > (VG[i]-VT)) * (k/2.0)*sq(VG[i]-VT) +
           (VD[i] <= (VG[i]-VT)) * k*((VG[i]-VT)*VD[i]-sq(VD[i])/2.0));

begin main
  assign VG[i=all] = VDD*(1.0*i/TMAX);
```

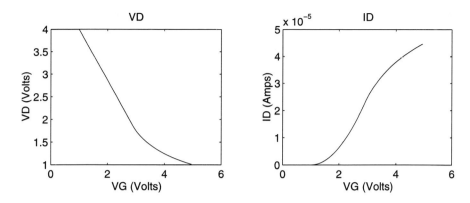

Figure 10.9. Characteristics of a MOSFET with an enhancement load.

```
   solve;
   write;
end
```

Depletion Load

The next example involves a depletion load. This case is more complicated to set up. The first file given is one attempt to handle the different conditions which can occur, which is only moderately successful. See Figure 10.10. The problem is that the idealized device model leads to a very awkward set of equations. A better result is obtained with more realistic equations. This is given in the file after this. Extra terms have been added to the equations used. (See the exercises.) These extra terms are needed to allow the computer to find V_D when the current is zero, for instance.

Input File mos3.sg

```
// FILE:  sgdir\ch10\mos3\mos3.sg
// This file contains a demonstration of how to
// model a MOSFET with a depletion load.

// Circuit Parameters
const TMAX = 100;
const VDD  = 5.0;
const VT = 1.0;
const k = 1.28e-5;
const VT2 = -3.0;
const k2 = 1.0e-5;
```

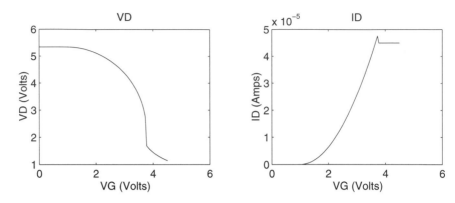

Figure 10.10. Characteristics of a MOSFET with a depletion load. The kinks in the curves are artificial and are due to the 'logical' statements in the input file.

```
var ID[TMAX];
var VG[TMAX];
var VD[TMAX];

set NEWTON ACCURACY  = 1.0e-10;

equ VD[i=0..TMAX-1]->
  ID[i] =  0.0 +  2e-6*VD[i] +
    (VDD<(-VT2 + VD[i]))*k2*( (-VT2)*(VDD-VD[i]) - sq(VDD-VD[i])/2.0 ) *
    (VD[i] >  (VG[i] -VT))
 +  (VD[i] > 0.0)*(VG[i] > VT)*(VDD>(-VT2 + VD[i]))*
    (VD[i] <= (VG[i] -VT)) *k*((VG[i]-VT)*VD[i]-sq(VD[i])/2.0);

equ ID[i=0..TMAX-1]->
  ID[i] =  0.0 +
    (VDD>=(-VT2 + VD[i]))*(k2/2.0)*sq(-VT2) +
    (VD[i] > 0.0)*(VG[i] > VT)*
    (VD[i] > (VG[i]-VT)) * (k/2.0)*sq(VG[i]-VT);

begin main
  assign VG[i=all] = 0.9* VDD*(1.01*i/TMAX);
  solve;
  write;
end
```

The previous example file had artificial corrections in it, to allow solutions to be obtained in a situation where the simple set of equations was indeterminate. The approach was only partially successful. This case allows an excellent illustration of what to do when a simple model fails. The proper approach is to look for the physical mechanism which has been left out. In reality, the equations are not indeterminate, and they are corrected below.   This example allows a weak dependence of I_D on V_D in saturation, which is more realistic than the simple model and leads to equations which are much easier to solve. The results from the improved model are shown in Figure 10.11.

Input File mos4.sg

```
// FILE:  sgdir\ch10\mos4\mos4.sg
// This file contains a demonstration of how to
// model a MOSFET with a depletion load, using slightly
// more realistic device models than mos3.sg.

// Circuit Parameters
const TMAX = 100;
const VDD  = 5.0;
const VT = 1.0;
const k = 1.28e-5;
const VT2 = -3.0;
const k2 = 1.0e-5;

var ID[TMAX];
var VG[TMAX];
var VD[TMAX];

set NEWTON ACCURACY  = 1.0e-10;

equ ID[i=0..TMAX-1]->
  ID[i] =  0.0 +
    (VDD<(-VT2 + VD[i]))*k2*( (-VT2)*(VDD-VD[i]) - sq(VDD-VD[i])/2.0 ) +
    (VDD>=(-VT2 + VD[i]))*k2/2.0 * sq(-VT2) ;

equ VD[i=0..TMAX-1]->
  ID[i] =  0.0 + ((VD[i] <= 0)+(VG[i]<VT))*1.0e-6*(VD[i]-VDD) +
    (VD[i] > 0.0)*(VG[i] > VT)*(
    (VD[i] < (VG[i]-VT)) * k * ((VG[i]-VT)*VD[i] - 0.5*sq(VD[i]) ) +
    (VD[i] >=(VG[i]-VT)) * (k/2.0)*sq(VG[i]-VT) * (1.0+0.05*(VD[i]-(VG[i]-VT)))
    );

begin main
  assign VG[i=all] = VDD*(1.01*i/TMAX);
  assign VD[i=all] = 0.2*VDD;
  assign ID[i=all] = 1.0e-6;
  solve;
  write;
end
```

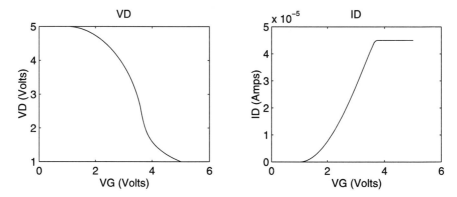

Figure 10.11. Characteristics of a MOSFET with depletion load , using a more realistic model which gives a better behaved solution.

CMOS Invertor

We now turn to a simple CMOS inverter, where the load is the complement of the driver MOSFET. The high degree of symmetry makes the distinction between load and driver largely meaningless. The symmetry should also be evident in the setting up of the equations. It is particularly interesting to notice the shape of I_D in Figure 10.12. The fact that the current is small in both the high- and low-voltage states (due to the symmetry) will minimize power dissipation in CMOS circuits. Compare this to all the preceding plots of I_D.

Input File mos5.sg

```
// FILE:  sgdir\ch10\mos5\mos5.sg
// This file contains a demonstration of how to
// model a MOSFET with a complementary load.

// Circuit Parameters
const TMAX = 100;
const VDD  = 3.0;
const VTN = 0.6;
const k = 6.0e-5;
const VTP = -0.6;
const k2 = 6.0e-5;

var ID[TMAX];
var VG[TMAX];
var VD[TMAX];

set NEWTON ACCURACY = 1.0e-6;


equ ID[i=0..TMAX-1]->
  ID[i] =  0.0    +
    (VG[i]<VDD+VTP)*(VD[i]<=VDD)*(
    (VD[i]>(VG[i]-VTP))*k2*((VDD-VG[i]+VTP)*(VDD-VD[i])-sq(VDD-VD[i]))/2.0) +
    (VD[i]<=(VG[i]-VTP))*k2/2.0 * sq(VDD-VG[i]+VTP)*(1.0+0.05*
                                  (VDD-VD[i]+(VG[i]-VTP))));

equ VD[i=0..TMAX-1]->
  ID[i] =  0.0 + ((VD[i] <= 0)+(VG[i]<VTN))*1.0e-2*(VD[i]-VDD) +
    (VD[i] > 0.0)*(VG[i] > VTN)*(
    (VD[i] < (VG[i]-VTN)) * k * ((VG[i]-VTN)*VD[i] - 0.5*sq(VD[i]) ) +
    (VD[i] >=(VG[i]-VTN)) * (k/2.0)*sq(VG[i]-VTN)*(1.0+0.05*
                                  (VD[i]-(VG[i]-VTN))));

begin main
  assign VG[i=all] = VDD*(1.01*i/TMAX);
  assign VD[i=all] = 0.2*VDD;
```

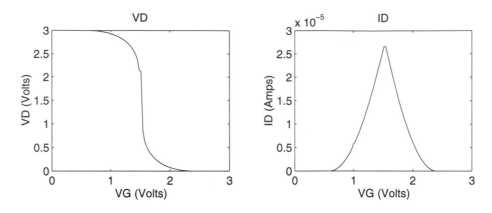

Figure 10.12. Characteristics of a MOSFET with a complementary load. The parameters in this case are for a low-voltage technology which might be needed for VLSI devices with very small dimensions.

```
   assign ID[i=all] = 1.0e-6;
   solve;
   write;
end
```

The next file is the same as the previous one but with slightly different values of V_{DD} and V_T. They are included to show the effect of changing the power supply on a CMOS technology. See Figure 10.13. Provided V_T is altered appropriately when V_{DD} is changed the curves can be made to look similar. If the same value of V_T was kept when V_{DD} was decreased from 5 to 3 V, the result would not be acceptable.

Input File mos6.sg

```
// FILE:  sgdir\ch10\mos6\mos6.sg
// This file contains a demonstration of how to
// model a MOSFET with a complementary load.

// Circuit Parameters
const TMAX = 100;
const VDD  = 5.0;
const VTN = 1.2;
const k = 6.0e-5;
const VTP = -1.2;
const k2 = 6.0e-5;

var ID[TMAX];
var VG[TMAX];
var VD[TMAX];
```

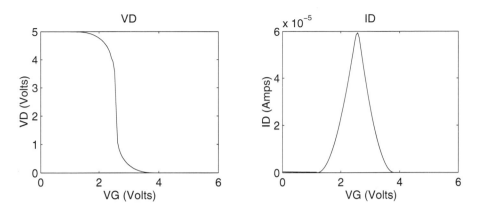

Figure 10.13. Characteristics of a MOSFET with a complementary load, with more standard device parameters corresponding to a higher-voltage technology.

```
set NEWTON ACCURACY   = 5.0e-4;

equ ID[i=0..TMAX-1]->
  ID[i] =  0.0     +
    (VG[i]<VDD+VTP)*(VD[i]<=VDD)*(
    (VD[i]>(VG[i]-VTP))*k2*((VDD-VG[i]+VTP)*(VDD-VD[i])-sq(VDD-VD[i])/2.0) +
    (VD[i]<=(VG[i]-VTP))*k2/2.0 * sq(VDD-VG[i]+VTP)*(1.0+0.05*
                            (VDD-VD[i]+(VG[i]-VTP))));

equ VD[i=0..TMAX-1]->
  ID[i] =  0.0 + ((VD[i] <= 0)+(VG[i]<VTN))*1.0e-2*(VD[i]-VDD) +
    (VD[i] > 0.0)*(VG[i] > VTN)*(
    (VD[i] < (VG[i]-VTN)) * k * ((VG[i]-VTN)*VD[i] - 0.5*sq(VD[i]) ) +
    (VD[i] >=(VG[i]-VTN)) * (k/2.0)*sq(VG[i]-VTN)*(1.0+0.05*
                            (VD[i]-(VG[i]-VTN))));

begin main
  assign VG[i=all] = VDD*(1.01*i/TMAX);
  assign VD[i=all] = 0.2*VDD;
  assign ID[i=all] = 1.0e-6;
  solve;
  write;
end
```

(The rest of the file is identical to mos5.sg.)

Invertors in Combination

mos7.sg uses two enhancement MOSFETs, each with a resistive load. Each constitutes a simple invertor, as was shown by mos1.sg. The output (drain voltage) of the first is fed into the input (gate voltage) of the second. See Figure 10.14.

Input File mos7.sg

```
// FILE:  sgdir\ch10\mos7\mos7.sg
// This file contains a demonstration of how to
// model two enhancement MOSFETs, one with its gate connected to the drain
// of the first and both having equal resistive loads RL
// connected to their drain.

// Circuit Parameters
const TMAX = 100;
const VDD  = 5.0;
const RL   = 5.0e+03;
const VT   = 1.2;
const k    = 5.0e-4;

var ID[TMAX];
var VG[TMAX];
var VD[TMAX];
var ID2[TMAX];
var VD2[TMAX];

set NEWTON ACCURACY = 1.0e-10;


equ VD[i=0..TMAX-1]->
  VD[i] = (VDD - RL*ID[i]);

equ VD2[i=0..TMAX-1]->
  VD2[i]= (VDD - RL*ID2[i]);

equ ID[i=0..TMAX-1]->
  ID[i] =  0.0 +
    (VD[i] > 0.0) * (VG[i] > VT)*(
    (VD[i] > (VG[i]-VT)) * (k/2.0)*sq(VG[i]-VT) +
    (VD[i] <= (VG[i]-VT)) * k*((VG[i]-VT)*VD[i]-sq(VD[i])/2.0));

equ ID2[i=0..TMAX-1]->
  ID2[i] = 0.0 +
    (VD2[i] > 0.0) * (VD[i] > VT)*(
    (VD2[i] > (VD[i]-VT)) * (k/2.0)*sq(VD[i]-VT) +
    (VD2[i] <= (VD[i]-VT)) * k*((VD[i]-VT)*VD2[i]-sq(VD2[i])/2.0));

begin main
  assign VG[i=all] = VDD*(1.0*i/TMAX);
  solve;
  write;
```

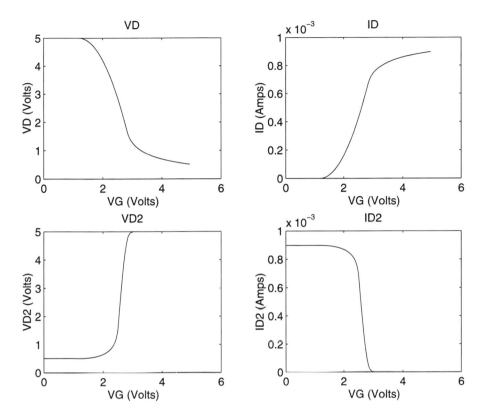

Figure 10.14. Results of modeling a circuit with two enhancement MOSFETs.

end

10.8   Summary

The MOSFET is the device which serves as the primary example throughout this document. A detailed simulation was presented in Chapter 6, which introduced the simulation software. Here we have described the MOSFET analytically to establish a more intuitive picture of how it works, and set up circuit models of some important MOSFET circuits and shown how they behave. In the problems below the analytic results are compared to simulation results. We return to MOSFETs again in power applications in Chapter 11.

PROBLEMS

1. Plot on the same figure the voltage versus position in a MOS capacitor, analytically and using a numerical solution of the nonlinear Poisson's equation.

2. Use the analytic expressions given in the text to plot the usual curves showing I_D versus V_G and V_D for a typical MOSFET. Repeat the exercise using a numerical model of the device. Compare the results.

3. Analytically estimate the electric field inside a MOSFET under the main normal operating conditions.

4. Using the electric field from the previous question, write a Monte Carlo program to follow carriers injected into a MOSFET through each terminal, under different bias conditions. Use the mean free path λ and collision frequency ν which can be estimated from data given in Chapter 4. How many collisions do you expect a typical particle to have before it leaves each junction where it is injected? How long will this take, in each case? Plot several trajectories, to confirm your estimates.

5. Obtain the small-signal characteristics g_d and g_m, analytically and numerically, for a standard MOSFET geometry. Compare the two sets of results.

6. Draw the circuits represented by the circuit simulations given in this chapter. Derive and explain the equations used. Suggest more realistic values for the device parameters used and rerun the simulations using those values. Plot the results.

7. Compare the two simulations of MOSFETs which employed a depletion load. What is the difference between the models used? Why is the second model appropriate, physically, and why is it more convenient to use in the simulation?

Chapter 11

Power Semiconductor Devices

Power semiconductors are an exciting and practically important area of research. They also provide an opportunity to use the ideas of the previous chapters in a context where device performance is not always well understood, and to consider the design of truly novel devices.

11.1 Introduction

This chapter extends the subject matter of the earlier chapters to the more difficult topic of power semiconductors, and considers the design of devices in detail. There are two main differences in the devices and the device theory which will be presented:

1. Small-signal analysis is not appropriate, since power devices are required to operate in a highly nonlinear fashion. After the (i, v) characteristics, we accordingly turn to switching characteristics of each device. We must also consider in detail how to model avalanche breakdown, which limits the operating range of many power devices. This is done in Section 11.2.

2. A different set of devices is tackled: power diodes and power MOSFETs have different structures than their low-power counterparts. In addition, a new class of devices, the insulated-gate bipolar transistors (IGBTs), is introduced which does not have a low-power equivalent.

After the section on the power diode, which is the first class of power device we consider, we include a presentation of how the simulation tools introduced here may be used in device design, and how the use of statistical methods may assist in optimizing the design process. The specific application is to a power diode. Since the topic of statistical methods for design is outside the mainstream of this text, and is not discussed elsewhere in the text, this section has been placed with the material which describes the class of device being designed, a power diode.

The most common power semiconductor devices are the following:

1. Power diodes resemble low-power diodes but have a more complicated doping profile than their low-power equivalents.

2. Power BJTs exist and are widely used but are probably less important than some of the other high-power devices and will not be discussed in detail here. A power BJT was modeled in Chapter 8.

3. Power MOSFETs have an extra drift region which is lightly doped next to the drain;

4. Thyristors or semiconductor controlled rectifiers (SCRs) are four layer npnp devices which do not have an exact low-power equivalent but are often represented as two interconnected BJTs. GTO (Gate Turn-Off) Thyristors have also been developed. Thyristors are perhaps the highest power-density devices.

5. Insulated-gate bipolar transistors (IGBTs) are also sometimes called (among other things) bipolar-MOS transistors, the names reflecting the fact that the device has features of BJTs and MOSFETs. An IGBT has a gate, a source and a drain but with a more complex set of layers between. Compared to the power MOSFET, which already has an extra drift region compared to a low-power MOSFET, the IGBT has an extra highly doped region next to the drain. The extra layer has the opposite doping to the layer which is adjacent to the drain in the equivalent power MOSFET.

Before outlining the devices we shall consider in detail the topic of breakdown. Methods of modeling breakdown are presented and illustrated in application to a pin diode. Power semiconductor devices are then considered in order of complexity. Section 11.3 discusses power diodes. It is followed by a detailed worked example of how statistical design techniques may be employed to optimize a semiconductor device. Since the purpose of simulation is largely to improve design of devices, this is a very important aspect of the techniques which are presented here. Sections 11.5, 11.6 and 11.7 handle power MOSFETs, thyristors and IGBTs, respectively. Section 11.8 discusses electrothermal simulations of power devices. These simulations solve the heat-transport equation in addition to the semiconductor equations. Section 11.9 summarizes the chapter.

11.2   Simulation of Breakdown in Semiconductors

Breakdown was discussed in Section 7.2.3, on the behavior of pn junctions. Because of the importance of breakdown for the operation of power devices, this section introduces and contrasts three alternate ways of simulating a semiconductor device in the vicinity of reverse breakdown, with the most complete model first and the most approximate last. It will be shown that all produce essentially the same results in a test case, while the third model is computationally much more efficient.

11.2.1   Basic Semiconductor Equations with Impact-Ionization Model

At first sight the most likely method of modeling the breakdown of semiconductor devices is to incorporate a carrier generation model due to impact ionization in the semiconductor continuity equations. This method requires the solution of three coupled nonlinear PDEs (the semiconductor equations) and will be referred to as the *basic semiconductor equations with impact ionization* (BSE-II) model. These equations were given in Chapter 4. They are used in the work described in this section, which serves to validate the less accurate models which are introduced later. (It is worth noting that the inclusion of the rapid generation of carriers in the continuity equation changes the way we can handle the continuity equation. The current now varies exponentially, between mesh points.)

11.2.2   Nonlinear Poisson Model

In general, both the electron and hole current densities J_n and J_p are small in a reverse-biased pn junction diode prior to breakdown. This fact can be exploited to derive the usual analytic expressions for the electron and hole concentrations as functions of the electrostatic potential. These lead to a highly simplified model which is the topic of this section.

This method is divided between two subsections. Subsection 11.2.3 derives the nonlinear Poisson (NP) model using the fact that the electron and hole current densities are very small. Subsection 11.2.4 discusses a method by which the breakdown voltage may be determined without direct knowledge of the electron and hole current densities. The extension of the method by a further simplification follows later.

11.2.3   Derivation of the Nonlinear Poisson Model

To derive the nonlinear Poisson equation, the analytical expressions for the electron and hole concentrations are substituted into the scaled Poisson equation. The resulting expression is given by Equation (11.1). Equation (11.1) may be solved in order to determine the electrostatic potential throughout the device. The electron and hole concentrations may then be determined. The computational requirements for the solution of the nonlinear Poisson equation are substantially (approximately nine times) smaller than the requirements for the semiconductor equations. There are many fewer nonzero elements, and much less fill, in the Jacobian matrix. The drawback of this method, however, is that the electron and hole current densities cannot be determined, since they are assumed to be zero in the derivation.

$$\lambda^2 \nabla \cdot (\epsilon_r \mathbf{E}) - (p_{ref} \exp(-V) - n_{ref} \exp(V) + C) = 0 \qquad (11.1)$$

where $p_{ref} n_{ref} = n_i^2$.

11.2.4   Calculation of the Breakdown Voltage

Since the electron and hole current densities cannot be determined by the NP model, an indirect method of determining the breakdown voltage must be employed. One such method is the evaluation of the electron and hole multiplication coefficients, which are defined by equations (11.2) and (11.3) [121] , respectively, and which were introduced in Chapter 7. Models for the ionization rates α_n and α_p are given by Equations (4.82) and (4.83) of Subsection 4.4.6. The multiplication coefficients determine the total number of secondary electron-hole pairs generated by impact ionization by one electron or hole when it traverses the distance from one end of the depletion region to the other. At breakdown, either the electron or hole multiplication coefficient (M_e or M_h) approaches infinity. The multiplication coefficients approach infinity when the integrals in their denominators (I_e and I_h) approach unity. These integrals are commonly referred to as ionization integrals and are given by Equations (11.4) and (11.5).

$$M_e = \frac{\exp\left\{\int_{x_n}^{x_p} [\alpha_n(x) - \alpha_p(x)]\, dx\right\}}{1 - \int_{x_n}^{x_p} \alpha_n(x) \exp\left\{\int_{x}^{x_p} [\alpha_p(x') - \alpha_n(x')]\, dx'\right\} dx} \tag{11.2}$$

$$M_h = \frac{\exp\left\{\int_{x_n}^{x_p} [\alpha_n(x) - \alpha_p(x)]\, dx\right\}}{1 - \int_{x_n}^{x_p} \alpha_p(x) \exp\left\{\int_{x_n}^{x} [\alpha_n(x') - \alpha_p(x')]\, dx'\right\} dx} \tag{11.3}$$

$$I_e = \int_{x_n}^{x_p} \alpha_n(x) \exp\left\{\int_{x}^{x_p} [\alpha_p(x') - \alpha_n(x')]\, dx'\right\} dx \tag{11.4}$$

$$I_h = \int_{x_n}^{x_p} \alpha_p(x) \exp\left\{\int_{x_n}^{x} [\alpha_n(x') - \alpha_p(x')]\, dx'\right\} dx. \tag{11.5}$$

In principle, the ionization integrals can be evaluated along any path through the depletion region. However, the critical ionization path is the trajectory of the electric-field line which yields the largest ionization integral. Hence, a practical measure of device breakdown is when either the electron or hole ionization integral is unity when evaluated along the critical ionization path. In practice, it is not necessary to evaluate both ionization integrals. The hole ionization integral is always larger than the electron ionization integral in p^+n junctions, and the opposite is true in n^+p junctions [122]. The accuracy of this breakdown calculation is dependent upon the accuracy with which the ionization integrals can be calculated and the accuracy of the models for ionization rates. Goud and Bhat [123] compared the ionization-rate models of several researchers and found the model of Overstraeten and DeMan [51] to be the most accurate for calculating the breakdown voltage. The critical ionization path is computed by computing the ionization integral along several electric-field lines.

11.2.5   Linear Poisson Model with Depletion-Region Logic

The nonlinear Poisson model may be further simplified by using the depletion approximation. The approximation assumes that in the depletion region, the electron and hole carrier concentrations are negligible compared to the ionized dopant concentration. It further assumes quasi-neutrality in the undepleted regions of the device. With these assumptions it is possible to model reverse-bias diodes with a linear Poisson equation coupled with 'depletion-region logic' to determine whether a region is depleted or not.

The next sections describe the linear Poisson model with depletion-region logic (LP-DRL), and is divided between two subsections. Subsection 11.2.6 presents a justification of the LP-DRL model. References will be given to a more rigorous analysis of this model. We describe the simulation results of the LP-DRL model. The equation specification for this model is identical to the equation specification for the NP model, except that the expressions for charge density are either zero or equal to the charge density due to the ionized impurities. Since Laplace's, and this version of Poisson's, equations are linear, one can solve the resulting system of equations with an iterative technique such as SOR.

11.2.6   Justification of Linear Poisson Model with Depletion-Region Logic

This subsection presents a justification of the LP-DRL model. As the first step, the discretized Poisson equation is solved assuming a charge density equal to the charge density of the ionized impurities. This is done node by node using an SOR solution algorithm. However, when a device is reverse biased, the electrostatic potentials in the regions around the anode and cathode are fixed by the anode and cathode potentials, respectively. Therefore, in the second step, if the computed electrostatic potential at a mesh point is greater than the cathode potential or less than the anode potential, it is 'clamped' to the appropriate contact potential. The validity of constricting the solution for the electrostatic potential between the contact potentials is discussed by Adler [124, 125, 126]. If the electrostatic potential were not clamped, local potential extrema would result and, in reality, free carriers would migrate to the extrema until the potential near the electrode readjusted itself to the appropriate contact potential. This clamping is known as depletion-region logic. The overall procedure is equivalent to solving Poisson's equation in the depletion region, with ρ set equal to the charge on the impurities, and Laplace's equation in the quasineutral bulk.

The algorithm described above is applied at each mesh point in the simulation domain in succession. This process is referred to as a sweep. The order in which the nodes are handled is not important as long as each node is updated during the sweep. After each sweep, the maximum change in the electrostatic potential is compared to some predetermined limit. If the maximum change in the electrostatic potential is greater than the predetermined accuracy, another sweep is performed. Once the electrostatic potential has been computed to sufficient accuracy, the ionization

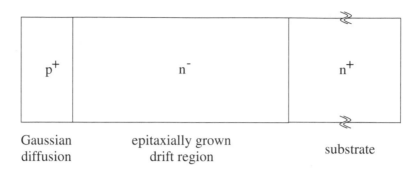

Figure 11.1. A schematic diagram of the one-dimensional pin diode used in the comparison of the simulation methods.

integrals are evaluated, as described in the previous section, in order to determine whether breakdown occurs.

11.2.7   Comparison of Breakdown Models

This section compares the results of the BSE-II, NP and LP-DRL models described in Sections 11.2.1, 11.2.2 and 11.2.5. The BSE-II model consists of the basic semiconductor equations (Poisson's equation, an electron continuity equation and a hole continuity equation) and includes a model for generation due to impact ionization. The NP model consists of a nonlinear Poisson equation. Instead of solving the continuity equation for the electron and hole concentrations, the NP model uses analytical expressions for the electron and hole concentrations. The LP-DRL model requires us to solve a linear Poisson equation and use depletion-region logic to determine which semiconductor regions are depleted of charge and which regions are quasi-neutral.

The geometry of the pin diode used in this section is shown in Figure 11.1. The device is restricted to one dimension in order to simplify the specification of the model. A higher-dimensional simulation requires additional boundary conditions which complicate the mesh specification and distract from the main features of the models. The issues which result from a higher-dimensional simulation will be discussed later. The doping profile of the pin diode is as follows: The substrate is heavily doped n type with a dopant concentration of 10^{19} donors per cubic centimeter. The lightly doped drift region is epitaxially grown upon the substrate and is doped with 10^{14} donors per cubic centimeter. The heavily doped anode is diffused into the drift region. A Gaussian profile is assumed with a junction depth of 10 microns and a surface concentration of 10^{19} acceptors per cubic centimeter.

The results of the simulations are shown in Figure 11.2.

The knee in the simulated (i, v) characteristics (obtained using the BSE-II model) is around -1435 volts. The NP model predicted a breakdown voltage of -1430.1 volts and the LP-DRL model predicted a breakdown of -1430.6 volts. All

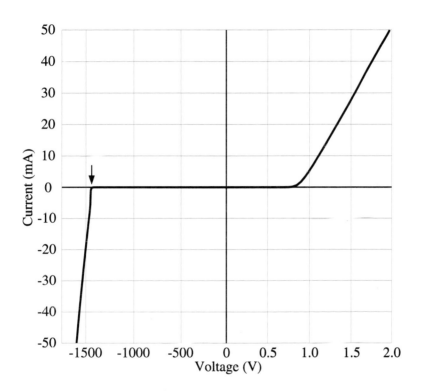

Figure 11.2. The current-voltage characteristics of the pin diode simulated using the BSE-II model. The arrow represents the breakdown voltage predicted by the NP and LP-DRL models.

of the methods are in good agreement. Comparison of the electrostatic-potential profiles further demonstrates the excellent agreement between the models. The maximum difference between the electrostatic potentials found using the NP and the BSE-II models is 1.06 volts, while the maximum difference between the electrostatic potentials found using the LP-DRL and the BSE-II models is 2.01 volts. The difference between the LP-DRL model and the more exact BSE-II model is slightly larger than that between the NP and the BSE-II models due to the fact that the LP-DRL model neglects the built-in potential of the pin diode.

In terms of computational resources, the BSE-II model requires significantly more than the NP or LP-DRL models. The LP-DRL model requires the least amount of computer resources, and empirical tests indicate that this method will converge regardless of the initial guess. The BSE-II model is the most sensitive to the accuracy of the initial guess.

This section has shown that all of the methods predict the same breakdown voltage to within a few volts in the single-dimensional test case. Since the simulation was one dimensional, the electric-field lines were straight, and 'each' path through the depletion region was identical. In a multidimensional case, the critical path has to be accurately determined. This requires very fine resolution in the regions where the electric field peaks, in order to resolve accurately the electric-field profile and calculate the ionization integral. The majority of the ionization tends to occur in a very small area. It should be verified that the mesh is fine enough, by checking that several cells contribute significantly to the ionization integral. The ionization integrals (and the calculation of M) are only valid in regions where there is no gain. The expressions would not apply in the base of a BJT. Carriers produced in the base by ionization have the same effect ass carriers injected into the base. They will attract several more carriers into the base. This is not accounted for in the equations.

11.3   Power Diodes

The simplest power electronic device is now introduced. Performance of this device has already been discussed and modeled in Section 7.2.4. More details of its structure and uses will be given here. In the next section the optimization of such a device is discussed, using the methods of the previous section to determine when breakdown occurs.

Several important power electronic applications require rectification. An ideal rectifier allows current to flow in one direction without a voltage drop and prevents current from flowing in the reverse direction regardless of the applied voltage. Such a rectifier would have current-voltage (i, v) characteristics as shown in Figure 11.3. The ideal rectifier would also have zero switching time. Unfortunately, it is not possible to fabricate an ideal rectifier.

The most common power rectifier is the power diode, which is also referred to as a pin diode. In contrast to an ideal rectifier, pin diodes are designed to block large yet finite voltages. The maximum blocking voltage, in which little reverse-bias current flows, is known as the breakdown voltage. The breakdown voltage is an extremely important rating of pin diodes. It determines the range of voltages in which the device may safely be operated. Exceeding the breakdown voltage may result in destruction of the device if the reverse-bias current is not limited by the external circuit.

As usual, we describe the structure of the device, its (i, v) characteristics and its switching characteristics. We then describe breakdown, followed by 'conductivity modulation'. The focus of this and the next sections is in part on the reverse-bias simulation of pin diodes for the purpose of determining the breakdown voltage. This simulation is done in Section 11.4. in the context of an overall optimization of the device structure, which is done using statistical techniques of experimental design. (Since this simulation requires a custom Numerical Algorithm Module, it is not included on the CD-ROM.) The next subsection reviews the basic ideas underlying design of pin diodes.

In real devices, diffused and epitaxially grown junctions are finite in extent and must terminate. Several techniques known as terminations have been developed to terminate junctions. An ideal termination would produce a device with a breakdown voltage equal to the hypothetical infinite plane-parallel breakdown voltage and require no additional real estate on the wafer. Section 11.4 briefly lists the various methods of termination. In the last section the optimization of pin diodes with one particular type of termination, floating-field rings, will be discussed. The optimization techniques specific to the field rings, presented in Section 11.4, may be extended to other types of terminations.

11.3.1 Theory and Operation of PIN Diodes

This section briefly discusses the theory and operation of pin diodes and is divided into four subsections. Subsection 11.3.2 describes the basic structure of pin diodes and their typical (i, v) characteristics. The next section considers their switching characteristics. Subsection 11.3.4 discusses the avalanche-breakdown mechanisms and the geometrical factors which affect the breakdown voltage. Subsection 11.3.5 discusses the phenomenon of conductivity modulation.

11.3.2 Basic Structure and (i, v) Characteristics

The (i, v) characteristics of a typical pin diode are shown in Figure 11.4. In contrast to the ideal rectifier, pin diodes have both a small forward voltage drop when conducting and a small reverse-bias current when blocking. In addition, pin diodes are not capable of blocking an infinite voltage. When a voltage greater than the breakdown voltage (BV) is applied across the device, it will conduct a large current. If the current is not limited by the external circuit, the device will dissipate excessive power and will be destroyed. Unlike low-power diodes, the forward characteristics of pin diodes do not exhibit an exponential behavior. This behavior is masked by the Ohmic resistance of the pin diode. The drift region in forward bias has its resistance decreased by carrier injection, but it still is substantial enough to be

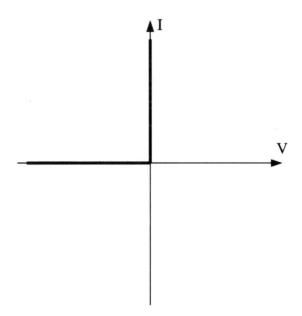

Figure 11.3. The current-voltage characteristic of an ideal rectifier.

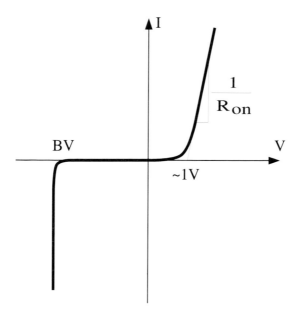

Figure 11.4. The current-voltage characteristic of a pin diode.

important. The resistance of the drift region appears in series with the pn junction. If the junction resembles an ideal diode when it is forward biased (in which case an ideal diode is often thought of as behaving like a short circuit), it is clear that the resistor in series with it determines the (i, v) curve in forward bias. The wafer and contacts also contribute resistance.

Since pin diodes are designed to conduct large on-state currents, it is important to make the on-state resistance as small as possible. This is achieved by heavily doping the diode's n-type substrate and p-type diffusion. However, since p^+n^+ diodes have small breakdown voltages due to tunneling, a lightly doped n^- region (commonly called an intrinsic or drift region) is epitaxially grown on top of the n^+ substrate, and the p^+ anode is diffused into the drift region, as shown in Figure 11.5. The purpose of the drift region is to absorb (to contain or form a buffer for) the depletion layer of the p^+n junction. Both the thickness and donor concentration of the drift region will determine the breakdown voltage of the device. The drift region supports a large reverse voltage V_{dr}, since the electric field $E \simeq V_{dr}/W$ can be kept below the critical value of the electric field for a larger V_{dr} if W, the width of the region where the voltage is dropped, is large. A heavily doped region would have a narrow depletion layer, as described in Section 7.1. The drift region is therefore lightly doped, to allow the depletion layer to be wide, and the drift region is made wide to allow room for a large depletion layer.

While the drift region supports substantial reverse bias voltages, it also in-

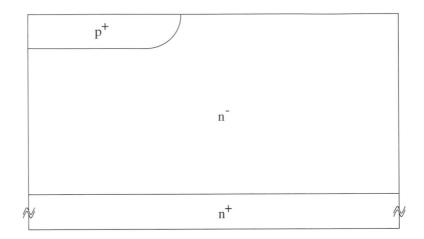

Figure 11.5. A cross-sectional view of a pin diode.

creases the resistance of the device, which would apparently lead to unacceptably large power dissipation when the device is conducting current. Fortunately, another mechanism, conductivity modulation, greatly reduces this apparent problem. Conductivity modulation was discussed in Section 7.2.4 and will be returned to in Subsection 11.3.5. Conductivity modulation is the increase in the conductivity due to injection of large numbers of carriers when the pn junction is forward biased.

The next section describes the switching characteristics of pin diodes. Power diodes may break down due to punch-through or avalanching. The following section will discuss the breakdown voltage which may be sustained before avalanche breakdown occurs.

11.3.3   Switching Characteristics

As for a low-power (digital) diode, we need to distinguish between turn-on and turn-off when discussing switching. To turn on, the device typically has to undergo a transition from reverse bias to forward bias. To switch to forward bias, the fixed charge near the reverse-biased junction has to be shielded. The stored charge in the diode is the exposed charge of the impurities inside the depletion layer. Shielding the charge means allowing majority carriers to reenter the region. The time this takes is likely to be limited mainly by the drift region, since the low charge density in the drift region will make the drift side of the pn junction wide. The time taken for charge to cross the drift side of the junction will be relatively long.

Once the junction becomes forward biased, then carriers begin to be injected into the drift region, one type across the junction and the other type pulled in from the other side to maintain charge neutrality. Holes from the (say) p^+ diffusion are injected across the junction into the drift region, where they diffuse towards the substrate. In addition, electrons from the n^+ substrate diffuse towards the p^+

diffusion. Since the drift region is resistive, as the forward current gets bigger, the voltage drop across the drift region also gets bigger, until conductivity modulation (i.e., the increased carrier density) decreases the resistance. The inductance of the wafer and its contact wires also creates a voltage $L\frac{di}{dt}$ which may be large if the current changes rapidly. These two components of the voltage, taken together, create a potentially large voltage overshoot during turn-on. Once the device is on, and the current through the device reaches a constant value, the inductive term goes to zero and the high conductivity brings the resistive voltage across the drift region back down, so the voltage drops again. (See the discussion of conductivity modulation, 7.2.4 and 11.3.5.)

The turn-off transient of a power (pin) diode in principle resembles the turn-off of a low-power diode. As long as there are excess carriers in the drift region, the junctions remain forward biased, so the junction voltages change only a little until the carrier densities drop enough to allow the junctions to become reverse biased. The depletion regions can then expand and a large reverse bias can build up. Before the device becomes reverse biased, the reverse current will cause some Ohmic voltage drop, which decreases the voltage slightly. The external circuit will attempt to maintain a large negative current, but once the carrier density becomes very small, the current drops to a very low value. The total time between the diode current going negative and returning to zero is called the reverse recovery time, t_{rr}.

The maximum reverse current flowing in this interval is called I_{rr}, and the total charge flowing in the time t_{rr} is called Q_{rr}, the reverse recovery charge. t_{rr} and Q_{rr} are often stated as functions of the rate of change of the reverse current, on the device specification.

11.3.4 Breakdown Voltage

When a pin diode is in its blocking state, almost all of the reverse bias voltage (V_r) is dropped across the lightly doped drift region. As V_r increases, the electric field in the drift region increases. As V_r approaches the breakdown voltage (BV), the peak electric field in the drift region (E_{max}) approaches the critical electric field (E_c) and substantial impact ionization begins, as illustrated in Figure 11.6. The value of V_r which causes E_{max} to approach E_c depends upon the doping profile of the p^+n junction.

Since most junctions in actual devices are formed via masked 'diffusions' of impurities, they will inevitably have some degree of curvature, as shown in Figure 11.7. The degree of curvature (specified by a radius of curvature) is dependent upon several factors, including the geometry of the diffusion mask and the duration and temperature of the diffusion drive. Unless the radius of curvature is much greater than the plane-parallel depletion width ($W_{c,pp}$), the depletion width of the actual device (with the curved junction) is spatially nonuniform. Furthermore, the electric field in the depletion region will have its largest magnitude where the radius of curvature is the smallest. In fact, the maximum electric field in the device with the curved pn junction (E_{max}) will be substantially larger than the electric field in a

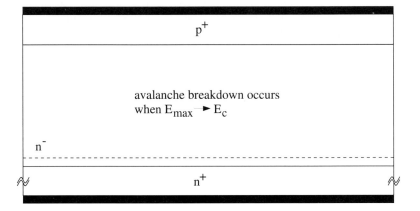

Figure 11.6. Breakdown in a plane-parallel pin diode.

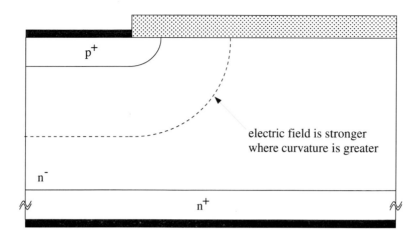

Figure 11.7. Breakdown in a pin diode with a diffused (hence curved) junction.

device with a (infinite) plane-parallel pn junction ($E_{max,pp}$) for the same value of reverse bias. Consequently, devices with curved pn junctions will have significantly lower breakdown voltages than the plane-parallel idealization.

11.3.5 Conductivity Modulation

A major portion of the power dissipated in a pin diode occurs when the device is forward biased and conducting large currents. As discussed above, in addition to the junction's forward voltage drop, a significant Ohmic drop occurs across the device's lightly doped drift region. Care must be exercised in estimating the power dissipation in the drift region, since the on-state resistance is significantly smaller than the apparent Ohmic value calculated on the basis of the geometric size and thermal-equilibrium carrier densities [127]. Large numbers of excess carriers (holes) are injected into the drift region from the heavily doped anode. At high injection levels where $\delta p > n_{no}$, the hole space charge is large enough to attract electrons across the n^+n^- interface. To maintain local charge neutrality, the excess electron and hole concentrations in the drift region are approximately equal. If the ambipolar diffusion length of the excess carriers is greater than the drift-region length W_d, the spatial distribution of excess carriers will be relatively flat, as in Figure 11.8. Furthermore, δn and δp are much greater than the thermally generated carrier concentrations, hence the on-state, high-level injection conductivity of the drift region is significantly greater than the equilibrium or low-level injection conductivity. This phenomenon is referred to as conductivity modulation. The presence of the drift layer also causes turn-on transients, as discussed in the literature [127, 128].

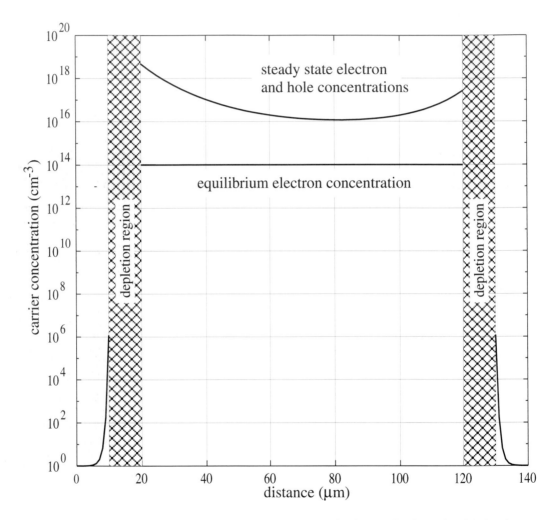

Figure 11.8. Conductivity modulation of a pin diode. Under high-level injection, the steady-state electron and hole concentrations are several orders of magnitude larger than their equilibrium concentrations. This, in turn, increases the on-state conductivity of the device.

11.4   Optimization of a Semiconductor Device

In this section we present a detailed investigation of how a semiconductor device may be optimized. The focus is on a particular power semiconductor device, a power diode. The special features of the optimization are, first, that it is done numerically using the same approaches used throughout the text, and second, that it employs statistical design techniques. These methods are of very general applicability and extend to many areas of semiconductor design, and beyond. The discussion is included at this point because it emphasizes the power diode as an example.

In order to increase the voltage-blocking capabilities of power devices, it is important to reduce the peak electric fields at the semiconductor surface [121]. Several termination techniques have been developed to this end: field-limiting rings (FLR) [125, 129, 130], field plates (FP) [131, 132, 133, 134, 135, 136, 137], the combination of both field-limiting rings and field plates (OFP-FLR) [138], lateral charge-controlled devices (RESURF) [139, 140] and junction-termination extension (JTE) [141, 142, 143]. Of these techniques, RESURF is attractive only for lateral devices and JTE is complicated [122]. Comparisons of the junction-termination extension techniques may also be found in the literature [138, 144, 145, 146, 147].

In this section, pin diodes with multiple FLRs will be optimized. While the details of the optimization techniques developed in this section are chosen so as to be applicable to FLR devices, they are not limited to FLR. Exactly the same general approach applies to any semiconductor design; and the particular details of the methods may be used to optimize devices with other terminations.

11.4.1   Optimization of PIN Diodes with Field-Limiting Rings

One of the most critical problems in the design of high-voltage devices is the optimization of the junction termination. The device termination has a serious impact on the device reliability as well as the size of the device [147]. Junction-termination techniques are discussed above. Previous research has noted the effects of variability of the process parameters upon the breakdown voltage of pin diodes with optimally placed field-limiting rings (FLRs) [130]. However, prior to [106] no systematic method of optimizing devices with multiple FLRs had been proposed. The focus of this section is the optimization of pin diodes with multiple FLRs. Two optimization techniques will be described and used to optimize the spacing between FLRs. Although these techniques are demonstrated with FLR terminations, they may also be applied to other types of junction terminations. FLR terminations have been chosen, since the fabrication of FLRs requires no additional processing steps; hence the cost of manufacturing pin diodes with FLR terminations is directly proportional to the size of the device. Thus, the optimal device is the one which satisfies the specified performance criteria (the specified breakdown voltage) using the least amount of wafer area. Other types of terminations, such as field plates and junction-termination extension, require additional processing steps which increase the fabrication costs. Since 'cost' data are not published in the literature, direct comparison between devices with different types of terminations is not possible; hence, we restricted ourselves to FLR terminations.

This section has seven more subsections. Section 11.4.1 contains the problem description. This section describes the structure (the pin diode with multiple FLRs) which is to be optimized as well as the constraints on the optimization. The next sections describe two methods of optimizing the FLR termination. Section 11.4.2 first discusses an iterative method. The next part of Section 11.4.2 describes a factorial design method; the following subsection compares the two methods of optimization, and the next describes the results of the optimization.

Problem Description

Planar-diffusion technology is commonly used to fabricate power devices. When a planar diffusion is formed using a rectangular diffusion window, a 'cylindrical' junction is formed at the straight edge of the diffusion window and a 'spherical' junction is formed at each of the corners of the diffusion window. The interior portion of the junction has the electric-field profile of a plane-parallel junction. However, the electric-field lines become crowded at the cylindrical and spherical junctions, resulting in a local increase in the electric field and a decrease of the breakdown voltage (see Figure 11.7). In many devices, such as DMOSFETs and IGBTs, the junction depth is only three to five microns. As the junction depth is decreased, the electric-field crowding becomes greater, resulting in a further decrease in the breakdown voltage.

One approach to reducing the electric-field crowding at the edges of a diffused junction is the strategic placement of FLRs. Field limiting rings are also referred to as floating-field rings, since the potentials of the rings are not fixed by direct electrical contact. Instead, the potentials of the FLRs are indirectly determined by the bias applied to the main junction. If the FLRs are placed within the depletion region of the main junction, their potentials will lie at intermediate values between the anode and cathode potentials. Since some of the electric-field lines will terminate upon the FLRs, the electric-field crowding at the main junction may be alleviated increasing the breakdown voltage of the main junction. This also has the effect of reducing the curvature of the depletion region, as shown in Figure 11.9.

One advantage of FLRs is that they are easy to fabricate and do not require any additional processing steps, since they can be fabricated at the same time as the main junction. The disadvantage of FLRs is that they require additional 'real estate', thus they increase the size of the device without enhancing its current handling capabilities. Furthermore, the effectiveness of FLRs is strongly dependent upon the placement of the FLRs as well as process-dependent factors such as the surface-charge density. Therefore, in order to maximize the breakdown voltage of a device, it is critical to judiciously choose the optimal placement of the FLRs. One of the most cost-effective tools in the design of devices with FLR terminations is computer simulation.

The next several subsections provide the necessary background for the opti-

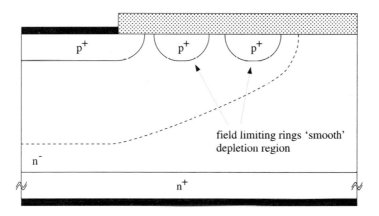

Figure 11.9. The field-limiting rings reduce the electric-field crowding at the main junction and 'smooth' the depletion region. This, in turn, increases the breakdown voltage of the main junction.

mization. The first subsection describes the geometry of the test device which will
be used in this section. In addition, the nomenclature and the parameters which
may be varied in order to optimize the breakdown voltage are clearly described.
Subsection 11.4.2 discusses the method of optimization; specifically, the simulation
method and the use of statistical design techniques are described. This subsection
then states the optimization objectives and constraints, and describes the design
process.

Geometry of Test Device

This subsection describes the geometry of the class of pin diodes whose breakdown
voltage will be maximized by optimal placement of FLRs. A schematic of the diode
is shown in Figure 11.10. The drift region is doped n type with a concentration of
10^{14} donors per cubic centimeter. The thickness of the drift region is 106 microns.
Since the electric field quickly vanishes in the heavily doped substrate (with 10^{19}
donors per cubic centimeter), only 10 microns of the substrate are included in the
simulation domain. The remaining portion of the substrate would merely add series
resistance to the device, which is not important in reverse-bias simulations. The
main junction, as well as the junctions formed by the FLRs, are assumed to have a
Gaussian doping profile. Their diffusion depth is 7.5 microns with a lateral diffusion
of 5.0 microns. The surface concentration of the diffusions is 10^{19} acceptors per cubic
centimeter. The radius of the main junction is 100 microns and the width of the
FLRs is 6 microns. The surface of the semiconductor is passivated with 5 microns
of silicon dioxide and 195 microns of silicon nitride. The parameters listed above
are consistent with the data listed in the current literature for pin diodes.

'No-flux' or 'reflective' boundary conditions for the normal electric field are used
in the simulation on the top, left and right edges, and the electrostatic potential on
the bottom edge is set to the cathode potential. The extent of the silicon nitride
passivation and the width of the device are chosen so that the artificial no-flux
boundary conditions on the top and right edge do not greatly perturb the solution
obtained for the physical system. The left no-flux boundary condition is exact, since
the device is symmetric about the left edge.

In order to optimize the design of a pin diode with FLRs, the spacing between
the successive FLRs has to be judiciously chosen such that the breakdown voltage
is maximized. The space between the main junction and the first FLR is W_{s1}, the
space between the first FLR and the second FLR is W_{s2} and so on. In general,
W_{si} is the space between the $(i-1)$th and the ith FLRs, where the zeroth FLR is
considered the main junction.

11.4.2   Method of Optimization

In order to calculate the breakdown voltage of a pin diode with FLRs, the linear
Poisson model with depletion-region logic (LP-DRL) is used, as described in Section 11.2.5. The LP-DRL solution algorithm clamps the voltage of neutral regions
near the n$^+$ cathode to the cathode potential and the neutral regions near the p$^+$

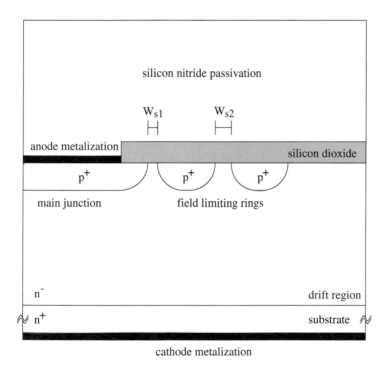

silicon nitride passivation

W_{s1} W_{s2}

anode metalization

silicon dioxide

p^+ p^+ p^+

main junction   field limiting rings

n^- drift region

n^+ substrate

cathode metalization

Figure 11.10. A schematic diagram of the 'test' pin diode used in the optimization simulation. The spacing between the main junction and the first field-limiting ring is W_{s1} and the spacing between the first and second field-limiting rings is W_{s2}.

anode to the anode potential. To accommodate the FLRs, however, the basic algorithm is generalized, since the voltages of the neutral portions of the FLRs are not known. One of the proper boundary conditions for the FLRs is that the net current into each FLR is zero. However, since the current densities are not determined by the LP-DRL model, a different means of calculating the potential of the FLRs is necessary. The approach used is that of Adler et al [125]. In order to satisfy the zero-net-current condition, the FLRs' junctions cannot be completely reverse biased or forward biased. The potential of each of the FLRs is equal to the lowest voltage (V_{low}) along its metallurgical ring-substrate junction minus a built-in potential which is approximately 0.85 volts for silicon devices.

The modified LP-DRL algorithm is as follows:

1. At the beginning of each sweep over the mesh, calculate the lowest voltage $V_{low}(i)$ along the metallurgical ring-substrate junction for each FLR i.

2. Next loop through each node j and compute the value of the electrostatic potential of this node $V_{new}(j)$ by solving the discretized Poisson equation at that node, using the current value of the electrostatic potential at node j's neighboring nodes.

    (a) If node j is part of the ith FLR and $V_{new}(j)$ is less than $V_{low}(i)$ - 0.85 then clamp $V_{new}(j)$ to the value $V_{low}(i) - 0.85$.

    (b) If node j is not part of a FLR then $V_{new}(j)$ should be bounded above and below by the anode and cathode potentials.

3. At the end of the sweep, the maximum change in the electrostatic potential ΔV_{max} is compared to a predetermined change. If ΔV_{max} is not sufficiently small, then another sweep is performed and the process described above is repeated.

Once the electrostatic potential is computed to sufficient accuracy, the hole ionization integrals (Equation (11.5)) are evaluated along the critical paths (electric-field lines which yield the maximum ionization integrals starting by each FLR including the main junction). Breakdown occurs when the hole ionization integral along one of the critical paths is unity. It has been empirically observed that the value of the ionization integral varies exponentially as a function of the reverse bias (MCD = $A \exp(B)$). Since the ionization integrals are actually the integrals of the multiplication coefficient's denominator, the value of the ionization integral has been abbreviated by the acronym 'MCD'. The multiplication coefficients and the ionization integrals are discussed in Subsection 11.2.4. The exponential dependence of the MCD may be exploited to construct a method which quickly converges on the breakdown voltage. Given two initial guesses for the breakdown voltage V_1 and V_2 and the corresponding largest values of the ionization integrals MCD_1 and MCD_2, the next estimate of the breakdown voltage V_3 may be calculated using Equation (11.6). The iteration described by Equation (11.6) is terminated when the largest MCD is within 5 percent of unity.

$$V_{i+1} = (V_i - V_{i-1}) \frac{MCD_i^{V_{i-1}/(V_i - V_{i-1})}/MCD_{i-1}^{V_i/(V_i - V_{i-1})}}{\ln(MCD_i/MCD_{i-1})} \qquad (11.6)$$

In order to accurately calculate the electrostatic potential and hence the break-down voltage, a suitable solution mesh is generated. The solution mesh contains regions of fine and coarse resolution, as shown in Figure 11.11. The mesh is very fine in the vicinities of the peaks in the electric field. Fine resolution is needed in these areas to accurately calculate the values of the hole ionization integrals along the critical paths, as discussed in Subsection 11.2.4. Furthermore, sufficient resolution is needed around the main junction and the FLRs in order to accurately determine the low-voltage points along the metallurgical ring-substrate junctions.

Figure 11.12 shows the graphical output of a typical simulation. In this figure, the silicon, silicon-dioxide and metal regions are shown. In addition, the metallur-gical junctions of the main junction and of the FLRs are shown. The lines which wrap around the junctions are equipotentials and the lines originating from each junction are the critical paths. The electrostatic potentials of the main junction and of the FLRs are labeled. In addition, several data files are generated in the simulation which log the potential of the main junction and FLRs, the peak electric field around each junction, the coordinates of the critical paths and the values of the hole ionization integral evaluated along each critical path.

Optimization Objective

pin diodes are rated to operate at a certain voltage. Thus the diode must be able to block voltages which are equal to the rated voltage plus a safety margin. In addition, the area of the device should be kept to a minimum in order to increase manufacturing yields. Since the area of the main junction is determined by the power rating of the diode, the only manner in which one can minimize the total area of the device is by minimizing the area of the junction termination (the area required by the FLRs).

If the number of FLRs is fixed at n, then the breakdown voltage may be maxi-mized by optimally choosing the spacing between adjacent FLRs. Any other spacing will lead to a decrease in the breakdown voltage. The maximum voltage achievable increases with n, provided the radius is not limited. Now suppose the area of the device is constrained to the area required by a device with n optimally placed FLRs. The question arises, could the breakdown voltage be further increased by adding additional FLRs within the same area? Empirical tests indicate that the breakdown voltage decreases upon addition of more than the n FLRs within the area required by the optimal design for n FLRs.

The above observation dictates the manner in which to design the smallest pin diode with a given voltage rating. For instance, suppose one wishes to design a pin diode rated at 900 V with a 30 percent safety margin. The optimal device has n optimally spaced FLRs and a breakdown voltage of 1170 volts or greater with the

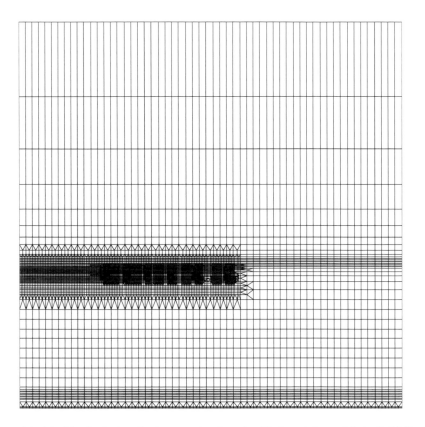

Figure 11.11. Typical simulation mesh of a pin diode with multiple field-limiting rings. The dimensions of the mesh are 445 microns wide and 317 microns high. The top 195 microns of the simulation domain is the silicon nitride passivation. The mesh is very fine in the vicinities of the field-limiting rings and the main junction.

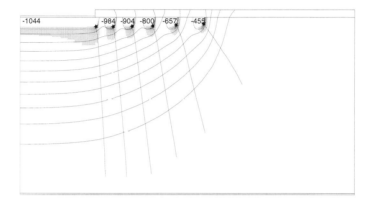

Figure 11.12. Graphical output of a typical simulation. The dimensions of the plot are 445 microns wide and 122 microns high. The plot does not show the silicon nitride passivation. The voltages of the neutral portion of the main junction and the field-limiting rings are labeled. The electrostatic potential contours, the critical paths and the locations of the peak electric field by each junction are shown. Grey areas denote regions where impact ionization occurs, with darker grey corresponding to higher levels of ionization.

smallest n. Therefore, the problem of optimizing a device is reduced to finding the optimal spacing of a given number of FLRs, and this is the topic of the next two sections.

We now turn to the method by which the optimization is performed.

Iterative Method

This subsection discusses an iterative method by which to optimize the spacings between a given number of FLRs in order to maximize the breakdown voltage. An alternative method is presented in the next section and a comparison of the methods in the following section.

Consider a pin diode with a single FLR. Although the FLR can be intuitively expected to increase the breakdown voltage of the main junction, its effectiveness depends upon its exact spacing from the main junction [121]. If the FLR is too far away from the main junction, the depletion region of the main junction will not be altered and the breakdown voltage of the main junction with the FLR will be the same as that of the junction without the FLR. On the other hand, if the FLR is too close to the main junction, the FLR's potential will be essentially the same as the main junction's potential. In this case, the electric-field crowding will not be alleviated. The crowding is merely shifted away from the main junction to the FLRs junction, hence the breakdown voltage of the pin diode and FLR will be roughly the same as the junction without the FLR.

The above observations can be stated in terms of the values of the hole ionization integrals (MCDs) evaluated along the critical paths in the vicinity of the main junction and FLR. If the FLR is far away from the main junction, the MCD of the main junction will be unity and the FLR MCD will be approximately zero at breakdown. If the FLR is too close to the main junction, the MCD of the main junction will approximately be zero and that of the FLR will be unity at breakdown. If the FLR is optimally placed such that the breakdown voltage of the pin and FLR in combination is maximized, then the MCD of the main junction and that of the FLR will both be unity at breakdown [121]. The above statement may be generalized to a pin with multiple FLRs. The values of the ionization integrals evaluated along each junction's critical path are simultaneously equal to unity in an optimally designed device [121].

Therefore, the differences between the MCDs of adjacent rings may be used as a guide to determine the optimal spacing between the FLRs. In the following discussion, the main junction will be considered to be the zeroth ring and the FLRs will be numbered consecutively starting with one. Hence MCD(0) is the value of the ionization integral evaluated along the critical path in the vicinity of the main junction, while MCD(i) is the value of the ionization integral evaluated along the critical path in the vicinity of the ith FLR. The difference between the ith and the $(i + 1)$th MCDs is defined as $\Delta\text{MCD}(i) = \text{MCD}(i + 1) - \text{MCD}(i)$. If $\Delta\text{MCD}(i)$ is positive, then the ith and the $(i + 1)$th FLRs are too close and the spacing between the FLRs should be increased. On the other hand, if $\Delta\text{MCD}(i)$ is negative then

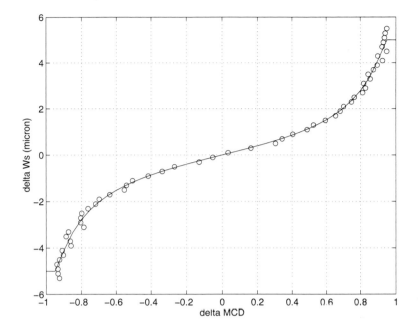

Figure 11.13. A plot of the spacing update function. The solid line is the fitted tangent function and the circles are the simulation data points.

the FLRs are too far apart and the distance between them should be decreased.

In order to determine the magnitude of the spacing adjustment, a pin diode with a single FLR was simulated first. The distance between the main junction and the FLR was varied and the resulting difference between the MCD of the main junction and that of the FLR was recorded (see Figure 11.13). The data was then fitted to a tangent function. The choice of a tangent was made on the basis of an examination of the data when plotted, as shown in Figure 11.13. The resulting expression is given by Equation (11.7). The ordinate (ΔMCD) has been translated so that the fitting function passes through the origin. ΔW_s given here is a predicted spacing-change designed to equalize the MCDs. The maximum allowable spacing update is set to 5 microns in order to avoid overshoot problems.

$$\Delta W_s = 1.38919 \tan\left(1.38412 \Delta \text{MCD}\right). \tag{11.7}$$

In summary the algorithm of the iterative method is as follows.

1. Choose an initial guess for the spacing between the FLRs, simulate the pin diode and compute the MCD corresponding to each junction (the main junction and the FLRs' junctions).

2. Compute the differences between the MCDs of successive junctions.

i	1	2	3	4	5	6
BV	990	1020	1032	1037	1035	1044
W_{s1}	6.0	6.3	6.7	6.9	7.0	7.0
W_{s2}	7.9	8.2	8.3	8.4	8.5	8.6
W_{s3}	10.5	10.5	10.6	10.6	10.8	10.5
W_{s4}	13.9	13.6	13.6	13.8	13.6	13.8
W_{s5}	18.0	19.1	19.4	19.3	19.3	19.6
MCD_0	0.34	0.53	0.72	0.81	0.84	0.94
MCD_1	0.50	0.76	0.82	0.86	0.82	0.97
MCD_2	0.65	0.83	0.90	0.90	0.89	0.86
MCD_3	0.67	0.86	0.87	1.00	0.73	0.92
MCD_4	0.49	0.84	0.98	0.90	0.82	0.88
MCD_5	0.99	1.02	0.92	0.92	0.96	0.96

Table 11.1. The convergence of the iterative optimization method for a pin diode with five field-limiting rings.

3. Adjust the spacing of the FLRs to obtain ΔMCD $= 0$ according to Equation (11.7).

4. Repeat this process until the optimal spacing is achieved.

   The above algorithm does not consider the interaction between nonadjacent FLRs. Consequently, the algorithm does not always converge to the optimal spacing between field rings. However, it does provide a spacing which is very close to optimal (the breakdown voltage is within a few percent of the optimal breakdown voltage). Furthermore, since it is not possible to fabricate devices with the desired spacings between FLRs achieved to arbitrary accuracy, the spacing between the field rings is rounded to the nearest tenth of a micron. This discrete spacing also introduces a small error. It is possible to slightly improve upon the design provided by this algorithm. However, the effort involved in finding the true optimal spacing is usually not warranted, since the simulation error due to the uncertainty in the process parameters (such as the surface charge) will overshadow the error of this algorithm.

   Table 11.1 shows several iterations of this method for a pin diode with five FLRs. The initial gap spacing was chosen by using the formula $W_{si} = 6x_i^2 + 6x_i + 6$ where $x_i = (i - 1) / (n - 1)$, where n equals the number of rings, which is five in this case. Since the initial guess was very good, the algorithm converged to a nearly optimum solution in six iterations. The appropriateness of the initial gap formula will be discussed in the next section.

Factorial Design Method

The previous section discussed an iterative method to design a near-optimal pin with multiple FLRs. The iterative method adjusted the spacing between consecutive rings based on the differences between MCDs. This section describes an alternative method of optimization based upon the method of steepest ascent [4]. The direction of steepest ascent is determined by a series of simulations, chosen using a two-level factorial design (FD).

An obvious choice for the factors of the FD are the spacings between the FLRs. However, as the number of rings becomes large, the number of simulations required to implement an FD (or even a partial FD) becomes prohibitive. In order to limit the number of factors to a reasonable number, an alternative approach is used. Since it has been observed that the spacing between the FLRs of an optimally designed pin diode monotonically increases as one moves away from the main junction, the spacing between the FLRs may be modeled as a well-behaved function $S(x)$ where x is proportional to the ring number minus one – see below. The spacing function may, in turn, be represented as the Taylor series $S(x) = s_0 + s_1 x + s_2 x^2 + \cdots$ where $s_i = (1/i!) \left(d^i S/dx^i \right)_{x=0}$. The spacing function may be approximated as an nth order polynomial by truncating the series. Then the coefficients of the series may be used as the factors of the FD.

Empirical tests have shown that a second-order polynomial is adequate for pin diodes with up to at least ten FLRs. The independent variable of the spacing function is defined by $x = (i-1)/(n-1)$ where i is the ring number ($i = 1$ for the first FLR, $i = 2$ for the second FLR, etc). and n is the number of FLRs. Since the spacing function is approximated by a second-order polynomial, this method will not provide the exact spacing of an optimally designed pin diode. The method does, however, provide a solution which is reasonably close to the optimum. Another advantage of this method is the ability to easily choose a good initial guess for the coefficients. As the number of rings increases, the distance between the main junction and the first FLR (W_{s1}) decreases. Furthermore, the distance between the last and the second-to-last rings (W_{sn}) is almost the same regardless of the number of rings provided that number is large. These two observations allow one to choose good initial guesses for the coefficients, since $W_{s1} = s_0$ and $W_{sn} = s_0 + s_1 + s_2$. Therefore, the coefficients which best fit the spacing function of a pin with three FLRs will be a reasonably good initial guess for a pin with ten FLRs.

The methodology of this approach is illustrated in Figure 11.14. First, an initial guess for the factors s_0, s_1 and s_2 is chosen which serves as the 'center point' of the first factorial design and as a simulation run. The factors of the remaining runs are chosen to be the vertices of a 'cube' which is generated about the center point by displacing the factors by the quantities δs_0, δs_1 and δs_2, respectively. The output variable of the simulations is the breakdown voltage of the pin diode's main junction. The direction of the steepest ascent is determined by approximating the gradient with first-order central differences about the center point. Once the direction of steepest ascent is determined, simulations are conducted along that

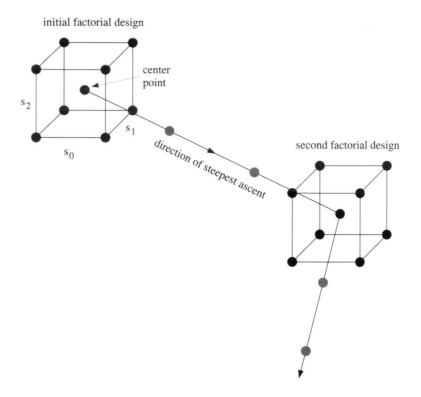

Figure 11.14. A schematic diagram of the method of steepest ascent using a factorial design to determine the gradient of steepest ascent. The cubes represent the factorial designs while the circles represent the simulation runs.

direction until the breakdown voltage of the main junction ceases to increase. At this point, another factorial design may be conducted about a new center point, or the iterative method may be used to further improve the design.

   Table 11.2 lists the results of a FD. The first entry in the table is the center point, which in this case is $s_2 = 6$, $s_1 = 7$ and $s_0 = 6$. The remaining entries in the table are the vertices of the 'design cube'. The displacement of each factor was unity, and the units of the factors are in microns. The 'H' character in the table represents a high value and the 'L' character represents a low value. Hence, the entry 'L', 'H', 'L' represents the run with $s_2 = 6 - 1 = 5$, $s_1 = 7 + 1 = 8$ and $s_0 = 7 - 1 = 6$. The direction of the steepest ascent is $-0.04508\hat{s}_2 - 0.2200\hat{s}_1 - 0.9745\hat{s}_0$. This concludes the section on the factorial design method. Results of the factorial design method are deferred to the sections which follow.

s_2	s_1	s_0	BV
-	-	-	1049.2
L	L	L	1062.8
L	L	H	1013.2
L	H	L	1059.9
L	H	H	1006.4
H	L	L	1077.1
H	L	H	1008.5
H	H	L	1045.0
H	H	H	1001.7

Table 11.2. The results of a designed simulation with six field limiting rings.

Comparison of the Optimization Methods

In the previous two sections, two methods of designing the optimal pin diode with FLRs were discussed. The first method, the iterative method, adjusted the spacings between adjacent rings based on the differences between MCDs. The second method, the factorial design method, utilized a factorial design and the method of steepest ascent to determine the coefficients of a polynomial approximation to an optimal spacing function. While the first method provides a 'better' design than the second method, it converges rather slowly if the initial guess is poor. If a 'good' initial guess is not known, the best method of design is a combination of the iterative and factorial design method. The factorial design method may be used to provide the iterative method with a good initial guess.

Results

This subsection discusses the results of the optimization described earlier in this section. The goal was to design a device which could block 1170 volts using the fewest FLRs. To this end, the device described in Subsection 11.4.1 was optimized using from one to ten FLRs. Figure 11.15 shows the breakdown voltage of an optimally designed device as a function of the number of FLRs. Nine rings were required to achieve a breakdown voltage of 1170 volts or higher. As seen in Figure 11.15, the breakdown voltage as a function of the number of FLRs tends to saturate. In order to further increase the breakdown voltage significantly, one must increase the junction diffusion depth. This, however, increases the processing time. Figure 11.16 shows a schematic of the optimally designed pin diode and the contours of electrostatic potential.

   This section addressed the design of a complex power semiconductor device. Two different design strategies were developed, the optimal strategy in general being a combination of the two. An optimal FLR termination was found for a maximum breakdown voltage of 1170 volts. The optimal device was found to have nine FLRs.

11.5   Power MOSFETs

A power MOSFET like any power device needs to be able to carry a large current when on, and block large voltages when off. Such devices are now available which are sometimes known as *vertically diffused MOS* (VDMOS) devices, since they are constructed using four layers arranged vertically. A schematic of a high-power MOSFET is shown in Figure 11.17.

   The drain contact is at the back (or bottom.) Both the gate and the source are at the top. Starting at the bottom, above the drain contact is the drain region. The drain region is heavily doped. For definiteness we shall take the drain region to be n^+, where the n indicates it is n-type and the + implies that it is heavily doped. Next comes an n^- (lightly doped n-type) drain drift region. Like the pin diode, the power MOSFET relies on the drift region to block reverse voltages. Above parts of the drift region is a p-type body and above parts of the p-type body are n^+ source

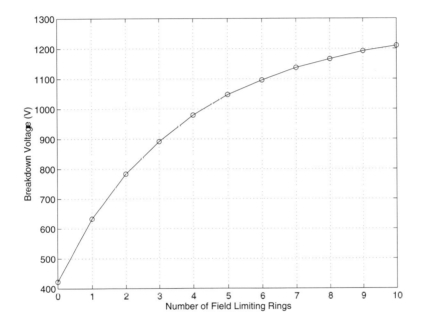

Figure 11.15. A plot of the breakdown voltage as a function of the number of field-limiting rings.

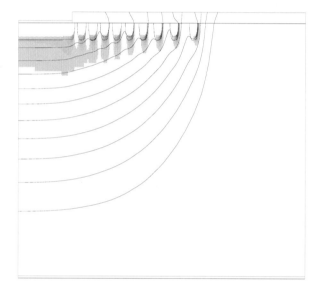

Figure 11.16. A schematic diagram of an optimally designed pin diode with nine field-limiting rings. The gap spacings in microns are $W_{s1} = 4.5$, $W_{s2} = 4.7$, $W_{s3} = 5.5$, $W_{s4} = 6.3$, $W_{s5} = 7.5$, $W_{s6} = 8.8$, $W_{s7} = 10.9$, $W_{s8} = 13.9$ and $W_{s9} = 19.6$.

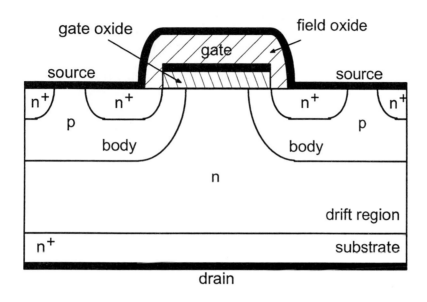

Figure 11.17. A schematic of a high-power metal oxide semiconducting field-effect transistor.

regions. The source contact covers the n^+ source region, but it also covers parts of the p-body, as will be explained shortly.

All of the regions except the drain region reach the top of the device in places and at least partially overlap the gate oxide. This allows the channel which is formed under the gate to make a connection which bridges the source, body and drift regions.

If the gate is positive relative to the source contact, then an n-type channel will be set up under the gate oxide. The channel connects the n^+ source region to the body and to the drift region. This allows current from the drain which can get into the drift region to flow into the channel and on through the body to the source.

Because the device contains an npn structure (treating the drain and the drift regions as a single n-type region, temporarily), there is a parasitic transistor in the structure that, if possible, must be prevented from turning on. As mentioned above, the body and source regions both touch the source contact. This shorting of body and source is intended to prevent the transistor from turning on. Since the source contact is connected to the p-body, there is a path from source contact to drain contact through a pn junction which forms a parasitic diode.

11.5.1 (i, v) **Characteristics**

The (i, v) characteristics of the power MOSFET are similar to those of low-power MOSFETs with modifications, which are, in principle, minor. To be more specific, the usual relationship between drain current I_D and the gate voltage (relative to the source voltage) V_G is obeyed at low values of I_D. At high I_D the electric field in the channel may reach a level at which velocity saturation occurs (as was discussed for short-channel MOSFETs in integrated circuits). In addition at high carrier densities, which can be found in a power device, the scattering between the carriers can decrease the mobility. The overall effect is that the relationship between I_D and V_G becomes linear instead of quadratic at high currents. As in the power diode, the presence of the drift region adds resistance which the forward current must pass through.

Breakdown is typically by avalanching, which is most likely to occur at the high curvature points where the body meets the drift region. The gate metallization extends over the drift region, which is perhaps not necessary for its main purpose of forming the channel. This allows it to act like the field rings in the pin diode described above and extend the depletion region sideways to minimize its curvature and to postpone breakdown. The overlap of the gate with the drift region also extends the channel into the drift region so as to increase the conductivity of the drift region in the on-state.

11.5.2 **Switching Characteristics**

The switching behavior of power MOSFETs may be understood in terms of equivalent-circuit models. The models emphasize the use of the MOSFET as a switch with the gate controlling the power supplied through the drain-source connection.

Capacitors are included between gate and drain, C_{gd}, and between gate and source, C_{gs}, to allow for the depletion layers (plus any parasitic capacitances). There are no excess minority carriers to be removed during switching, as opposed to in a BJT. Since the excess minority carriers limit the switching speed of a BJT, a MOSFET can switch faster.

The Ohmic model is used when $V_D < V_G - V_T$, in which case a resistor r_{ds} is included between drain and source. The current i_d from drain to source is then $i_d = V_D/r_{ds}$. Otherwise i_d depends on V_G, and in a power MOSFET the dependence may be more nearly linear than the usual (low-power) quadratic dependence. The voltages are referenced to the source. If $V_G - V_T > 0$ an n-channel is formed at the source. If $V_G - V_T > V_D$, the channel is formed at the drain (and in between) as well, since $(V_G - V_D) - V_T > 0$. In power switching applications $V_G \gg V_T$, so $V_G > V_D$ is the condition for there to be a channel at the drain. (If the gate is positive relative to the drain, it attracts electrons to the region beneath the drain end of the oxide layer and forms a channel there.) If there is a channel all the way from source to drain it behaves as a resistor, although most of the resistance of the device comes from the drift region which is in series with the channel. There is no channel at all at the source when $V_G < V_T$ (cutoff) so $I_D = 0$. I_D is quadratic in $V_G - V_T$ when there is a channel which is pinched-off at the drain end of the channel (active mode).

The behavior of the device during switching should be examined using these circuit models within the context of particular external circuits.

Power MOSFET with Resistive Load

The first example which will be given will have a resistive load. We will then examine the difficult case of a load including a diode and an inductor. The circuit modeled in this file is like the standard MOSFET invertor circuit, with a load resistor R_L between the drain and the power supply, as set up in the file pmos1.sg. In addition, in this example there is a resistor R_G at the gate. Since this is a time-dependent problem, the MOSFET's internal capacitances are included. In addition to the capacitances due to the gate oxide, C_{gd} and C_{gs}, there is a capacitor C_{ds} between drain and source.

V_G is found from the equation for i_g, which is composed of the currents through the two capacitors C_{gd} and C_{gs}. i_g is found from the voltage across the resistor R_G at the gate, which is $V - V_G$, so $i_g = (V - V_G)/R_G$. The drain voltage V_D is found from the power-supply voltage V_{dr} minus the voltage dropped across the load resistor, $R_L i_d$. The most complex equation in the file is that for the drain current i_d. This consists of the currents through the two capacitors attached to the drain, C_{gd} and C_{ds}, and of a part which only flows when $V_G > V_T$. This equivalent circuit is similar to the MOSFET model used by Ramshaw [148].

```
// FILE:  sgdir\ch11\pmos1\pmos1.sg
// This file contains a demonstration of how to
// model a MOSFET with a resistive load.
```

```
// Circuit Parameters
const TMAX = 4000;          // Max. number of time steps
const dt = 1.0e-13;         // Time Step
const Vdr = 150.0;
const VGM = 10.0;
const VTH = 1.8;
const RDS = 0.1;
const G = 1.0 ;
const CGS = 2.5e-9;         // Capacitance between Gate and Source
const CGD = 1.0e-9;         // Capacitance between Gate and Drain
const CDS = 1.5e-9;         // Capacitance between Drain and Source
const RG = 0.01;
const RL = 10.0;
const L = 0.0;

var t[TMAX];
var ID[TMAX];
var VD[TMAX];
var IG[TMAX];
var VG[TMAX];
var VL[TMAX];
var IL[TMAX];
var IDi[TMAX];
var V[TMAX];

set NEWTON ACCURACY  = 4e-2;


equ VG[i=1..TMAX-1]->
  IG[i] = CGS*(VG[i]-VG[i-1])/dt + CGD*( (VG[i]-VD[i])-(VG[i-1]-VD[i-1]))/dt ;

equ IG[i=all]->
  VG[i] = V[i] - RG*IG[i] ;

equ VD[i=all]->
  VD[i] = Vdr  - RL*ID[i] ;

equ ID[i=1..TMAX-1]->
  ID[i] =  0.0    +
    (VG[i] > VTH)* {G*(VG[i]-VTH) +  VD[i]/RDS} +
    CGD*((VD[i]-VG[i])-(VD[i-1]-VG[i-1]))/dt   +
    CDS*(VD[i]-VD[i-1])/dt ;


begin main
  assign VG[i=0] = 0.0 ;
  assign ID[i=0] = 0.0 ;
  assign t[i=all] = i*dt ;
```

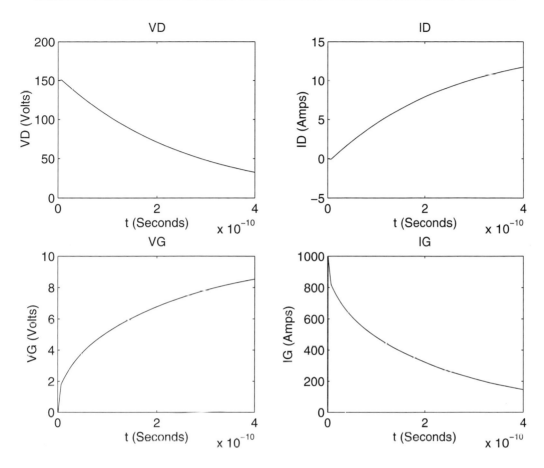

Figure 11.18. The performance of a MOSFET with a resistive load.

```
    assign V[i=10..TMAX-1] = VGM ;
    solve;
    write;
end
```

Power MOSFET with Inductive Load; 'Long-Term' Behavior

We will examine the MOSFET with an inductive load in two parts. First its long term behavior will be considered, where the internal MOSFET capacitances can be neglected and the focus is on the inductor. Then we shall assume the inductor has a constant current in it to look at the short-term behavior.

In these files most of the same equations are used as before, although the dependent variable is different in some cases. In this circuit there is an inductor L in series with a resistor R_L, and this combination is in parallel with a diode. This combination is attached to the drain of the MOSFET. An equation for the diode current, i_{Di}, is added, which depends on the diode voltage V_{Di}. The inductor current i_L is the diode current plus the drain current. The inductor voltage V_L is given by L times the rate of change of i_L. The diode voltage V_{Di} is (minus) $V_L + i_L R_L$.

```
// FILE:  sgdir\ch11\pmosdilt\pmosdilt.sg
// This file contains a demonstration of how to
// model a MOSFET switching a load with a diode and inductor.
// This file looks at long time scales on which the 30-kHz
// switching can interact with the inductor.
// The capacitor time scales are extremely short compared to
// the times involved here so the capacitors are treated as open circuits.

// Circuit Parameters
const TMAX = 200;              // Max. number of time steps
const TH= TMAX/2;              // Half of Max. Time
const TINT = TMAX/20;          // Switching Interval
const dt = 3.0e-4/TMAX;        // Time Step
const Vdr  = 150.0;
const VGM = 20.0;
const VTH = 1.8;
const RDS = 0.1;
const G = 1.0 ;
const CGS = 0.0;
const CGD = 0.0;
const CDS = 0.0;
const RG = 0.02;
const VT = 0.025;
const Is = 1.0e-12;
const V1 = 0.6;
const V2 = 0.7;
const RL = 5.0;
const L = 1.0e-3;
const ILM = Vdr/(RL + RDS) ;

var t[TMAX], ID[TMAX], VD[TMAX], IG[TMAX], VG[TMAX];
var VL[TMAX], IL[TMAX], VDi[TMAX], IDi[TMAX], V[TMAX];

set NEWTON ACCURACY  = 1.0e-4;

equ IG[i=1..TMAX-1]->
  IG[i] = CGS*(VG[i]-VG[i-1])/dt + CGD*((VG[i]-VD[i])-(VG[i-1]-VD[i-1]))/dt;

equ VG[i=all]->
  VG[i] = V[i] - RG*IG[i] ;

equ VD[i=all]->
  VD[i] = Vdr  + VDi[i] ;
```

```
equ ID[i=1..TMAX-1]->
  ID[i] =  0.0 + (VG[i]>VTH)*{G*(VG[i]-VTH)+VD[i]/RDS} +
    CGD*{(VD[i]-VG[i])-(VD[i-1]-VG[i-1])}/dt +
    CDS*{VD[i]-VD[i-1]}/dt ;

equ IDi[i=all]->
  IDi[i] = (VDi[i] <= V1)*Is*(exp(V1/VT)-1.0) +
           (VDi[i] >  V1)*Is*(exp(V1/VT)-1.0 +
            exp(V2/VT)*(VDi[i]-V1)/VT) ;

equ IL[i=0..TMAX-1]->
  IL[i] = ID[i] + IDi[i] ;

equ VL[i=1..TMAX-1]->
  VL[i] = L*(IL[i] - IL[i-1])/dt ;

equ VDi[i=all]->
  VDi[i] =  - (VL[i] + IL[i]*RL ) ;

begin main
  assign IG[i=0] = 0.0 ;
  assign ID[i=0] = 0.0 ;
  assign VL[i=0] = 0.0 ;
  assign t[i=all] = i*dt ;
  assign V[i=TINT..2*TINT]   = VGM ;
  assign V[i=3*TINT..4*TINT] = VGM ;
  assign V[i=5*TINT..6*TINT] = VGM ;
  assign V[i=7*TINT..8*TINT] = VGM ;
  assign V[i=9*TINT..10*TINT]= VGM ;
  assign V[i=11*TINT..12*TINT]=VGM ;
  assign V[i=13*TINT..14*TINT]=VGM ;
  assign V[i=15*TINT..16*TINT]=VGM ;
  assign V[i=17*TINT..18*TINT]=VGM ;
  assign V[i=19*TINT..TMAX-1] =VGM ;
  solve;
  write;
end
```

Power MOSFET with Inductive Load; Short-Term Behavior

We now examine the short-term behavior (by which we mean the inductor acts like a constant current source on this time scale) of this same sort of circuit. The inductor current is held at i_{LM}. The diode current is thus $i_{Di} = i_L - i_d = i_{LM} - i_d$, and the value of i_{Di} controls V_{Di}. V_D is found from the equation for i_d when $i_d > i_{LM}$, and from $V_D = V_{Di} + V_{dr}$ when $i_d < i_{LM}$.

```
// FILE:  sgdir\ch11\pmosdist\pmosdist.sg
```

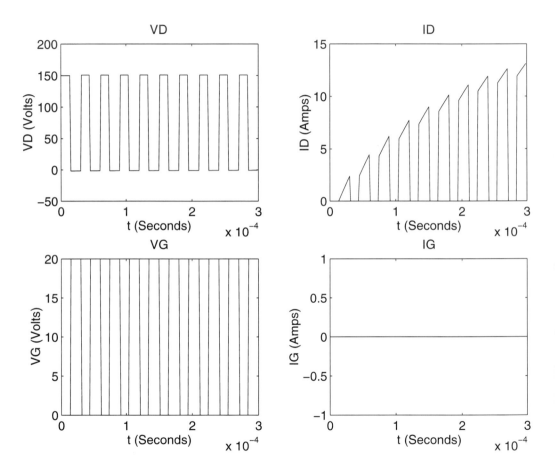

Figure 11.19. Various voltages and currents for a MOSFET used as a switch, with a diode in the load, as described in the text.

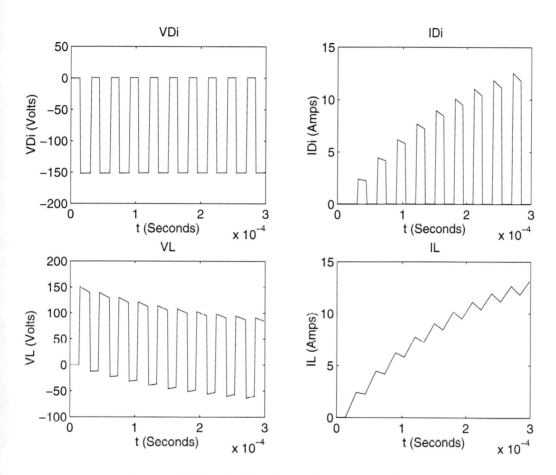

Figure 11.20. Continuation of the previous figure.

```
// This file contains a demonstration of how to
// model a MOSFET with a load consisting of a diode and an inductor.
// This file handles short-time behavior, so the inductor acts as
// a current source.

// Circuit Parameters
const TMAX = 500;                       // Maximum Number of Time Steps
const dt = 2.4e-11;                     // Time Step
const Vdr  = 150.0;
const VGM = 10.0;
const ton = 2;
const VTH = 1.8;
const RDS = 0.1;
const G = 1.0 ;
const CGS = 3.0e-9;                     // Capacitance, G to S
const CGD = 2.0e-10;                    // Capacitance, G to D
const CDS = 3.0e-9;                     // Capacitance, D to S
const RG = 0.5;
const taug = RG*(CGS + CGD);
const VT = 0.025;
const Is = 1.0e-10;
const V1 = 0.0;
const V2 = 0.75;
const RL = 10.0;
const ILM = Vdr/(RL + RDS);
const ts1 = taug*VTH/VGM + ton*dt;


var t[TMAX];
var ID[TMAX];
var VD[TMAX];
var IG[TMAX];
var VG[TMAX];
var IL[TMAX];
var VDI[TMAX];
var IDI[TMAX];
var V[TMAX];

set NEWTON ACCURACY  = 1.0e-10;


equ VG[i=1..TMAX-1] ->
  IG[i] = CGS*(VG[i]-VG[i-1])/dt + CGD*( (VG[i]-VD[i])-(VG[i-1]-VD[i-1]))/dt ;

equ IG[i=all] ->
  VG[i] = V[i] - RG*IG[i] ;

equ VD[i=0..ton]->
```

```
  VD[i] = Vdr ;

equ VD[i=ton..TMAX-1] ->
  (ID[i] <= ILM) * {VD[i] - (VDI[i] + Vdr)} =
  (ID[i] >  ILM) * [-ID[i] +  (t[i] > ts1) *
    {sq(VG[i] - VTH)/(1.0 + sq(VG[i] - VTH))}*{G*(VG[i] - VTH) + VD[i]/RDS} +
    CGD*{(VD[i] - VG[i]) - (VD[i-1] - VG[i-1])}/dt +
    CDS*{VD[i] - VD[i-1]}/dt];

equ ID[i=1..TMAX-1]->
  (ID[i]>ILM)*(ID[i]-ILM) = (ID[i] <= ILM) * [-ID[i] +
    (t[i] > ts1) * {sq(VG[i]-VTH)/(1.0 + sq(VG[i]-VTH))} *
    (G*(VG[i]-VTH) + VD[i]/RDS) +
    CGD*((VD[i]-VG[i])-(VD[i-1]-VG[i-1]))/dt +
    CDS*(VD[i] - VD[i-1])/dt];

equ VDI[i=all]->
  (IDI[i] < 0.0) * (VD[i]-(VDI[i]+Vdr)) =
  (IDI[i]>=0.0)  * [-IDI[i] + Is*exp(V2/VT) * (VDI[i]-V1)/VT];

equ IL[i=0..TMAX-1]->
  IL[i] = ILM ;

equ IDI[i=all]->
  IDI[i] = IL[i] - ID[i] ;

begin main
  assign VG[i=0] = 0.0 ;
  assign ID[i=0] = 0.0 ;
  assign t[i=all] = i*dt ;
  assign V[i=ton..TMAX-1] = VGM ;
  solve;
  write;
end
```

11.6   Thyristors

Thyristors provide switches that latch. The original version of the thyristor can be turned on by the gate but cannot be turned off by the gate. It has a pnpn structure, with once again a lightly doped drift region. See Figure 11.22.

The drift region can accommodate the depletion layer of either of the junctions next to it, when that junction is blocking an applied reverse voltage. Its terminals are called the gate, cathode and anode. Its pnpn structure and external connections differ from a BJT in that there is an extra layer next to the anode (as well as in that the drift region is different from the equivalent layer in a BJT). Starting at the

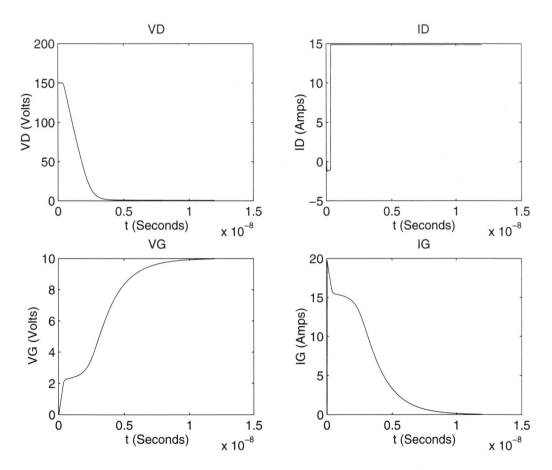

Figure 11.21. Short-term behavior of the circuit with the MOSFET used as a switch.

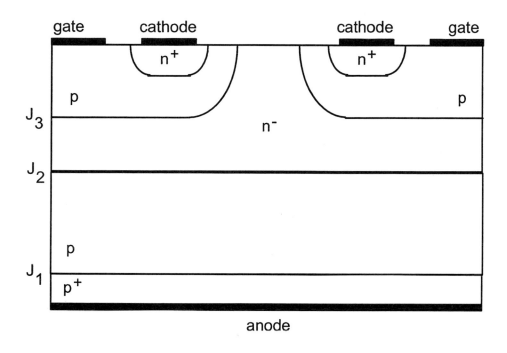

Figure 11.22. A schematic of a thyristor.

anode and choosing to begin with a p-type layer, there is first a p^+ layer followed by a p layer, which together form the extra feature that this device possesses relative to a BJT. The n^- layer that follows is the first feature common to the BJT and the thyristor, but it is thicker and much more lightly doped than it would be in a BJT. All the layers described so far are flat. The next layer, which is p-type, extends in one location up to the top of the device where it makes contact with the gate electrode. In other places the p-type layer is covered by n^+ regions which contact the cathode.

11.6.1 (i, v) Characteristics

The thyristor is often thought of, first as a one-dimensional structure laid out as:

1. The anode A,

2. a p-type region p_1,

3. a junction J_1,

4. an n^- region (the drift region) n_1,

5. a junction J_2,

6. a p-type region P_2 which is connected to the gate,

7. a junction J_3,

8. an n^+ region n_2 and

9. the cathode (denoted by a K for the German word Kathode).

The junctions J_1, J_2 and J_3 each have an n-type region on one side and a p-type region on the other side (so the boundary between p^+ and p does not count as a junction).

Second, the low-frequency equivalent circuit is drawn as a pair of pnp and npn transistors with the emitter of the pnp forming the anode of the thyristor and the emitter of the npn forming the cathode of the thyristor. The base of the npn is the gate of the thyristor, but this is connected to the collector of the pnp device and the collector of the npn is connected to the base of the pnp.

In reverse bias the first and third junctions J_1 (p below n^-) and J_3 (p below n^+) are reverse biased but the second junction J_2 (n^- below p) is forward biased. The doping on either side of J_3 (p below n^+) is high so it will break down quickly. The low doping of the drift-region means that voltage is blocked primarily on the drift region side of J_1 (p below n^-). As usual the limit on the voltage is set by avalanche breakdown. On the other hand, in forward bias junction J_2 (n^- below p) is reverse biased, while the other two junctions are forward biased. On one side of J_2 (n^- below p) is the drift region, which again does the voltage blocking.

When the thyristor is forward biased but blocking voltage, the two transistors are in active mode and obey the Ebers-Moll equations. Solving these indicates that a condition can arise where the anode current becomes very large. This in turn indicates that the thyristor is about to pass into the stable on-state. The forward on-state is a second region of stable forward behavior, in addition to the forward blocking state. In the forward on-state the impedance of the device is much lower than in the forward blocking state. The two regions are connected by an unstable region on the (i, v) curve. The cause of the transition can be understood in terms of the αs of the two transistors, which grow because the base thicknesses of both transistors grow when the depletion regions around J_2 (n^- below p) grow. More physically, if the gate current is large and positive enough, electrons will be injected across the forward-biased junction J_3 (p below n^+) from n_2 into p_2 to supply the carriers needed for that current. However, many of those electrons will make it to junction J_2 (n^- below p) and be swept into the drift region (since all the junctions are biased to push electrons from cathode to anode, in forward bias). The electrons have two effects in the drift region: they make it more negative, so the junction has to get wider to expose more positive ions to compensate; and they attract holes from p_1, which constitute extra current and which flow back through J_2 (n^- below p) into p_2. There the holes attract more electrons from n_2, and so on. The high levels of minority-carrier injection going on in the forward on-state lead to high conductivity modulation. The transistors are in saturation in this state. The on-state can now be terminated only by the external circuit limiting the anode current long enough for the stored charge to decay by recombination and by being swept out. (Gate turn-off thyristors also exist where the gate can turn off the device, but these will not be discussed here.)

11.6.2 Switching Characteristics

The switching characteristics of the thyristor must be studied in the context of an external circuit, using the models given above. Since $I = f(V)$ is not single valued, current boundary conditions are needed. Voltage boundary-conditions underspecify the problem.

11.7 IGBTs

The insulated gate bipolar transistor resembles the power MOSFET but with an extra p^+ layer next to the drain contact. In fact it has features in common with BJTs and MOSFETs. Each time an extra layer is added to an existing simpler structure, there is a tendency to create new classes of parasitic of increasing complexity. It might seem that the device which had an extra layer added to it would be the parasitic in the new device, but that depends on how the new device is intended to function. The IGBT has inside it an npnp structure which can function as a thyristor, but is not designed to do so.

The n^- region and the layers above it have the same names as in the MOSFET.

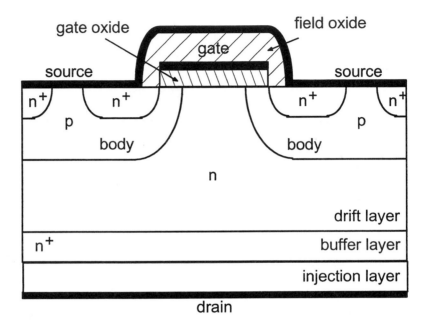

Figure 11.23. A schematic of an insulated gate bipolar transistor.

There is an n^+ source region, a p-body and an n^- drift region. The n^+ layer below the drift region is called a buffer layer and the p^+ region is the injecting layer. The essential structure of the IGBT would not be altered by removing the n^+ buffer layer above the p^+ injecting layer. Instead of combined n^+ and n^- layers there would be only an n^- layer. The buffer layer reduces the on-state voltage drop, because a depletion layer in it will be narrow due to the high doping level. For the same reason the buffer layer decreases the reverse-blocking voltage. Finally, the buffer layer shortens the turn-off time.

The pn^- junction blocks forward voltage when the device is off. The p^+n^+ junction (or if there is no n^+ region, then the p^+n^- junction) blocks reverse voltages. The pn^- junction and the p^+n^- junction (if it is present) both block large voltages because they employ the lightly doped drift region on one side of the junction in both cases. Including the n^+ region removes the p^+n^- junction and lowers the reverse-blocking voltage to a very low value.

IGBTs have higher courrent densities than MOSFETs but slower switching times. IGBTs also have an additional voltage drop across the added junction.

11.7.1 (i, v) **Characteristics**

The variation of I_D with V_G is like a power MOSFET. I_D varies with V_D in forward bias like a BJT but it is controlled by V_G instead of the base current. In fact the IGBT is designed to resemble a MOSFET with conductivity modulation used to control the drain drift region. The minority-carrier injection needed for conductivity modulation of the drift region is done by the p^+ (injecting) layer. By reducing the resistance of the drift region in the on-state the injecting layer allows much larger currents to flow than in the MOSFET. Otherwise the device behaves like a MOSFET with regard to the effect of V_G (measured relative to the source voltage), which determines whether there is a channel. The voltage at the other end of the channel where the drain would be in a normal MOSFET controls whether the channel is pinched off or not. The other end of this channel is separated from the drain of the IGBT, however, by the drain-drift region and a junction from an n-region to a p-region.

This discussion does allow equivalent-circuit models to be formulated. The equivalent circuit can be expected to include a MOSFET. The source of the MOSFET and the source of the IGBT coincide, as do the gates of the two devices. The drains of the IGBT and of the MOSFET are not the same, however, and the source has a connection to the drain of the IGBT through a pnp structure. This structure is represented as a BJT between the IGBT source and drain.

The MOSFET in the equivalent circuit has an n-type source and an n-type drain. The drain of this MOSFET is in the n^- drift region, but to allow for the resistance of the drift region a resistance is included in series with the MOSFET's drain. The other edge of the n^- region is the base of the BJT in the equivalent circuit, so the resistor is shown connected to the base of the BJT.

11.8   Electrothermal Simulation of Power Devices

Electrothermal simulation of power devices includes the energy-transport equation along with the carrier-transport equations (and Poisson's equation). The purpose of this kind of simulation is to show how heating of the device can alter the device performance and ultimately contribute to certain failure modes.

The equations which are used were described in Chapter 4, in Section (4.3.2), Equations (4.35) and (4.36). The quantities appearing in the equations are defined in that section. Additional information on the transport coefficients is given in Sections (4.4.9), (4.4.10) and (4.4.11).

11.9   Summary

This chapter has reviewed in outline some of the power semiconductor devices which are in use. It has emphasized the physical operation of each device rather than taking the more usual approach of considering the devices as sets of lumped circuit elements in the equivalent circuits. Another feature of this chapter which is noteworthy is the introduction of statistical design methods for optimization of semiconductors. Mixed mode simulation, which allows these detailed models to be used in realistic circuits and shows the device functioning in complex situations, is the topic of Chapter 12.

It is suggested that the BJT and MOSFET simulations be adapted to model thyristors and IGBTs. A large number of mesh points is needed, because the devices have large areas, and (for instance) the MOSFET in the IGBT needs the channel to be accurately modeled.

PROBLEMS

1. Write out the analytic forms of the equations used to describe the behavior of the power MOSFET when it has a combination of an inductor and a diode as its load. Sketch the circuit that goes with these equations. Are the choices of dependent variables in each equation appropriate, and why?

2. This question involves solution of the heat-transport equation, using 'propagators'. Set up a uniform rectangular mesh in the Cartesian coordinates (x, y), in the region $x = 0$ to $x = L_x$ and $y = 0$ to $y = L_y$, with $L_x = 1.5L_y$. The heat source is equal to H_0 in the region $x = L_x/3$ to $x = 2L_x/3$ and $y = L_y/4$ to $y = L_y/2$. The temperature is T_0 on the boundary of the region. Find the temperature using the steady-state propagator for the diffusion equation (which is the same as that for Poisson's equation). Use the method of images to approximately include the boundary condition.

3. Repeat the above question, using a time-dependent finite-difference scheme to follow the time evolution to steady state. Start with $T = T_0$ everywhere. What is the time scale τ for the temperature relaxation, in terms of the thermal conductivity κ and the specific heat C_V ? What is the appropriate time step Δt in terms of these quantities and the mesh size Δx ?

4. Set up a circuit model of a thyristor, representing it as two bjts. Plot the (i, v) characteristics of the thyristor.

Part IV

Advanced Topics

Chapter 12

Mixed-Mode Simulations

This chapter will be devoted primarily to simulations of time-dependent problems in which semiconductor devices are interacting with external circuits. The external circuit will normally be represented by lumped circuit-elements. The device itself is modeled at the microscopic level in a true mixed-mode simulation. The difference in the types of model leads to the term *mixed-mode.* The interaction between these different levels of modeling causes some difficulties, which will be addressed in what follows. The introductory material in this chapter is not truly mixed mode. We begin by using lumped circuit-elements to show how the circuits are modeled. Later sections build on this to replace the lumped elements with better models.

In this chapter we shall stress the need to build into a simulation the capability to reproduce the behaviors which we know are likely to occur. This may seem circular, but the ultimate guide to what is correct is experiment; our setting up of simulations should therefore rely on experiment to guide us rather than a purely mathematical development.

(The circuits in this and earlier chapters are not accompanied by circuit diagrams for the most part. Drawing the circuit diagrams is an exercise given at the end of the chapter.)

12.1   Time-Dependent Circuits

To introduce the issues arising in device modeling when the device is included in an external circuit, we shall begin by modeling circuits. The first circuits will be very simple RLC circuits. Then diodes and other devices will be introduced, represented as lumped circuit elements. After showing how the devices behave in general terms, we then progressively increase the level of sophistication of the description of the devices in the circuit.

A Series RLC Circuit

A series RLC circuit is set up in rlc.sg.

Input File rlc.sg

```
// FILE:  sgdir\ch12\rlc\rlc.sg
// This file contains a demonstration of how one can implicitly
// model a time-varying RLC circuit.

// Circuit Parameters
const TMAX = 200;
const L   = 1.0e-2;
const R   = 1.0;
const C   = 1.0e-2;
const Vo = 1.0;
const PI = 3.1415927;
const omega = 2.0*PI*50.0; // Angular Frequency
const dt = 0.1*PI/omega;   // Time Step

var Vtot[TMAX],V[TMAX],t[TMAX],I[TMAX];
var VL[TMAX],VR[TMAX],VC[TMAX];

equ VR[i=1..TMAX-1]->
  V[i] = VR[i]+VL[i]+VC[i];

equ I[i=1..TMAX-1]->
  VR[i] = R*I[i];

equ VC[i=1..TMAX-1]->
  0.5*(I[i]+I[i-1]) = C/dt*(VC[i]-VC[i-1]);

equ VL[i=1..TMAX-1]->
  0.5*(VL[i]+VL[i-1]) = +L*(I[i]-I[i-1])/dt;

begin main
  assign VR[i=0] = 0.0 ;
  assign  I[i=0] = 0.0 ;
  assign VC[i=0] = 0.0 ;
  assign t[i=all] = i*dt;
  assign V[i=all] = Vo*sin(omega*(i*dt));
  assign VL[i=0] = V[i];
  solve;
  assign Vtot[i=all] = VR[i]+VL[i]+VC[i];
  write;
end
```

The results of running this file are shown in figure 12.1.

As expected, after an initial transient the circuit settles down to a sinusoidal steady state, with the voltages across the devices exhibiting the appropriate phase differences. To understand the magnitudes, we can calculate the impedances of the circuit elements. From the input file, $R = 1\Omega$, $C = 10^{-2}$F and $L = 10^{-2}$H.

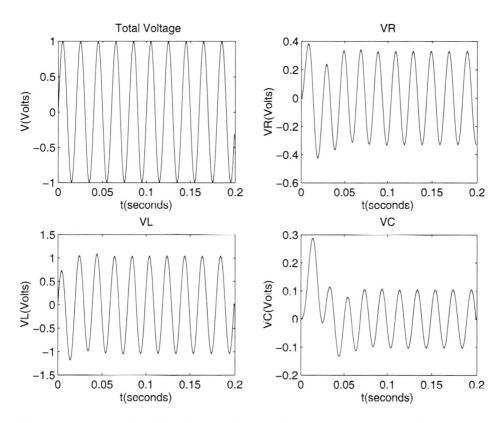

Figure 12.1. Results of simulation of an *RLC* circuit, showing it settling down to a sinusoidal steady state after an initial transient.

The angular frequency is $\omega = 100\pi$, so the impedances are $Z_L = \omega L = \pi$ and $Z_C = \frac{1}{\omega C} = \frac{1}{\pi}$. The overall circuit impedance is $Z_{TOT} = \sqrt{1.0 + (\pi - \frac{1}{\pi})^2} \simeq 3.0\Omega$, so for an applied voltage of 1.0V we expect a current of about 0.33A. The current is not shown, but since $R = 1.0\Omega$, V_R has the same magnitude, as can be seen from figure 12.1. The inductor voltage should be larger by a factor of π, and the capacitor voltage should be smaller by a factor of π. These expectations are also confirmed by the numerical results.

Diode Rectifier Circuit I

The effect of replacing the capacitor with a diode to obtain a simple rectifier is examined using the file rld.sg.

Input File rld.sg

```
// FILE:  sgdir\ch12\rld\rld.sg
// This file contains a demonstration of how one can implicitly
// model a time-varying RL+ diode circuit.

// Circuit Parameters
const TMAX = 200;
const L   = 5.0e-4;
const R   = 1.0;
const Is  = 1.0e-6;              // Diode Saturation Current
const VT = 0.03;                // Thermal Voltage
const Vo = 10.0;
const PI = 3.1415927;
const omega = 50*(2.0*PI);      // Angular Frequency
const dt = 7.0e-3*2.0*PI/omega; // Time Step
const phase = 0;                // Phase

var Vtot[TMAX];
var V[TMAX];
var t[TMAX];
var I[TMAX];
var VL[TMAX];
var VR[TMAX];
var VD[TMAX];

// Circuit Equations
equ VR[i=1..TMAX-1]->
  (I[i]>=0)*(V[i] - VR[i] - VL[i] - VD[i]) =
  (I[i]<0)*(VR[i] - R*I[i]);

equ I[i=1..TMAX-1]->
  (I[i]>=0)*(VR[i] - R*I[i]) =
  (I[i]<0)*(I[i] - Is*(exp(VD[i]/VT) - 1.0));
```

```
equ VD[i=1..TMAX-1]->
  (I[i]>=0)*(I[i] - Is*(exp(VD[i]/VT) - 1.0)) =
  (I[i]<0)*(V[i] - VR[i] - VL[i] - VD[i]);

equ VL[i=0]->
  VL[i] = V[i] ;

equ VL[i=1..TMAX-1]->
  VL[i] = +L*(I[i]-I[i-1])/dt;

begin main
  assign VR[i=0] = 0;
  assign I[i=0]   = 0;
  assign VD[i=0] = 0;
  assign t[i=all] = i*dt;
  assign V[i=all] = Vo*sin(omega*(i*dt)+phase);
  solve;
  assign Vtot[i=all] = VR[i]+VL[i]+VD[i];
  write;
end
```

The results are plotted in figure 12.2, and to numerical accuracy show the expected rectification and the effect of the inductor to keep the current positive as long as possible.

This problem is much harder to solve, and the values that are used for the inductance L and the load voltage V_o were kept small to help the convergence of the solution. It was helpful to introduce several conditional statements to get around some problems, which are caused by the interaction of the diode and the inductor. It is not appropriate to go into this problem in more detail yet, since similar issues arise frequently in mixed-mode simulations. The problem will be discussed carefully when microscopic simulations of the diode's interior behavior are done.

The best way to handle sudden changes in behavior involves examining the physical causes of the change and designing a numerical approach which can respond to the change. For example, a diode goes from its on-state to its off-state when the excess charge has been removed. By tracking the microscopic behavior and monitoring the excess charge, it will be possible to determine when the diode is about to switch. This will allow time steps to be taken which are suitably small when the physical situation dictates that it is necessary.

Diode Rectifier Circuit II

Two illustrations of using a diode model in a slightly different circuit follow. The circuit itself, like the one just studied, comes from power electronics. They both are simple rectifiers. The diode will be modeled first using the usual analytic diode model, the results being shown in Figure 12.3. We then use a simple 'piece-wise linear' (PWL) model. See Figure 12.4.

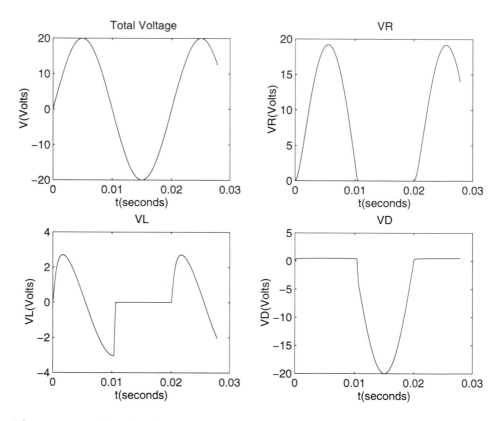

Figure 12.2. Simulated behavior of a circuit which uses an inductor and a diode to achieve rectification. Most of the voltage which is applied is dropped across the resistor (when the diode is conducting) or across the diode (when it is blocking). The diode voltage is less than a volt when it is conducting. The convergence of this run is marginal, so the input file given uses $V_0 = 10$ volts instead of the 20 shown here.

Input File rcd.sg

```
// FILE:  sgdir\ch12\rcd\rcd.sg
// This file contains a demonstration of how one can implicitly
// model a time-varying RC+ diode circuit.

// Circuit Parameters
const TMAX = 200;
const C  = 1.0e-2;
const R  = 10.0;
const Is  = 1.0e-6;              // Diode Saturation Current
const VT = 0.03;                // Thermal Voltage
const Vo = 10.0;
const PI = 3.1415927;
const omega = 50*(2.0*PI);      // Angular Frequency
const dt = 3.0e-2*2.0*PI/omega; // Time Step
const phase = 0;                // Phase

var Vtot[TMAX], V[TMAX], t[TMAX], I[TMAX], VC[TMAX], VD[TMAX];


equ I[i=1..TMAX-1]->
  (I[i]>=0)*(I[i] - VC[i]/R -C*(VC[i]-VC[i-1])/dt) =
  (I[i]<0)*(I[i] - Is*(exp(VD[i]/VT) - 1.0));

equ VD[i=1..TMAX-1]->
  (I[i]>=0)*(I[i] - Is*(exp(VD[i]/VT) - 1.0)) =
  (I[i]<0)*(V[i] - (VC[i-1]+ (I[i]-VC[i]/R)*dt/C) - VD[i]);

equ VC[i=1..TMAX-1]->
  VC[i] = V[i]-VD[i];


begin main
  assign t[i=all] = i*dt;
  assign V[i=all] = Vo*sin(omega*(i*dt)+phase);
  assign VC[i=0] = 0;
  assign I[i=0]  = 0;
  assign VD[i=0] = 0;
  solve;
  assign Vtot[i=all] = VC[i]+VD[i];
  write;
end
```

The results of this simulation are shown in figure 12.3.

The PWL model will be shown next. It is capable of converging for much higher applied voltages than the more 'correct' model, because the linearized equations are much better behaved than the full set, which have exponential factors in them.

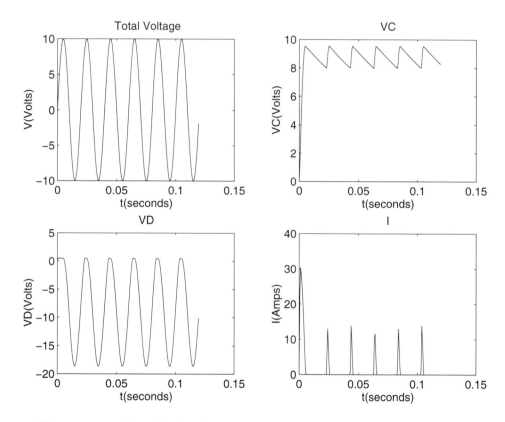

Figure 12.3. Rectification by a circuit using a capacitor and a diode. This particular model is difficult to use because of the exponential dependence of the diode current on the voltage.

Input File rcd2.sg

```
// FILE:  sgdir\ch12\rcd2\rcd2.sg
// This file contains a demonstration of how one can implicitly
// model a time-varying RC+ diode circuit, using a piece-wise
// linear model.

// Circuit Parameters
const TMAX = 200;
const C  = 1.0e-2;
const R  = 10.0;
const Is  = 1.0e-12;            // Diode Saturation Current
const VT = 0.025;              // Thermal Voltage
const Vo = 200.0;
const PI = 3.1415927;
const omega = 50*(2.0*PI);     // Angular Frequency
const dt = 3.0e-2*2.0*PI/omega; // Time step
const phase = 0;               // Phase
const V1 = 0.7;                // Breakpoint in PWL model
const IMIN = Is;               // Breakpoint in PWL model

var Vtot[TMAX],V[TMAX],t[TMAX],I[TMAX];
var VC[TMAX],VD[TMAX];

equ I[i=1..TMAX-1]->
  (VD[i]>=V1)*(I[i] - Is*(exp(V1/VT)*(VD[i]-V1)/VT + 1.0)) =
  (VD[i]< V1)*(I[i]-IMIN);

equ VC[i=1..TMAX-1]->
  VC[i] = VC[i-1] + (I[i]-VC[i]/R)*dt/C   ;

equ VD[i=1..TMAX-1]->
  VD[i] = V[i]-VC[i]  ;

begin main
  assign VC[i=0] = 0.0 ;
  assign  I[i=0] = 0.0 ;
  assign VD[i=0] = 0.0 ;
  assign t[i=all] = i*dt;
  assign V[i=all] = Vo*sin(omega*(i*dt)+phase);
  solve;
  assign Vtot[i=all] = VC[i]+VD[i];
  write;
end
```

Results from this piece-wise linear model are shown in figure 12.4.

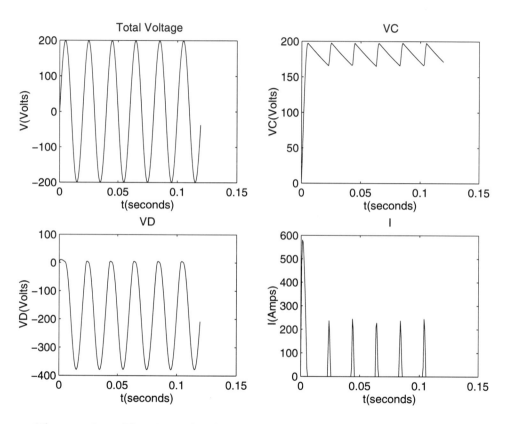

Figure 12.4. The same circuit as the previous figure, but this time a piece-wise linear diode model was used, which makes convergence easier.

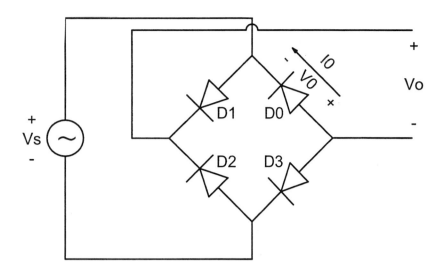

Figure 12.5. A schematic of a single phase diode bridge.

12.2 Circuit Boundary Conditions for Device Models

In this section we use some of the device models from earlier chapters, but now the devices are embedded in external circuits so the external circuit provides feedback and determines the boundary conditions on the device model. The first example is of a four-diode bridge circuit. After that we provide a rather sophisticated simulation of a BJT (modeled using Ebers-Moll equations) in a circuit containing an inductor and a pin diode (modeled at a microscopic level).

12.2.1 A Diode Bridge Circuit

A single-phase diode bridge is shown schematically in Figure 12.5.

A simple 'mixed-mode' simulation of a diode bridge circuit is set up in the file mm01.sg which follows. The input file mm01.sg is based on the simulation of a pin diode contained in pin01.sg.

The bridge is laid out in a square. The corners of the square are at top and bottom and left and right. Each face of the square has a diode in it, and all the diodes point to the left. Diode zero is on the top right, and the diode numbering runs counterclockwise. The bottom corner of the square is grounded. The applied voltage V_s is connected to the top corner. The output voltage V_o is the voltage on the left minus the voltage on the right.

The potential is initialized using the equilibrium potential, but with an offset added to D1 and D2 so the potential is the same where their contacts are connected. Each device can be simulated separately to get initial conditions in general. One contact of D2 and D3 is set to zero and one contact of D0 and D1 is set to V_s, according to the above description of the circuit. Where D0 meets D3 the potentials must be equal, as where D1 meets D2. Since the output is open-circuited, the currents must be equal in magnitude; $I_1 = -I_2$ and $I_0 = -I_3$.

The step size must be chosen to be small enough to allow the simulation to easily converge. Ideally an adaptive scheme would be used which kept $\frac{dV_s}{dt}$ roughly constant. Time-independent equations are used, however, since the frequency is low.

In the simulation which follows, the diodes are identical so the four diodes can be included using an array index, j, which runs from 0 to 3.

Input File mm01.sg

```
// FILE:   sgdir\ch12\mm01\mm01.sg
// mesh constants ... these correspond to the data in mesh.dat
const NX  = 37;                     // number of x-mesh points
const LX  = NX-1;                   // index of last mesh point
const Wd  = 25.0e-4;                // device width
const Na  = 5.0e+19;                // peak acceptor concentration (cm^-3)
const Nd  = 1.0e+14;                // substrate donor concentration (cm^-3)

// physical constants and built-in potential
const T   = 300.0;                  // ambient temperature (K)
const e   = 1.602e-19;              // electron charge (C)
const kb  = 1.381e-23;              // Boltzmann's constant (J/K)
const eo  = 8.854e-14;              // permittivity of vacuum (F/cm)
const eSi = 11.8;                   // relative permittivity of Si
const eps = eSi*eo;                 // permittivity of Si (F/cm)
const Ni  = 1.25e10;                // intrinsic concentration of Si (cm^-3)
const Vt  = kb*T/e;                 // thermal voltage (V)
const Vbi = Vt*ln(Na*Nd/sq(Ni));    // built-in potential (V)

// scaling constants
const x0  = Wd;                     // distance scaling (cm)
const T0  = T;                      // temperature scaling (K)
const V0  = kb*T0/e;                // electrostatic-potential scaling (V)
const C0  = max(Na,Nd);            // concentration scaling (cm^-3);
const D0  = 35.0;                   // diffusion-coefficient scaling (cm^2/s)
const u0  = D0/V0;                  // mobility-coefficient scaling (cm^2/V/s)
const R0  = D0*C0/sq(x0);          // recombination-rate scaling (cm^-3/s)
const t0  = sq(x0)/D0;             // time scaling (s)
const E0  = V0/x0;                  // electric-field scaling (V/s)
const J0  = e*D0*C0/x0;            // current-density scaling (A/cm^2)
const L0  = V0*eo/(e*sq(x0)*C0);   // Laplacian scaling constant

// 1D Cartesian discretized gradient operator
```

```
func grad(f1,f2,<h>)
  return (f2-f1)/h;

// 1D Cartesian discretized divergence operator
func div(f1,f2,<h>)
  return (f2-f1)/h;

// equilibrium electron concentration of uniformly doped silicon
func Neq(<N>)
  assign temp = [abs(N)+sqrt(sq(N)+4*sq(Ni))]/2.0;
  return (N>=0)*temp + (N<0)*sq(Ni)/temp;

// equilibrium hole concentration of uniformly doped silicon
func Peq(<N>)
  assign temp = [abs(N)+sqrt(sq(N)+4*sq(Ni))]/2.0;
  return (N<=0)*temp + (N>0)*sq(Ni)/temp;

// unscaled electron mobility
func MUn(<N>,<T>)
  const Al = 1430.0,  Bl = -2.20;
  const Ai = 4.61e17, Bi = 1.52e15;
  assign MUl = Al*pow((T/300.0),Bl);
  assign MUi = {Ai*T*sqrt(T)/N}/{ln(1+Bi*sq(T)/N)-Bi*T*T/(N+Bi*sq(T))};
  assign X   = sqrt(6.0*MUl/MUi);
  return MUl*{1.025/[1+pow(X/1.68,1.43)]-0.025};

// unscaled hole mobility
func MUp(<N>,<T>)
  const Al = 495.0,   Bl = -2.20;
  const Ai = 1.00e17, Bi = 6.25e14;
  assign MUl = Al*pow((T/300.0),Bl);
  assign MUi = {Ai*T*sqrt(T)/N}/{ln(1+Bi*sq(T)/N)-Bi*T*T/(N+Bi*sq(T))};
  assign X   = sqrt(6.0*MUl/MUi);
  return MUl*{1.025/[1+pow(X/1.68,1.43)]-0.025};

// scaled recombination
func R(n,p)
  const tn = 1.0e-6/t0;            // scaled electron lifetime
  const tp = 1.0e-6/t0;            // scaled hole lifetime
  const ni = Ni/C0;               // scaled intrinsic concentration
  return (n*p-sq(ni))/[tp*(n+ni)+tn*(p+ni)];

// scaled electric field
func E(V1,V2,<h>)
  return -grad(V1,V2,h);

// scaled electron-current density
func Jn(n1,n2,E,<un1>,<un2>,<h>)
```

```
  const Ut  = Vt/V0;                   // scaled thermal voltage
  assign un  = ave(un1,un2);
  assign dV  = E*h/(2*Ut);
  assign n   = n1*aux2(dV)+n2*aux2(-dV);
  assign dndx = aux1(dV)*grad(n1,n2,h);
  return un*(n*E+dndx);

// scaled hole-current density
func Jp(p1,p2,E,<up1>,<up2>,<h>)
  const Ut  = Vt/V0;                   // scaled thermal voltage
  assign up  = ave(up1,up2);
  assign dV  = E*h/(2*Ut);
  assign p   = p1*aux2(-dV)+p2*aux2(dV);
  assign dpdx = aux1(dV)*grad(p1,p2,h);
  return up*(p*E-dpdx);

// set numerical algorithm parameters
set NEWTON DAMPING    = 3;
set NEWTON ACCURACY   = 1.0e-12;
set NEWTON ITERATIONS = 100;
set LINSOL ALGORITHM  = GAUSSELIM;
set LINSOL FILL       = INFINITY;


// declare variables
var x[NX], C[NX], V[NX,4], n[NX,4], p[NX,4], un[NX], up[NX], h[LX];
unknown V[1..LX-1,all], V[0,0], V[LX,1], V[LX,2], V[0,3];
unknown n[1..LX-1,all];
unknown p[1..LX-1,all];

// scaled Poisson's equation
equ V[i=1..LX-1,j=all] ->
  L0*div(eSi*E(V[i-1,j],V[i,j],h[i-1]),
         eSi*E(V[i,j],V[i+1,j],h[i]),
         ave(h[i],h[i-1])) - (p[i,j]-n[i,j]+C[i]) = 0.0;

// scaled electron continuity equation
equ n[i=1..LX-1,j=all] ->
  +div(Jn(n[i-1,j],n[i,j],E(V[i-1,j],V[i,j],h[i-1]),un[i-1],un[i],h[i-1]),
       Jn(n[i,j],n[i+1,j],E(V[i,j],V[i+1,j],h[i  ]),un[i],un[i+1],h[i  ]),
       ave(h[i],h[i-1])) -
  R(n[i,j],p[i,j]) = 0.0;

// scaled electron continuity equation
equ p[i=1..LX-1,j=all] ->
  -div(Jp(p[i-1,j],p[i,j],E(V[i-1,j],V[i,j],h[i-1]),up[i-1],up[i],h[i-1]),
       Jp(p[i,j],p[i+1,j],E(V[i,j],V[i+1,j],h[i  ]),up[i],up[i+1],h[i  ]),
       ave(h[i],h[i-1])) -
```

```
   R(n[i,j],p[i,j]) = 0.0;

// boundary conditions of the diode bridge
equ V[i=0 ,j=0] -> V[0,0]  = V[0,3];

equ V[i=LX,j=1] -> V[LX,1] = V[LX,2];

equ V[i=LX,j=2] -> // BC:  I2 = -I1
  +(Jn(n[LX-1,2],n[LX,2],E(V[LX-1,2],V[LX,2],h[LX-1]),un[LX-1],un[LX],h[LX-1])+
    Jp(p[LX-1,2],p[LX,2],E(V[LX-1,2],V[LX,2],h[LX-1]),up[LX-1],up[LX],h[LX-1]))=
  -(Jn(n[0,1],n[1,1],E(V[0,1],V[1,1],h[0]),un[0],un[1],h[0])+
    Jp(p[0,1],p[1,1],E(V[0,1],V[1,1],h[0]),up[0],up[1],h[0]));

equ V[i=0,j=3]  -> // BC:  I3 = -I0
  +(Jn(n[0,3],n[1,3],E(V[0,3],V[1,3],h[0]),un[0],un[1],h[0])+
    Jp(p[0,3],p[1,3],E(V[0,3],V[1,3],h[0]),up[0],up[1],h[0]))=
  -(Jn(n[LX-1,0],n[LX,0],E(V[LX-1,0],V[LX,0],h[LX-1]),un[LX-1],un[LX],h[LX-1])+
    Jp(p[LX-1,0],p[LX,0],E(V[LX-1,0],V[LX,0],h[LX-1]),up[LX-1],up[LX],h[LX-1]));


begin SetUpMesh
  open "mesh.dat" read;
  file goto "mesh points";
  file read x;
  file goto "dopant concentration";
  file read C;
  close;
end


var Veq[NX], neq[NX], peq[NX];

begin InitializeVariables
  // read equilibrium values into 1D "storage" arrays
  open "mm01eq.dat" read;
  file goto "V";
  file read Veq;
  file goto "n";
  file read neq;
  file goto "p";
  file read peq;
  close;

  // initialize the 2D arrays
  assign V[i=all,j=0] = Veq[i];
  assign V[i=all,j=1] = Veq[i] - Veq[0];
  assign V[i=all,j=2] = Veq[i] - Veq[0];
  assign V[i=all,j=3] = Veq[i];
```

```
    assign n[i=all,j=all] = neq[i];
    assign p[i=all,j=all] = peq[i];

    // initialize the other 1D arrays
    assign h[i=all] = x[i+1]-x[i];
    assign un[i=all] = MUn(abs(C[i]),T);
    assign up[i=all] = MUp(abs(C[i]),T);
end

begin ScaleVariables
    // NOTE:  the V, n, and p variables are already scaled since the
    //        equilibrium simulation did not unscale the variables
    //        before writing them to the equilibrium data file.
    assign x[i=all] = x[i]/x0;
    assign h[i=all] = h[i]/x0;
    assign C[i=all] = C[i]/C0;
    assign un[i=all] = un[i]/u0;
    assign up[i=all] = up[i]/u0;
end


const PI = 3.14159265358979323846;
const Fs = 60.0;                      // frequency of ac supply (Hz)
const As = 5.0;                       // amplitude of ac supply (V)
const ws = 2*PI*Fs;                   // angular frequency of ac supply
const Nt = 500;                       // number of steps
const dt = 1.0/Fs/Nt;                 // time step

var Vs, Vo, t;

begin main
    // setup simulation and solve for equilibrium
    call SetUpMesh;
    call InitializeVariables;
    call ScaleVariables;
    solve;

    // null the output file
    open "mm01.out" write;
    close;

    // simulate one cycle of an ac sine wave
    assign t = 0.0;
    while (t < (Nt+0.5)*dt) begin
        assign Vs = As*sin(ws*t);
        assign V[i=LX,j=0] = Vs/V0;
        assign V[i=0 ,j=1] = Vs/V0;
        solve;
```

```
      assign Vo = (V[LX,1]-V[0,0])*V0-2*Vbi;
      open "mm01.out" append;
      file write t,Vs,Vo;
      close;
      assign t = t+dt;
  end
end
```

This file makes use of the equilibrium solution obtained from mm01eq.sg, which is the next input file.

```
// FILE:  sgdir\ch12\mm01\mm01eq.sg
// mesh constants ... these correspond to the data in mesh.dat
const NX  = 37;                  // number of x-mesh points
const LX  = NX-1;                // index of last mesh point
const Wd  = 25.0e-4;             // device width
const Na  = 5.0e+19;             // peak acceptor concentration (cm^-3)
const Nd  = 1.0e+14;             // substrate donor concentration (cm^-3)

// physical constants and built-in potential
const T   = 300.0;               // ambient temperature (K)
const e   = 1.602e-19;           // electron charge (C)
const kb  = 1.381e-23;           // Boltzmann's constant (J/K)
const eo  = 8.854e-14;           // permittivity of vacuum (F/cm)
const eSi = 11.8;                // relative permittivity of Si
const eps = eSi*eo;              // permittivity of Si (F/cm)
const Ni  = 1.25e10;             // intrinsic concentration of Si (cm^-3)
const Vt  = kb*T/e;              // thermal voltage (V)
const Vbi = Vt*ln(Na*Nd/sq(Ni)); // built-in potential (V)

// scaling constants
const x0  = Wd;                  // distance scaling (cm)
const T0  = T;                   // temperature scaling (K)
const V0  = kb*T0/e;             // electrostatic-potential scaling (V)
const C0  = max(Na,Nd);          // concentration scaling (cm^-3);
const D0  = 35.0;                // diffusion-coefficient scaling (cm^2/s)
const u0  = D0/V0;               // mobility-coefficient scaling (cm^2/V/s)
const R0  = D0*C0/sq(x0);        // recombination-rate scaling (cm^-3/s)
const t0  = sq(x0)/D0;           // time scaling (s)
const E0  = V0/x0;               // electric-field scaling (V/s)
const J0  = e*D0*C0/x0;          // current-density scaling (A/cm^2)
const L0  = V0*eo/(e*sq(x0)*C0); // Laplacian scaling constant


// 1D Cartesian discretized gradient operator
func grad(f1,f2,<h>)
  return (f2-f1)/h;


// 1D Cartesian discretized divergence operator
```

```
func div(f1,f2,<h>)
  return (f2-f1)/h;

// equilibrium electron concentration of uniformly doped silicon
func Neq(<N>)
  assign temp = [abs(N)+sqrt(sq(N)+4*sq(Ni))]/2.0;
  return (N>=0)*temp + (N<0)*sq(Ni)/temp;

// equilibrium hole concentration of uniformly doped silicon
func Peq(<N>)
  assign temp = [abs(N)+sqrt(sq(N)+4*sq(Ni))]/2.0;
  return (N<=0)*temp + (N>0)*sq(Ni)/temp;

// unscaled electron mobility
func MUn(<N>,<T>)
  const Al = 1430.0,  Bl = -2.20;
  const Ai = 4.61e17, Bi = 1.52e15;
  assign MUl = Al*pow((T/300.0),Bl);
  assign MUi = {Ai*T*sqrt(T)/N}/{ln(1+Bi*sq(T)/N)-Bi*T*T/(N+Bi*sq(T))};
  assign X   = sqrt(6.0*MUl/MUi);
  return MUl*{1.025/[1+pow(X/1.68,1.43)]-0.025};

// unscaled hole mobility
func MUp(<N>,<T>)
  const Al = 495.0,   Bl = -2.20;
  const Ai = 1.00e17, Bi = 6.25e14;
  assign MUl = Al*pow((T/300.0),Bl);
  assign MUi = {Ai*T*sqrt(T)/N}/{ln(1+Bi*sq(T)/N)-Bi*T*T/(N+Bi*sq(T))};
  assign X   = sqrt(6.0*MUl/MUi);
  return MUl*{1.025/[1+pow(X/1.68,1.43)]-0.025};

// scaled recombination
func R(n,p)
  const tn = 1.0e-6/t0;              // scaled electron lifetime
  const tp = 1.0e-6/t0;              // scaled hole lifetime
  const ni = Ni/C0;                  // scaled intrinsic concentration
  return (n*p-sq(ni))/[tp*(n+ni)+tn*(p+ni)];

// scaled electric field
func E(V1,V2,<h>)
  return -grad(V1,V2,h);

// scaled electron-current density
func Jn(n1,n2,E,<un1>,<un2>,<h>)
  const Ut  = Vt/V0;                 // scaled thermal voltage
  assign un  = ave(un1,un2);
  assign dV  = E*h/(2*Ut);
  assign n   = n1*aux2(dV)+n2*aux2(-dV);
```

```
    assign dndx = aux1(dV)*grad(n1,n2,h);
    return un*(n*E+dndx);

// scaled hole-current density
func Jp(p1,p2,E,<up1>,<up2>,<h>)
    const Ut  = Vt/V0;                    // scaled thermal voltage
    assign up  = ave(up1,up2);
    assign dV  = E*h/(2*Ut);
    assign p   = p1*aux2(-dV)+p2*aux2(dV);
    assign dpdx = aux1(dV)*grad(p1,p2,h);
    return up*(p*E-dpdx);

// set numerical algorithm parameters
set NEWTON DAMPING    = 3;
set NEWTON ACCURACY   = 1.0e-12;
set NEWTON ITERATIONS = 100;
set LINSOL ALGORITHM  = GAUSSELIM;
set LINSOL FILL       = INFINITY;


// declare variables
var x[NX], C[NX], V[NX], n[NX], p[NX], un[NX], up[NX], h[LX];

// scaled Poisson's equation
equ V[i=1..LX-1] ->
  L0*div(eSi*E(V[i-1],V[i],h[i-1]),eSi*E(V[i],V[i+1],h[i]),ave(h[i],h[i-1])) -
  (p[i]-n[i]+C[i]) = 0.0;

// scaled electron continuity equation
equ n[i=1..LX-1] ->
  +div(Jn(n[i-1],n[i],E(V[i-1],V[i],h[i-1]),un[i-1],un[i],h[i-1]),
       Jn(n[i],n[i+1],E(V[i],V[i+1],h[i ]),un[i],un[i+1],h[i ]),
       ave(h[i],h[i-1])) -
  R(n[i],p[i]) = 0.0;

// scaled electron continuity equation
equ p[i=1..LX-1] ->
  -div(Jp(p[i-1],p[i],E(V[i-1],V[i],h[i-1]),up[i-1],up[i],h[i-1]),
       Jp(p[i],p[i+1],E(V[i],V[i+1],h[i ]),up[i],up[i+1],h[i ]),
       ave(h[i],h[i-1])) -
  R(n[i],p[i]) = 0.0;


begin SetUpMesh
  open "mesh.dat" read;
  file goto "mesh points";
  file read x;
  file goto "dopant concentration";
```

```
  file read C;
  close;
end

begin InitializeVariables
  assign h[i=all] = x[i+1]-x[i];
  assign n[i=all] = Neq(C[i]);
  assign p[i=all] = Peq(C[i]);
  assign V[i=all] = Vt*ln(n[i]/Nd);
  assign un[i=all] = MUn(abs(C[i]),T);
  assign up[i=all] = MUp(abs(C[i]),T);
end

begin ScaleVariables
  assign x[i=all] = x[i]/x0;
  assign h[i=all] = h[i]/x0;
  assign V[i=all] = V[i]/V0;
  assign C[i=all] = C[i]/C0;
  assign n[i=all] = n[i]/C0;
  assign p[i=all] = p[i]/C0;
  assign un[i=all] = un[i]/u0;
  assign up[i=all] = up[i]/u0;
end

begin main
  call SetUpMesh;
  call InitializeVariables;
  call ScaleVariables;

  // solve for equilibrium and write the results to the output file
  solve;
  open "mm01eq.dat" write;
  file label "V";
  file write V;
  file label "n";
  file write n;
  file label "p";
  file write p;
  close;
end
```

This can be run using the DOS script runmm01.bat, or the UNIX script file runmm01, which are given in tables. The output from the simulation is shown in figure 12.6.

12.2.2   BJT with Inductive Load and Shunt Diode

The schematic in Figure 12.7 shows a BJT switching an inductive load.

The next mixed-mode simulation demonstrates how the BJT (Ebers-Moll model)

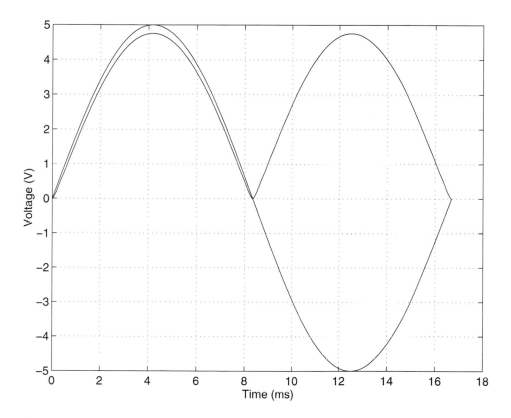

Figure 12.6. The results of the diode bridge simulation. The applied voltage is the continuous sinusoid. The rectified voltage is the output, which is seen to be slightly smaller in magnitude than the applied voltage, due to the voltage drop in the diode.

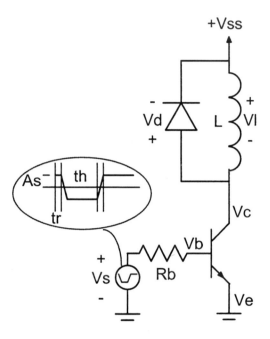

Figure 12.7. A schematic of a BJT switching an inductive load.

Script Files runmm01 and runmm01.bat

runmm01.bat

```
echo off
cls
rem generate equilibrium solution
sgxlat -p15 mm01eq.sg
call sgbuild sim mm01eq
order mm01eq.top mm01eq.prm
mm01eq -vs2
del mm01eq.exe
   (etc.)
del mm01eq.prm
rem translate, compile, link and run the diode bridge mixed mode simulation
sgxlat -p15 mm01.sg
call sgbuild sim mm01
order mm01.top mm01.prm
mm01 -vs1
del mm01.exe
   (etc.)
del mm01.prm
```

runmm01

```
#!/bin/sh
# generate equilibrium solution
sgxlat -p15 mm01eq.sg
sgbuild sim mm01eq
order mm01eq.top mm01eq.prm
mm01eq -vs2
rm mm01eq
  (etc.)
rm mm01eq.prm
# translate, compile, link and run the diode bridge mixed mode simulation
sgxlat -p15 mm01.sg
sgbuild sim mm01
order mm01.top mm01.prm
mm01 -vs1
rm mm01
  (etc.)
rm mm01.prm
```

is used here, where a BJT is switching an inductive load with a shunt 1D pin diode. The inductor and the diode are connected in parallel between the power supply V_{ss} and the collector of the BJT (the diode with its forward direction toward the power-supply connection). The emitter of the BJT is grounded. The base of the

BJT is connected to a voltage source V_S through the base resistor R_B.

This is a very interesting simulation. It has quite sophisticated time-step-size control, since the rise and fall of the input signal both need to be resolved with better resolution than the flat part of the square wave. Also, when the diode's stored charge is removed and the diode starts to reverse bias, a small time step is needed. It also shows how circuit values can be put in a data file for easy modification. This simulation shows the BJT model going from forward active (FA), to off, to FA, to saturation, and back to FA. It shows both the turn-on and turn-off transient of the PIN. The turn-off transient is particularly difficult to simulate. (It also demonstrates that the current through the inductor cannot instantaneously change.) See figures 12.8 to 12.11. The figure captions explain the results observed.

A summary of the events taking place can be given in terms of four phases.

PHASE I: The BJT is in forward active mode (the Base-Emitter junction is forward-biased, the Base-Collector junction is reverse-biased). The current $I_c = 3$, $I_l = 3\beta I_b$. Since $\frac{dI_l}{dt} = 0$, $V_l = 0$ and the diode is in equilibrium.

PHASE II: The base drive source starts to quickly ramp down, i.e. I_b decreases, which causes I_c to decrease. Since I_l cannot change instantaneously, some of the inductor current starts to flow through the diode. Since the current in the diode increases rapidly, there is a forward voltage overshoot. The voltage overshoot is caused by the intrinsic region, which is highly resistive until injected carriers can lower its resistivity (a process known as conductivity modulation).

PHASE III: The BJT is off (both junctions are reversed biased). All of the inductor current now flows through the diode. The inductor current decays with a time constant determined by the inductor-diode circuit.

PHASE IV: The base drive source starts to quickly ramp up. We then expect the diode to be forced to go from its on-state to a blocking state (a process known as reverse recovery). To do this, the stored (minority) charge which was put into the intrinsic region earlier must be removed. The removal is achieved by means of a reverse current and by recombination.

With a circuit like this, it is important to think carefully about how to simulate it. If one did not know that the large reverse current flowing through the diode would abruptly change once the stored charged in the device was removed, then one would take too large a time step and the simulation would diverge. One can anticipate this event by watching the voltage across the PIN. This strategy was exploited in the simulation. Unlike the case where the BJT turns off, and the simulation makes use of an increase in the time step immediately after the input signal finishes its transition and the BJT turns off, in this phase the simulation must postpone increasing the time step after the BJT turns on so that it can capture the sudden transition which occurs when the stored charge in the pin is removed.

The input file mm02.sg which performs this simulation is given next.

```
// FILE:  sgdir\ch12\mm02\mm02.sg
// mesh constants ... these correspond to the data in mesh.dat
const NX  = 176;                    // number of x-mesh points
```

```
const LX  = NX-1;                    // index of last mesh point
const Wd  = 160.0e-4;                // device width
const Nd  = 5.0e+19;                 // peak acceptor concentration (cm^-3)
const Ne  = 1.0e+14;                 // epitaxial donor concentration (cm^-3)
const Ns  = 1.0e+19;                 // substrate donor concentration (cm^-3)

// physical constants and built-in potential
const T   = 300.0;                   // ambient temperature (K)
const e   = 1.602e-19;               // electron charge (C)
const kb  = 1.381e-23;               // Boltzmann's constant (J/K)
const eo  = 8.854e-14;               // permittivity of vacuum (F/cm)
const eSi = 11.8;                    // relative permittivity of Si
const eps = eSi*eo;                  // permittivity of Si (F/cm)
const Ni  = 1.25e10;                 // intrinsic concentration of Si (cm^-3)
const Vt  = kb*T/e;                  // thermal voltage (V)
const Vbi = Vt*ln(Nd*Ns/sq(Ni));     // built-in potential (V)

// scaling constants
const x0  = Wd;                      // distance scaling (cm)
const T0  = T;                       // temperature scaling (K)
const V0  = kb*T0/e;                 // electrostatic-potential scaling (V)
const C0  = max(Nd,Ns);              // concentration scaling (cm^-3);
const D0  = 35.0;                    // diffusion-coefficient scaling (cm^2/s)
const u0  = D0/V0;                   // mobility-coefficient scaling (cm^2/V/s)
const R0  = D0*C0/sq(x0);            // recombination-rate scaling (cm^-3/s)
const t0  = sq(x0)/D0;               // time scaling (s)
const E0  = V0/x0;                   // electric-field scaling (V/s)
const J0  = e*D0*C0/x0;              // current-density scaling (A/cm^2)
const L0  = V0*eo/(e*sq(x0)*C0);     // Laplacian scaling constant


// 1D Cartesian discretized gradient operator
func grad(f1,f2,<h>)
  return (f2-f1)/h;

// 1D Cartesian discretized divergence operator
func div(f1,f2,<h>)
  return (f2-f1)/h;

// electron lifetime
func LTn(<N>)
  const LT0  = 3.95e-4;
  const Nref = 7.1e15;
  return LT0/(1.0+N/Nref);

// hole lifetime
func LTp(<N>)
  const LT0  = 3.52e-5;
```

```
     const Nref = 7.1e15;
     return LT0/(1.0+N/Nref);

  // scaled electron mobility
  func MUn(<N>,<T>,pn)
     const Al = 1430.0,  Bl = -2.20;
     const Ai = 4.61e17, Bi = 1.52e15;
     assign NU  = abs(N)*C0;
     assign pnU = (pn>0.0)*pn*sq(C0)+(pn<0.0);
     assign MUl = Al*pow((T/300.0),Bl);
     assign MUi = {Ai*T*sqrt(T)/NU}/{ln(1+Bi*sq(T)/NU)-Bi*T*T/(NU+Bi*sq(T))};
     assign MUc = {2e17*T*sqrt(T/pnU)}/ln(1+8.28e8*sq(T)*pow(pnU,-1/3));
     assign X   = sqrt(6*MUl*(MUi+MUc)/(MUi*MUc));
     return MUl*{1.025/[1+pow(X/1.68,1.43)]-0.025}/u0;

  // scaled hole mobility
  func MUp(<N>,<T>,pn)
     const Al = 495.0,   Bl = -2.20;
     const Ai = 1.00e17, Bi = 6.25e14;
     assign NU  = abs(N)*C0;
     assign pnU = (pn>0.0)*pn*sq(C0)+(pn<0.0);
     assign MUl = Al*pow((T/300.0),Bl);
     assign MUi = {Ai*T*sqrt(T)/NU}/{ln(1+Bi*sq(T)/NU)-Bi*T*T/(NU+Bi*sq(T))};
     assign MUc = {2e17*T*sqrt(T/pnU)}/ln(1+8.28e8*sq(T)*pow(pnU,-1/3));
     assign X   = sqrt(6*MUl*(MUi+MUc)/(MUi*MUc));
     return MUl*{1.025/[1+pow(X/1.68,1.43)]-0.025}/u0;

  // scaled recombination
  func R(n,p,<tn1>,<tp1>)
     const tn = 4.9e-5/t0;
     const tp = 4.2e-5/t0;
     const Cn = 2.8e-31/[D0/sq(C0*x0)];// scaled electron Auger coefficient
     const Cp = 1.2e-31/[D0/sq(C0*x0)];// scaled hole Auger coefficient
     const ni = Ni/C0;                 // scaled intrinsic concentration
     const pn = sq(ni);                // scaled intrinsic concentration squared
     assign Rsrh = (n*p-pn)/[tp*(n+ni)+tn*(p+ni)];
     assign Raug = (n*p-pn)*(Cn*n+Cp*p);
     return Rsrh+Raug;

  // scaled electric field
  func E(V1,V2,<h>)
     return -grad(V1,V2,h);

  // scaled electron-current density
  func Jn(n1,n2,p1,p2,E,<C1>,<C2>,<h>)
     const  Ut   = Vt/V0;             // scaled thermal voltage
     assign un1  = MUn(C1,T,n1*p1);
     assign un2  = MUn(C2,T,n2*p2);
```

```
  assign un    = ave(un1,un2);
  assign dV    = E*h/(2*Ut);
  assign n     = n1*aux2(dV)+n2*aux2(-dV);
  assign dndx  = aux1(dV)*grad(n1,n2,h);
  return un*(n*E+dndx);

// scaled hole-current density
func Jp(n1,n2,p1,p2,E,<C1>,<C2>,<h>)
  const  Ut    = Vt/V0;               // scaled thermal voltage
  assign up1   = MUp(C1,T,n1*p1);
  assign up2   = MUp(C2,T,n2*p2);
  assign up    = ave(up1,up2);
  assign dV    = E*h/(2*Ut);
  assign p     = p1*aux2(-dV)+p2*aux2(dV);
  assign dpdx  = aux1(dV)*grad(p1,p2,h);
  return up*(p*E-dpdx);


// set numerical algorithm parameters
set NEWTON DAMPING    = 3;
set NEWTON ACCURACY   = 1.0e-12;
set NEWTON ITERATIONS = 100;
set LINSOL ALGORITHM  = GAUSSELIM;
set LINSOL FILL       = INFINITY;


// declare variables
var x[NX], C[NX], V[NX], n[NX], p[NX], tn[NX], tp[NX], h[LX], Ad, Vd, Id;
var nlast[NX], plast[NX], dt, t;

// scaled Poisson's equation
equ V[i=1..LX-1] ->
  L0*div(eSi*E(V[i-1],V[i],h[i-1]),eSi*E(V[i],V[i+1],h[i]),ave(h[i],h[i-1])) -
  (p[i]-n[i]+C[i]) = 0.0;

// scaled electron continuity equation
equ n[i=1..LX-1] ->
  +div(Jn(n[i-1],n[i],p[i-1],p[i],E(V[i-1],V[i],h[i-1]),C[i-1],C[i],h[i-1]),
       Jn(n[i],n[i+1],p[i],p[i+1],E(V[i],V[i+1],h[i ]),C[i],C[i+1],h[i ]),
       ave(h[i],h[i-1])) -
  R(n[i],p[i],tn[i],tp[i]) = (n[i]-nlast[i])/(dt/t0);

// scaled hole continuity equation
equ p[i=1..LX-1] ->
  -div(Jp(n[i-1],n[i],p[i-1],p[i],E(V[i-1],V[i],h[i-1]),C[i-1],C[i],h[i-1]),
       Jp(n[i],n[i+1],p[i],p[i+1],E(V[i],V[i+1],h[i ]),C[i],C[i+1],h[i ]),
       ave(h[i],h[i-1])) -
  R(n[i],p[i],tn[i],tp[i]) = (p[i]-plast[i])/(dt/t0);
```

```
// current boundary condition
equ V[i=0] -> Id/Ad/J0 =
  Jn(n[0],n[1],p[0],p[1],E(V[0],V[1],h[0]),C[0],C[1],h[0])+
  Jp(n[0],n[1],p[0],p[1],E(V[0],V[1],h[0]),C[0],C[1],h[0]);


// declaration of simulation variables
var Af, Ar, Ifo, Iro, Ifc, Irc;
var Vb, Ve, Vc, Ib, Ic, Ie;
var Rb, L, Vl, Il, Illast, Vs, Vss;
var tr, th, ta, rr, As;

// declaration of simulation equations ...
// unknown variables are determined from equation headers
equ Ifc -> Ifc = Ifo*{exp((Vb-Ve)/Vt)-1.0};
equ Irc -> Irc = Iro*{exp((Vb-Vc)/Vt)-1.0};
equ Ie  -> Ie  = Ifc-Ar*Irc;
equ Ic  -> Ic  = Af*Ifc-Irc;
equ Ib  -> Ib  = Ie-Ic;
equ Id  -> Id  = Il-Ic;
equ Il  -> Vl  = L*(Il-Illast)/dt;
equ Vb  -> Ib  = (Vs-(Vb-Ve))/Rb;
equ Vd  -> Vd  = Vbi+(V[0]-V[LX])*V0;
equ Vl  -> Vl  = -Vd;
equ Vc  -> Vss = Vl+(Vc-Ve);

begin SetUpDiodeMesh
  open "mesh.dat" read;
  file goto "mesh points";
  file read x;
  file goto "dopant concentration";
  file read C;
  close;
end

begin InitDiodeVars
  // read the scaled values of the V, n, and p equilibrium profiles
  // from the pin02 simulation
  open "Veq.dat" read;
  file read V;
  close;
  open "neq.dat" read;
  file read n;
  close;
  open "peq.dat" read;
  file read p;
  close;
```

```
// initialize the other variables and scale the variables
assign x[i=all] = x[i]/x0;
assign C[i=all] = C[i]/C0;
assign h[i=all] = x[i+1]-x[i];
assign tn[i=all] = LTn(abs(C[i]))/t0;
assign tp[i=all] = LTp(abs(C[i]))/t0;
assign Vd = 0.0;
assign Id = 0.0;
end

begin InitCircuitVars
  open "mm02.dat" read;
  file read Af,Ar,Ifo,Vb,Ve,Vss,Rb,L,Ad,As,tr,th,rr;
  close;
  assign Iro = Af*Ifo/Ar;
  assign Vs  = As;
  assign Vc  = Vss;
  assign Ifc = Ifo*{exp((Vb-Ve)/Vt)-1.0};
  assign Irc = Iro*{exp((Vb-Vc)/Vt)-1.0};
  assign Ie  = Ifc-Ar*Irc;
  assign Ic  = Af*Ifc-Irc;
  assign Ib  = Ie-Ic;
  assign Vl  = 0.0;
  assign Il  = Ic;
end

begin UpdateLastVariables
  assign nlast[i=all] = n[i];
  assign plast[i=all] = p[i];
  assign Illast = Il;
end

var temp1,temp2;
begin ComputeAdjustmentInterval;
  assign temp1 = log(th/tr)/log(rr);
  assign temp2 = rr;
  assign ta = 0.0;
  while (temp1 > 1.0) begin
    assign ta = ta+temp2;
    assign temp2 = temp2*rr;
    assign temp1 = temp1-1.0;
  end
  assign ta = ta*tr/100.0-th/100.0;
end

var hold;
begin main
```

```
// initialize the simulation variables
call SetUpDiodeMesh;
call InitDiodeVars;
call InitCircuitVars;
call UpdateLastVariables;

// NOTE:   dt has to be large to simulate steady state but it cannot
//         be too large, since Vl = -Vd = L*(Il-Illast)/dt and Vd is
//         not strictly zero in equilibrium because of numerical
//         round-off errors.  For example, if Vd is -1e-17, L = 1e-3,
//         and dt = 1e2, then Il needs to be 1e86.  This will cause
//         havoc in the numerical solver.
assign t    = 0.0;
assign dt   = 1.0e3;
solve;
call UpdateLastVariables;
write;

// clear output file
open "mm02.out" write;
close;

// ramp base drive current to maximum voltage and hold
assign hold = 1;
assign dt = tr/1000.0;
call ComputeAdjustmentInterval;
write;
do begin
  // set the base drive voltage, solve the simulation, and update
  // the last arrays and variables
  if      (t < tr)     assign Vs = As*max(-1.0,1.0-2.0*t/tr);
  else if (t < th)     assign Vs = -As;
  else if (t < th+tr)  assign Vs = As*min(1.0,2.0*(t-th)/tr-1.0);
  else                 assign Vs = As;
  solve;
  call UpdateLastVariables;

  // write the circuit values to the output file
  open "mm02.out" append;
  file write t,dt,Vs,Vb,Vc,Vd,Vl,Ib,Ic,Il,Id;
  close;

  // a small time step is needed until the stored charge in the diode
  // is removed ... once the diode becomes sufficiently reverse biased
  // it is safe to take a larger time step.  The variable hold signals
  // when it is safe to increase the time step.
  if (Vd < -0.975*Vss)
    assign hold = 0;
```

Script File runmm02.bat

```
echo off
cls

sgxlat -p15 mm02.sg
call sgbuild sim mm02
order mm02.top mm02.prm
mm02 -vs1 -vl1 -Lmm02.log
group t dt Vs Vb Vc Vd Vl Ib Ic Il Id <mm02.out >mm02.tab
del mm02.exe
del mm02.top
del mm02.res
del mm02.cpp
del mm02.h
del mm02.prm
```

```
    // increment the time and adjust the time step
    assign t = t+dt;
    if       (t < tr)     assign dt = min(tr/100.0,dt*rr);
    else if (t < th-ta)  assign dt = min(th/100.0,dt*rr);
    else if (hold > 0.5) assign dt = max(tr/100.0,dt/rr);
    else                 assign dt = min(th/100.0,dt*1.1);
  end while (t < 2.0*th);
end
```

The DOS script runmm02.bat which runs this simulation is given in a table.

12.3   Summary

This chapter has considered the issue of simulation of devices when the devices are embedded in external circuits. The interaction with the external circuit causes additional complications, which were pointed out. To simulate these circuits it is necessary to make sure that the simulation has built into it the capability to describe the behaviors which we know are likely to take place. In other words, we need to provide the benefit of an intuitive understanding of the device and its function in the circuit when we set up the simulation, if that simulation is to be effective. This is a reasonable state of affairs. Simulations which use standard numerical techniques in a 'brute-force' way are rarely optimal. In situations where complex behaviors can occur, and especially when those behaviors happen on widely different time scales, it is imperative to construct an optimal simulation which reflects the real behavior by building it into the simulation explicitly; otherwise the simulation is unlikely to be successful.

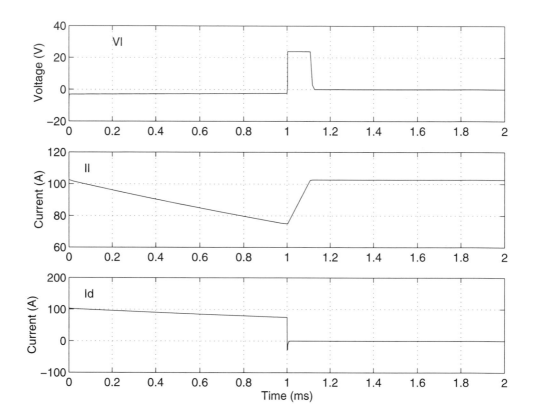

Figure 12.8. The inductor and diode current and voltage waveform over an entire switching cycle. The diode voltage waveform is not shown, since $V_d = -V_L$. In the middle plot, of the inductor current, we see an initial ramp down because the BJT is off; when the BJT turns back on, the current ramps back up to its original value. It should be possible to see that $V_L = L\frac{dI}{dt}$. Energy is being dissipated in the diode, which acts as a resistor, after its turn-on transient. The turn-off transient of the diode is explained in figure 12.11.

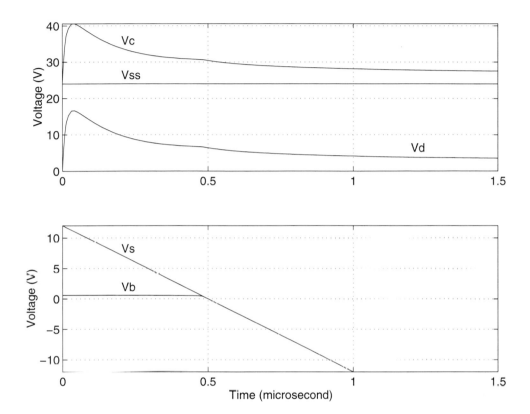

Figure 12.9. The BJT turn-off transient, and pin turn-on transient. At first, V_C is larger than the supply voltage; the inductor is acting as an energy source. Since the inductor time constant is long, compared to the the switching time of the BJT, the current in the inductor is roughly constant. Since the inductor current cannot change instantaneously, the current, which was flowing through the BJT, must now flow through the pin diode. The curves change slope at 0.5 microseconds, when all the current which was flowing in the BJT has been transferred to the diode. Since the current in the pin diode is ramped up rapidly, there is a large voltage across the pin until its resistance is reduced by conductivity modulation. If a simple circuit model of the diode was used, this turn-on transient would be neglected.

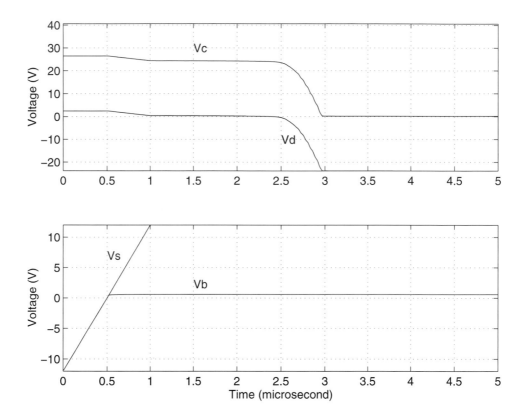

Figure 12.10. Shows the turn-on transient of the BJT. The diode voltage goes from a positive value (forward bias) to a reverse-bias value over a finite time, because the charge stored in the diode has to be removed before the diode can turn-off.

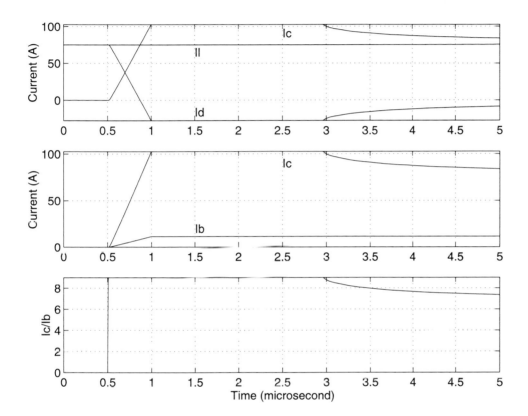

Figure 12.11. Shows the turn-off transient of the diode. A very large negative diode current flows for a considerable time, until all the stored charge is removed from the diode's drift region. Once the stored charge is removed, the reverse current goes to its saturation value. Once the diode is discharged, the BJT goes to saturation, as shown by the fact that I_c/I_b decreases to a value less than β_F. The BJT stays saturated until the inductor current ramps up to its maximum value.

PROBLEMS

1. Sketch the circuits corresponding to the input files for circuit problems given in this chapter. Derive the equations used and comment on their applicability.

2. Draw the circuit for the diode bridge example in the text. Rewrite the simulation to include a resistive load. This will require modification of the current boundary conditions. Then add a shunt capacitor to simulate a simple ac-dc converter.

3. Modify the diode bridge simulation to operate with diode number three replaced by a diode with half the doping level everywhere. Repeat the simulations given.

4. The schematic below indicates a circuit which is to be used to make either a NOR gate or a NAND gate.

```
    + VDD
      |
      |
      /
      \ R
      /
      |
    -----
  |   X   |
    -----
      |
      ---
       -
```

A NOR gate can be implemented, by using two MOSFETs in parallel with each other as element 'X'.

A NAND gate can be implemented by using two MOSFETs in series.

Set up a mixed mode simulation of each of these cases, and simulate the performance of the gates, as the input voltage is varied.

5. The following input file, op.sg, simulates an operational amplifier used in a Wien bridge oscillator. Derive the finite-difference equations which are used. Analyze the time step which is allowed. Until the resistor R_2 is modified, the problem is linear. Run the simulation with a fixed value of R_2 for the initial transient while the voltage is building up to a level where the value of R_2 will start to switch. Then run the simulation with a small time step for two cycles, starting from the end of the first simulation you did, allowing the simulation to determine the value of R_2. Plot the results.

```
// FILE:  sgdir\ch12\op\op.sg
//
// Wien Bridge Oscillator Circuit, using an Operational Amplifier
//
```

```
//                                  R2
//                              ---/\/\/\---
//                  R1              |          |
//              ---/\/\/\-------|\         |
//              |               VB | \        |
//              +                  |  >---------- Vo
//              Vs             VA | /         |
//              -                  --|/         |i1(down)
//              |                  |            |
//              ---             |--|  |-/\/\/\-
//              -               |  C     R
//                          -------
//                  |           |
//              C ---       > R
//              ---          >
//                  |           |
//                          -------
//                              |
//                             ---
//                              -
```

```
// Circuit Parameters
const R1 - 1.0e+4;
const R2 = 20300;
const R  = 1.0e+4;
const C  = 1.6e-8;
const Rsh= 1.0e+3;
const dt=0.01*(R*C);      // Time Step
const NT=2000;            // Number of Time Steps
const LT=NT-1;            // Maximum value of Time Index

var vo[NT],vA[NT],i1[NT],vC[NT],vS[NT],R2e[NT],t[NT];

equ R2e[i=1..LT]-> R2e[i]= ((vo[i-1]-vA[i-1])*(vo[i-1]-vA[i-1]) < 225.0) *
                           (R2 - Rsh) + Rsh ;
equ vo[i=1..LT] -> vo[i] = (R2e[i] + R1)/R1 * vA[i-1] + vS[i-1] ;
equ i1[i=1..LT] -> i1[i] = (vo[i] - vA[i-1] - vC[i-1])/R ;
equ vA[i=1..LT] -> vA[i] = dt/C * (i1[i] - vA[i-1]/R) + vA[i-1] ;
equ vC[i=1..LT] -> vC[i] = vC[i-1] + i1[i] * dt/C ;

begin main
  assign t[i=all] = i*dt ;
  assign vS[i=0..5] = 5.0 ;
  solve;
end
```

6. One of the main issues to be dealt with in setting up a complex simulation, such as is common in mixed-mode simulations, is how to deal with simulations that have very

different time scales. The strategy we shall develop involves solving the problem on the fast time scale, either by averaging over fast motions or by finding an analytic solution which is valid on the 'fast' time scale. Once we know what happens on the fast time scale, we may be able to use the analytic solution or the average behavior to solve the problem on the slow timescale. A classic example of this in a nonlinear problem is the 'nonlinear oscillator'. The case we shall examine was originally studied because it was important for rotating machines. The 'method of averaging' was subsequently applied to a wide range of problems [7]. Suppose an object is at a radius r and an angle θ, and it is rotating so that $\frac{d\theta}{dt} \simeq \omega$, where ω is constant and $\frac{dr}{dt} \simeq 0$. However, its exact equations of motion are

$$\frac{d\theta}{dt} = \omega + \alpha r \cos(\theta) \tag{12.1}$$

and

$$\frac{dr}{dt} = \alpha r \sin(\theta) \tag{12.2}$$

where α is very small. Solve these equations numerically for $\alpha = 0.1$ and several initial values of (r, θ), where r is between about 0.5 and 1.5. Follow the motion for about ten gyrations of the object.

7. Now divide the two variables (r, θ) into an average part and an oscillating part, so that $r = \bar{r} + \tilde{r}$, where $\bar{r}$ is the average part and $\tilde{r}$ the oscillating part of r. Similarly, $\theta = \bar{\theta} + \tilde{\theta}$. The equations for θ and r can be solved approximately to obtain the average and the oscillating parts, 'order by order' in α. That is, the lowest order includes all the terms with magnitude α^0. α does not appear in them. The first order terms are proporional to α. The second order terms are proportional to α^2 and so on. For example, the average part of θ is just $\bar{\theta} = \theta_0 + \omega t$, to order zero in alpha, while $\bar{r} = r_0$ to order zero. All oscillating quantities are zero to this order. Taking the lowest-order solution for r and θ and substituting it into the above equations, we obtain

$$\frac{d\theta}{dt} = \omega + \alpha r_0 \cos(\theta_0 + \omega t) \tag{12.3}$$

and

$$\frac{dr}{dt} = \alpha r_0 \sin(\theta_0 + \omega t) \tag{12.4}$$

where θ_0 and r_0 are the initial values for the particular orbit. On the right hand side, terms which involve α^2 have been omitted. Solve these equations for the oscillating parts of θ and r.

8. Since r and θ can be divided into parts which oscillate and parts which do not, and since the oscillating parts can be written as sin or cos of $n\omega t$, where n is an integer, then it should be possible to average $\frac{dr}{dt}$ and $\frac{d\theta}{dt}$ over one cycle to find the average part of each of these quantities. Integrate the exact equations carefully over one cycle and find the rate of change of the average quantities. Then use the rates of change of the average quantities to integrate the average quantities for N_c cycles, where N_c is a large integer. Use a time step equal to exactly N_s cycles, where N_s is a small integer. Reevaluate the averages over a cycle. Check to see how much the average has changed. If the change is small, make N_s larger; if the change is large, make N_s smaller. Take another step of N_s cycles, and so on. Plot the average values of r and

θ as you integrate them. Compare the results to those from the first problem, where the full motion was integrated in detail.

9. Take the solution for the oscillating parts of θ and r and substitute into the right-hand sides of the original equations. Divide each equation into a part which oscillates and a part which does not, keeping only terms up to and including order α^2. Integrate the parts which do not oscillate, for about ten cycles of the motion.

Chapter 13

Kinetic Transport Models

This chapter presents models for kinetic simulation of particles, with specific examples of applications to carrier behavior and to ion transport during ion implantation. The sections of the chapter are arranged in the order which makes the kinetic simulations clearest.

Section 13.1 is an overview of kinetic theory; Section 13.2 describes Monte Carlo particle simulations; in Section 13.3 are given the equations governing the motion of charged particles in semiconductors, upon which the MC method and the other methods which follow rely. Section 13.4 introduces integral equation methods in kinetic theory. We are interested specifically in (Section 13.5) methods for finding the distribution directly and (Section 13.6) methods to find the scattering rate of particles, from which the distribution can also be found. The integral equation methods are basically straightforward, although they may be unfamiliar.

In some circumstances (which are described below), it is no longer adequate to use the fluid description of transport. The cases we shall discuss here where the model fails are ion implantation and hot-electron behavior in semiconductors.

In fluid calculations the densities of particles such as the carriers are the main dependent variables that are calculated. As we shall see, the density by itself is often not enough to describe the particles' behavior. Frequently, the densities are obtained using average fluxes in the particle-conservation equation. These fluxes are found from expressions which are written in terms of the diffusion coefficient and possibly the mobility. Fluid equations fail in two situations which are really the same:

1. When the mean free path of the particles is no longer short compared to the spatial scales we are interested in, or

2. the collision time is no longer short compared to the time scales we need to resolve.

These are frequently the same condition, since the time scale can often be found from the spatial scale by dividing it by the particle velocity.

When these occur, the fluid description and the concepts that go with it may not even be applicable. Then a kinetic description of the particles' behavior is called for.

Kinetic simulations are generally computationally difficult to do accurately, and expensive in the sense that they require very large amounts of computational resources. For these reasons, the remarks made in Chapter 12, on mixed-mode, about the importance of constructing optimal numerical representations of real systems apply here too. 'Brute-force' numerical calculations are too slow for kinetic simulations, but worse than this, they are not accurate. An optimal simulation must be set up that has the relevant physical properties built into the simulation explicitly. This, in turn, means that we must use our understanding of the problem to model it so that the simulation captures its essential features.

13.1   Kinetic Simulations

Kinetic models not only describe the particles in terms of the density, the average flow velocity, and so on (which is the information contained in the solution of the fluid equations, which include the continuity equation), but also provide a description of how the particles are distributed with respect to their individual velocities. Kinetic models can be divided into two main types, at least in terms of their computational implementation. First, particle simulations represent the behavior of a system of particles by having the computer follow the motion of a large number of simulation particles. ('Simulation particles' is the name given to the fictional entities which the computer 'follows'.) Monte Carlo calculations are of this type; see Section 13.2. The behavior of the simulation particles is intended to sample that of the real particles. Following the particles' motion means integrating the equations of motion over time, keeping track of the position and velocity of each particle.

The second kinetic simulation method involves calculation of the distribution function of the particles. The distribution function is a density, but it is the density not only per unit of physical space but also per unit of velocity. For example, if we are using the position x and the velocity v_x (which together are called the phase space (x, v_x)), then the distribution function $f(x, v_x)$ is the number of particles in a unit of x near x, which are also in a unit of v_x near v_x. In other words, the number of particles between x and $x + dx$ which are also between v_x and $v_x + dv_x$ is $f(x, v_x)\, dx\, dv_x$.

The distribution function obeys a kinetic equation, which is basically an equation stating that particles are conserved as they flow around in phase space. Conservation is local in phase space, meaning not only that particles are conserved but also that they do not jump around suddenly except in the case of collisions, which make the particles' velocities change abruptly.   We define a collision to be an interaction of particles which happens too fast for us to resolve the details. This means that it is very fast compared to the time scales which we intend to use when we solve the kinetic equation, so it appears to be sudden within the solution. (The collision would appear to be a slow, thermodynamically reversible process to an observer who was fast enough to witness the details of what we choose to call a collision.) Finally, any ionization or recombination of the particles also leads to a failure to conserve particles. The conservation equation must therefore include a collision term which moves particles around in velocity and a source term which can add (or remove) particles.

As mentioned above, ion implantation and hot-electron transport in semiconductors provide examples of important processes where fluid equations are not adequate and where a kinetic description of the particles is essential. During ion implantation a high-energy beam of ions is injected into a solid (in our case a semiconductor). The injection is usually in a direction nearly normal to the solid surface. The ions are slowed down and deflected sideways by collisions with nuclei in the solid and with electrons. Fluid equations, such as those described in Chapter 3, do not furnish enough detail about the particles, such as their angular distribution of velocities (which is mostly pointing in the direction of the incoming beam) or their distribution in energy (which is not at all like a thermal distribution) to describe the behavior of the implanted ions. Instead Monte Carlo methods (which when applied to the simulation of particle kinetic behavior are a subset of particle simulations) or direct solutions of the kinetic equation have been used to model ion implantation.

Hot electrons are important in very small semiconductor devices. They may cause damage in VLSI devices (see Chapter 10, on MOS devices). Some novel devices are designed to exploit the capabilities of a hot group of electrons, since they move much faster than a group of thermal electrons and could therefore be used to achieve a more rapid response from the device. The reason for their presence in small devices is that a moderate potential difference ΔV applied to the ends of a very small semiconducting region, of length L, can lead to a very large electric field of the order of $\Delta V/L$, which will accelerate the electrons rapidly. Looked at another way, the applied potential multiplied by the charge on the electron is equal in magnitude to the energy the electron can pick up in passing through the region. If the mean free path λ is comparable to the size of the region, so that $\lambda \geq L$, some of the electrons may cross the region and retain all or most of their energy. Since most of them are presumably not being deflected sideways by collisions either, given that λ is not small compared to L, they may form a high-energy beam of particles. If particles have velocities which are mostly in the same direction, then their distribution of velocities is highly anisotropic. Anisotropic means that it is not the same in all directions; if the distribution of velocities were the same in all directions it would be said to be isotropic. The distribution is also far from being an equilibrium (or thermal) distribution. In these regards the hot electrons are like the implanted ions; their velocities are not isotropic nor are their energies thermally distributed. These features rule out the use of fluid equations, which assume isotropy and a thermal distribution of energies, among other things.

In both cases, of ion implantation [149] and hot-carrier transport [150], the most common method in use in simulations is the Monte Carlo (MC) method. When applied to kinetic theory of particles the MC method becomes a variant of the particle simulation method, meaning that large numbers of simulation particles are followed in this approach. Monte Carlo methods are relatively straightforward to understand and to set up. The results of MC calculations can be difficult to interpret, because the simulation particles that are followed are only a sample of the real particles. There may be too few simulation particles to give adequate statistics for some purposes. The basic ideas of MC methods are described in the next section, followed by the equations of motion that are needed when an MC method is used in application to semiconductor problems. Then integral equation methods will be introduced.

13.2 Monte Carlo Particle Simulation

The Monte Carlo method, as applied to the behavior of particles, consists of following the particles, treating their collisions as random events [151]. The randomness in the method is the origin of the name, providing a link to games of chance. In between collisions, the particle's motion is described by equations of motion, which are generally of the form

$$m\frac{d\mathbf{v}}{dt} = q\mathbf{E} \tag{13.1}$$

and

$$\frac{d\mathbf{x}}{dt} = \mathbf{v} \tag{13.2}$$

where m is the mass and q the charge of the particle; $\mathbf{v}$ is its velocity and $\mathbf{x}$ its position. These equations are stepped forward in time, until a collision occurs, through time steps of Δt. The time step may be a constant or it may be varied during the run. The determination of when a collision will be said to have occurred is the next topic to be discussed.

There are two aspects to deciding when a collision occurs in an MC calculation. First, we need to know the probability of a collision happening. Second, we must decide when a collision actually occurs, from that probability. If the particle travels a distance dl, then, provided dl/λ is very small, the chance of a collision in that step is approximately $p = dl/\lambda$. The exact expression for the probability of a collision while traveling a finite distance l is

$$p = 1 - \exp(-\frac{l}{\lambda}) \tag{13.3}$$

Equation 13.3 is valid for large or small values of l and gives the above result for very small l/λ. One way to include collisions is to take steps which are small compared to λ, and, after each step, calculate $p = dl/\lambda$ (since both dl and λ may vary during the particle's motion). A random-number generator is used to provide a random number r, which is uniformly distributed between zero and one. If r is less than p, then a collision is said to have occurred. r will be less than p a fraction of the time equal to p, since the random-number generator returns the number r, which is distributed uniformly in the correct range.

If a collision was indicated, the particle's velocity is updated to reflect the effects of the collision. The equations of motion are then integrated further, until the next collision, and so on.

An alternative way to determine when the collision occurs, which is applicable when λ is constant, relies on the cumulative probability. The cumulative probability is in this case the probability of having a collision before traveling a distance l. The probability of not colliding while traveling the distance l is $\exp(-\frac{l}{\lambda})$, which is the basis for the expression given above for the probability of colliding, which is one minus this expression. The cumulative probability of colliding before reaching the distance l is denoted $P = 1 - \exp(-\frac{l}{\lambda})$. P is also uniformly distributed between zero and one. If the random number r is generated and set equal to P, the corresponding distance l is found by inverting the equation obtained. $r = 1 - \exp(-\frac{l}{\lambda})$, so that $\exp(-\frac{l}{\lambda}) = 1 - r$. Since $1 - r$ is also uniformly distributed between zero and one, we can replace $1 - r$ by r, and we find that

$$l = -\lambda \ln r. \tag{13.4}$$

l is the distance traveled until the next collision. If we know l, then we do not need to test for a collision after each time step. Instead, after each collision we generate a new value of r and recalculate l. We next include a collision when the particle has gone a total distance equal to this new l since its last collision.

By following many particles in this way, either simultaneously or one after another, a representative sample is built up of the behaviors which can be expected from the real particles. How to extract useful information from that sample, and the reliability of the sample, are complex issues in their own right [151].

13.3   Motion of Charged Particles in a Semiconductor

The main difficulty which generally occurs when using the MC method, described above, is due to the statistical nature of the information it provides. On the other hand, the main difficulty, which is associated with the two particular physical situations we wish to consider here, is the calculation of the collision frequencies (or equivalently the mean free paths) and the particles' motion between collisions, which we refer to as the 'ballistic' motion. In the case of carriers in a semiconductor, a very valuable summary of the necessary data for setting up kinetic simulations in the most important semiconductor materials was given by Ferry [152]. Collision rates for implanted ions are discussed in [153] and in Chapter 4.

The ballistic, or collisionless, part of the motion is also made quite complicated in both of these examples by the crystal lattice. Those of the implanted ions which are moving along certain preferred directions or channels parallel to the crystal planes have greatly reduced drag associated with their ionic collisions. Channeling can significantly increase the ion range over the value it would have in an amorphous crystal, in which there would be no channels.

Turning to the carrier behavior, the energy E of an electron in the crystal depends on the value of its wavevector or k-vector. The group velocity of the packet of waves, which make up the electron, is given in terms of its angular frequency ω

and of k by

$$v_g = \frac{\partial \omega}{\partial k}. \tag{13.5}$$

This is the one-dimensional result. Written as a vector this is

$$\mathbf{v}_g = \nabla_{\mathbf{k}}\omega \equiv \left(\frac{\partial \omega}{\partial k_x}, \frac{\partial \omega}{\partial k_y}, \frac{\partial \omega}{\partial k_z}\right) \tag{13.6}$$

where the operator $\nabla_{\mathbf{k}}$ is defined by this equation.

If we use the de Broglie relation $E = \hbar\omega$ to relate the electron's energy to its angular frequency, the electron group velocity is thus

$$\mathbf{v}_g = \frac{1}{\hbar}\nabla_{\mathbf{k}}E. \tag{13.7}$$

The group velocity is the rate of change of the particle's position, $\mathbf{v}_g = \frac{d\mathbf{x}}{dt}$. When describing the electrons in the crystal, the other appropriate independent variable to use in addition to x is the wavenumber k_x (and the other components of the position and wavevector). The rate of change of the k-vector is obtained by setting the rate of change of the crystal momentum, which is defined to be $\mathbf{p}_c = \hbar\mathbf{k}$, equal to the Lorentz force, so that

$$\frac{d\mathbf{p}_c}{dt} = \hbar\frac{d\mathbf{k}}{dt} = e(\mathbf{E} + \mathbf{v}_g \times \mathbf{B}). \tag{13.8}$$

These equations for the electron's motion are known as the semiclassical equations.

To solve the semiclassical equations one must know the band structure, that is, one must know how E depends on $\mathbf{k}$, so a knowledge of the details of the lattice is again required.

13.4 Integral Equation Methods in Kinetic Theory

The solution of these two problems by nonstatistical techniques will now be described. These techniques use methods which are, in principle, based on using integral equations to describe the transport. Nonstatistical simply implies that we are not sampling the real particles by means of simulation particles. Instead, we calculate the density in cells, which are spread throughout phase space.

To see how the integral equations arise, consider a mesh which is divided into cells. The density in any given cell of the mesh at the end of a time step (for example) gets there from a number of different initial cells where the density was at the start of the time step. The total density put into the cell is the sum of the contributions from all of the initial cells. That density is a sum on a discrete mesh made up of finite cells. When the cell size is allowed to go to zero, the sum is replaced by an integral over the initial locations the particles came from at the start of the time step. In this sense, the method uses integral equations, although, in practice, we are summing contributions from finite cells.

The advantage of these methods is that they combine high efficiency with high accuracy. Both are achieved in part because the methods allow particles to take large 'steps', as we shall see.

13.5   Distribution Function Found from the Convected Scheme

The first calculation we describe is of the particle distribution function. If the distribution in question is the carrier distribution, then the phase space will normally consist of the independent variables $(\mathbf{x}, \mathbf{k})$. The usual equations of motion in these variables are the semiclassical equations given in the previous section. The discussion that follows is in terms of $(\mathbf{x}, \mathbf{v})$, but the use of $(\mathbf{x}, \mathbf{k})$ is exactly the same. The much-cited, but rarely used, 'iterative method' of Rees [154],[155] used a convolution integral to solve the integral form of the kinetic equation. The technique which follows, which we call a 'convected scheme' (CS), is more intuitive and usually more accurate.

The calculation of the distribution function in a cell at time $t + \Delta t$ is done by summing contributions from different cells where particles began at time t, the start of the time step Δt. Different physical effects which redistribute density on the mesh within the duration Δt of the time step are accounted for one after another. We begin with the ballistic motion, which is the motion in the absence of collisions. The effects of the collisions are included in a second step.

13.5.1   The Ballistic Move

To describe the distribution function, we divide the phase space into cells. For simplicity the phase space can be thought of as being (x, v_x). A cell might have width Δx in x and width Δv_x in v_x. As usual, the cells of the mesh are fixed in place. To describe how the particles get distributed by the ballistic move, we also choose to introduce moving cells. The moving cells go where the (fixed) mesh cells would go if they could flow along at the same velocity as the particles inside them. There is no numerical limitation on the size of the time step which is allowed. The numerical Courant limit on finite-difference schemes does not apply to this method. It is always necessary, however, to limit Δt to allow the appropriate physical behavior to be resolved. As was discussed in the section on the use of propagators in transport theory, the propagators allow large time steps, and the large time step reduces numerical diffusion.

The density is in a set of fixed cells at time t. It gets put back into the fixed cells at time $t + \Delta t$, although usually with different amounts in each fixed cell. Where the moving cells go in the time step Δt indicates where the density should be put at the end of the step. See Figure 13.1.

As particles move around during the time step Δt, changing their positions in x and in v_x, the faces of each moving cell follow the trajectories of the particles. Particles which were along the faces of the fixed mesh cell at the start of the time step can be thought of as markers, which indicate where the moving cell goes. The moving cells overlap a number of fixed mesh cells at the end of the time step. The particles are split between final fixed cells according to where the moving cell goes.

In the simplest version of this scheme, the area of each moving cell which falls into each fixed cell, is used to determine how much of the density in the moving cell is also in that fixed cell. Suppose some initial fixed cell i has a density N_i in it at time t. The moving cell, which coincided with fixed cell i at the start of the step, typically overlaps several fixed cells (which may include the initial cell i) including a fixed cell f at time $t + \Delta t$. The fraction of the moving cell which overlaps the fixed final cell is F_i^f, after the moving cell has been allowed to move during Δt. A fraction F_i^f of the density N_i will be placed in the final fixed cell f. The density which is put in cell f, from cell i, is thus $F_i^f N_i$. The total density placed in f is

$$N_f = \sum_i F_i^f N_i. \tag{13.9}$$

The sum over initial cells includes the final cell itself, since particles often stay in the same cell during a time step. Since the fractions of the particles starting in the initial cell i, and either staying there or going elsewhere, must sum to one, the number of particles in that initial cell is exactly conserved. Conservation of particles, locally in phase space, is therefore achieved by making sure that the sum over final cells f of the fractions F_i^f is exactly one for each initial cell i. See Figure 13.1.

Often it is useful to design the scheme to conserve energy as well as particles. In that case the kinetic energy E_i of particles in each spatial cell i (which is not necessarily the initial cell now) is found from conservation of energy. First, we calculate the positions of the front and back of the moved cell, denoted x_f and x_b, and the kinetic energy of the moved cell in each spatial cell i it overlaps. Then we can use the overlap of the moving cell with the fixed cells (in space and energy, at the end of the time step) to find how many particles to put back into each final fixed cell. Overlap in energy means that if a fraction equal to 0.9 of the moving cell is in the range of energies associated with energy cell k, then the fraction 0.9 of the particles in the moving cell will be put into energy cell k, when they are replaced on the mesh. This is equivalent to saying that particles are replaced, on the velocity or energy mesh, so that total energy is conserved. Using overlap in energy in the replacement 'rule' ensures that the mean energy is correct for the particles from that initial cell which are put back in any particular spatial cell. This is discussed further, in Figure 13.1 and Figure 13.2.

There is an obvious advantage to having the kinetic energy E_i of the moving cell exactly match one of the energies E_j on the energy mesh. If this happens, all the particles in the moving cell within a given spatial cell go to the same energy cell, so there is no numerical diffusion along the energy axis. (This can be quite important, since diffusion in energy can cause some particles to move up in energy and give artificially high rates of inelastic processes, such as ionization.) If the electric field is constant, then the change in particle energy between successive spatial cells is $\Delta E_i = \mp q E_x \Delta x$. The spacing between energy cells can be chosen to match this; $E_{j+1} = E_j \pm \Delta E_i$.

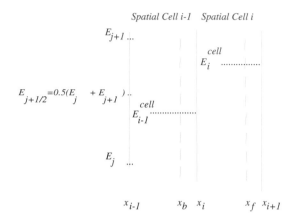

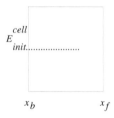

Figure 13.1. Schematic showing how an initial cell, at the lower left, moves during a time step. The final positions of its front and back edges and the mean kinetic energies which the particles from the initial cell have in each of the final fixed spatial cells are indicated. The fraction of the density to the left of x_i, which is equal to $(x_i - x_b)/(x_f - x_b)$, goes into spatial cell $i - 1$. The rest of the density goes into spatial cell i. The same procedure is used to redistribute the particles in energy or velocity. If we wish to conserve energy, we make energy the velocity-space variable. We use the same sort of overlap rule for energy as was just described for use in space. The kinetic energy of the center of the moving cell in spatial cell $i - 1$ is E_{i-1}^{cell}. In spatial cell i the moving cell center has kinetic energy E_i^{cell}. These kinetic energies are found from conservation of energy. The density from the moving cell in each spatial cell i is shared between the energy cells at E_j and E_{j+1} so as to obtain the correct mean kinetic energy E_i^{cell} in that spatial cell. See Figure 13.2.

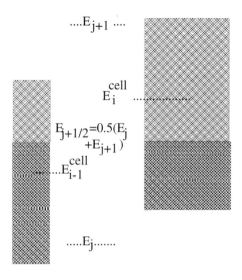

Figure 13.2. The distribution of particles from the same initial cell according to their kinetic energy, in spatial cells $i-1$ and i. Particles are shared between the energy cells so that the total energy of the batch of particles which are put back in each spatial cell is conserved. If the energy cells have uniform spacing (which is not always convenient), particles with kinetic energy above $E_{j+1/2}$ are placed in the energy cell at E_{j+1}, and those with kinetic energy below $E_{j+1/2}$ are placed in the energy cell at E_j.

The density in all the initial cells is replaced on the mesh by reading the densities into an empty array, which will represent the new densities. After this step, collisions can be included. Many variants of the ballistic move described here have been developed for use in different circumstances. The details of the methods are too lengthy to give here but can be found in the literature [8].

13.5.2   The Collision Operator

Collisions change the velocities of the particles, but not their positions - at least, not directly. If their velocity is changed, then their position will eventually be changed as a result of the different velocity, but not immediately. The collision operator only moves particles in velocity, therefore.

The collision operator consists of two probabilities. One is the probability of having a collision of a given type. This is just the collision frequency multiplied by the time step, or $\nu \Delta t$, at least approximately and provided $\nu \Delta t$ is small. This must be calculated based on the local densities which affect ν and on Δt. The particles that collide are then removed from the cell they were in; the fraction removed is equal to the probability of the collision. Of those which are removed by the collision, we now need to know what is the probability of their subsequently going to each other velocity cell (at the same spatial location). The fraction of the particles which scatter in velocity cell a, having had a collision of a given type, and then go to velocity cell b (at the same spatial location) depends only on the velocities associated with cells a and b and on the type of collision. It can be calculated once at the start of the simulation and used later, since it is the same for each spatial cell and at all times. The sum over all cells b for a given initial cell a must be one, to conserve particles.

Once all initial cells have been looped over, the time step is complete and a new time step can be begun, all these steps being repeated in each subsequent time step until the maximum run time is reached.

13.5.3   An Alternative Approach to Kinetic Calculations

An alternative approach to the direct calculation of the distribution function using the CS, is to use the CS to calculate the scattering rate in each phase-space cell of the mesh. The calculation begins with an initial guess for the scattering rate in each phase-space cell of the mesh. Those particles are propagated along trajectories until they scatter. The number scattering per second into each cell from the trajectory starting in any given initial cell represents the contribution of that initial cell to the total source (scattering) rate in that phase-space cell. The CS allows an accurate determination of the phase-space cells where particles are replaced and guarantees conservation of particles and energy in complex phase-space geometries. The calculation is iterated to steady state. Once the scattering rate is known, the distribution function is found from the scattering rate, using the CS to construct the distribution in each cell. The advantages of working with the scattering rate (reduced numerical diffusion and memory requirements) are described in the next section. Only the reduction in numerical diffusion applies to the method described in this subsection.

13.5.4   Self-Consistent Calculations

In a problem involving charged particles, it is often important to use an electric field that is found consistently with the density of charged particles. The density of charged particles can be found from the distribution function by integrating (i.e. summing) over all velocities. This density can then be inserted into the total free charge which appears in Poisson's equation. This allows the electrostatic potential V and the electric field to be found, for use during the next time step.

When V is found self-consistently, the time step Δt is limited to be a small fraction of the period of a 'plasma oscillation'. The plasma oscillation is an oscillation of the plasma in which the electric field provides the restoring force. The lightest charged species being followed has the fastest oscillation. By calculating V self-consistently, we make plasma oscillations possible in the simulation. This in turn means that we must be able to resolve the plasma oscillation, which means using a time step smaller than the oscillation period. (If we do not limit Δt in this way, the plasma particles will artificially heat up and the simulation will eventually diverge.)

13.5.5   Charged Particles in a Constant Electric Field

A simple example of a kinetic simulation based on these ideas will now be given. This example file is set up to describe the motion of charged particles falling under the influence of a constant electric field. They undergo collisions which reduce their velocity to a low value, close to zero. The particles are injected into the simulation region from the left of the region, at a moderate velocity, and move to the right. Only positive velocities are included in this simulation, which is done on a rectangular mesh in (x, v_x) space.

This example does not make use of the capabilities of the SGFramework software in a significant way. It could just as well have been implemented using a standard Fortran or C program. In fact part of the example could have been done more generally in a standard program. SGFramework does not support general integer variables, hence array indices cannot be calculated from expressions within the program. The possible changes in the cell indices within a time step are 'hard coded' into this example. Two possibilities are allowed for the distance traveled in the ballistic move, as should be clear from the file. One is a step of less than a cell width, and the other is a step of between one and two cell widths.

Input File cs.sg

```
// FILE:  sgdir\ch13\cs\cs.sg
// DESCRIPTION:  This simulation finds a distribution function,
//               using the convected scheme (CS).
```

```
CONST NX = 11;                 // number of x-mesh points
CONST NV = 65;                 // number of v-mesh points
CONST NVB2 = NV/2 ;            // NV by 2.
CONST LX = NX - 1;            // index of the last x-mesh point
CONST LV = NV - 1;            // index of the last v-mesh point
CONST DX = 2.5 / LX;          // mesh spacing in the x-direction
CONST DT = 0.1;               // time step
CONST VMAX = 2.0*DX/DT;       // maximum velocity
CONST DV = VMAX / LV;         // mesh spacing in the v-direction
CONST C = 1.0;                // velocity
CONST PI = 3.14159;           // pi
CONST NU = 3.0 ;              // collision frequency
CONST TMAX = 10.0;            // maximum time
CONST IS = 0;                 // location of source on the x-axis
CONST JS = LV/4 ;             // location of source on the v-axis
CONST JSPR=44 ;               // spread of the source along v-axis
CONST DELV = 1.4*DV;          // acceleration in dt
CONST JINT = 1 ;              // integer part of DELV/DV;
CONST NUDT = NU*DT;           // fraction of particles which collide in dt


// the t variable stores the time
// the F array stores the distribution function
//     F[0,*,*] is the amplitude at time = t - 1 * DT
//     F[1,*,*] is the amplitude at time = t

var t,dlv,FP[NX,NV],FCL[NX,NV],F[2,NX,NV],DELX[NV];
var TEMP[NV,NV],TEMP2[NV-2,NX-3];
var X[NX],VX[NV];

//
//  The main equation finds the new F, at h=1, from the old F at h=0.
//

begin main
  // initialize variables
  assign X[i=all] = i*DX ;       // position array
  assign VX[j=all]= j*DV ;       // velocity array
  assign t = 0.0;
  assign F[h=all,i=all,j=all] = 0.0 ;
  assign DELX[j=0..NVB2] = DV*j*DT/DX ;
  assign DELX[j=NVB2+1..LV] = DV*j*DT/DX - 1.0 ;
  assign dlv=DELV/DV - JINT;

  // loop from t = 0 to TMAX with a time step of DT
  while (t <= TMAX) begin
    assign F[h=0,i=IS..IS+1,j=JS..JS+JSPR] =
              1.0e+12/(sq(j-JS+1)) ;  // Set F at the inlet.
```

```
 assign FCL[i=all,j=all] = 0.0 ;
 assign FP[i=all,j=all]  = 0.0 ;

 assign FCL[i=all,j=all] = NUDT*F[0,i,j] ;

 assign F[h=0,i=1..LX,j=1..LV] = F[h,i,j] - FCL[i,j] ;

 assign FP[i=all,j=1..LV] = FP[i,j-1] + FCL[i,j] ;

 assign F[h=1,i=2..LX-1,j=2..NVB2]  =        // Update F where deltax < dx
     F[h-1,i-1,j-JINT-1]*DELX[j]*dlv
    +F[h-1,i-1,j-JINT]*DELX[j]*(1.0-dlv)
    +F[h-1,i,j-JINT-1]*(1.0-DELX[j])*dlv
    +F[h-1,i,j-JINT]*(1.0-DELX[j])*(1.0-dlv)    ;

 assign F[h=1,i=2..LX-1,j=NVB2+1..LV-1] =   // Update F  where deltax > dx
     F[h-1,i-2,j-JINT-1]*DELX[j]*dlv
    +F[h-1,i-2,j-JINT]*DELX[j]*(1.0-dlv)
    +F[h-1,i-1,j-JINT-1]*(1.0-DELX[j])*dlv
    +F[h-1,i-1,j-JINT]*(1.0-DELX[j])*(1.0-dlv) ;

 assign F[h=1,i=0..LX-1,j=0..1] = F[h,i,j] + 0.5*FP[i,LV] ; //replace scttrs.
 assign F[h=0,i=0..LX,j=0..LV] = F[h+1,i,j] ;       // Update "old" distrn.
 assign F[h=1,i=0..LX,j=0..LV] = 0.0 ;              // Zero out "new" dstn.

 assign t = t + DT ;
end
// write the solution to the result file
assign TEMP[i=0..LX-3,j=0..LV] = F[0,i+2,j];
assign TEMP[i=0..LX-3,j=0..LV] = log(0.001+TEMP[i,j]) ;
assign TEMP2[i=0..LV-2,j=0..LX-3] = TEMP[j,i] ;
write;
end
```

The distribution F found in this way is shown next. The logarithm of the distribution is plotted in Figure 13.3, since F itself varies by many orders of magnitude across the mesh.

13.6   Kinetic Simulations Based on the Scattering Rates

In some physical situations, it is convenient to simulate kinetic phenomena using a different formulation than a direct calculation of the distribution function. In the method outlined in this section, the quantity being calculated in each cell of the mesh is the rate at which particles scatter in that cell. From the scattering rate, and in some cases with extra information about the angular distribution of the velocities of the scattered particles, we can find the distribution function [156],[157],[158] . Later in the section we show how ion implantation may be described in this way.

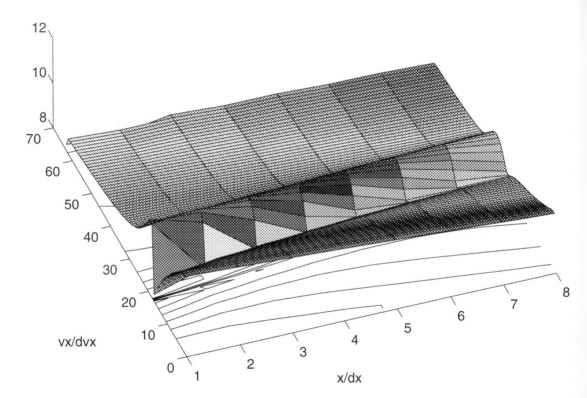

Figure 13.3. The distribution function is the density per unit velocity (or per unit energy) as well as per unit length. Since the distribution varies by many orders of magnitude over the velocity axis, and sometimes along the spatial axis as well, the logarithm of the distribution is shown. The 'tail' of the distribution (the most energetic particles, which usually are low in density) may be important, despite its low density, since the energetic particles may ionize other particles or damage the crystal lattice. The distribution of particles shown here was found using a propagator method (the convected scheme) and corresponds to particles which are accelerating (for instance in an electric field) and undergoing collisions which reduce their velocity to zero.

One reason why this is a useful approach is that specifying the scattering rate may be a very compact way to summarize the particle behavior. A second reason is that in this method each 'step' the particles take corresponds to the distance between successive collisions. This is a relatively large step, so numerical diffusion is reduced and the calculation itself is speeded up.

Suppose that each scattered particle has an equal chance of going in any direction after it has a collision. This means that all the particles which have a collision in any given cell of the mesh are isotropically distributed in velocity. The total distribution of particles in a cell labeled k is made up of particles coming from a number of other cells labeled i. That distribution is not necessarily isotropic, even though the particles scattering in each cell i were isotropic when they left the cell i where they scattered. For example, if the scattering only happens in one cell i and the particles are isotropic when they leave cell i, the (total) distribution of particles in some other cell k (which all came from cell i in this example) is not isotropic because all the particles are traveling away from cell i. This example suggests how a relatively simple situation, in which all the particles are isotropic when they leave the cell in which they scattered, when viewed differently in terms of the total distribution in each cell appears to be a great deal more complex, since the total distributions are likely to be anisotropic. The anisotropic total distribution requires much more data to specify its form, so the description of the particle kinetics will be more compact if it is given in terms of the (isotropic) scattering rates.

The advantage of using the scattering rates is most pronounced when the angular distribution of the scattered particles is relatively simple. If the angular distribution after scattering is isotropic, then the whole distribution function can be specified in terms of the scattering rate in each spatial and energy cell. This rate is represented by one number per cell on a combined space and energy mesh. But the whole mesh in phase space has two more dimensions than this, the extra dimensions corresponding to two angles which specify the direction of motion. These angles were not needed to specify the scattering rate when the scattering was isotropic. In general, the distribution might be highly anisotropic after each scattering event. In the case of anisotropic scattering, we would need to specify the density of particles in each cell in space, in energy and in addition in the two angles, using a total of six independent variables. In the case of isotropic scattering, the use of the scattering rate as the primary dependent variable lets us get away with four independent variables instead of six. This decrease in the number of independent variables means that the mesh is four dimensional instead of six dimensional and has hundreds or thousands of times fewer cells in which the dependent variable (the scattering rate) must be calculated. In a complex physical problem where the spatial mesh is already large, this is a very important benefit.

The situation, to which we intend to apply these ideas, is the transport of ions during ion implantation. The angular distribution of scattered particles is actually highly anisotropic, but we will be able to describe that distribution in a relatively simple form. The advantage of the method will thus be retained, in that it will be possible to express the scattering rate along with the angular distribution of scattered particles using a relatively compact array of numbers.

To obtain the scattering rates, we need to calculate several sets of probabilities. These probabilities are used to find how many of the particles which scatter in any given cell of the mesh have their next scatter in each cell of the mesh. The calculation of the probabilities will be discussed later. First the use of the probabilities to find scattering rates in steady state will be outlined. (See Figure 13.4.)

The probability of scattering next in cell i having just scattered in cell i' (which may be the same cell) is denoted as $P_{i'}^i$. $R_{i'}$ is the rate of scattering in cell i'. A fraction of these scatters equal to $P_{i'}^i$ lead to scatters in i. This means that in steady state the scattering rate $R_{i'}$ in cell i' contributes a rate of scattering equal to $P_{i'}^i R_{i'}$ in cell i. All the other initial cells i' also contribute to the scattering rate in cell i, so summing the rate over the initial cells i' gives the total scattering rate in cell i:

$$R_i = \sum_{i'} P_{i'}^i R_{i'}. \tag{13.10}$$

If particles are being injected in some cell from outside the system being simulated, then their rate of scattering must be added to this. Starting with an initial guess at the rates (which might be zero except for the scattering caused by the external injection, for example) this equation is iterated until steady state is reached.

To give a simple example, if 10^{14} particles per second scatter in cell 1 and if $P_1^2 = 0.01$, so that a fraction equal to 0.01 of these scatter next in cell 2, then in steady state cell 1 contributes 10^{12} scatters per second to the total rate of scattering in cell 2.

The reason for a particle going from an initial cell to a final cell could be that it moves in space, or it could be a collision that changes the particle's velocity. The way the probability is used in the iterations does not depend on the process that is described by the probability, except that the probabilities for different processes are stored separately and are used one after another within each iteration of the equation for the scattering rate.

13.6.1   Ion Motion Inside a Crystal

The physical process of ion motion and slowing down inside a crystal is described in detail in the literature [149],[159]. The subject is somewhat difficult to grasp from the articles which give the details of the physical models, and we shall review it here.

In ion implantation the physical processes that must be included are:

1. A ballistic motion of ions between collisions with the ions in the crystal lattice. The probability of going in a ballistic move from a scatter in the initial cell (c', E') to a subsequent scatter in the final cell (c, E'), at the same energy E' but a different final location c' from the initial location c, is denoted as $T_{bal}(c, E'; c', E')$. T_{bal} stands for the Transition probability in the ballistic move.

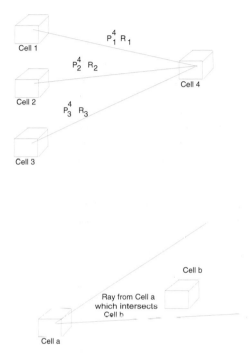

Figure 13.4. Calculation of the total scattering rate in a cell. The scattering at rate R_1 in cell 1 contributes a scattering rate $P_1^4 R_1$ in cell 4, since P_1^4 is the probability that a particle which just scattered in cell 1 will scatter next in cell 4. All the cells contribute to the scattering in each other cell. The lower part of the figure shows the 'ray' from cell a which intersects cell b, as described in the discussion of the calculation of the probabilities.

2. The electron drag, which lowers the ion's energy during the ballistic motion. The probability of going from energy E' to some other energy cell at energy E due to electron drag is denoted $T_{ele}(E'; E)$. This probability also depends on the distance traveled in space while the drag is acting on the ion.

3. When the ions collide with lattice ions, they again lose energy and are deflected in angle.

Several different approaches to modeling ion implanation have been employed in the past. The most similar to that used here was a solution of the kinetic equation done by Giles and Gibbons [160] . The method we employ relies on obtaining transition probabilities between cells of the mesh, which describe the probability of being moved from one cell of the mesh to another in various processes. The way these probabilities can be found, for the case of ion implantation, will be described next.

13.6.2   Ballistic Motion

Just like in the kinetic simulation described in the previous section, the collisions themselves are in one regard relatively easy to handle numerically, because they change the particle's velocity but not its position. If we know the energy E' of the particle immediately before it had the collision, then the probability of its going to any energy E as a result of the collision depends on E and E' and the type of collision. This probability does not depend on the position where the collision occurred. The probabilities of the possible angular deflections when the energy changes from E' to E are also known, so the new angular distribution of velocities at energy E can be found from the old distribution at energy E'. This will be explained in detail below.

The main difficulty with the numerical simulation is caused by the spatial movement of the particles. In three spatial dimensions there is a very large number of cells in the mesh. Each axis is divided into a large number of increments. Suppose each axis has N_a increments; then the mesh has $N_m = N_a^3$ cells. Since $N_a \geq 20$, the mesh will have $N_m \geq 10^4$ cells. The problem with this is that we need to know the probability of going from any of these cells to any other of the cells. There are N_m^2 ways to pair the cells with each other, if it makes a difference which cell is the first member of the pair, and half this number of ways of it does not. This means there are of order N_m^2 different probabilities of particles going from some cell to some other cell (or to the same cell). Calculating and storing $N_m^2 \sim N_a^6$ different numbers is cumbersome. If $N_a \geq 20$, then $N_m^2 \geq 10^8$. We need a way to find the probabilities which is fast and which does not require storage of all of these numbers.

To find the probability of a particle traveling from one cell to another, we need to know the likelihood of the particle going in any direction, the mean free path of the particle at any point along its path and the geometry of the cells. To demonstrate what is involved, we can describe how the probability can be found approximately, using a Monte Carlo (or similar) method.

One relatively straightforward way to find the probability of a particle going from an initial cell i' to a final cell i is to use a computer to 'launch' a large number of simulation particles from the initial cell i'. If we can find how many of the particles launched in i' have their next collision in i, then dividing that number by the total number launched gives the fraction going from i' to i, which is the probability we want.

The particles' launch positions should be spread out uniformly in the initial cell. The direction of particle motion should be chosen so that the correct fraction of the particles go in each direction. Finally, the distance they travel before the next collision must be found using the mean free path λ, which may vary from point to point. If λ is constant, then the distance traveled can be found by setting

$$s = -\lambda \ln r, \tag{13.11}$$

where r is a random number which is distributed uniformly from zero to one. If λ is not constant, then a fraction of the particles equal to ds/λ will be said to collide in each short distance ds, provided that $ds/\lambda \ll 1$. (An exact expression for the fraction colliding is $(1 - \exp(-ds/\lambda))$.) To choose 'randomly' whether an event has occurred, when the event has a probability p of occurring, we call the random-number generator and obtain a random number r which (as before) is uniformly distributed between zero and one. Then if $r < p$ we say the event occurred; otherwise it did not.

The initial positions, the launch angles, and the distances traveled can be chosen either by using random numbers or by making sure we launch the correct fraction of the particles in each range of space and angle, and that the right fraction collide in each range of distance traveled. For example, if the initial cell is between $x_{i'}$ and $x_{i'} + \Delta x$, then the particles' initial positions in x could be set equal to $x_{i'} + r\Delta x$, where r is our usual random number, which is distributed uniformly between zero and one. Alternatively, one could choose a small number α (preferably such that $1/\alpha$ is an integer) and launch a fraction α of the particles at $x = x_{i'}$, a fraction α at $x = x_{i'} + \alpha\Delta x$, a fraction α at $x = x_{i'} + 2\alpha\Delta x$, and so on all the way across the cell (and spreading the particles out similarly in y and z). In the same way, if λ is constant we can choose the position of the next collision by setting $s = -\lambda \ln r$. Alternatively, we can put a fraction α of the particles between $s = 0$ and $s = -\lambda \ln(1 - \alpha)$; another fraction α between $s = -\lambda \ln(1 - \alpha)$ and $s = -\lambda \ln(1 - 2\alpha)$, and so on.

These calculations can in principle also be done by analytic integration, but this is difficult in general and probably impossible if λ is not constant. It should be clear that the calculation of the probabilities by launching particles is straightforward but likely to be very time-consuming, even for a fast computer, for large numbers of cells. We now need to consider ways to find the probabilities, which are much more efficient than these straightforward approaches.

There are several ways to simplify this problem, which are applicable in different circumstances. If the scattering is isotropic (which means that there are equal chances of particles going in all directions) and the mean free path λ is constant, and if there is spatial symmetry in the region modeled, it is possible to express the probabilities in a relatively compact form [156]. The spatial invariance of the probabilities, which in a symmetric case such as this depend only on the difference in the cell positions in the symmetry coordinate and not otherwise on the cell positions in that coordinate, can be used to simplify the calculation of the probabilities and to reduce the amount of memory needed to store them. If the scattering is nearly isotropic and the mean free path varies by a moderate amount, then a good approximation to the probabilities can also be found in a compact form [157]. A method for finding the probabilities efficiently which works for arbitrary λ and arbitrary angular distributions after scattering, on a regular Cartesian mesh, has been developed for use in ion implantation and will be explained next.

Ballistic Transition Probability; General Case

The calculation of the transition probabilities will now be explained, for a case which is rather general and is limited primarily in using a uniform rectangular mesh. The probabilities will be found, first, in what could be thought of as an infinite mean free path limit. From these, it will be possible to find the probabilities for arbitrary finite mean free paths. Now if λ is very large, the probability of colliding in a distance ds is $p = ds/\lambda \ll 1$. Instead of calculating ds/λ as $\lambda \to \infty$, we shall instead (in effect) calculate ds for each final cell. We can then find ds/λ for finite λ by dividing by λ. In practice, what it means to find ds is to find the average path length, which particles coming from all locations in an initial cell i' will spend in cell i, as well as the chance of their going in the direction from i' to i.

Starting in an initial cell, we divide the spherical angles (θ, ϕ) into a mesh of cells of width $\Delta\theta$ in θ and $\Delta\phi$ in ϕ. Then some angular cell might run from $n\Delta\theta$ to $(n + 1)\Delta\theta$ and from $m\Delta\phi$ to $(m + 1)\Delta\phi$. The angular distribution of the particles which scattered in the spatial cell will indicate what fraction of the scattered particles are traveling within this angular cell, from $n\Delta\theta$ to $(n + 1)\Delta\theta$ and between $m\Delta\phi$ and $(m + 1)\Delta\phi$, as they leave the spatial cell. The angular cell defines a 'ray'. We need to know which other spatial cells are overlapped by this ray, and how much it overlaps them. The bigger the overlap of a spatial cell with the ray, in both angle and distance along the ray, the greater the chance that particles which are traveling within the ray will have a collision in that cell.

To find the chance of particles in our 'ray' colliding in a particular Cartesian cell, we can suppose that the Cartesian cell is divided into a large number of very small Cartesian cells, so that almost all of the small cells are either completely inside or completely outside the ray. Suppose a particle is in the ray. We want to know if that particle is likely to scatter in a small cell labeled α. The chance of a particle, that is in the ray, intersecting our small cell, α, is proportional to the fraction of the ray which goes through cell α. This fraction is equal to the solid angle subtended by the cell (at the origin of the ray) divided by the total solid angle of the ray. This is equivalent to the fraction found as described next, by going out along the ray to a distance r_α where the small cell is located. At r_α one then finds the area of the ray. Next, one projects the small cell onto that area, as it appears from the origin of the ray, and finds the fraction of the area of the ray which is covered by the projection of the small cell. The fraction of the ray going through the small cell α will be denoted f_α^{angle}.

The average distance a particle in the ray, which is moving radially outward, has to go to cross the small cell α (provided that it is in the range of angles which intersect cell α) is found next. This average distance is called $\bar{ds}_\alpha$. The chance that a particle which starts to cross the small cell α then collides in α is $\bar{ds}_\alpha/\lambda$. The overall probability that a particle in the ray, which succeeds in reaching radius r_α, will be both in the right part of the ray to cross small cell α (which has probability f_α^{angle}), and will then collide in the cell (which has probability $\bar{ds}_\alpha/\lambda$) is the product of these probabilities. That is, if a particle makes it to radius r_α in this ray, it has a probability of

$$p_\alpha^c = f_\alpha^{angle} \frac{\bar{ds}_\alpha}{\lambda} \tag{13.12}$$

of colliding in the small cell α.

If the probabilities p_α^c are all very small, we can add them to find the probability of the particle colliding in a group of these small cells. In this way, if a 'large' Cartesian cell β partially overlaps our ray, we can find which 'small' Cartesian cells are in both the ray and the 'large' Cartesian cell. We can then find the probability of a collision happening in that group of Cartesian cells, by summing the probabilities of a collision in each of that group of small cells:

$$p_\beta^c \simeq \sum_\alpha p_\alpha^c \tag{13.13}$$

(provided that p_β^c is also small). Then

$$p_\beta^c \simeq \sum_\alpha f_\alpha^{angle} \frac{\bar{ds}_\alpha}{\lambda}. \tag{13.14}$$

Since we want to be able to use this information for all possible values of λ, we actually store the values after multiplying through by λ:

$$p_\beta^{c\infty} = \lambda p_\beta^c \simeq \sum_\alpha f_\alpha^{angle} \bar{ds}_\alpha. \tag{13.15}$$

To find the probability that particles, which scattered in initial cell i', have their next scatter in each other cell i, we first divide the particles into groups, according to which ray they are placed in after their last scatter, which was in cell i'. How the angular distribution is found will be described later. Of the particles in a ray, the fraction which collide in the first Cartesian cell is $p_\beta^{c\infty}/\lambda$, using the $p_\beta^{c\infty}$ and the λ for that cell. That fraction of the particles is removed from the number in the ray. The fraction of the remaining particles colliding in the next cell is similarly $p_\beta^{c\infty}/\lambda$

evaluated in the next cell. That fraction of the particles which remain in the ray is removed, and so on.

The reasons why this procedure is effective, for a Cartesian mesh, are:

1. it works for arbitrary angular distributions and arbitrarily varying mean free path (as long as λ is considerably bigger than the Cartesian cell size); and

2. the 'infinite mean free path' probabilities $p_\beta^{c\infty}$ depend only on the coordinates of the final cell relative to the initial cell. For example, they depend only on the initial cell's x-coordinate x' and the final cell's x-coordinate x in the combination $x - x'$. They do not depend on x' and x separately. The situation is the same for the y- and z-coordinates. The $p_\beta^{c\infty}$ can be written

$$p_\beta^{c\infty} \equiv p_\beta^{c\infty}(x - x', y - y', z - z'). \qquad (13.16)$$

This means that we can store these probabilities as three-dimensional arrays. If no simplification of this kind had been possible, we would have been forced to use six-dimensional arrays, which would have been much less efficient.

The cell centers are at positions $(i\Delta x, j\Delta y, k\Delta z)$, where (i, j, k) are integers and $(\Delta x, \Delta y, \Delta z)$ are the cell dimensions along the corresponding axes. Then $x - x' = (i - i')\Delta x = \Delta i \, \Delta x$. The probabilities are thus stored in terms of the integers, as $p_\beta^{c\infty} \equiv p_\beta^{c\infty}(\Delta i, \Delta j, \Delta k)$. The high degree of symmetry in a Cartesian mesh means that only positive (and zero) values of $(\Delta i, \Delta j, \Delta k)$ need be stored, since the values for the cases with negative integer separations can be found from these.

13.6.3   Angular Distribution

The angular distribution of scattered particles is found from the angular distribution the particles had before they collided, and from their deflection in a collision. For example, suppose we know the initial velocities of all the particles that collide in a certain spatial cell. Suppose we know the fraction of their initial momentum, denoted by $h(E', E)$, that will be retained on average by particles with initial energy E' that scatter to final energy E in a particular type of collision. (This may or may not be true in reality.) If each of N particles that end up with energy E had (different) initial velocities $\mathbf{v}'$, then the mean final velocity in energy cell E will be taken to be

$$\mathbf{v} = \sum_N \frac{h(E', E)\, \mathbf{v}'}{N}. \qquad (13.17)$$

If all we know is the mean velocity at this energy and the energy itself, we can impose an angular distribution of the form

$$f = A + B \cos\theta \qquad (13.18)$$

where $\theta = 0$ is along the direction of the mean velocity. A and B are chosen to give the correct number of particles and mean velocity. From this distribution we can find the number of particles within each ray, by rotating the coordinates to line up with a set of spherical coordinates with (say) $\theta' = 0$ along the z-axis, and then by integrating the distribution over the ray.

13.6.4   Collisions of Injected Ions

There are two major collision processes to be accounted for: nuclear scattering of the injected ions, and electron drag on the ions.   Nuclear scatters which deflect the ions through a very small angle are extremely numerous, but each such scatter has little effect. These small-angle collisions have the effect of causing a drag which can be handled numerically in the same way as the electron viscosity.   Both effects slow the ions down without changing the direction of their motion substantially. A table $T_{drag}(E', c'; E, c)$ can be set up to record the energy E an ion has in each Cartesian cell c it crosses for a given initial energy E' in the initial cell c'. The table T_{drag} allows for energy losses due to the electron drag and the small-angle nuclear scatters. The mean free path $\lambda(E)$ for the larger-angle nuclear scatters depends on the energy E, so T_{drag} must be used to find λ so that we know the mean free path during the ballistic move.

If the material is amorphous, this completes the outline of the steps involved in the setting up of the various probabilities which are needed. If the material is crystalline, there are crystal planes along which ions may travel in which the ions have a reduced probability of suffering nuclear scatters. Ions traveling close enough to a crystal plane have a reduced chance of scattering off nuclei in the crystal. These ions are said to be 'channeled'.   Some range of angles around the crystal axis will typically allow ions to be more or less effectively channeled [153].

13.6.5   Implementation of the Transition Matrix

Figure 13.5 contains a flow chart showing how the 'transition matrix' method is used in a simulation of the injection of ions into a crystal. The ions are followed while they slow down, through multiple generations of scatters. The calculation is set up so that each energy shell can be iterated in order, from the highest to the lowest. Eventually, after some number of iterations, a given higher energy shell will be cleared of ions, since some number of generations of scatters has reduced the ions' energies enough to put them into lower shells. At this point, the calculation proceeds to deal with the next lower shell, and so on.

13.7   Summary

In this chapter we reviewed several effective methods for numerical simulation of problems arising in semiconductor and device behavior, which require a kinetic description of particles. The methods presented here for the kinetic simulation of particle behavior are all designed to be 'optimal' in the sense of being based on the real physical behavior of the particles.

Implantation Model Flow Chart

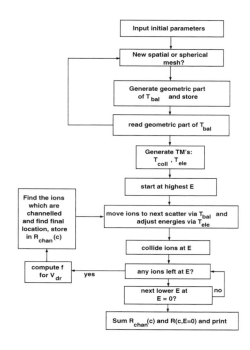

Figure 13.5. Flow chart showing the steps in a 'transition matrix' simulation of ion injection into a crystal.

PROBLEMS

1. Suppose that in the redistribution of particles illustrated in Figure 13.1 the spacing of the energy cells E_j was nonuniform. Sketch the equivalent diagram to Figure 13.1, showing how particles are redistributed so as to conserve energy, in a case where the energy of the kth energy cell is proportional to the index k, $E_k = E_0 + \alpha k$. Suppose the initial cell was in the fifth energy cell from the bottom, E_5, and that in the 'final' spatial cell at the end of the time step the kinetic energy is between E_{15} and E_{16}. Choose α so that the spacing $E_{16} - E_{15}$ is about twice as large as $E_6 - E_5$.

2. These examples are intended to illustrate the ideas of propagators in transport theory. The kind of transport discussed in this chapter is the transport of particles, but now we will use an example of transport of people - by means of London Transport.

   Suppose I catch a train at Harrow, and want to go to Paddington station. Having poor vision I cannot read the names of the stations. The stations along the route are:

   Harrow, Kenton, South Kenton, North Wembley, Wembley Central, Stonebridge Park, Harlesden, Willesden Junction, Kensal Green, Queen's Park, Kilburn Park, Maida Vale, Warwick Avenue and Paddington.

   Assign each of these stations an index i, so that the probability of my being at that station can be referred to by the index.

   Each time the train stops I have a 20 percent probability of getting off. Once off, I will take any train leaving the station I am at, including trains going back the way I just came, with equal probability.

   Whichever station I am at, I have to leave going to one of the neighbors. There are as many states that I can be in when I leave each station as there are neighboring stations (that is, two) since I have to leave going to one of them. What is the probability that at any station I keep on going the way I came in? (This is the probability I stay on the train, plus the probability that I got off, but got onto another train going in the same direction.) The probability of changing direction must be one minus this.

   Sketch a schematic showing the possible routes I can take for the first N_t trips, by drawing an arrow from Harrow to Kenton on the first line; arrows from Kenton to South Kenton and from Kenton back to Harrow on the second line; and so on. (Do not allow travel past the 'boundaries' at Harrow and Paddington.) Draw the schematic for $N_t = 4$. Calculate by hand the value of $P_s[i = 4, N_t = 4]$, the probability of having $i = 4$ when $N_t = 4$. What does this number represent?

3. At each station I can be in one of two states. Set up an array to store the probability that I am going north or south at each station, $P_d[id, is]$, where $id = 0$ might be south and $id = 1$ might be north. is labels the station and runs from zero to thirteen. For each state, the probability of my going to any other state is known. The probabilities can be used to find what my chance is of being in any state at any time. If I am in any given 'old' state with probability $P_d[id, is]$, and my chance of switching to another state is P_{sw}, then my chance of going to the new state next and of having been in the old state immediately before is $P_d[id, is] P_{sw}$. This probability must be added to the probability that I got into the new state by any other means, in that last trip I took, to find the overall probability of my being in the particular new state after

some number of trips. Thus if $P_d[0, 5]$ is 0.01 after 10 trips, and the chance of going to state $[0, 6]$ next is a little more than 0.8, then this particular combination of events gives me a chance of about 0.008 of being in state $[0, 6]$ next. This must be added to my chance of getting there by any other method, to find the actual probability of being there after 11 trips.

It may be easier to understand this process if we think about a large number of travelers, so that we can calculate the actual number of people in each state. Then suppose a thousand soccer fans get on the train at Harrow. There could be many plausible reasons for their failing to notice the signs at the stations. The array we calculate could now be called $N[id, is]$, the number of fans in each state. Then if $P_d[0, 5]$ was 0.01 after 10 trips, that means that a fraction equal to 0.01 of the thousand fans or $N[0, 5] = 10$ fans are at station 5 (Wembley Central, which is probably where they ought to get off) going south after 10 trips. If they have a probability of 0.8 of going to state $[0, 6]$ after one more trip, that means that 8 of those 10 continue to station 6 (Stonebridge Park) and go south from there. To those 8 have to be added any other fans who end up making the same trip south from station 6, but who got there by some other means.

What is my chance of getting to Paddington after 20 trips? (Or alternatively, how many soccer fans make it? A fractional number must be understood as giving a probability of any individual making it, when divided by the total number of fans.) Assume I do not realize where I am, so I may leave.

4. What is the probability of my being there if I manage to stop once I get there? Repeat the above for a journey of 50 trips.

Chapter 14

Related Work

This book has described the design, development and implementation of a highly flexible tool for numerical solution of large-scale computational problems in applied physics, the Simulation Generation Framework (SGFramework). The design of the framework is the result of investigating which aspects of applied physics simulations are generic. Some features might be common to many simulations, and might also be implemented in a general form in a multipurpose simulation framework. We set out to design a tool which could automate the time-consuming and error-prone tasks involved in constructing complex applied physics simulations. The SGFramework has been used to simulate many complex problems in engineering. During this process, several additional key problems were identified and the framework was extended to handle these problems. The issues addressed by SGFramework include:

1. the automatic generation and refinement of irregular meshes,

2. the specification and symbolic differentiation of discretized partial differential equations (PDEs) and boundary conditions on irregular meshes,

3. the specification of complex parameter models, and

4. the generation of efficient computer code.

The main purpose of this book is to describe the simulation of semiconductor devices. To this end, we have outlined the analytic treatment of several important devices. We have then contrasted this with numerical simulations of those devices, or deferred the comparison to the exercises.

More advanced simulation issues were then introduced, including mixed-mode simulations and the use of simulations in combination with statistical methods in the design of semiconductor devices. As our worked example of the design process we considered the avalanche breakdown of pin diodes with field-limiting rings (FLRs). Two new methods of optimizing the FLR terminations were described, and results were presented. The optimization techniques were applied to pin diodes, but statistical design coupled with simulation is a technique which has wide applicability.

The novel aspects of this work include the recognition of the need for, and the implementation of, a simulation tool which can integrate all of the relevant components needed for the efficient simulation of applied physics problems and the application of this framework. The simulation tool clearly has applications to areas other than semiconductors, and this section will outline some which have been developed to date.

14.1 Dry Etching of Semiconductors

Several applications to semiconductor processing, and particularly to 'dry etching' of semiconductors, have been implemented. Dry etching uses a plasma to transfer a pattern from a 'mask' onto the semiconductor. The plasma etches the parts of the semiconductor surface which are accessible through openings in the mask. The advantage of plasma etching is that the plasma tends to cut almost straight into the surface, whereas chemical etching cuts sideways as well as inward. This directional etching is known as 'anisotropic' etching, while the chemical etching tends to be 'isotropic'.

The reason for the directional etching is that plasma etching is largely due to positive ions from the plasma. The positive ions are accelerated toward the surface by a strong electric field in a region close to the surface known as a 'sheath'. The electric field in the sheath attracts ions to the surface (in nearly all cases) and is in the direction normal to the surface. This field is set up because fast-moving electrons reach the surface before the ions can get there in significant numbers, and they charge the surface negatively. Only when the surface is charged negatively does the electric field repel electrons and attract positive ions enough to make the fluxes of each be equal so that a steady state can be reached. This sheath is very like the field in a pn junction, the function of which is to repel the mobile majority carriers on either side of the junction.

SGFramework has been used extensively in describing the uniformity and dose of ion fluxes striking semiconductor surfaces during processing. Multiple physical models are needed to describe a processing plasma. The main physical models are frequently implemented in separate 'modules' which must be able to communicate with each other and obtain self-consistent solutions. For example, the plasma ions are often described using fluid equations similar to those outlined for the carrier densities in this book. These fluid equations are solved in one 'module' of the code, using one SGFramework input file. The electrons cannot always be handled in this way, since their behavior is quite complex. A more sophisticated, 'kinetic' model of the electrons is often called for. The electrons' kinetic equation is solved using a second SGFramework input file. The electrostatic fields which govern the behavior of both species are found in another module, using the densities of all the charged species in Poisson's equation. In many cases it may also be necessary to calculate radio-frequency or microwave electromagnetic fields which are used to heat the plasma electrons. These fields will have to be found using another module, and so on.

A highly sophisticated, modular kinetic simulation of an inductively coupled plasma was set up in SGFramework. The results were shown to give excellent agreement with parameter scalings obtained experimentally [161]. One major stumbling block was that the vacuum electric fields predicted by SGFramework did not agree with experiment. This prompted considerable controversy, which was resolved by the removal from the experiment of a pyrex plate which had not been included in the specification of the experiment which was furnished. (Once the pyrex plate was removed, the vacuum fields as measured and calculated were in close agreement.)

SGFramework has also been used to describe microwave irradiation of ceramic materials and the interaction of this radiation with ionic defects [162][163].

Chapter 15

The SGFramework User's Manual

The user's manual given here provides the syntax of the mesh and equation specification files, and explains the commands needed to use them. (This manual is not a tutorial; it is intended as a reference manual. The text contains an extensive tutorial on the setting up and running of simulations.)

In summary, the procedure to generate and run a simulation is given below (followed by a similar one to construct a mesh).

1. Translate the equation specification file with the SGFramework translator:

`sgxlat filename.sg`

2. Compile and link with the Numerical Algorithm Module:

`sgbuild sim filename`

3. (If necessary) order the unknown elements to reduce sparse-matrix fill:

`order filename.top filename.prm`

4. run the simulation:

`filename`

If the simulation uses an irregular mesh, you must construct it before executing the above procedure.

1. parse the mesh specification file:

`mesh filename.sk`

2. construct the initial mesh:

`sggrid filename.xsk`

3. build the mesh refinement program:

`sgbuild ref file_ref`

4. run the mesh refinement program:

`file_ref`

(Each time SGFramework opens a window to plot the mesh, it is necessary to close the window before the next line will be executed.) These commands all make use of SGFramework Executables. The SGFramework Executables are invoked from a command line. This means that Windows95 and NT users must first open an MS-DOS prompt before giving these commands to run the SGFramework Executables.

Table: Syntax of Syntax Statements

A character enclosed in single quotes represents the literal character. Hence, ',' means one literally types a comma.

The asterisk character (*) means zero or more of the item or set that preceeds it.

The plus character (+) means one or more of the item or set that preceeds it.

Items enclosed by vertical lines are optional.

Items enclosed by brackets form a set. For instance, [A-Z] mean one can type either an A, B, C, ..., or Z.

Items enclosed by braces form a group. Groups are treated as a single item.

Keywords appear in boldface type.

Italized text represents a previous defined entity.

In UNIX, a shell must be opened. The commands are explained in Section 15.4, on SGFramework Executables. The first parts of this manual, on the other hand, are devoted to explaining how to set up the files which are used by the SGFramework Executables.

The text to this point is full of examples of input files. The first of these is the 'Game of Life' in Section 1.5. This gives (and explains) the commands needed to run an existing input file (if that file does not need a mesh to be generated by SGFramework). Section 1.5 is the appropriate point to begin learning to use SGFramework. The examples progress through simple electrostatics problems, beginning with Section 2.1. The use of irregular meshes is introduced in Section 2.8. Section 2.8 is the appropriate place to begin learning about the mesh refinement program and the commands required to use it. The examples in the text are all explained where they are presented, and the easiest way to start is to read the sections which were indicated above, from the first few chapters.

Throughout this chapter there are definitions of the syntax which is appropriate for the various SGFramework commands and functions. The meaning of the syntax statements is summarized in a table of 'syntax syntax'.

This chapter is divided into four sections: The Syntax and Grammar of the Equation Specification File 15.1, The Syntax and Grammar of the Mesh Specification File 15.2, Interfacing the Equation and Mesh Specification Files 15.3, and SGFramework Executables 15.4.

15.1 The Syntax and Grammar of the Equation Specification File

This section describes the syntax and grammar of equation specification files. It is divided into twelve subsections, each of which describes a particular part of the SGFramework language. The topics covered in this section are Comments 15.1.1, Numbers 15.1.2, Strings 15.1.3, Identifiers 15.1.4, Operators 15.1.5, Constants 15.1.6, Variables and Arrays 15.1.7, Functions 15.1.8, Constraint Equations 15.1.9, Procedures 15.1.10, Conditional Statements 15.1.10, and Numerical Algorithm Parameters 15.1.11. Some of the subsections are further divided. The Functions subsection describes internal, user-defined, and external functions, whereas the Procedures sub-section describes assignment statements, single-word statements, file input/output statements, subroutine execution statements, conditional statements and looping statements.

15.1.1   Comments

Comments may be placed anywhere in the source file and are ignored by the equation specification file parser. Comments start with the // characters and terminate at the end of the line. It is a recommended practice to use comments to document a specification file. Proper use of comments will enable other users to easily determine the purpose of the simulation. Examples of comments can be found in the examples given in the text.

15.1.2   Numbers

Both integer and floating-point numbers are supported in equation specification files. Integers are numbers without a fractional part. Floating-point numbers contain both an integral and a fractional part.

Integers

Integers consist of an optional plus or minus sign followed by one or more digits. The range of integer numbers is dependent upon the implementation of the host system's C++ compiler. At a minimum, the integer range is guaranteed to include the numbers -32768 to 32767 inclusive. The syntax of integers is as follows.

$|[+,-]|$ *digit* $+$ where *digit* is the set [0-9].

The numbers $2, -5, 12536, -32768$, and 23 are examples of allowed integers. The number 40000 may be interpreted as an integer or it may be interpreted as a floating-point number if it exceeds the C++ compiler's integer range. It is suggested that users adhere to the integer range of -32768 to 32767 regardless of their C++ compiler, in order to generate portable simulations which will run correctly on multiple platforms. To avoid writing simulations where certain numbers may be interpreted as integers on one computer and as floating-point numbers on another computer, add a decimal point and a zero (e.g. 40000.0) to integral numbers outside the range of -32768 to 32767. This will guarantee that these numbers are always interpreted as floating-point numbers.

Floating-Point Numbers

Floating-point numbers consist of an optional plus or minus sign followed by one or more digits which are optionally followed by a period and one or more digits. Scientific (exponential) notation is supported. SGFramework implements floating-point numbers with double precision and with a guaranteed minimum range of -1.7×10^{-308} to 1.7×10^{308} with 15-digit precision. The syntax of floating-point numbers is as follows.

$|[+,-]|\,digit+|[.]\ digit+|[E,e][+,-]||\ digit+|$

The numbers 1.0, -2.3, $1e1$, 3.14159, $-2.0e-3$, and $4.23e+23$ are examples of floating- point numbers. Although numbers such as 5 and -10 are valid floating-point numbers as far as syntax is concerned, they will be treated as integers, since they are within every C++ compiler's integer range.

Integer and Floating-Point Expressions

Mathematical expressions may be divided into two categories: integer expressions and floating-point expressions. A mathematical expression which consists entirely of integers and operators will evaluate to an integer. Fractions are dropped in integer division. A mathematical expression which contains a floating-point number or a function will evaluate to a floating-point number. Consider the following examples.

$9/2$ evaluates to 4

$9/2.0$ evaluates to 4.5

$log(100)$ evaluates to 2.0

$2*2*2*2$ evaluates to 16

The first expression evaluates to the integer 4, since the expression consists entirely of integers and the division operator. The second expression, however, evaluates to the floating-point number 4.5, since it contains a mixture of an integer and a floating-point number. The third expression evaluates to the floating-point number 2.0, since it contains the *log* function. The fourth expression evaluates to the integer 16, since it consists entirely of integers and the multiplication operator.

In addition, integer and floating-point expressions may be subdivided into two further categories: constant expressions and variable expressions. Constant expressions are expressions that the SGFramework translator can evaluate to a constant at translation time. Constant expressions consist of numbers, constants, operators, internal functions and user-defined functions. Variable expressions are expressions that the SGFramework translator cannot evaluate to a constant at translation time. They contain at least one variable, array element, function argument, or external function. As a final note, integer expressions are always constant, since they cannot contain variables, array elements, function arguments, or external functions.

15.1.3   Strings

A string is a collection of one or more characters enclosed by double quotes. Every displayable ASCII character except the double quote (") character is a valid string character. Strings may not be broken over multiple lines in a specification file. The syntax of strings is as follows.

$[\ "\ ]\ character+[\ "\ ]$ where *character* is any printable ASCII character except for the double quote character.

Examples of strings are as follows.

```
"Hello,"
"Welcome to SGFramework."
```

```
"This is an example of a valid string."
```

15.1.4   Identifiers

Identifiers are used as constant, variable, array and user-defined function names. Certain identifiers are reserved as keywords. Since keywords have a special meaning, they should not be used as constant, variable, array and user-defined function names. Doing so will result in an error. Furthermore, SGFramework contains over 40 internal functions. The names of internal functions also should not be used as constant, variable, array and user-defined function names. Consult the internal function section for the names of the internal functions.

Naming Convention

Identifiers must begin with a letter, which is optionally followed by one or more letters or digits.

```
voltage, Electron, hOLe, cf4
```

are examples of valid identifiers, whereas

```
_voltage, elec_hole, 9lives, cf4+
```

are examples of invalid identifiers. Identifiers are not case sensitive unless the case-sensitive command switch is specified. Only the first seven characters of an identifier are guaranteed to be significant. Therefore the identifiers CH2OHCHOHCH2OH and CH2OHCHOHCH2NH2 may be identical and should not be used. The syntax of identifiers is as follows.

   letter letter , *digit* ∗ where *letter* is the set [A-Z,a-z].

Keywords

Keywords are not case sensitive. Keywords may not be used as identifiers, since the translator interprets them in special ways. Using a keyword in place of an identifier will usually result in a syntax error. SGFramework reserves the following identifiers as keywords.

   'accuracy',
   'algorithm',
   'all',
   'and',
   'append',
   'assign',
   'begin',
   'call',
   'close',
   'comment',

'const',
'coordinates',
'damping',
'diagonal',
'divisions',
'do',
'edge',
'else',
'end',
'exit',
'equ',
'file',
'fill',
'func',
'gausselim',
'goto',
'if',
'ilu',
'infinity',
'iterations',
'known',
'label',
'length',
'linear',
'linsol',
'logarithmic',
'main',
'maximum',
'minimum',
'newton',
'no',
'none',
'not',
'open',
'or',
'over',
'pcg',
'point',
'preconditioner',
'read',
'real',
'rectangles',
'refine',
'region',

Table: Operator Precedence

Operations	Associativity	Precedence
(unary) +, (unary) −, not	right-to-left	highest
*, /	left-to-right	
+, −	left-to-right	
>, <, >=, <=, ==, <>	left-to-right	
and, or, xor	left-to-right	lowest

'return',
'set',
'signedlog',
'solve',
'sor',
'sort',
'unknown',
'var',
'while',
'write',
'xor',
'yes', and
'zero'.

15.1.5 Operators

SGFramework supports the addition, subtraction, multiplication, division, negation and exponentiation operators. The precedence of these operators is shown in the table.

The unary plus, unary minus, addition, subtraction, multiplication, and division operators (+, −, *, /) behave as expected. The greater than, less than, greater than or equal to, less than or equal to, equal to, and not equal to operators (>, <, >=, <=, ==, <>) evaluate to 0 or 1 if both the left and right operands are integer expressions, or 0.0 or 1.0 if either the left or right operands are floating-point expressions. If the expression is true, unity is the result; while if the expression is false, zero is the result. The not, and, or and exclusive or (xor) operators likewise return unity if the expression is true and zero if the expression is false. However, these operators assume that nonzero operands are true. Hence the expression '5 and 3' evaluates to 1. Left-to-right associative operators with the same precedence are evaluated from left to right. Right- to-left associative operators with the same precedence are evaluated from right to left. Consider the following examples.

$4 + 5 * -6$ evaluates as $4 + (5 * (-6)) = 4 + -30 = -26$
$3 - 4 - 7.0$ evaluates as $(3 - 4) - 7.0 = -1 - 7.0 = -8.0$
$3 > 2 and 4 > 9$ evaluates as $(3 > 2) and (4 > 9) = 1 and 0 = 0$
5 xor not $4 + 2$ evaluates as 5 xor $((not 4) + 2) = 5$ xor $(0 + 2) = 0$

In the first example, the operations are performed in the following order: negation (unary minus), multiplication and addition. In the second example, the left subtraction operation is performed first followed by the right subtraction operation. Notice the result is a floating point number as the last operator is the subtraction of an integer and a floating-point number. In the next example, the greater than operators are evaluated before the and operator. The first greater than operation is true, hence unity is the result. The second greater than operation is false, hence zero is the result. The fourth example uses both boolean and mathematical operators. The not operator has highest precedence, so it is evaluated first. Since its operand is true (nonzero) it evaluates to zero (false). Next the addition operator is evaluated and finally the exclusive or operator.

15.1.6   Constants

A constant is a value that is fixed - it does not change during the simulation. Proper use of constants can make a specification file more readable and manageable.

Declaring Constants

To declare a constant one must type the keyword 'CONST' followed by an identifier, an equal sign, an expression which evaluates to a constant and a semicolon. Multiple constants may be declared in the same constant statement. The syntax of the constant statement is as follows.

> CONST *identifier* '=' *const expr* |',' ... *identifier* '=' *const expr* | ';'

Examples of constant statements are as follows:

```
CONST T  = 300;       // temperature (K)
CONST Ks = 11.8;      // dielectric constant of Si
CONST k  = 1.381e-23; // Boltzmann's constant (J/K)
CONST q  = 1.602e-19; // charge of an electron (C)
CONST Vo = k*T/q;     // potential scaling (V)

CONST MIN = 2, MEAN = 8, MAX = 16;
CONST TRUE = 1, FALSE = 0;
CONST A = 20, B = sq(A), C = ln(B);
```

A constant expression is an expression that evaluates to a constant. All previously defined constants and user-defined functions as well as numbers, operators and internal functions may appear in constant expressions. Variables and array elements, on the other hand, cannot appear in constant expressions, since their values need not remain constant throughout a simulation. This is not to say that variables or array elements cannot be initialized to a value which does not change during a simulation. However, in general the values of variables and array elements may change during a simulation. Since the translator has no knowledge of which variables and array elements remain constant and which change, all variables and array elements are excluded from constant expressions. See the following sections for more information concerning variables, arrays, user-defined functions and internal functions.

Using Constants

It is advantageous to use constants when the same value appears many times throughout a specification file. Consider the following specification file which solves Poisson's equation on a nine-by-nine square grid with uniform spacing. The electrostatic potential is held at zero on the boundary and 10 (volts) at the center.

```
var     V[9,9];
unknown V[1..7,1..7];
known   V[4,4];

equ V[i=1..7,j=1..7] ->
  V[i-1,j] + V[i,j-1] - 4.0*V[i,j] +
  V[i+1,j] + V[i-1,j] = 0.0;

begin main
  assign V[i=4,j=4] = 10.0;
  solve;
  write;
end
```

Suppose we decide that the simulation is not accurate enough. To remedy this problem, we wish to more than double the number of grid points. Let us choose a twenty-one by twenty-one grid. Since the dimensions of the array V are explicitly set to nine, we need to manually change all of the integer quantities in this simulation, i.e., the nines to twenty-ones, the sevens to nineteens, and the fours to tens. If we ran this simulation and decided we needed still more accuracy, we would have to repeat this process again. A better way to write this simulation is to define a constant which specifies the dimensions of the array. Consider the following simulation.

```
const DIM = 9;         // number of x and y grid points
const Vcenter = 10.0;  // voltage at center of grid (volts)

// define array V and make all array elements unknowns
// except for those at the center and on the boundary
var     V[DIM,DIM];
unknown V[1..DIM-2,1..DIM-2];
known   V[DIM/2,DIM/2];

// Poisson's equation
equ V[i=1..DIM-2,j=1..DIM-2] ->
  V[i-1,j] + V[i,j-1] - 4.0*V[i,j] +
```

```
  V[i+1,j] + V[i-1,j] = 0.0;

// initialize center voltage, solve Poisson equation and
// write the results
begin main
  assign V[i=4,j=4] = Vcenter;
  solve;
  write;
end
```

Now if we ran this simulation and decided we needed more accuracy (suppose a twenty-one by twenty-one grid), we would only have to change the value of the constant DIM. Although the two specification files are equivalent in function, the second specification file is advantageous. It is much easier to change the number of grid points in the second simulation than the first. Furthermore, the lack of comments in the first specification file make its purpose more difficult to determine.

15.1.7 Variables and Arrays

Variables are blocks of memory that store floating-point numbers whose value may change during a simulation. Every variable must be declared before it can be used in a SGFramework specification file. Failure to declare a variable before using it will result in an error. Array variables, usually called arrays, are a group of variables that are referenced by a common identifier and one to three indices. The use of arrays can greatly simplify the writing of specification files by reducing redundancy.

Declaring Variables

It is easy to declare variables in SGFramework. To declare one or more variables, type the keyword 'VAR' followed by one or more identifiers, separated by commas. As with all SGFramework statements, the variable declaration statement must end with a semicolon. The syntax of the variable declaration statement is as follows.

VAR *identifier* |',' ... *identifier* | ';'

Examples of variable declaration statements follow.

```
VAR voltage;
VAR ElectronConcentration, HoleConcentration;
VAR GasMileage, OdometerReading, TotalDistance;
VAR CF4,H2O,CaCl;
VAR A,B,C;
```

It should be emphasized that only the first seven characters of variable names are guaranteed to be significant. Therefore the following declaration statements may result in errors.

```
VAR ElectronLifetime, ElectronMobility;
VAR ElectrostaticPotential, ElectrostaticField;
```

Declaring Arrays

Declaring arrays is very similar to declaring variables. The only difference is that the array size must also be specified. To declare an array, type the keyword 'VAR', an identifier and the number of items to be stored, with the number surrounded by square brackets. Like variables, multiple arrays may be declared in the same statement with the arrays separated by commas. The statement must end with a semicolon. The syntax of the array declaration statement is as follows.

VAR *identifier* '[' *int expr* ']' |',' ... *identifier* '[' *int expr* ']'| ';'

Examples of array declaration statements are as follows:

```
VAR DaysOfTheWeek[7], MonthsOfTheYear[12];

VAR V[20], n[20], p[20];
```

In the above examples the size of the array was specified by an integer. However, any mathematical expression that evaluates to an integral constant is also acceptable. (See subsection 15.1.2 for more information on integral constant expressions.) Furthermore, multidimensional arrays (arrays which are referenced by more than one index) may be declared. The maximum number of dimensions an array may have is three. The syntax of the multidimensional array declaration statement is as follows.

VAR *identifier* '[' *int expr* |',' ... *int expr* | ']' |',' ... *identifier* '[' *int expr* |',' ... *int expr* | ']'| ';'

Variables, single-dimensional arrays and multidimensional arrays may all be declared in the same statement. The following examples demonstrate this point.

```
CONST DIM     =  9;
CONST HOLES   = 18;
CONST PLAYERS = 15;

VAR V[DIM];
VAR Scores[PLAYERS, HOLES];
VAR Points[4*DIM-6,3+10/4], Series, Strikes[PLAYERS];
```

The number of elements (size) in each array dimension must be greater than or equal to one. If the size of any dimension of an array evaluates to zero, a negative integer or a floating-point number, then an error will occur. As a final note, consider the first array declaration in the above examples. An array named V is declared with 9 members (since DIM is a constant equal to 9). The individual members of this array are referred to as V[0], V[1], V[2], ... V[7], V[8]. When referencing an array element the index may be between zero and the size of the array dimension minus one. For multidimensional arrays each index may be between zero and that dimension's size minus one. SGFramework follows the C++ language array indexing convention, which is different from those for languages such as Pascal and Fortran.

Specifying Unknown Variables

Typically not all of the variables or array elements declared in a specification file are unknown quantities whose value is determined by solving the equations. Consider the two-dimensional Poisson's equation file listed in Subsection 15.1.6. The center mesh point, array element $V[DIM/2, DIM/2]$, remains set to 10.0 volts throughout the simulation. Likewise the elements at mesh points on the perimeter of the mesh remain at zero volts throughout the simulation.

Because of this it is necessary to tell the SGFramework numerical algorithm modules which variables and array elements are the unknown quantities, for which a solution must be obtained. In order to declare variables or array elements as unknown, these variables and array elements must first be declared as variables using the VAR statement and then tagged as unknown variables using the UNKNOWN statement. Furthermore, 'unknown variables' may be untagged (converted back to known variables) by the KNOWN statement. The syntax of these statements is as follows.

UNKNOWN *identifier* |',' ... *identifier* | ';';

UNKNOWN *identifier* '[' *range* |',' ... *range* | ']' |',' ... *identifier* '[' *range* |',' ... *range* | ']'| ';'

KNOWN *identifier* |',' ... *identifier* | ';';

KNOWN *identifier* '[' *range* |',' ... *range* | ']' |',' ... *identifier* '[' *range* |',' ... *range* | ']'| ';'

Although it is not shown above, both variables and arrays may be present in the same UNKNOWN or KNOWN statement. Arrays require a range (or a collection of ranges) that specify which array elements should be tagged or untagged as known and unknown. The syntax of a range is as follows.

int expr |.. *int expr* |: *int expr* ||

all

The first integer expression is the range's starting value, the second integer expression is the range's ending value, and the third integer expression is the range's step value. All of the integer expressions must evaluate to an integer constant. If only the start value is specified, then range spans only that value. If the step value is omitted, then it defaults to plus or minus one. It is not allowable to specify a start value and a step value without the range's ending value, since such a range is meaningless. Examples of the KNOWN and UNKNOWN statements can be found in the example specification file given in Subsection 15.1.6 above and in the following examples. If the keyword 'ALL' is specified, then the range loops through all elements, i.e., its starting value is zero, its ending value is the size of the array dimension minus one, and its step is unity.

```
VAR Cell[3,3,3], Total, Time, x;
UNKNOWN Cell[0..2,1,0..2], Total, Time, x;
KNOWN Cell[1,0..2,1], Time;
```

After these KNOWN and UNKNOWN statements are processed, the following

Table: Internal Functions by Type

Group	Internal Function List
Trigonometric	sin, cos, tan, csc, sec, cot
Inverse Trigonometric	arcsin, arccos, arctan, arccsc, arcsec, arccot
Hyperbolic	sinh, cosh, tanh, csch, sech, coth
Inverse Hyperbolic	arcsinh, arccosh, arctanh, arccsch, arcsech, arccoth
Exponential and Logarithmic	exp, log, pow10, log, pow, sq, sqrt, inv
Miscellaneous	ave, bern, aux1, aux2, erf, nsdep, ngdep
Special	abs, sign, nonneg, step, min, max

variables and array cells are unknown variables: $Total, x, Cell[0,1,0], Cell[0,1,1],$ $Cell[0,1,2], Cell[1,1,0], Cell[1,1,2], Cell[2,1,0], Cell[2,1,1]$ and $Cell[2,1,2]$. Note that subsequent KNOWN and UNKNOWN statements may override the results of previous statements.

15.1.8   Functions

SGFramework supports three types of functions: internal, user-defined, and external functions. Internal functions are built into the SGFramework language and need not be declared. User-defined functions are defined by the user in the specification file. External functions are declared in the specification but are implemented in some programming language, compiled, and then linked to the source code generated by SGFramework.

Internal Functions

SGFramework contains over forty internal functions which are available to every specification file for use. All SGFramework functions accept integral or floating-point expressions as arguments and return floating- point values. The internal functions can be divided into seven groups: trigonometric functions, inverse trigonometric functions, hyperbolic functions, inverse hyperbolic functions, exponential and logarithmic functions, miscellaneous functions and special functions whose derivatives are singular. The following table lists the internal functions according to their type; the table lists the syntax and a description of the internal functions in alphabetical order.

Since the max, min, nonneg, sign and step functions have singular derivatives and since SGFramework does not implement the partial derivatives of the nsdep and ngdep functions, it is recommended that these functions are not used in the constraint equations which are solved by SGFramework using a Newton method. (At a minimum, the user should be aware of the numerical issues resulting from the use of these functions in such equations.) If these functions are present in the constraint equations and have arguments containing unknown variables or array elements, then the Jacobian matrix will have singularities which may cause undesirable behavior during the numerical solution of the equations. Examples of the

Table: Syntax/Description of Internal Functions

$\mathrm{abs}(x)$	returns the absolute value of the argument x
$\mathrm{arccos}(x)$	returns the inverse cosine of the argument x
$\mathrm{arccosh}(x)$	returns the inverse hyperbolic cosine of the argument x
$\mathrm{arccot}(x)$	returns the inverse cotangent of the argument x
$\mathrm{arccoth}(x)$	returns the inverse hyperbolic cotangent of the argument x
$\mathrm{arccsc}(x)$	returns the inverse cosecant of the argument x
$\mathrm{arccsch}(x)$	returns the inverse hyperbolic cosecant of the argument x
$\mathrm{arcsec}(x)$	returns the inverse secant of the argument x
$\mathrm{arcsech}(x)$	returns the inverse hyperbolic secant of the argument x
$\mathrm{arcsin}(x)$	returns the inverse sine of the argument x
$\mathrm{arcsinh}(x)$	returns the inverse hyperbolic sine of the argument x
$\mathrm{arctan}(x)$	returns the inverse tangent of the argument x
$\mathrm{arctanh}(x)$	returns the inverse hyperbolic tangent of the argument x
$\mathrm{aux1}(x)$	returns $x/sinh(x)$
$\mathrm{aux2}(x)$	returns $1/\left(1+e^{x}\right)$
$\mathrm{bern}(x)$	returns $x/\left(e^{x}-1\right)$
$\mathrm{ave}(x,y)$	returns average value of x and y
$\mathrm{csc}(x)$	returns the cosecant of the argument x
$\mathrm{csch}(x)$	returns the hyperbolic cosecant of the argument x
$\mathrm{cos}(x)$	returns the cosine of the argument x
$\mathrm{cosh}(x)$	returns the hyperbolic cosine of the argument x
$\mathrm{cot}(x)$	returns the cotangent of the argument x
$\mathrm{coth}(x)$	returns the hyperbolic cotangent of the argument x
$\mathrm{erf}(x)$	returns $\frac{2}{\sqrt{\pi}}\int_{0}^{x}\exp(-s^{2})ds$, the error function of the argument x
$\mathrm{exp}(x)$	returns the exponential of the argument x, i.e. e to the x
$\mathrm{inv}(x)$	returns the reciprocal of the argument x
$\mathrm{ln}(x)$	returns the natural log of the argument x
$\mathrm{log}(x)$	returns the log to base 10 of the argument x
$\mathrm{max}(x,y)$	returns x if $x>y$, otherwise y is returned
$\mathrm{min}(x,y)$	returns x if $x<y$, otherwise y is returned
$\mathrm{ngdep}(x,y,W,a_{x},a_{y})$	if $\hat{x}=|x|-W/2$ is positive, returns $\exp\left(-a_{x}\hat{x}^{2}-a_{y}y^{2}\right)$; otherwise returns 1.0
$\mathrm{nonneg}(x)$	returns 1.0 if $x\geq 0.0$, otherwise 0.0 is returned
$\mathrm{nsdep}(x,W,Dt)$	returns $0.5\left[erf\left(\frac{W/2+x}{2\sqrt{Dt}}\right)+erf\left(\frac{W/2-x}{2\sqrt{Dt}}\right)\right]$
$\mathrm{pow}(x,y)$	returns x to the power y, i.e. x^{y}
$\mathrm{pow10}(x)$	returns 10 to the power x, i.e. 10^{x}
$\mathrm{sec}(x)$	returns the secant of the argument x
$\mathrm{sech}(x)$	returns the hyperbolic secant of the argument x
$\mathrm{sign}(x)$	returns 1.0 if $x>0.0$, returns 0.0 if $x=0.0$ and -1.0 if $x<0.0$
$\mathrm{sin}(x)$	returns the sine of the argument x
$\mathrm{sinh}(x)$	returns the hyperbolic sine of the argument x
$\mathrm{sq}(x)$	returns the square of the argument x
$\mathrm{sqrt}(x)$	returns the square root of the argument x
$\mathrm{step}(x)$	returns 1.0 if $x\geq 0.0$, otherwise 0.0 is returned
$\mathrm{tan}(x)$	returns the tangent of the argument x
$\mathrm{tanh}(x)$	returns the hyperbolic tangent of the argument x

use of internal functions can be found in the specification files listed in this manual. This applies to virtually all the semiconductor simulations, beginning with the simulation listed in Chapter 7, in the input file pn01.sg.

User-Defined Functions

In many simulations it is very convenient to declare and use application-specific functions. It is typical to implement parameter models as user-defined functions. For instance, parameter models for the electron and hole mobilities are implemented as user-defined functions in the simulations presented in this book. Once declared, user-defined functions may be used everywhere that internal functions are allowed. User-defined functions consist of a header and a body, as discussed below.

User-Defined Function Headers To declare a user-defined function header, one must type the keyword FUNC followed by an identifier, and a parenthesized list of function arguments. The identifier following the FUNC keyword is the user-defined function's name. The argument list consists of a comma-separated list of identifiers. Each identifier is the name of a function argument. Each argument may optionally be enclosed by a pair of less than '<' and greater than '>' signs.

FUNC '(' |'<'| *identifier* |'>'| |',' ... |'<'| *identifier* |'>'|| ')'

Normally, each user-defined function is symbolically differentiated with respect to each function argument. Both the function and its partial derivatives are then coded in the output file. The partial derivatives are needed, since functions, in general, may be used in equations, which in turn are differentiated with respect to each unknown variable and array element. The equations are differentiated to obtain expressions that evaluate the elements of the Jacobian matrix. If, for some reason, the user does not wish to generate the ith partial derivative, then the user can enclose the ith argument between less than and greater than signs. There are at least two reasons why one might not want to generate a partial derivative of a user-defined function. First, one may know that the partial derivative always evaluates to zero. Second, one may not want to include the contribution of a partial derivative when evaluating the elements of the Jacobian matrix. For instance, in semiconductor simulations, the distance between mesh points is frequently a user-defined function argument. Since this distance is usually constant with respect to the simulation's unknown variables, there is no need to symbolically differentiate the function with respect to this argument or to evaluate this partial derivative. Hence this argument is one which is often enclosed within less than and greater than signs.

User-Defined Function Bodies The body of a user-defined function consists of zero or more function statements, followed by a return statement. Two types of function statements are allowed: constant statements and assignment statements. The syntax of function constant statements is identical to the syntax of the constant statements described in Section 15.1.6. The expression which defines the constant statement must evaluate to a constant. Hence, all previously defined 'general' constants and user-defined functions as well as numbers, operators, and internal functions may appear in constant expressions. Variables and array elements, on the other hand, cannot appear in a constant expression. In addition, the constant expression may contain the function constants which have been previously defined in the function in which the constant statement appears. One or more constants may be defined in a single function constant statement. The syntax of the function constant statement is as follows.

   CONST *identifier* '=' *const expr* |',' ... *identifier* '=' *const expr* | ';'

   The only difference between general constants (constants declared outside of a function) and function constants is the scope of the constant. General constants may be used from their point of declaration to the end of the file. Function constants may be used only from their point of declaration to the end of the function in which they are defined. The names of function constants need not be unique throughout the file. The bodies of several user-defined functions may define the same constant. The value of the constant may also be different in any or all of the user- defined functions in which it is defined. Second, the name of a function constant may be the same as the name of a general constant. For instance, consider the following specification file.

```
const COUNT = 9;
var a,b,c;

func MyFunc(x,y)
  const TOTAL = COUNT + 3;
  const COUNT = TOTAL;
  return (x < y) * COUNT;

begin main
  assign a = COUNT, b = 10;
  assign c = MyFunc(a,b);
end
```

   The first statement declares the COUNT constant whose value is 9. The second statement declares three variables. Next the function MyFunc is declared. MyFunc declares two function constants: TOTAL and COUNT. TOTAL is equal to COUNT plus 3, that is, 12. Note that COUNT, in this case, refers to the general constant COUNT whose value is 9. SGFramework does this because the function constant COUNT has not been declared yet. The second function statement defines a function constant named COUNT and assigns it the value of TOTAL, which is 12. MyFunc's return value is COUNT if x is less than y, or 0 is x is greater than or equal to y. Note that COUNT, in this case, refers to the function constant COUNT whose value is 12, not the general constant COUNT whose value is 9. Since functions return floating-point numbers, MyFunc actually returns the value 12.0 or 0.0. The integer constant is automatically converted to a floating-point number by the return statement. Lastly, let us look at the main procedure. The first statement assigns the variable a the value of the general constant COUNT, which is 9, and assigns the variable b the value 10. The variable c is assigned the value of MyFunc whose arguments are a and b. Since the function arguments x and y correspond to variables a and b, and x is less than y, the variable c is assigned the value 12.0, which is the (floating-point) value of the function constant COUNT.

The second type of optional body statement is a function assignment statement. Function assignment statements declare temporary variables that may be used in the function in which they are declared. The syntax of the function assignment statement is the keyword 'ASSIGN' followed by an identifier that is the variable's name, an equal sign, a function assignment expression and a semicolon.

ASSIGN *identifier* '=' *func expr* |',' ... *identifier* '=' *func expr* | ';'

The function assignment expression may contain constants, numbers, internal variables, previously declared user-defined and external functions, internal functions, the function arguments and mathematical operators. One or more function variables may be declared in a single function assignment statement by comma-separating the declarations and terminating the last declaration with a semicolon.

The function body must end with a return statement. The syntax of the return statement consists of the keyword RETURN followed by a return expression and a semicolon.

RETURN *func expr* ';'

The syntax of the return expression is identical to that of the function assignment expression. The value of the return expression is the value which is returned by the user-defined function. If the expression evaluates to an integral value, the value is automatically converted to a floating-point number.

Consider the three following user-defined functions.

```
func ave3(x, y, z)
  assign sum = x + y + z ;
  return sum / 3.0 ;

func grad(y1, y2, <h>)
  return (y2 - y1) / h ;

function AreaOfCircle(r)
  const PI = 3.141592654 ;
  return PI*sq(r) ;
```

The first function returns the average of its three arguments. It does this by first summing its three arguments and storing the result in a function variable. Then it returns one-third of the sum.

The second function returns the finite-difference approximation to the (one-dimensional) gradient. The arguments y_1 and y_2 will typically be the values of some scalar field at adjacent mesh points, and h is the distance between the mesh points. Since the location of the mesh points is often fixed, throughout the duration of the simulation, the derivative of the grad function with respect to the argument

h is not needed. To instruct the simulation not to evaluate this particular partial derivative, the h argument is enclosed by '<' and '>'.

The last function returns the area of a circle with radius r. To do this the function defines the local (function) constant PI. It then returns PI times the square of the radius.

External Functions

In addition to internal and user-defined functions, SGFramework supports external functions. External functions can be used anywhere that internal or user-defined functions can be used, except in constant expressions. External functions are functions whose bodies are not defined in the specification file. Instead the body of an external function is implemented in a programming language (such as C, C++, Fortran or Pascal.) Consequently the user must compile and link external functions with the SGFramework-generated code and a numerical algorithm module (NAM). Since implementing these functions requires a thorough understanding of code and data structures generated by the SGFramework translator, this topic will not be discussed further here. External functions are scarcely used in the simulations presented in this book, and, when they are used, the source-code implementation is provided and is well-documented.

15.1.9   Constraint Equations

The constraint equations usually form a system of linear or nonlinear equations which a SGFramework numerical algorithm module solves to obtain the values of the unknown variables. If one is modeling a physical system, these equations constitute a mathematical description of physical laws (or approximations to those laws) as well as fixed relationships (constraints) between the system variables. For instance, if a system of masses and springs is modeled, application of Newton's and Hooke's laws provides the simulation's constraint equations. If the electrostatic interaction of a system of point charges is modeled, application of Coulomb's law provides the simulation's constraint equations. If the governing equations are partial differential equations, the constraint equations usually will be the corresponding discretized form of the PDEs. (We use the words 'constraint equations' here in a manner which is not entirely consistent with the notion of constraint equations as used in Lagrangian and Hamiltonian mechanics.)

The syntax of the constraint-equation statement is as follows.

EQU *header* '->' expr $=$ *expr* ';'

The use of mesh labels within constraint equations is described in Section 15.2, and especially in Subsection 15.3.1, after mesh labels have been introduced.

Constraint Equation Headers

The constraint-equation header consists either of a variable name or an array name followed by one to three loops enclosed in brackets. The syntax of the header is as follows.

 identifier identifier '[' *loop* |',' ... *loop* | ']'

The syntax of a loop is similar to that of a range with the addition of an index identifer. The syntax of a loop is as follows.

 identifier '=' *range*

Loop indices are similar to variables with the following exceptions. First, loop indices store integer quantities, whereas variables store floating-point quantities. Second, the scope of a loop index is limited to the equation statement in which it appears. There are three implications of the second statement. First, the name of a loop index may be reused in subsequent equations. Second, if two or more loops appear in a header, then the indices must be unique. Third, indices may appear in the range expression of subsequent loops of the same statement but not in the range expressions of the loop in which the index is defined.

The equation headers serve two purposes. First, the headers may specify which variables and array elements are unknown. If KNOWN and UNKNOWN statements do not appear in the specification file, SGFramework uses the equation headers to determine which variables are unknown. Each variable and array element specified by the array header is assumed to be unknown in the absence of KNOWN and UN-KNOWN statements. However, if any KNOWN or UNKNOWN statements appear in the file, all declared variables and array elements are assumed to be known unless explicity tagged as unknown by an UNKNOWN statement. For more information about KNOWN and UNKNOWN statements, refer to Subsection 15.1.7.

The main purpose of equation headers is to associate equations with unknown variables. If the header is a variable, then the equation is associated with that variable. Consequently, the variable which appears in the header should not be explicitly tagged as a known variable via a KNOWN statement. Furthermore, if a KNOWN or UNKNOWN statement appears in the specification file, header variables should explicitly be tagged as unknown. If an equation is associated with a known variable, then SGFramework will issue a warning that the equation is useless.

Frequently, one wishes to solve a system of discretized partial differential equations when simulating a physical system. Often the resulting discretized equations can be represented by a couple of template equations, i.e., equations which have the same form but different array indices. Consider the SGFramework specification file of Subsection 15.1.6. This specification file numerically solves Poisson's equation on a nine by nine rectangular mesh. The electrostatic potential, V, is held constant at the center of the mesh and its perimeter. Consequently, the V array has 81 elements, 48 of which are tagged as unknown variables. Since there are 48 unknowns in this specification file, there must be 48 constraint equations as well. These equations are as follows.

```
V[0,1] + V[1,0] - 4 * V[1,1] + V[2,1] + V[1,2] = 0
V[0,2] + V[1,1] - 4 * V[1,2] + V[2,2] + V[1,3] = 0
V[0,3] + V[1,2] - 4 * V[1,3] + V[2,3] + V[1,4] = 0
```

$$\cdots$$

```
V[6,7] + V[7,6] - 4 * V[7,7] + V[8,7] + V[7,8] = 0
```

Since all of these constraint equations have the same form and differ only in the values of their array indices, they may be represented by the following template (indexed) constraint equation.

```
V[i-1,j] + V[i,j-1] - 4 * V[i,j] + V[i+1,j] + V[i,j+1] = 0
```

Both the indices i and j loop through the values one through seven, i.e., $i = 1..7$ and $j = 1..7$. Consequently the header for this constraint equation would be $V[i=1..7, j=1..7]$. Since the electrostatic potential is known at the center, there is no equation associated with the array element $V[4,4]$.

When SGFramework processes equation statements, it associates that equation with either a variable or one or more array elements. An equation is tagged 'useless' if SGFramework could not associate the equation with one or more unknown variables and/or array elements. This warning is usually generated if (1) the variable or at least one array element with which one wishes to associate the equation is not specified as unknown or (2) the variable or array elements are already associated with a previous equation. It is important to recognize that SGFramework associates a variable or array element with the first equation it encounters that specifies that variable or array element in its header. Consider the following example.

```
var x[10];
unknown x[all];

equ x[j=0..7] -> x[j] = 5.0;
equ x[j=8..9] -> x[j] = 3.0;
equ x[j=7..8] -> x[j] = 4.0;
```

The last equation is 'useless', since the equation associated with array element $x[7]$ is the first equation and the equation associated with array element $x[8]$ is the second.

Constraint Equation Expressions

Constraint-equation expressions are mathematical expressions consisting of previously defined constants, variables, array elements, external functions, internal functions and user-defined functions as well as operators and the index variables of the loops that are specified in the constraint equation header. Consider the following system of equations.

```
x - y - z = 0
x - y + z = 2
x + y + z = 6
```

The above equation can easily be implemented as follows:

```
var x, y, z;
equ x -> x - y - z = 0 ;
equ y -> x - y + z = 2 ;
equ z -> x + y + z = 6 ;
```

It worth noting that in the above system of equations, any of the equations could be associated with any of the variables. In general, this is not true. One cannot associate a variable or array element with an equation in which the variable or array element does not appear in the equation's expression.

15.1.10   Procedures

Procedures are needed for several reasons in a simulation. Usually it is necessary to initialize variables and array elements prior to solving the specified equations in order to determine the values of the unknown variables and/or array elements. (By default, all variables and/or array elements are initialized to zero.) In addition it is often desirable to solve the equations for different boundary conditions and to output simulation results to data files.

Procedures are the mechanism by which users can initialize their variables and/or array elements, tell the numerical algorithm modules to solve their specified equations and write their simulation results to data files. Procedures are each specified by a unique name and contain one or more procedural statements. The syntax for procedures is:

begin *identifier*
procedural statement ';'
procedural statement ';'
...
end

Procedures start with a header that consists of the keyword 'begin', followed by an identifier which serves as the procedure's name. Procedures terminate with the keyword end. Between the header and terminator are one or more procedural statements. In addition, SGFramework supports external procedures. External procedures are directly coded in a programming language such as C/C++, Fortran or Pascal. External procedures are specified by generating a procedure with no body (i.e., a procedure with a header directly followed by the terminating keyword 'end'.)

Every simulation must have at least one procedure, the main procedure. The syntax of the main procedure is identical to that of the generic procedure described above, except for the header. The main procedure's header consists of the keyword 'begin' followed by the keyword 'main'. Upon execution of a SGFramework simulation, the simulation initializes its internal data structures and numerical algorithm module. After this start-up code is complete, program control is turned over to the main procedure. Each statement in the main procedure is executed, branching to and returning from subroutines (other procedures) as necessary. When the end of the main procedure is reached, the simulation calls its clean-up code and the simulation terminates.

Between the procedure's header and terminator are one or more procedural statements. There are six types of procedural statements: assignment statements, single-word statements, file input/output statements, subroutine execution statements, conditional statements and looping statements. Each of these statement types will be discussed in detail.

Assignment Statements

Assignment statements are used initialize and/or modify the values of variables and array elements. By default the values of all simulation variables and array elements are set to zero upon invoking the simulation. This default may be overridden by means of assignment statements for some or all of the variables and array elements. Assignment statements may also modify the values of variables and array elements in the middle of simulation to simulate time varying boundary conditions. Finally, assignment statements may calculate quantities based upon other variables and elements. For instance, assignment statements are used in the semiconductor simulations

1. To initialize variables and array elements,

2. To modify the values of boundary conditions such as contact voltages which provide the interface between regions where the physical equations are solved, and

3. To calculate quantities such as terminal currents in terms of variables and array elements.

The difference between assignment statements in procedures and those in functions is as follows. Assignment statements in functions both declare and initialize variables that are visible only in the function where they are declared. Assignment statements in procedures do not declare new variables or array elements. They are strictly used to modify the values of existing variables and array elements that are visible to the entire program, i.e. those variables and array elements which have been declared in **VARIABLE** statements. The syntax of ASSIGNMENT statements is as follows.

ASSIGN *identifier* = *expr* ';'

ASSIGN *identifier* '[' *loop* |',' ... *loop* | ']' = *expr* ';'

Procedural assignment statements begin with the keyword 'assign' followed by an identifier, which is the name either of a previously declared variable or of a previously declared array element. If the identifier is the name of an array element, then it must be followed by a comma-separated list of loops enclosed in brackets. The loops specify which array elements are included in the assignment statement. See Section 15.1.9 for more information concerning loops. Whether the identifier is the name of a variable or of an array, the statement is completed by an equal sign, an expression and a semicolon. The expression may be any mathematically valid formula composed of previously declared constants, variables, array elements, functions as well as numbers, internal functions, and operators (as well as loop indices, if applicable).

Examples of assignment statements can be seen throughout this book in numerous specification files. For example, in the 'Game of Life' specification file that is listed in the introduction, the following assignment statements were used:

```
assign t=0.0 ;

assign AO[i=is,j=js+1]   = 1 ;       // launch stop light
assign AO[i=is-1..is+1,j=js] = 1 ;   //

assign AO[i=ig,j=jg+2]   = 1 ;       // launch glider
assign AO[i=ig+1,j=jg+1] = 1 ;       //

assign AO[i=ig-1..ig+1,j=jg] = 1 ;   //

assign A[i=all,j=all] = AO[i,j] ;

assign COUNT[i=1..LX-1,j=1..LY-1] =
  AO[i-1,j-1] + AO[i-1,j] + AO[i-1,j+1] +
  AO[i,j-1]   +             AO[i,j+1]   +
  AO[i+1,j-1] + AO[i+1,j] + AO[i+1,j+1];

assign A[i=1..LX-1,j=1..LY-1] =
  (COUNT[i,j]==3) or (COUNT[i,j]+AO[i,j]==3);
  assign AO[i=all,j=all] = A[i,j] ;
assign t = t+dt ;
```

These statements are largely self-explanatory. Some assignment statements used above are for a single variable, some refer to a single element of an array, and some apply to a specified range of elements of an array. (To see how this fits into the overall input file, see Section 1.5.)

Single-Word Statements

There are three procedural statements that consist of single keywords: solve, write, and exit. The syntax is the keyword followed by a semicolon.

    solve ';'
    write ';'
    exit ';'

The solve statement causes the simulation to invoke the appropriate numerical analysis algorithm to solve the specified equations to obtain the values of the unknown variables and array elements. The equations are solved, starting by using the current values of all of the simulation variables and array elements. The values of the unknown variables and array elements are modified as a consequence of invoking the solve statement.

The write statement dumps a snapshot of all the variables and array elements to a binary data file, referred to as the result file. If a write statement is not explicitly present in the simulation, an implicit write statement is appended after the last statement in the main procedure. It is possible to override the behavior of the write statement to dump a snapshot of selected variables and array elements by writing code to customize the SGFramework default behavior. Data in the SGFramework data files may be extracted and viewed using certain SGFramework tools such as 'extract' and 'triplot'. The extract command is discussed in Section 15.4.8. The triplot command is discussed in the section on Graphical Output, 15.4.10.

The exit statement causes the simulation to call its clean-up code and then terminates the simulation.

Input and Output

In many circumstances, it is convenient to initialize variables and/or array elements from numbers stored in data files rather than computing their values from a formula. It may also be expedient to write the values of certain variables and/or array elements to ASCII data files rather than dumping the values of all variables and/or arrays to a binary result file. Towards this end, SGFramework provides several commands to accomplish the input and output needs of most users.

Data Files   SGFramework data files are ASCII files which can contain three types of data: numerical data (floating-point numbers), labels and comments. The most prevalent type of data in SGFramework data files is numerical data. Numerical data is always treated as floating-point numbers, since this data either will initialize variables and/or array elements or was generated by writing the values of variables and/or array elements. Numerical data need not have any special format. It is only required that each number be separated by whitespace (spaces, tabs, new lines, etc.).

Labels mark a particular place in the data file in the same fashion as tabs mark a book. When reading numerical data from an input file, one can instruct the simulation to search for a particular label and read data starting from that point in the input file. This feature is useful when several simulations use the same data file to initialize their variables and/or array elements. Common data can be stored at the beginning of the input file and simulation-specific data can be tagged with a label and stored at the end of the data file.

Data files may also have comments. Like SGFramework mesh and equation specification files, comments begin with two slash characters '//' and end at the end of the line. Comments are ignored when reading numerical data and searching for labels. Comments serve no other purpose than to annotate the data file for human readability.

Opening Data Files   In order to read data from or write data to an ASCII data file, it is necessary first to open the file. The syntax for the open file statement is as follows. The file open command begins with the keyword OPEN followed by a string and a file mode. The command is terminated by a semicolon.

OPEN ["] *file name* ["] *file mode* ';'

The string following the 'OPEN' keyword specifies the full name of the file, i.e., path, name, and extension. The file mode is one of the following keywords: 'READ', 'WRITE', and 'APPEND'. The READ file mode opens the data file for input of data from the file. The WRITE and APPEND file modes open the data file for output to the file. The differences between WRITE and APPEND are apparent only when trying to open an existing data file. WRITE will erase and overwrite the contents of the data file, whereas APPEND does what its name implies: it appends the new data to the end of the existing data.

File Pointer   When a file is opened for data input from the file, a pointer, referred to as the file pointer, is positioned at the beginning of the file. When the simulation reads numerical data from the file, it reads the first number that is located at or following the file pointer. After it reads the value, the file pointer is advanced past the data which was just read. Similarly, when the simulation is searching for a label in a data file, it starts its search at the file pointer. If the search key is found, the file pointer is updated to the position which immediately follows the label. If the label is not found, the file pointer will point to the end of the data file. To locate a search key which is located prior to the file pointer's current location, the file must first be closed and then reopened. Reopening the data file causes the file pointer to be reset to the beginning of the file.

When a file is opened for data output to the file, the file pointer is located at either the beginning or the end of the file, depending upon the mode in which the file was opened. If the data file is opened with the WRITE mode, then the file pointer is located at the beginning of the file. If the file is opened with the APPEND mode, then the file pointer is located at the end of the file (unless, of course, the file does not exist, in which case the beginning and the end of the file are the same). Whenever numerical data, comments, or labels are written to the output file, the information is written at the current location of the file pointer. The file pointer is then updated to point to the end of the information that was written.

Reading from a Data File   Once a file is open for input from the file, i.e., opened with the READ file mode, numerical data from the file can be read into simulation variables and array elements via the FILE READ command. The syntax for this command is as follows. The FILE READ command begins with the keywords 'FILE' and 'READ' and is followed by a comma-separated list of identifiers. The command is terminated by a semicolon. The identifiers in the list are the names of variables and arrays into which the numerical data in the data file will be read. If an identifier refers to the name of a simulation variable, only one number is read from the data file. If, however, an identifier refers to the name of a simulation array, then several numbers are read from the data file and stored sequentially in the elements of the array. It is not possible to read data into a select range of array elements. Data is read from the current position of the file pointer, and the file pointer is updated upon execution of this command.

FILE READ *identifier* |',' ... *identifier* | ';'

Writing to a Data File   Once a file is open for output to the file, i.e., opened with the WRITE or APPEND file mode, the values of simulation variables and array elements may be written to the data file via the FILE WRITE. The syntax for this command is similar to the syntax of the FILE READ command. The FILE WRITE command begins with the keywords 'FILE' and 'WRITE' and is followed by a comma-separated list of identifiers. The command is terminated by a semicolon. The identifiers in the list are the names of variables and arrays whose values will be written to the data file. If an identifier refers to the name of a simulation variable, its value will be written to the data file. If, however, an identifier refers to the name of a simulation array, then the values of each of its array elements will be sequentially written to the data file. It is not possible to write data from a selected range of array elements. Data is written to the current position of the file pointer, and the file pointer is updated upon execution of this command.

FILE WRITE *identifier* |',' ... *identifier* | ';'

Closing a Data File   Once data has either been read from or written to a data file, the file needs to be closed before another file can be opened. It is a recommended practice to always close a data file regardless of whether another file will be opened. The syntax of the close file statement is very simple. The command starts with the keyword 'CLOSE' and is terminated by a semicolon.

CLOSE ';'

Subroutine Execution Statements

As stated in the introduction to the procedure subsection, every SGFramework simulation must have a main procedure. When a SGFramework simulation is invoked, an initialization procedure is first executed. After the simulation has initialized its internal data structures, it then transfers program control to the main procedure.

Many simulations have additional procedures. Procedures provide a convenient mechanism of grouping a series of related statements into a unit, which makes the equation specification file more modular. A procedure can call another procedure via a call statement. Typically, the statements in a procedure are executed sequentially. When a call statement is encountered, program control is transferred to the procedure that is called. Once all of the statements in the called procedure have been executed, program control is transferred back to the calling procedure. The syntax of the call statement is as follows.

CALL *identifier* ';'

Call statements begin the keyword 'CALL', followed by an identifier and a semicolon. The identifier is the name of a procedure which is to be executed.

Consider the following example.

```
var A, B, C;
```

```
begin MyProc1
  assign C = A + B;
end

begin MyProc2
  assign A = A + 1.0;
  assign B = B - 2.0;
  call MyProc1;
end

begin main
  assign A = 3.0;
  assign B = 7.0;
  call MyProc2;
  assign C = 2.0 * C;
end
```

After this simulation is invoked and it completes its initialization procedure, program control is transferred to the main procedure. The statements in the main procedure are sequentially executed. Hence, the first two assignment statements are executed which initialize variables A and B to 3.0 and 7.0 respectively. The next statement calls the procedure MyProc2. Hence the statements in procedure MyProc2 begin to execute sequentially. The first two assignment statements in MyProc2 reassign the values of variables A and B to 4.0 and 5.0 respectively. The next statement in procedure MyProc2 calls the procedure MyProc1. Consequentially, the procedure MyProc1 is executed which assigns the variable C the value of 9.0 (the sum of variables A and B.) Since procedure MyProc1 only has one statement, program control returns to the procedure that called procedure MyProc1, i.e. procedure MyProc2. Since procedure MyProc2 does not have any more statements after its call statement, program control returns to the procedure which called MyProc2, i.e. the main procedure. Execution in the main procedure continues after the call statement. Hence, the assignment statement which reassigns variable C to twice its value is executed. At the end of this simulation, the variables A, B, and C have the values of 4.0, 5.0, and 18.0 respectively.

Conditional Statements

All of the statements discussed thus far, with the exception of procedure calls, execute sequentially. In other words, the statements are executed in the order in which they appear in the main procedure. Often it is desirable to execute different statements depending upon the result of some computation or condition. Conditional statements provide such a mechanism. SGFramework supports IF-THEN and IF-THEN-ELSE conditional statements. By combining several conditional statements, it is possible to construct ELSE-IF conditional statements as well. The syntax for these statements is as follows.

IF '(' *condition* ')' THEN *statement*
IF '(' *condition* ')' THEN *statement* ELSE *statement* .

IF-THEN statements begin with the keyword 'IF' and are followed by a parenthesized condition. After the condition, the keyword 'THEN' is specified followed by one SGFramework procedural statement. Conditions are simply mathematical expressions. The condition is treated as being TRUE if it evaluates to a nonzero value and FALSE if it evaluates to zero. The procedural statement is executed only if the condition is TRUE. IF-THEN-ELSE statements have a similar syntax. Instead of the statement terminating with a procedural statement, the keyword 'ELSE' and another procedural statement are appended. In IF-THEN-ELSE statements, if the condition is TRUE, the first procedural statement is executed (the one after the THEN keyword). If the condition is FALSE, the second procedural statement is executed (the one after the ELSE keyword).

It is often desirable to conditionally execute several procedural statements rather than one procedural statement. To accommodate this feature, SGFramework allows several statements to be treated as one by enclosing the statements between the keywords 'BEGIN' and 'END'. All statements discussed in this section thus far are valid procedural statements. Furthermore, conditional statements and branching statements (which will be discussed next) are also valid procedural statements. For instance, consider the following example.

```
VAR a, b, c;

IF      ( a > b ) THEN
   ASSIGN c = +1.0;
ELSE IF ( a < b ) THEN
   ASSIGN c = -1.0;
ELSE
   ASSIGN c = 0.0;
```

This example compares the values of variables a and b and stores the result in variable c. If a is greater than b, then the result is positive unity. If a is less than b, then the result is negative unity. If a and b are equal, then the result is zero. Notice that in this example, an IF-THEN-ELSE statement is used as the procedural statement of another IF-THEN-ELSE statement. If the condition 'a is greater than b' is true, then the assignment statement 'c is set to positive unity' is executed. If the condition 'a is greater than b' is not true, then the statement that begins with 'if a is less than b' is executed.

A slight variation of the use of conditional statements within conditional statements is illustrated in the next example.

```
VAR a, b, c;

ASSIGN c = 0.0;
IF      (a > 0.0) THEN
```

```
    BEGIN
      IF      (b > 0.0) THEN ASSIGN c = +1.0;
      ELSE IF (b < 0.0) THEN ASSIGN c = -1.0;
    END
ELSE IF (a < 0.0) THEN
    BEGIN
      IF      (b > 0.0) THEN ASSIGN c = -1.0;
      ELSE IF (b < 0.0) THEN ASSIGN c = +1.0;
    END
```

These statements check the sign of variables a and b and store the result in variable c. If either a or b is zero, then the result is zero. If a and b are both positive or both negative numbers, then the result is positive unity; otherwise, the result is negative unity. In this case, we wanted the first IF-THEN-ELSE statement to execute if a is greater than zero and the second IF-THEN-ELSE statement to execute if a is less than zero. Therefore, these conditional statements had to be enclosed by the 'BEGIN' and 'END' keywords. Otherwise, it would not be clear whether the first 'ELSE' keyword should be associated with the first or second 'IF' keyword.

Looping Statements

The last type of procedural statements are looping statements. Looping statements allow any procedural statement (or a group of procedural statements if they are enclosed by the BEGIN and END keywords) to be executed repeatedly as long as some condition remains TRUE. SGFramework supports two types of looping statements: WHILE statements and DO-WHILE statements. The syntax of these commands is as follows.

WHILE '(' *condition* ')' *statement*

DO *statement* WHILE '(' *condition* ')' ';'

WHILE statements begin with the keyword 'WHILE' and are followed by a parenthesized condition and a procedural statement. As long as the condition remains TRUE, then the procedural statement will be executed over and over. If the condition becomes FALSE, then the procedural statement is skipped and the procedural statement which immediately follows the entire WHILE statement is executed.

DO-WHILE statements begin with the keyword 'DO' and are followed by a procedural statement, the keyword 'WHILE', a parenthesized condition and a semi-colon. Like WHILE statements, the procedural statement will be executed over and over as long as the condition remains TRUE. The difference between these looping statements is apparent when the condition is FALSE prior to executing the looping statement. In the case of a WHILE statement, the procedural statement will never be executed if the conditional is initially FALSE. However, with the DO-WHILE statement, the procedural statement will be executed once. Hence the procedural statement of a WHILE statement may be executed zero or more times and the procedural statement of a DO-WHILE statement may be executed one or more times.

In most cases, the procedural statement of a loop will consist of several procedural statements enclosed by the 'BEGIN' and 'END' keywords. At least one of the enclosed statements should affect the value of the condition; otherwise, the loop will never end. SGFramework omits FOR loops, since these may be implemented with a *while* statement. The following example will loop through the statements between the 'BEGIN' and 'END' keywords ten times. Note, however, that the loop variable i is a floating-point number and not an integer.

```
ASSIGN i = 1.0;
WHILE (i < 10.1) BEGIN
  statement
  statement
    ...
  statement
  ASSIGN i = i + 1.0;
END
```

15.1.11   Numerical Algorithm Parameters

Equation specification files may also specify several parameters. Parameters are divided into three categories: nonlinear solution algorithm parameters, linear solution algorithm parameters, and mesh scaling parameters. Parameters are specified via SET statements. To understand the parameters specified in this example, it is important to understand how a nonlinear system of algebraic equations is solved. It is recommended that the reader review section 5.4 and the sections which follow.

Nonlinear Solution Algorithm Parameters

There are three nonlinear solution algorithm parameters. These parameters specify the maximum number of Newton iterations, minimum accuracy with which to solve the simulation's equations, and the maximum amount of damping that can be applied to the Newton update vector. This subsection assumes the reader is familar with the Newton algorithm.

The parameter statement that specifies the maximum number of Newton iterations begins the keywords 'SET', 'NEWTON', and 'ITERATIONS'. These keywords are followed by an equal sign, an integer, and a semi-colon. The integer specifies the maximum number of Newton iterations. The numerical algorithm module will continue to iterate, as long as this maximum number of iterations is not exceeded and the simulation equations have not been solved to the minimum specified accuracy (and no error has been detected). The syntax of this parameter statement is as follows.

   SET NEWTON ITERATIONS '=' *integer* ';'

The parameter statement that specifies the minimum accuracy with which to solve the simulation's equations begins with the keywords 'SET', 'NEWTON', and 'ACCURACY'. Following these keywords is an equal sign, a floating-point (real) number, and a semicolon. The floating-point number specifies the minimum accuracy. The syntax of this parameter statement is as follows.

SET NEWTON ACCURACY '=' *real* ';'

The parameter statement that specifies the maximum amount by which to damp the Newton update vector begins with the keywords 'SET', 'NEWTON', and 'DAMPING'. These keywords are followed by an equal sign, an integer, and a semicolon. The maximum amount of damping is given by the expression 2.0^{-i} where i will be set equal to the integer mentioned in the previous sentence. The syntax of this parameter statement is as follows.

SET NEWTON DAMPING '=' *integer* ';'

Linear Solution Algorithm Parameters

There are seven linear solution algorithm parameters. These parameters specify the linear solution algorithm, the maximum matrix fill level, the conjugate gradient preconditioner, the maximum number of linear solution iterations, the minimum accuracy of the linear solution algorithm, the value of the overrelaxation parameter, and whether to zero the Newton update vector prior to executing the linear solution algorithm.

The parameter statement that specifies the linear solution algorithm begins with the keywords 'SET', 'LINSOL', and 'ALGORITHM'. These keywords are followed by an equal sign, the keyword 'SOR', 'PCG', or 'GE', and a semicolon. If the linear solution algorithm parameter is set to 'SOR', then the successive overrelaxation algorithm is used. If this parameter is set to 'PCG', then the preconditioned conjugate gradient algorithm is used. Finally, if this parameter is set to 'GE', then the Gaussian elimination algorithm is used. The syntax of this parameter statement is as follows.

SET LINSOL ALGORITHM '=' SOR ';'
SET LINSOL ALGORITHM '=' PCG ';'
SET LINSOL ALGORITHM '=' GE ';'

The parameter statement that specifies the maximum levels of matrix fill begins with the keywords 'SET', 'LINSOL', and 'FILL'. These keywords are followed by an equal sign, an integer, and a semicolon. The integer specifies the maximum matrix fill level. This parameter is only used when the preconditioned conjugate gradient with an ILU preconditioner or Gaussian elimination algorithm is used as the linear solution solver. The syntax of this parameter statement is as follows.

SET LINSOL FILL '=' *integer* ';'
SET LINSOL FILL '=' INFINITY ';'

The parameter statement that specifies the conjugate gradient preconditioner begins with the keywords 'SET', 'LINSOL', and 'PRECONDITIONER'. These keywords are followed by an equal sign, the keyword 'NONE', 'DIAGONAL', or ILU, and a semicolon. If this parameter is set to 'NONE', then no preconditioner is used.

If this parameter is set to 'DIAGONAL', then the preconditioner is the matrix whose diagonal elements are $d_{ii} = \partial F_i/\partial x_i$ (where F_i represents the ith equation and x_i represents the associated variable) and whose off diagonal elements are zero. If this parameter is set to 'ILU', then the preconditioner is the incomplete LU factored Jacobian matrix. This parameter is only used when the preconditioned conjugate gradient is the linear solution solver. The syntax of this parameter statement is as follows.

SET LINSOL PRECONDITIONER '=' NONE ';'
SET LINSOL PRECONDITIONER '=' DIAGONAL ';'
SET LINSOL PRECONDITIONER '=' ILU ';'

The parameter statement that specifies the maximum number of linear solution iterations begins with the keywords 'SET', 'LINSOL', and 'ITERATIONS'. These keywords are followed by an equal sign, an integer, and a semi colon. The integer specifies the maximum number of linear solution iterations. This parameter is only used when the successive overrelaxation or preconditioned conjugate gradient algorithm is the linear solution solver. The syntax of this parameter statement is as follows.

SET LINSOL ITERATIONS '=' *integer* ';'

The parameter statement that specifies the minimum linear solution accuracy begins with the keywords 'SET', 'LINSOL', and 'ACCURACY'. These keywords are followed by an equal sign, a floating-point (real) number, and a semicolon. The floating-point number is the minimum linear solution accuracy. This parameter is only used when the successive overrelaxation or preconditioned conjugate gradient algorithm is the linear solution solver. The syntax of this parameter statement is as follows.

SET LINSOL ACCURACY '=' *real* ';'

The parameter statement that specifies the value of the overrelaxation parameter begins with the keywords 'SET', 'LINSOL', and 'OVER'. These keywords are followed by an equal sign, a floating-point (real) number, and a semicolon. This parameter is only used when the successive overrelaxation algorithm is the linear solution solver. The syntax of this parameter statement is as follows.

SET LINSOL OVER '=' *real* ';'

The parameter that specifies whether the Newton vector update is zeroed prior to executing the linear solution algorithm begins with keywords 'SET', 'LINSOL', and 'ZERO'. The keywords are followed by an equal sign, the keyword 'YES' or 'NO', and a semicolon. If the value of this parameter is 'YES', then the Newton update vector is zeroed prior to calling the linear solution algorithm. If the value of this parameter is 'NO', then the Newton update vector is not zeroed prior to calling the linear solution algorithm. This parameter is only used when the successive overrelaxation or preconditioned conjugate gradient is the linear solution algorithm. Note, the Newton update vector serves as an initial guess for the indirect linear solution solvers. The syntax of this statement is as follows.

SET LINSOL ZERO '=' YES ';'
SET LINSOL ZERO '=' NO ';'

Mesh Scaling Parameters

There is only one mesh scaling parameter. This parameter sets the mesh distance scaling. The mesh distance scaling parameter begins with the keywords 'SET' and 'DIST'. These keywords are followed by an equal sign, a constant expression, and a semicolon. The constant expression evaluates to the value of the mesh distance scaling. The syntax of this parameter statement is as follows.

SET DISTANCE '=' *const expr* ';'

This concludes the discussion of the syntax and grammar of equation specification files. Following is a discussion of the syntax and grammar of mesh specification files.

15.2 The Syntax and Grammar of the Mesh Specification File

SGFramework provides the ability to automatically generate and refine hybrid rectangular/triangular two-dimensional meshes based on the specification contained in an input file. The algorithms and associated data structures are discussed in Chapter 6. This section focuses on describing the syntax and grammar of the mesh specification file. This section is divided into several subsections: An Overview of Mesh Specification Files 15.2.1; Comments, Numbers, Identifiers and Constants 15.2.2; Coordinates 15.2.3; Points 15.2.4; Edges 15.2.5; Regions 15.2.6; Labels; 15.2.7; Refinement Statements 15.2.8; Mesh Parameters 15.2.9; and the Element Refinement Criteria 15.2.10.

15.2.1 An Overview of the Mesh Specification File

SGFramework mesh specification files are usually divided into two sections: the mesh skeleton and the mesh refinement criteria. The mesh skeleton contains the minimum amount of information necessary to accurately describe the mesh. The mesh skeleton consists of a hierarchical declaration of points, edges and regions. SGFramework performs several data consistency checks and is designed so that additional checks may be implemented easily. The mesh refinement criteria specify the minimum and maximum spacing between nodes, the minimum and maximum numbers of refinement divisions and the conditions under which to refine (divide) triangles and/or rectangles. The naming of all of the mesh quantities is important, since it is the mechanism which links the mesh specification and equation specification files. The equation specification files were previously referred to as the specification files.

A very simple mesh skeleton file 'sq.sk' was given in Section 2.8. The discussion there would serve as a simple introduction to this topic. As a more complete example, consider the MOSFET mesh specification file listed in Section 10.4. The file is duplicated below, for convenience.

```
// mos.sk
// mesh constants
```

```
const WDEV  = 2.0e-4;      // device width          | (cm) | segment 1
const DDEV  = 2.0e-4;      // device depth          | (cm) | segment 2
const WOX   = 1.4e-4;      // oxide width           | (cm) | segment 3
const DOX   = 0.2e-6;      // oxide depth           | (cm) | segment 4
const WCONT = 0.2e-4;      // contact width         | (cm) | segment 5
const DRECT = 0.2e-4;      // depth of rect. region | (cm) | segment 6


//        |--5--|    |-----------3-----------|    |--5--|
//
//                  A-----------------------B          -
//                  |\\\\\\\\\\\\\\\\\\\\\\\|          4
//    -   C-----D---E-----------------------F---G-----H   -
//    6   |        |                        |      |   |
//    -   |        I-----------------------J       |   |
//        |                                        |   |
//        |                                        |   2
//        |                                        |   |
//        |                                        |   |
//        |                                        |   |
//        K----------------------------------------L   -
//
//        |--------------------1--------------------|

// define points
point pA = ((WDEV-WOX)/2, DOX), pG = (WDEV-WCONT, 0.0);
point pB = ((WDEV+WOX)/2, DOX), pH = (WDEV, 0.0);
point pC = (0.0, 0.0),          pI = ((WDEV-WOX)/2, -DRECT);
point pD = (WCONT, 0.0),        pJ = ((WDEV+WOX)/2, -DRECT);
point pE = ((WDEV-WOX)/2, 0.0), pK = (0.0, -DDEV);
point pF = ((WDEV+WOX)/2, 0.0), pL = (WDEV, -DDEV);


// define edges
edge eAB = GATE   [pA, pB] (WOX/20, 0.0);
edge eDE = NOFLUX [pE, pD] (1.0e-7, 0.5);
edge eEF = SISIO2 [pE, pF] (WOX/20, 0.0);
edge eFG = NOFLUX [pF, pG] (1.0e-7, 0.5);
edge eIJ =        [pI, pJ] (WOX/20, 0.0);
edge eCD = DRAIN  [pC, pD] (WCONT/8, 0.0);
edge eAE = NOFLUX [pE, pA] (1.0e-7, 0.5);
edge eGH = SOURCE [pG, pH] (WCONT/8, 0.0);
edge eBF = NOFLUX [pF, pB] (1.0e-7, 0.5);
edge eCK = NOFLUX [pC, pK] (WCONT/8, 0.5);
edge eEI =        [pE, pI] (1.0e-7, 0.5);
edge eHL = NOFLUX [pH, pL] (WCONT/8, 0.5);
edge eFJ =        [pF, pJ] (1.0e-7, 0.5);
edge eKL = SUB    [pK, pL] (WDEV/5, 0.0);


//define regions
```

```
region r1 = SIO2 {eAE, eEF, eBF, eAB} RECTANGLES;
region r2 = SI   {eEI, eIJ, eFJ, eEF} RECTANGLES;
region r3 = SI   {eCK, eKL, eHL, eGH, eFG, eFJ, eIJ, eEI, eDE, eCD};


// define coordinate labels
coordinates x, y;


// physical constants and properties of Si and SiO2
const T     = 300.0;                 // operating temperature
const e     = 1.602e-19;             // electron charge          (C)
const kb    = 1.381e-23;             // Boltzmann's constant     (J/K)
const e0    = 8.854e-14;             // permittivity of vacuum   (F/cm)
const eSi   = 11.8;                  // dielectric constant of Si
const eSiO2 = 3.9;                   // dielectric constant of SiO2


// doping constants
const NS = 1.0e16;                   // substrate doping         (cm^-3)
const NC = 1.0e19;                   // contact doping           (cm^-3)
const WDIFF = (WDEV-WOX)/2;          // diffusion width          (cm)
const DDIFF = 0.25e-4;               // diffusion depth          (cm)
const DT    = 1.0e-11;               // diffusion coef. * time   (cm^2)


// doping profile
refine C (SignedLog, 3.0) = (y <= 0.0) * { -NS              +
   (NC+NS) * nsep(x,     2*WDIFF,DT) * nsep(y,2*DDIFF,DT) +
   (NC+NS) * nsep(WDEV-x,2*WDIFF,DT) * nsep(y,2*DDIFF,DT) };


// set min/max edge spacing and min/max refinement levels
set minimum length = sqrt(e0*eSi*(kb*T/e)/e/abs(C));
set maximum length = 1.0;
set minimum divisions = 0;
set maximum divisions = 20;
```

15.2.2   Comments, Numbers, Identifiers and Constants

Equation and mesh specification files have similarities in their syntax and grammar. Both of these specification files allow comments to be placed anywhere in the source file. Numbers, identifiers, and constants are also common between equation and mesh specification files.

Comments in mesh specification files are ignored by the mesh specification file parser. Comments start with the // characters and terminate at the end of the line. It is a recommended practice to use comments to document and annotate mesh specification files. Comments are used liberally in the MOSFET mesh specification file. For example, a series of comments towards the beginning of the example file provide a schematic of the MOSFET mesh. Furthermore, comments are used to explain the meaning of the constants declared in the file.

Mesh specification files also support integer and floating-point numbers. Integer and floating-point expressions are also supported with one exception. Since mesh specification files do not have variables, all mathematical expressions evaluate to a constant. For more information about numbers, refer to Subsection 15.1.2.

Identifiers are used in mesh specification files as the names of constants, points, edges, regions, labels, coordinates, and refinement criteria parameters. As with equation specification files, certain identifiers are reserved as keywords. For more details about identifiers, refer to Subsection 15.1.4.

Finally, constants are supported in mesh specification files. The syntax of constants in mesh specification files is identical to the syntax of constants in equation specification files. The proper use of constants can make a mesh specification more readable and managable. It is good practice to define geometrical quantities such as the device width and depth as constants as is done in the sample mesh specification file. Defining these quantities as constants allows users to easily modify the device geometry by simply changing the value of the constant. For more information about constants, refer to Subsection 15.1.6.

15.2.3 Coordinates

In order to access the coordinate values of the mesh points, one must first name the coordinates using a coordinate statement. Unlike any other mesh specification file statement, only one coordinate statement is allowed in a mesh specification file. The syntax of the coordinate statement is as follows.

COORDINATES *identifier* ',' *identifier* ';'

See Section 15.3.1 for details of how the coordinates of the mesh points are passed to the equation specification file.

In the MOSFET example, the coordinates are labeled 'x' and 'y', since the MOSFET geometry is specified in the Cartesian coordinate system. If the mesh specification file does not contain a coordinate statement, SGFramework labels the coordinates 'x' and 'y'.

15.2.4 Points

The basic building blocks of mesh specification files are points. Edges are defined by points, and regions are defined by edges. A point is defined by two coordinates. The coordinates are enclosed by parentheses. Each point is given a name which is an identifier. In the example given here, point names consist of the lowercase letter 'p' followed by an uppercase letter. For instance, the first point is named pA and the last point is named pL. In general, the names of objects must begin with a letter and may be followed by one or more letters and/or digits. For more information about naming points, refer to the identifier Subsection 15.1.4. The syntax of the point definition is as follows.

POINT *identifier* '=' '(' *const expr* ',' *const expr* ')' |',' ... *identifier* '=' '(' *const expr* ',' *const expr* ')'| ';'

15.2.5   Edges

Edges in SGFramework mesh specification files may either be straight line segments or circular arcs. A straight line-segment is defined by its two endpoints. A circular arc is defined by its two endpoints and a third point on the arc. In each case the points defining the edge are enclosed by brackets. The edges in the example MOSFET mesh specification file are all straight line-segment edges. Each edge is given a name. In this example, edge names consist of the letter 'e' followed by two letters which indicate the points at the ends of the edges. This naming, however, is not dictated by the language. In general, the name of edges may be any legal identifier (see Subsection 15.1.4).

In addition, an edge may be given a label. For example, in the MOSFET example file, edge eAB is labeled GATE and edge eCD is labeled DRAIN. Edges without labels are considered internal edges. Points on an internal edge may be adjusted to improve the quality of the mesh, whereas points on a labeled edge may not be nudged. A detailed explanation of labels will be presented in Subsection 15.2.7.

The edge definition may also include an initial node spacing and a grade. The grid-generation program will use these parameters to control the insertion of points on the edge. If these parameters are omitted, the grid-generation program will compute appropriate values for them. For example, edge eAE is a line-segment that connects points pE and pA. Edge eAE has an initial node spacing of 10 angstroms (10^{-7} cm) and a grade of 0.50. This declaration causes the grid generation program to insert a point on the edge that is 10 angstroms from point pE. The next point will be spaced 15 angstroms from the previous point or 25 angstroms from point pE, so that the distance between the first and second nodes is 50 percent larger than the distance between the first point and point pE.

The syntax of the edge definition is

EDGE *identifier* '=' |*identifier* | '[' *identifier* ',' *identifier* |',' *identifier* | ']' |'(' *const expr* ',' *const expr* ')'| |',' ... *identifier* '=' |*identifier* | '[' *identifier* ',' *identifier* |',' *identifier* | ']' |'(' *const expr* ',' *const expr* ')'|| ';'

where the two or three identifiers in the square brackets define the endpoints, and, in the case of a curved circular edge, a third point on the edge. (The third point on the curve is the middle identifier). Curved edges should not exceed a quarter circle.

15.2.6   Regions

Regions are defined by a list of three or more edges. The list is enclosed between braces. The list of edges must form a simple closed curve. Regions include both the curve and its interior. The interiors of regions cannot overlap; however, regions may share boundaries (edges). Each region has a name. However, unlike edges where labels are optional, regions must be given a label. In this example, region names consist of the letter 'r' followed by a number. In general, region names may be any valid identifier (see Subsection 15.1.4). The edges must be listed in counterclockwise order in the definition of the region.

By default, regions are divided into triangular elements. Rectangular regions may be divided into rectangular elements if the opposite sides of the region have the same initial spacing and grade. The edges of rectangular regions must be parallel to the coordinate axes. There is one more constraint on rectangular regions. The opposite edges must 'point' in the same direction. For example, consider horizontal edges of a rectangular region. If the points are ordered from left to right on one of the edges, the points on the other edge must follow the same ordering. The user may specify rectangular elements by inserting the keyword 'RECTANGLES' at the end of the region statement.

The syntax of the region definition is

REGION *identifier* '=' *identifier* '{' *identifier* |',' ... *identifier* | '}' |RECTANGLES| |',' ... *identifier* '=' *identifier* '{' *identifier* |',' ... *identifier* | '}' |RECTANGLES|| ';'

where the identifiers in the brackets form a list of edges.

15.2.7   Labels

Labels provide links between the mesh specification and equation specification files. The distinction between names and labels is that names must be unique whereas the same label can be used to identify several edges and/or regions. For example, consider the second and third regions. These regions have different names ('r2' and 'r3') but share the same label ('SI'). Equation specification files refer to edges and regions by their labels (such as 'GATE' and 'DRAIN') while the mesh specification refers to points, edges and regions by their names (such as 'eAB' and 'eCD'). Labels refer to a collection of points. For instance, the label 'GATE' refers to all of the points on the edge 'eAB' whereas the label 'SI' refers to all of the points in the regions 'r1' and 'r2' and on their boundaries. Interfacing mesh and equation specification files with labels will be discussed in more detail in Subsection 15.3.2.

15.2.8   Refinement Statements

As mentioned in the overview, mesh specification files consist of two parts: a mesh skeleton and mesh refinement criteria. The mesh skeleton consists of a hierarchy of points, edges, and regions. From this information, the SGFramework mesh generation program constructs an initial grid. Often, this grid needs to be refined by dividing the triangular and/or rectangular elements in certain subdomains. The refinement criteria are specified by refinement functions.

The syntax of refinement functions is as follows.

REFINE *identifier* '(' *identifier* ',' *const expr* ')' '=' *expr* ';'

Refinement functions consist of a name, two refinement parameters and a body. The function name is declared after the keyword 'refine'. The refinement parameters are enclosed by parentheses and follow the function's name. The first refinement parameter is the refinement measure (which may be LINEAR, LOG or SLOG) and the second is the refinement distance. The body is the expression which follows the equal sign and may be a function of position. In this MOSFET example, the body of the refinement statement is the device's doping profile. The refinement measure may be linear, log or signed log, as defined by equations (15.1), (15.2) and (15.3) respectively.

$$M_{\text{lin}}(x) = x \qquad\qquad (15.1)$$

$$M_{\text{log}}(x) = \log(x) \qquad\qquad (15.2)$$

$$M_{\text{slog}}(x) = sign(x) * \log(1.0 + |x|) \qquad\qquad (15.3)$$

In order to determine whether a triangular or rectangular element should be refined, the mesh refinement program will evaluate the refinement functions at each of the element's vertices. If the difference in measure between any two vertices exceeds the refinement distance, the element is refined. For example, consider the MOSFET example. The refinement measure is signed log and the refinement distance is 4. In order to determine if a triangular element with vertices (x_1, y_1), (x_2, y_2) and (x_3, y_3) should be refined, the mesh refinement program would first evaluate the mesh refinement function C at each vertex ($C_i = \text{C}(x_i, y_i)$ for $i = 1, 2, 3$). It would then check the measure between the vertices ($M_{12} = |M_{\text{slog}}(C_1) - M_{\text{slog}}(C_2)|$, $M_{23} = |M_{\text{slog}}(C_2) - M_{\text{slog}}(C_3)|$ and $M_{13} = |M_{\text{slog}}(C_1) - M_{\text{slog}}(C_3)|$). If $M_{12} > 4$, $M_{23} > 4$ or $M_{13} > 4$, then the triangle would be refined. Multiple mesh refinement functions may be declared. The aforementioned procedure is performed using each refinement function. The results of these tests are logically OR'ed together. Thus an element is refined if any one of the tests signifies that it should be refined.

15.2.9   Mesh Parameters

Mesh specification files may also specify several parameters such as the minimum and maximum number of divisions and the minimum and maximum lengths of edges. Each triangular or rectangular element in the initial grid is assigned a division level of zero. If an element is refined, then the resulting elements are assigned the division level of their parent plus one. Thus elements formed from the division of a level-zero element would have division level one and elements formed from the division of a level one element would have division level two. The minimum and maximum division parameters set the minimum and maximum number of divisions through which an element may be refined. The syntax of the statements that set these parameters is as follows.

SET MINIMUM DIVISIONS '=' *const expr* ';';
SET MAXIMUM DIVISIONS '=' *const expr* ';';

The minimum and maximum edge lengths may be functions of position. Furthermore, the maximum edge length may be specified independently in both coordinates. The minimum edge length in the MOSFET example is set to the Debye length. The maximum edge length is set to one micron in this example. The syntax of the these parameters is as follows.

SET MINIMUM LENGTH '=' *expr* ';';
SET MAXIMUM LENGTH '=' *expr* ';';
SET MAXIMUM *identifier* LENGTH '=' *expr* ';';

Note that the identifier in the last syntax statement is the name of a coordinate label. Furthermore, the expressions which define the minimum and maximum lengths may be functions of position.

15.2.10   Element Refinement Criteria

The following rules are used in the order shown to determine whether an element should be refined. Once a decision has been reached, the remaining rules are not invoked for that element. An element can be refined only once on each pass over the mesh.

1. If two or more of the element's edges have been divided by the refinement of adjacent elements, then the element is divided.

2. If an element's division level is less than the minimum division level specified in the mesh input file, then the element is divided.

3. If the element's division level is greater than the maximum division level specified in the mesh input file, then the element is not divided.

4. If any of the lengths of the element's edges is larger than the maximum edge length specified in the mesh input file, then the element is divided.

5. If all of the lengths of the element's edges are smaller than the minimum edge length specified in the mesh input file, then the element is not divided.

6. If the criteria based on the refinement functions are not satisfied as discussed above, then the element is divided.

15.3   Interfacing the Equation and the Mesh Specification Files

In order to implement a finite-difference scheme, one needs to know how the mesh nodes are connected. In other words, each node must know who are its neighbors. This problem is trivial with rectangular meshes, since the simulation variables may be stored in multidimensional arrays which reflect the mesh geometry. Since, in general, the nodes of irregular meshes are not stored in any predictable order, it is not convenient to store the simulation variables in multidimensional arrays. With irregular meshes, the simulation variables are usually stored in one-dimensional arrays. In addition, one usually explicitly stores the mesh connectivity (lists of neighboring nodes for each node).

This section discusses the SGFramework language constructs that are used to interface the mesh and equation specification files. Specifically, this section describes the equation specification file extensions, such as mesh connectivity functions, that allow a user to write a simulation using an irregular mesh. This section is divided into five subsections: Importing an Irregular Mesh 15.3.1, Using Labels in Equation Specification Files 15.3.2, Mesh Connectivity Functions 15.3.3 and Mesh

Geometry Functions 15.3.4, Mesh Summation Functions 15.3.5, and Precomputed Functions 15.3.6.

15.3.1   Importing an Irregular Mesh

In order to write a simulation using an irregular mesh, the equation specification file must first import the mesh via the mesh statement. The syntax of the mesh statement is as follows.

MESH *string* ';'

The mesh statement consists of the keyword 'mesh' followed by a string and a semicolon. The string is the name of the mesh to import.

Importing a mesh does several things. First of all, it imports all of the constants in the mesh specification file from which the mesh was generated. In order words, it allows the author of equation specification files to use the constants that are defined in the mesh specification file of meshes they import. For example, the example MOSFET mesh specification file declares the constants WDEV and DDEV. These constants define the width and depth of the MOSFET respectively. Any equation specification file that imports the mesh generated from the MOSFET mesh specification file can use constants WDEV and DDEV in the statements that follow the mesh statement.

Secondly, importing a mesh implicitly declares three additional constants: NODES, EDGES, and ELEMENTS. These constants define the number of nodes, edges, and elements (both triangular and rectangular) that are present in the imported mesh. These constants are often used to declare the number of elements in an array. For instance, suppose we wanted to write a simulation that computes the electrostatic potential of a MOSFET whose geometry and doping profile is defined by the example MOSFET mesh specification file. To do so, we would first import the MOSFET mesh using the mesh statement. We would then declare an array which would store the electrostatic potential at each node on the mesh. To do this, we need to know how many nodes the mesh contains. Thus, we use the NODES constant as shown in the following statements.

```
mesh "mosfet.msh";
var V[NODES];    // array V stores the electrostatic potential
```

Thirdly, importing a mesh imports all of the labels of the mesh specification file from which the mesh was created. Labels represent a collection or list of nodes. For instance, importing the the mesh generated from the example MOSFET, would import the labels GATE, NOFLUX, SISIO2, DRAIN, SOURCE, SUB, SI02, and SI. The label GATE represents a list of nodes which are on the edge labeled GATE. The label SI02 represents a list of nodes which are in and on the boundary of the region labeled SI02. The use of labels in equation specification files is discussed in Subsection 15.3.2.

Finally, importing a mesh imports some arrays. All of these imported arrays are one-dimensional and the number of elements in each array is equal to the number of mesh nodes. At least two arrays will be imported. The names of these two arrays will be the names given to the coordinate labels in the mesh specification file via the coordinate statement. The values of the elements of these arrays will be the positions of the mesh nodes. Furthermore, an additional array will be imported for each refinement function present in the imported mesh's specification file. The names of these arrays will be the names given to the refinement functions. The values of the elements of these arrays will be the bodies of the refinement functions evaluated at each mesh point. For example, the example MOSFET mesh specification file labels its coordinates 'x' and 'y'. In addition, this file defines one refinement function named 'C' whose body computes the doping profile. When meshes that are generated from this mesh specification file are imported, three arrays will be imported whose names are 'x', 'y', 'C'. The values of the elements of these arrays will be the x position, y position, and dopant concentration at each node in the mesh.

One word of caution needs to be given. Authors of equation specification files that import meshes should not declare constants, variables, functions, etc. whose names are identical to the names of the imported constants, labels, or arrays. If one does do this, a symbol redeclaration error will be generated.

15.3.2 Using Labels in Equation Specification Files

Several equation specification file statements such as known and unknown statements, equation statements, and assignment statements require a range or a loop when used in conjunction with arrays. The ranges and loops specify a group of indices that specify to which array elements the statements apply. (see 15.1.7 and 15.1.9).

For instance, consider the example MOSFET equation specification file. Suppose we want to determine the electrostatic potential throughout the device. As our first step, we import the mesh and declare an array to store the value of the electrostatic potential at each node (see Subsection 15.3.1). As our next step, we declare all the elements of the electrostatic potential array unknown except for the nodes on the gate, source, and drain contacts. As our final step, we assign the values of the elements of the electrostatic potential array at the gate, source, and drain contacts equal to the constants Vgate, Vsource, and Vdrain. The following example shows the statements which accomplish these steps.

```
mesh "mosfet.msh";

const Vgate = 2.0;
const Vsource = 0.0;
const Vdrain = 0.0;

var V[NODES];     // array V stores the electrostatic potential
unknown V[all];
known V[GATE], V[SOURCE], V[DRAIN];
```

```
begin main
  assign V[i=GATE] = Vgate;
  assign V[i=SOURCE] = Vsource;
  assign V[i=DRAIN] = Vdrain;
end
```

The above example illustrates the use of labels in equation specification files. Notice that labels, which are defined in the mesh specification file, are used as ranges. This is possible because ranges represent a group of nodes. One may ask, 'Why not use a traditional range which specifies a starting index, an ending index, and a step value?' The answer is three-fold. First, since we do not know which nodes reside on the gate, source, and drain contacts, we cannot define a range by specifying a starting index, an ending index, and a step value. Second, even if we did know which nodes reside on these contacts, in general, we still could not specify the starting index, ending index, and a step value because the nodes on any edge or region are not guaranteed to be sequential. Third, suppose we did know which nodes reside on the contacts and they could be represented by a range that is specified by a starting index, an ending index, and a loop. We still would not want to use this approach, since we would have to change the ranges' starting indices, ending indices and step values every time we modified the mesh.

Upon importing an irregular mesh, simulation authors may use labels as ranges in any equation specification statement that requires a range (or loop, since a loop contains a range). The syntax of a range is extended as follows.

int expr |*.. int expr* |*: int expr* ||

all

identifier

Hence, ranges may be used in known and unknown statements, equation headers, and assignment statement headers. Ranges provide a convenient way to specify a group of indices. Furthermore, they allow users to change certain characteristics of the mesh without having to change the equation specification file that imports the mesh. For instance, we could change the depth of the example MOSFET mesh specification file by changing the value of the WDEV constant. Since this change does not modify the mesh labels, no change would be required in a properly written equation specification file that imports the MOSFET mesh.

15.3.3   Mesh Connectivity Functions

As mentioned in the introduction to this section, the nodes of irregular meshes, in general, are not stored in any easily predictable order. Therefore, it is necessary to explicitly store the mesh connectivity. SGFramework provides two functions to access this information: node and edge. The syntax of these functions is as follows.

NODE '(' *int expr* ',' *int expr* ')'

EDGE '(' *int expr* ',' *int expr* ')'

The mesh connectivity functions begin with the keyword 'NODE' or 'EDGE' and are followed by a left parenthesis, an integer expression, a comma, another integer expression, and a right parenthesis. The first integer expression specifies a mesh node while the second integer expression specifies a neighbor index. Examples of mesh connectivity functions are node(i,j) and edge(i,j). The function node(i,j) returns an index to the jth neighbor of the ith node. The function edge(i,j) returns an index to the edge which connects the ith node to its jth neighbor. This function does not return the edge which connects the ith and jth node. Note that these functions are exceptional, in that they return an integer number as opposed to a floating-point quantity. Hence, node and edge functions may appear in the index expressions of arrays. For instance, the expression

```
V[i] - V[node(i,j)]
```

computes the difference between the values of array elements that correspond to the ith node and its jth neighbor.

15.3.4 Mesh Geometry Functions

It is often desirable to know geometrical quantities such as the distance between the ith node and its jth neighbor. One could compute this quantity via the following expression.

```
sqrt(sq(x[i]-x[node(i,j)])+sq(y[i]-y[node(i,j)]))
```

This, however, requires two differences to be computed and three function calls. Since the distances between a node and its neighbors are often used in simulations, these values have been precomputed and can be accessed via the edge length function.

SGFramework provides several functions that access certain precomputed geometrical quantities of irregular meshes: the edge length function (elen), the integration edge length function (ilen), the integration area function (area), and the partial integration area function (area). The syntax of each of these functions is as follows.

    ELEN '(' *int expr* ',' *int expr* ')'
    ILEN '(' *int expr* ',' *int expr* ')'
    AREA '(' *int expr* ')'
    AREA '(' *int expr* ',' *int expr* ')'

The first integer expression in each function evaluates to a node. The second integer expression (if it exists) evaluates to a neighbor index. To understand these functions, consider Figure 15.1. This figure illustrates the quantities that the mesh geometry function returns. Consider the node that is represented by the solid circle. An integration 'box' is formed by connecting the perpendicular bisectors of the edges that connect the node to its neighbors. The perpendicular bisectors are refered to as integration edges. The elen(i,j) function returns the length of the edge that connects the ith node and its jth neighbor (not the length of the edge that connects the ith

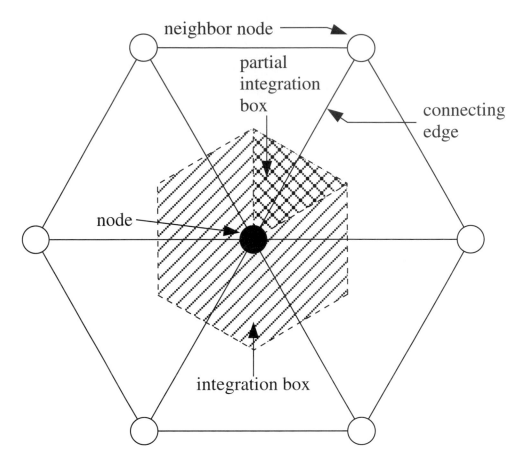

Figure 15.1. A schematic diagram illustrating the edge length, integration edge length, the integration area, and the partial integration area.

and jth nodes). The ilen(i,j) function returns the length of integration edge that perpendicular bisects the edge that connects the ith node and its jth neighbor. The area(i) function returns the area of the integration box that contains the ith node. Finally, the area(i,j) function returns the partial area of the integration box that is formed by the triangle whose vertices consist of the ith node and the endpoints of the integration edge discussed in the description of the $ilen(i, j)$ function.[1]

---

[1] The figures in this chapter are reprinted from Computer Physics Communications, Vol. 93, "Strategies for mesh-handling and model specification within a highly flexible simulation framework", 179-211, 1995, with permission from Elsevier Science - NL, Amsterdam, the Netherlands.

15.3.5 Mesh Summation Functions

It is often desirable to evaluate and sum an expression over a group of nodes. The SGFramework language provides two summation functions: the label summation function (lsum) and the node summation function (nsum). This subsection is divided into three parts. The first part discusses the label summation function. The second part describes the node summation function. The third part describes which summation statements may be nested inside other summation statements.

Label Summation Function

Labels represent a group of nodes. The label summation function loops over a group of nodes represented by a label, evaluates an expression at each node, and sums the results. The syntax of the label summation function is as follows.

LSUM '(' *identifier* ',' *identifier* ',' *expr* ')'

Label summation functions start with the keyword 'LSUM' followed by three comma-separated parameters enclosed by parentheses. The first parameter is an identifier. This identifier is the name of an index whose value is set to the index of the summation's current node. The second parameter is another identifier. This identifier is the name of a label that specifies the group of nodes the summation loops over. The last parameter is a mathematical expression. This expression is evaluated and summed at each node of the specified label. Usually this expression 'is a function of' the specified index. By 'is a function of', we mean that the specified index appears in the mathematical expression.

Node Summation Function

Whereas the label summation functions loops over a group of nodes represented by a label, the node summation function loops over the neighbors of a specified node. The syntax of the node summation function is as follows.

NSUM '(' *int expr* ',' *identifier* ',' *identifier* ',' *expr* ')'

NSUM '(' *int expr* ',' *identifier* ',' ALL ',' *expr* ')'

Node summation functions start with the keyword 'NSUM' followed by four comma-separated parameters enclosed by parentheses. The first parameter is an integer expression that evaluates to the node whose neighbors are looped over. The second parameter is an identifier. This identifier is the name of an index whose value is set to the index of the summation's current node. The third parameter is either an identifier which is the name of a label or the keyword 'ALL'. If this parameter is a label, then it filters the nodes over which the summation loops. Only those neighbors that are part of the group represented by the specified label are included in the summation. If this parameter is the keyword 'ALL', then all of the neighbors are included in the summation. The last parameter is a mathematical expression. This expression is evaluated and summed at each allowed neighbor. Usually this expression 'is a function of' the node given by the first parameter and the index specified by the second parameter. By 'is a function of', we mean that these indices appear in the mathematical expression.

Nesting Summation Functions

The last parameter of both summation functions is a mathematical expression. In general, one is not allowed to nest summation functions. In other words, one can not include another summation function in a summation function's expression. The only exception to this rule is that node summation functions may be nested in label summation functions.

To illustrate summation functions, consider a simulation which imports a mesh generated from the example MOSFET mesh specification file. Suppose this simulation solves for the MOSFET's electrostatic potential (V), electron concentration (n), and hole concentration (p) at each node of the mesh. The simulation declares user-defined functions which return the electron and hole current densities. The following code computes the MOSFET drain's current per unit length.

```
mesh "mosfet.msh";

func Jn(n1,n2,V1,V2,<h>)
...

func Jp(p1,p2,V1,V2,<h>)
...

var V[NODES], n[NODES], p[NODES];
...

begin main
  call Initialize;
  solve;
  assign I = lsum(i,DRAIN,nsum(i,j,SI,ilen(i,j)*
    {Jn(n[i],n[node(i,j)],V[i],V[node(i,j)],elen(i,j))+
     Jp(p[i],p[node(i,j)],V[i],V[node(i,j)],elen(i,j))}));
end
```

15.3.6   Precomputed Functions

Since a significant portion of the simulation time may be spent evaluating the elements of the Jacobian matrix, it is desirable to make this process as efficient as possible. One manner in which this may be accomplished is by eliminating redundant calculations by storing intermediate values in memory. To this end, the SGFramework language provides a mechanism that allows user-defined functions (and their partial derivitives with respect to their arguments) to be precomputed and stored in arrays for quick retrieval. Precomputation of user-defined functions is performed prior to constructing the Jacobian matrix or the function vector. It should be noted that function precomputation is only supported in simulations where SGFramework meshes are imported. This subsection is divided into two parts. The first part describes how to specify precomputed functions and the second part describes how to use precomputed functions.

Specifying Precomputed Functions

After a user-defined function has been declared, it may be tagged for precomputation via precompute statements. Not all user-defined functions may be precomputed. Precomputation is limited to user-defined functions that compute a value at a mesh node or a mesh edge. Specifying precomputation for each of the types of user-defined function will be discussed individually.

Specifying Node Precomputed Functions Precompute statements for user-defined functions that compute a value at a node begin with the keyword 'PRECOMPUTE' and are followed by an 'at' (@) symbol, the keyword 'NODE', an identifer, an arrow, a function, and a semicolon. The identifier is the name of a node index which may be used by specified user-defined function's arguments. The syntax of these precompute statements is as follows.

PRECOMPUTE '@' NODE *identifier* '->' *function* ';'

Examples of precompute statements for user-defined functions that compute a value at a node are given below.

```
var V[NODES], n[NODES], p[NODES], C[NODES], tn[NODES], tp[NODES];

func MUn(<N>,<T>,pn)
   ...
   return MUl*{1.025/[1+pow(X/1.68,1.43)]-0.025}/u0;

func R(n,p,<tn1>,<tp1>)
   ...
   return Rsrh+Raug;

precompute @ NODE i -> MUn(C[i],T,n[i]*p[i]);
precompute @ NODE i -> R(n[i],p[i],tn[i],tp[i]);
```

The first precompute statement specifies that the user-defined function MUn should be evaluated at node i with arguments N, T, and pn equal to C[i], T, and n[i]*p[i] respectively. The last precompute statement specifies that the user-defined function R should be evaluated at node i with arguments n, p, tn1, and tp1 equal to n[i], p[i], tn[i], and tp[i] respectively.

Specifying Node Precomputed Functions Precompute statements for user-defined functions that compute a value at a mesh edge begin with the keyword 'PRECOMPUTE' and are followed by an 'at' (@) symbol, the keyword 'EDGE', an identifer, a parenthesized list of two comma-separated identifiers, the keyword 'ODD' or 'EVEN' an arrow, a function, and a semicolon. The first identifier is the name of a edge index. The second and third identifiers, which are parenthesized, specify the two nodes on the ends of the edge. These identifiers may be used by a specified user-defined function's arguments. The keywords 'ODD' or 'EVEN' specify that the user-defined function is either an odd or even function with respect to the order of the nodes. The syntax of these edge precompute statements is as follows.

PRECOMPUTE '@' EDGE *identifier* '(' *identifier* ',' *identifier* ')' ODD '->'
function ';'
PRECOMPUTE '@' EDGE *identifier* '(' *identifier* ',' *identifier* ')' EVEN '->'
function ';'

Examples of precompute statements for user-defined functions that compute a
value at an edge are given below.

```
var V[NODES], n[NODES], p[NODES], C[NODES], tn[NODES], tp[NODES];

func E(V1,V2,<h>)
  return -grad(V1,V2,h);

func Jn(n1,n2,E,un1,un2,<h>)
  ...
  return un*(n*E+dndx);

precompute @ EDGE i (j,k) ODD -> E(V[j],V[k],elen);
precompute @ EDGE i (j,k) ODD ->
Jn(n[j],n[k],@E(i,j,k),@MUn(j),@MUn(k),elen);
```

The first precompute statement specifies that the user-defined function E is computed at edge i whose nodes are j and k with arguments V1, V2, and h equal to
V[j], V[k], and elen respectively. The keyword 'ELEN' represents the length of edge
i. Note this function is odd, because E(V[j],V[k],elen) = -E(V[k],V[j],elen). The
last precompute statement specifies that the user-defined function Jn is computed
at edge i whose nodes are j and k with arguments n1, n2, E, un1, un2, and h equal
to n[j], n[k], @E(i,j,k), @MUn(j), @MUn(k), and elen respectively. The particular notation for @E(i,j,k), @MUn(j), and @MUn(k) will be explained in the next
section.

Using Precomputed Functions

Once a function has been specified for precomputation, the precomputed values of
the function may be used in the body of equation statements and in other precompute statements. The syntax of precomputed functions is as follows.

'@' *identifier* '(' *int expr* ')'
'@' *identifier* '(' *int expr* ',' *int expr* ',' *int expr* ')'

The first syntax statement is for node precomputed functions. It begins with
an 'at' (@) symbol and is followed by an identifier, and a parenthesized integer
expression. The identifier is the name of the node precomputed function and the
integer expression evaluates to the node index.

The second syntax statement is for edge precomputed functions. It begins with
an 'at' (@) symbol and is followed by an identifier, and a parenthesized list of
three comma-separated integer expressions. The identifier is the name of the edge
precomputed function and the integer expressions evaluate to the edge and node
indices.

For example, consider the last precompute statement below.

```
precompute @ EDGE i (j,k) ODD ->
Jn(n[j],n[k],@E(i,j,k),@MUn(j),@MUn(k),elen);
```

The third, fourth, and fifth arguments of the user-defined function Jn are @E(i,j,k), @MUn(j), and @MUn(k). @E(i,j,k) represents the precomputed value of the user-defined function E evaluated at edge i with nodes j and k. @MUn(j) and @MUp(k) represent the precomputed values of the user-defined functions MUn and MUp evaluated at nodes j and k respectively.

As an example of how precomputed functions may be used in an equation statement, consider the SGFramework code below.

```
equ n[i=SI] ->
  +nsum(i,j,all,{@Jn(edge(i,j),i,node(i,j))}*ilen(i,j)) -
@R(i)*area(i) = 0.0;
```

@Jn(edge(i,j),i,node(i,j)) represents the precomputed value of the user-defined function Jn evaluated at edge edge(i,j), whose nodes are i and node(i,j).

This concludes the discussion of interfacing the equation and mesh specification files. We will now turn our attention to the executable programs and scripts which comprise the SGFramework.

15.4   SGFramework Executables

This section briefly describes the executable programs and scripts that form the SGFramework. (The SGFramework executables are invoked from a command line, so Windows95 and NT users must open an MS-DOS prompt to run the SGFramework executables, whereas in UNIX, a shell must be opened.) It should be apparent that it is all too easy to set up simulations which will fail, for various reasons. If a simulation does not converge, this may lead to overflows or to domain errors. These will often be reported as such, or they may cause the simulation to crash.

The SGFramework is schematically represented by Figure 15.4. The circles in the flowchart represent executable programs and scripts (batch files). This section is divided into several subsections. Each subsection describes an executable or script, lists its command-line options, and enumerates the warning and error messages that the executable may generate.

15.4.1   Build Script

The mesh parser and SGFramework translator produce C++ code. The code these programs generate must be compiled and linked with either the mesh refinement or SGFramework numerical algorithm module libraries. The SGFramework build script (sgbuild) facilitates the compiling and linking of the code generated by SGFramework programs.

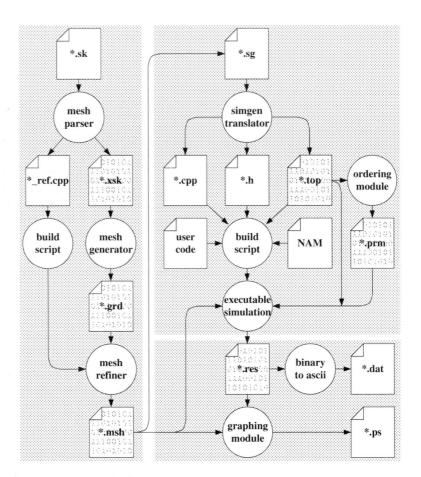

Figure 15.2. The overall flowchart for the SGFramework.

Command-Line Options

The SGFramework build script accepts two command-line arguments. The first argument is either 'ref' or 'sim' and the second argument is the name of the C++ source code to compile and link minus the extension. For instance, to compile the source code file `pin_ref.cpp` into a mesh refinement program, we would invoke the build script as follows.

```
sgbuild ref pin_ref
```

If the SGFramework build script is invoked with no or inappropriate command-line arguments, the following help information is displayed.

```
SGFramework Build Script    Version 1.0  Copyright (c) 1997 Kevin M. Kramer
* USAGE: sgbuild ref filename
*        sgbuild sim filename
*
* DESCRIPTION:
* sgbuild ref filename compiles and links the mesh refinement program
* sgbuild sim filename compiles and links the simulation program
*
* REMARKS:
* filename is the mesh refinement or simulation C++ source file name
* without the '.cpp' extension.
```

15.4.2 Mesh Parser

In any simulation which uses a complex mesh, we employ a mesh specification file to specify both the mesh and the way in which it is to be refined. The mesh specification file, which has an .sk extension, is parsed by the mesh program. This program generates a binary file with an .xsk extension and a C++ source file, which when linked with the mesh refinement library via the SGFramework build script, creates a mesh refinement program.

Command-Line Options

If the mesh parser (mesh) is invoked with no arguments or the -h command-line argument, the following help information is displayed.

```
SGFramework Skeleton Parser  Version 1.0  Copyright (c) 1995 Kevin M. Kramer
syntax:  mesh [options] file
 -h        display help screen on the standard output device
 -p##      floating-point precision where ## = 1 to 16 (default = 6)
```

Warning and Error Messages

The mesh parser may generate the following warning and error messages. The format of the messages will be '[X] (row, col) message' where X is either a W for warning, E for error, or F for fatal error; and row and col specify the row and column index of the place where the error occurred. Sometimes the actual error appears in the statement which precedes the one that is specified by the row and column indices. Because an error may cause several errors to follow, it is recommended that one fix the errors in the order they appear. In the explanation of the warning and error messages, all messages are referred to as errors even though they may only be warnings.

string cannot span multiple lines – Strings must begin and end with a double quote, both characters being on the same line.

unexpected end of file – The equation specification file ended unexpectedly. A common cause of this error is failure to specify the keyword 'END' at the end of the main procedure.

? expected – This error is caused by syntax errors. The question mark will be a list of valid characters or keywords that may appear at the specified position in the input file.

redefinition of symbol ? – The specified symbol has already been defined. For instance, one cannot define a constant named COUNT, and then declare a variable named COUNT.

symbol ? not defined – The specified symbol has not been defined. This error usually occurs when the name of a constant, function, variable, array or procedure has been misspelled.

? is not a ? symbol – The specified symbol is not of the required type. For instance, if an array is indexed by an expression enclosed in parentheses, rather than brackets, this error will occur, since the translator expects functions to be followed by a parenthesized list of expressions.

? symbol ? not declared – Because of a prior error, the specified symbol of the specified type has not been declared. This error will usually cause several additional errors to follow.

? not declared – This error is generated when the specified item (usually a parameter) has not been declared due to an error. This error usually occurs when a set statement attempts to initialize an unknown parameter.

coordinates not declared – This error is generated when the coordinates have not been generated due to an error.

? does not evaluate to a constant – The specified expression does not evaluate to a constant.

? does not evaluate to positive constant – The specified expression does not evaluate to a positive constant.

? variable does not evaluate to positive integer constant – The specified expression does not evaluate to a positive integer constant.

function ? has the wrong number of arguments – The specified function has the wrong number of arguments.

division by zero in ? – This error is caused when an expression or quantity is divided by another expression or quantity that evaluates to zero.

domain error in ? function – This error is caused when a function is called with arguments whose values are outside the domain of the function. For instance, taking the square root of a negative number will cause a domain error.

overflow in ? function – This error is generated when a precision overflow is generated.

underflow in ? function – This error is generated when a precision underflow is generated.

integer overflow – This error is caused when an integer expression evaluates to a value that is too large to represent as an integer.

cannot open file ? – The specified file cannot be opened.

cannot subdivide region ? into rectangular elements – The specified region cannot be divided into rectangular elements. Verify that the opposite edges have equal node spacing and that the edges are parallel to the coordinate axes.

15.4.3   Mesh Generator

The initial coarse grid is generated and plotted (unless specified) from a parsed mesh specification file. If a plot is generated, it is necessary to close the window before the next command can be accepted. Parsed mesh specification files have an .xsk extension.

Command-Line Options

If the mesh generator (sggrid) is invoked with no arguments or the -h command-line argument, the following help information is displayed.

```
SGFramework Grid Generator   Version 1.0  Copyright (c) 1995 Kevin M. Kramer
syntax:  gridgen [options] file               * = default
  -h        display this help screen
  -Oxxx     create ASCII output file named xxx
  -ee     * display errors on the standard error device
  -eo       display errors on the standard output device
  -t+     * exhaustive search for best triangulation
  -t-       nonexhaustive search wherever appropriate
  -i+       increase interior spacing
  -i-     * do not increase interior spacing
  -c+     * use centroid triangulation
  -c-       do not use centroid triangulation
  -g###     grade (1.5 to 2.5, default = 1.5)
  -s###     maximum mesh spacing
  -m###     minimum permissible quality (default 0.35)
  -qe###    number of edge enhancement iterations (default 2)
  -qn###    number of node enhancement iterations (default 2)
  -qb###    number of edge followed by node enhancement iterations
(default 2)
  -vr     * visualize mesh rating
  -vn       no graphics
```

```
-ps        generate a postscript file named grid.ps
-r1        graphics screen resolution 480 x 640
-r2      * graphics screen resolution 800 x 600
-r3        graphics screen resolution 1024 x 768
-aspect    keep x and y aspect ratio when plotting
```

Warning and Error Messages

The mesh generation may generate the following warning and error messages.

out of memory – If this error message occurs, try closing some applications and try again.

file not found – The parser cannot locate the input file, verify the name and path of the input file and try again.

cannot create or open file – This error is caused when either the user does not have permission to open or create a file on the specified directory or the disk is full.

file write error – This error is caused when either the user does not have permission to write on the specified directory or the disk is full.

directory name too long – This error is caused when the directory name is too long. To remedy this error use a shorter directory name and try again.

all triangulation heuristics failed – The mesh generator is not able to triangulate the mesh. Verify that each region is specified by a simple, closed counterclockwise curve, the regions do not overlap, and that the segment spacing changes gradually (several short segments are not next to a long segment).

15.4.4   Mesh Refiner

In order to refine a mesh, the user must do two things. First the mesh refinement program must be compiled and linked. Second, the mesh refinement program must be executed. The mesh refinement program may be compiled and linked using the SGFramework build script. The mesh refinement source file is generated by the mesh parser. The name of the mesh refinement source file is '**xxxx_ref.cpp**' where 'xxxx' is the first four characters of the mesh specification file from which the source file was generated. The mesh refinement program will generate a binary mesh file whose name is the name of the mesh specification file with a .msh extension. It may also plot the refined mesh (and the window containing the plot must be closed before the next command can be entered.)

Command-Line Options

If a mesh refinement program is invoked with no arguments or the -h command-line argument, the following help information is displayed.

```
SGFramework Mesh Refiner     Version 1.0  Copyright (c) 1995 Kevin M. Kramer
syntax:  refine [options]                            * = default
  -h         display this help screen
  -Oxxx      create ASCII output file named xxx
```

```
 -cyl        cylindrical geometry
 -dmin###    minimum number of divisions
 -dmax###    maximum number of divisions
 -dadj###    divide triangles of level < ### if adj. triangle will be
split
 -rm?=lin    use linear measurement for refinement variable ?
 -rm?=log    use logarithmic measurement for refinement variable ?
 -rm?=slog   use signed logarithmic measurement for refinement variable ?
 -rd?=###    maximum delta for refinement variable ?
 -qe###      number of edge enhancement iterations (default 2)
 -qn###      number of node enhancement iterations (default 2)
 -qb###      number of edge followed by node enhancement iterations
(default
2)
 -m###       minimum rectangular splitting angle in degrees (default 0)
 -g###       maximum grade between rectangular elements (default
infinite)
 -s###       maximum ratio between length and width (default infinite)
 -aspect     keep x and y aspect ratio
 -vr       * visualize mesh rating
 -vn         no graphics
 -yz         do not apply refinement statements to elements above y = 0
 -ps         generate a postscript file named mesh.ps
 -bw         generate grayscale plot
 -r1         graphics screen resolution 480 x 640
 -r2       * graphics screen resolution 800 x 600
 -r3         graphics screen resolution 1024 x 768
```

Warning and Error Messages

The Warning and Error Messages are:

cannot open file – This error is generated when the specified file could not be opened.

corrupt grid file — This error is generated when the grid (initial mesh) file is corrupt. To remedy this error, regenerate the initial mesh.

cannot create file – This error is generated when the specified file cannot be created.

15.4.5   SGFramework Translator

In order to generate a SGFramework simulation, a user must first translate an equation specification file via the SGFramework translator (using the command sgxlat). Then the user must compile and link the translator's output with a numerical algorithm module via the SGFramework build script (sgbuild).

Command-Line Options

If the SGFramework translator (sgxlat) is invoked with no arguments or the -h command-line argument, the following help information is displayed.

```
SGFramework Translator      Version 1.0  Copyright (c) 1995 Kevin M. Kramer
syntax:  sgxlat [options] file
 -h       display help screen on the standard output device
 -p##     floating-point precision where ## = 1 to 16 (default = 6)
 -nc      no coupling between equations
```

Warning and Error Messages

The SGFramework translator may generate the following warning and error messages. The format of the messages will be '[X] (row, col) message' where X is either a W for warning, E for error, or F for fatal error; and row and col specify the row and column index of the place where the error occurred. Sometimes the actual error appears in the statement which precedes the one that is specified by the row and column indices. Because an error may cause several errors to follow, it is recommended that one fix the errors in the order they appear. In the explanation of the warning and error messages, all messages are referred to as errors even though they may only be warnings.

string cannot span multiple lines – Strings must begin and end with a double quote, both characters being on the same line.

unexpected end of file – The equation specification file ended unexpectedly. A common cause of this error is failure to specify the keyword 'END' at the end of the main procedure.

? expected – This error is caused by syntax errors. The question mark will be a list of valid characters or keywords that may appear at the specified position in the input file.

redefinition of symbol ? – The specified symbol has already been defined. For instance, one cannot define a constant named COUNT, and then declare a variable named COUNT.

symbol ? not defined – The specified symbol has not been defined. This error usually occurs when the name of a constant, function, variables, array or procedure has been misspelled.

? is not a ? symbol – The specified symbol is not of the required type. For instance, if an array is indexed by an expression enclosed in parentheses, rather than brackets, this error will occur, since the translator expects functions to be followed by a parenthesized list of expressions.

? symbol ? not declared – Because of a prior error, the specified symbol of the specified type has not been declared. This error will usually cause several additional errors to follow.

? statement not declared – Because of a prior error, the specified statement of the specified type has not declared. This error may cause several additional errors to follow.

user-defined function ? not declared – Because of a prior error, the specified user-defined function was not declared. This error will usually cause several additional errors to follow.

? does not evaluate to a constant – The specified expression does not evaulate to a constant.

array ? has too many dimensions – The specified array has too many dimensions. This error is generated if an array is declared that has over three dimensions.

dimension ? of array ? does not evaluate to a integer constant – The specified dimension of the specified array does not evaulate to an integer constant.

dimension ? of array ? is not an integer expression – The specified dimension of the specified array is not an integer expression.

dimension ? of array ? must be greater than zero – The specified dimension of the specified array must evaulate to an integer that is greater than zero. In other words, one cannot declare an array with zero or a negative number of elements.

dimension ? of array ? is out of bounds – The specified dimension of the specified array is out of bounds. In other words, the expression evaluates to an integer which is less than zero or greater than the number of elements in the specified dimension minus one.

array ? has the wrong number of dimensions – The specified array has the wrong number of dimensions. For instance, if a two-dimensional array is declared, the use of this array must be accompanied by a bracketed list of two-index expressions.

function ? has the wrong number of arguments – The specified function has the wrong number of arguments.

? value of index ? does not evaluate to an integer constant – the start, end, or step value of the specified index does not evaluate to an integer constant.

? value of index ? is illegal – the start, end, or step value of the specified index is illegal.

equation corresponding to ? ? is useless – The equation corresponding to the specified variable or array is not used because the variable or the unknown elements of the specified array are already associated with one or more equations.

equation corresponding to ? ? has no pivot – The equation corresponding to the specified variable or array has no pivot. This error results when the variable or array elements specified by the equation header do not appear in the equation's expression.

no main procedure statements specified – The equation specification does not have the mandatory main procedure.

no equation statements specified – No equation statements are declared in the equation specification file yet a solve statement is present.

division by zero in ? – This error is caused when an expression or quantity is divided by another expression or quantity that evaluates to zero.

domain error in ? function – This error is caused when a function is called with arguments whose values are outside the domain of the function. For instance, taking the square root of a negative number will cause a domain error.

overflow in ? function – This error is generated when a precision overflow is generated.

underflow in ? function – This error is generated when a precision underflow is generated.

integer overflow – This error is caused when an integer expression evaluates to a value that is too large to represent as an integer.

corrupt or missing mesh file – The specified mesh file is either corrupt or missing. To remedy this error, regenerate the mesh file.

near singularity encountered – This error is generated when the absolute value of a pivot is less than the specified 'near singularity' quantity.

unknown ? does not have a corresponding equation – The unknown variable or array element does not have a corresponding equation. Either add an equation or equations to the equation specification file or declare this variable or array element to be known.

underspecified system of equations – The system of equations is underspecified. A common cause of this error is when one equation is just a multiplicative constant times another equation.

overrelaxation parameter not in range [1.0, 2.0] – The overrelaxation parameter is not within its required range of 1.0 to 2.0.

symbol ? is not allowed in user-defined function body – The specified symbol is not allowed in the body of a user-defined function. A common cause of this error is using a variable or array element in the function's body.

function ? may not call itself – This error is called when a function tries to recursively call itself. Recursion is not supported by the SGFramework translator.

a file must be open for ? prior to using this statement – This error is generated when the specified statement is used prior to opening an input or output file.

file already open, must issue a close statement – The SGFramework only supports one open file at a time. If the user tries to open a second file prior to closing the first file, this error will be generated.

cannot open file ? – The specified file cannot be opened.

the data file has not been closed – An open data file has not been closed prior to exiting the simulation.

unexpected end of data file – The end of a data file has been unexpectedly reached. A common cause of this error is searching for a label that does not exist in the data file.

syntax error in data file – A syntax error was encountered in the data file.

could not find label ? in data file – The specified label was not found in the data file.

main procedure already declared – An additional main procedure is present in the equation specification file.

? is not an integer expression – The specified expression is not an integer expression.

? function cannot contain another ? function – The first specified function cannot contain the second specified function. This error is usually caused by trying to nest a summing function inside another summing function.

irregular mesh has not been declared – This error is generated when a mesh connectivity function or mesh geometry function is used but a mesh has not been imported.

the ? operator cannot be used in edge precompute statements – The specified operator cannot be used in edge precompute functions.

cannot determine if argument is odd or even w/ respect to its indices – This error is generated when the SGFramework translator cannot determine if an argument of a function is odd or even with respect to its indices.

15.4.6   Ordering Module

In order to increase the speed and reduce the memory requirements of SGFramework simulations, a user should order a simulation's unknown variables and array elements via the SGFramework minimum degree ordering program (order). The ordering program requires two command-line arguments: a SGFramework topology file and a SGFramework permutation file. The SGFramework topology file is generated by the SGFramework translator. The SGFramework permutation file will be generated by the ordering program.

Command-Line Options

If the minimum degree ordering program (order) is invoked with no arguments or the -h command-line argument, the following help information is displayed.

```
Minimum Degree Order   Version 1.0  Copyright (c) 1995 Kevin M. Kramer
usage:  order [options] topfile permfile
  -h     display this help screen
  -q     run quietly
  -Pxxx generate a SMMS permutation vector named xxx
```

Warning and Error Messages

The Warning and Error messages are:

cannot open file ? – This error is generated when the specified file cannot be opened.

cannot create file ? – This error is generated when the specified file cannot be created.

15.4.7   SGFramework Simulations

SGFramework simulations are built by compiling and linking the code generated by the SGFramework translator with an appropriate numerical algorithm module.

Command-Line Options

If the simulation is invoked with the -h command-line argument, the following help information is displayed.

```
SGFramework Simulation, SimGen Copyright (c) 1994 K. M. Kramer
 -h       display this help screen
 -vs0     run quietly, i.e. do not output convergence information, etc.
 -vs1     output abbreviated convergence information
 -vs2     output full convergence information
 -vl1     output abbreviated convergence information to log file
 -vl2     output full convergence information to log file
 -z+      zero the dx vector prior to each iteration
 -z-      do not zero the dx vector prior to each iteration
 -Lxxx    generate log file xxx
 -i##     maximum number of Newton iterations
 -a##     minimum Newton accuracy
 -d##     minimum error-reducing damping factor (min. df = 2^-##)
 -e##     maximum percentage change per Newton iteration
 -s##     near singularity value
 -li##    maximum number of SOR, PCG, etc. iterations
 -la##    maximum SOR, PCG, etc. accuracy
 -lf##    maximum fill level for Gaussian elimination (## > 5 = infinity)
 -lr##    over-relaxation parameter
 -lsor    use successive over-relaxation algorithm
 -lge     use Gaussian elimination algorithm
 -p##     ASCII output precision for real numbers
 -n##     use ## norm (## = inf for infinite norm)
```

Warning and Error Messages

The Warning and Error messages are:

cannot open file ? – This error is generated if the specified file cannot be opened.

cannot find label ? in data file ? – This error is generated if the specified label cannot be found in the specified file.

syntax error in data file ? – This error is generated when a syntax error is encountered in the specified data file.

unexpected end of file in data file ? – This error occurs if the end of the specified data file is unexpectedly encountered. A common cause of this error is searching for a label that either does not exist or is present only before the location of the file pointer.

near singularity encountered – This error is generated when the absolute value of a pivot is less than the specified 'near singularity' quantity.

15.4.8   Extract Program

The binary-to-ASCII extraction program (extract) is used to extract data from a binary SGFramework result file. The data is written to ASCII files which are labeled by the name of the array whose value they contain. All variables are stored by the main program in a single file. The extension of these data files is determined by which snapshot of data is extracted. The command-line syntax of the extract program is as follows.

extract filename.res i

The number i is an integer that indicates that the ith set of data which was written to the .res file is to be extracted. Examples of shell scripts set up to perform the extraction are given in the text - see Section 1.5.

15.4.9 Group Program

The data file format program (group) is used to format unformatted data files generated by a SGFramework simulation. Often, several variables are repeatedly written to a data file. For instance, the mixed-mode simulation listed in Section 12.2.2 writes the variables t, dt, Vs, Vb, Vc, Vd, Vl, Ib, Ic, Il, and Id to the data file mm02.out at each time step via the file write statement. Since the file write statement outputs each variable on its own line, it is very difficult to analyze the data file mm02.out. A portion of the mm02.out file is shown below.

```
0

1e-09

12

0.5988724

24

4.572742e-07
```

The first number, 0, corresponds to t; the second number, 10^{-9}, corresponds to dt, and so on. It would be very hard to determine to what variable the hundredth or thousandth number corresponded. However, by using the group program, one can organize the data file into a list of columns. For example, consider the following invocation of the group program.

```
group t dt Vs Vb Vc Vd Vl Ib Ic Il Id <mm02.out >mm02.tab
```

The above example will group the data in the file mm02.out into eleven columns named t, dt, Vs, Vb, Vc, Vd, Vl, Ib, Ic, Il, and Id. (The command-line arguments are taken as the names of the columns, hence the number of command-line arguments is the number of columns into which the data is organized.) The SGFramework data file is redirected to the standard input device via the < symbol. The output, which is the organized (grouped) data, is redirected to the file mm02.tab. A portion of mm02.tab is shown below. Note that only the first five columns are shown.

t	dt	Vs	Vb	Vc
0.00000e+00	1.00000e-09	1.20000e+01	5.98872e-01	2.40000e+01

```
1.00000e-09   1.50000e-09   1.19760e+01   5.98818e-01   2.57290e+01
2.50000e-09   2.25000e-09   1.19400e+01   5.98736e-01   2.77730e+01
4.75000e-09   3.37500e-09   1.18860e+01   5.98613e-01   3.05299e+01
8.12500e-09   5.06250e-09   1.18050e+01   5.98427e-01   3.40187e+01
1.31875e-08   7.59375e-09   1.16835e+01   5.98146e-01   3.73361e+01
2.07812e-08   1.00000e-08   1.15013e+01   5.97718e-01   3.92992e+01
```

15.4.10 Graphical Output

The SGFramework provides mechanisms by which to view and print SGFramework meshes. Both the mesh generation program and the mesh refinement programs are capable of generating mesh plots. Furthermore, if these programs are invoked with the -ps command-line argument, these programs can generate postscript files of the mesh. These postscript files may be viewed and printed with postscript viewing applications.

Because it is not possible, in general, to solve semiconductor problems on regular meshes, and because most plotting software is only useful for regular meshes, SGFramework has its own plotting capabilities. The SGFramework surface plotter (triplot) is capable of plotting the results of a SGFramework simulation that used a mesh specification file. triplot requires one command-line argument, the name of a SGFramework result file without the extension. In addition, the user may specify an array to plot. The array must be one-dimensional and have as many elements as the mesh has nodes. If the name of an array is not specified on the command line, the user will be prompted to enter an array. Finally, the user may specify optional command-line arguments. For instance, -log plots the log of the values of the specified array (to be precise, sign(x)*log(abs(x))) , -az specifies the azimuthal viewing angle of the plot, etc.

The SGFramework also contains another graphing program (sgplot). sgplot is similar to triplot in that it generates surface plots on irregular meshes. However, sgplot can also produce contours and it provides axes with tick marks. Invoke sgplot with the -h command line argument for instructions.

To make a postcript file of the results, the command line option -ps should be specified. The postcript file will have the same name as the simulation, with a '.ps' on the end. In this book, for example, this command was used to plot the voltage from the simulation file pn05.sg. The command thus read

```
triplot pn05 V -ps
```

In this case, the SGFramework surface plotting program will generate a postscript file named pn05.ps. If there is more than one set of data (say from multiple time steps) the -i### command-line option should be specified to choose which data set to use. For instance, -i1 uses the first data set, -i2 the second, and so on. If the -i### command line option is not specified, the default is to use the first data set. This plotting routine makes use of a SGFramework mesh file, so it will not work in cases where we did not generate such a file.

Command-Line Options

If the SGFramework surface plotter (triplot) is invoked with no arguments or the
-h command-line argument, the following help information is displayed.

```
SGFramework Surface Plotter  Version 1.0  Copyright (c) 1995 Kevin M. Kramer
syntax:  refine filename [array] [options]                    * =
default
   -h         display this help screen
   -i###      use the ###'th data set (default 1)
   -az###     azimuthal angle (default 30)
   -el###     elevation angle (default 60)
   -ps        generate a postscript file
   -wf        wire frame
   -bw        black and white (grayscale) plot
   -nm        do not plot the mesh
   -lin     * use a linear scale
   -log       use a logarithmic scale
   -r1        graphics screen resolution 480 x 640
   -r2      * graphics screen resolution 800 x 600
   -r3        graphics screen resolution 1024 x 768
```

Warning and Error Messages

cannot open file ? – This error is generated when the specified file cannot be opened.

Bibliography

[1] Sun Tzu. *The Art of War*. Oxford University Press, Oxford, 1963.

[2] J.G. Frazer. *The Golden Bough*. Avenel, New York, 1981.

[3] U. Eco. *Foucault's Pendulum*. Harcourt Brace Jovanovich, San Diego, 1988.

[4] G.E.P. Box, W.G. Hunter, and J.S. Hunter. *Statistics for Experimenters, An Introduction to Design, Data Analysis, and Model Building*. Wiley Series in Probability and Mathematical Statistics. Wiley, New York, 1978.

[5] D. Brin. *Glory Season*. Bantam Books, New York, 1993.

[6] P.M. Morse and H. Feshbach. *Methods of Theoretical Physics*. McGraw-Hill, New York, 1953.

[7] W.D. D'haeseleer, W.N.G. Hitchon, J.D. Callen, and J.L. Shohet. *Flux Coordinates and Magnetic Field Structure*. Springer-Verlag, Berlin, 1991.

[8] G.J. Parker, W.N.G. Hitchon, and J.E. Lawler. Numerical solution of the boltzmann equation in cylindrical geometry. *Physical Review E*, 50:3210–3219, 1994.

[9] S. Selberherr. *Analysis and Simulation of Semiconductor Devices*. Springer-Verlag, Wien, 1984.

[10] Z. Yu and R.W. Dutton. Sedan iii - a generalized electronic material device analysis program. Stanford electronics laboratory technical report, Stanford University, July 1985.

[11] R.C. Jaeger and F.H. Gaensslen. Simulation of impurity freezeout through numerical solution of poisson's equation and application to mos device behavior. *IEEE Transactions on Electron Devices*, ED-27:914–920, May 1980.

[12] R. Stratton. *Physical Review*, 126:2002, 1962.

[13] K. Blotekjaer. *IEEE Transactions on Electron Devices*, ED-17:38, 1970.

[14] W. Quade, M. Rudan, and E. Scholl. Hydrodynamic simulation of impact-ionization effects in p-n junctions. *IEEE Transactions on Computer-Aided Design*, 10(9):1287–1293, 1991.

[15] E. Fatemi, J. Jerome, and S. Osher. Solution of the hydrodynamic device model using high-order nonoscillatory shock capturing algorithms. *IEEE Transactions on Computer-Aided Design*, 10(2):232–243, 1991.

[16] C.L. Gardner, P.L. Lanzkron, and D.J. Rose. A parallel block iterative method for the hydrodynamic device model. *IEEE Transactions on Computer-Aided Design*, 10(9):1187–1192, 1991.

[17] C.T. Sah, P.C.H. Chan, C.K. Wang, R.L.Y. Sah, K.A. Yamakawa, and R. Lutwack. Effect of zinc impurity in silicon solar-cell efficiency. *IEEE Transactions on Electron Devices*, ED-28(3):304–313, 1981.

[18] J.T. Watt. PhD thesis, Stanford University, 1989.

[19] C. Jacoboni, C. Canali, G. Ottaviani, and A.A. Quaranta. A review of some charge transport properties of silicon. *Solid State Electronics*, 20.77–89, 1977.

[20] K. Seeger. *Semiconductor Physics*. Springer, Wien, 1973.

[21] L. Elstner. *Phys. Status Solidi*, 17:139, 1966.

[22] E. Conwell and V.F. Weisskopf. Theory of impurity scattering in semiconductors. *Physical Review*, 77:388–390, 1950.

[23] H. Brooks. Scattering by ionized impurities in semiconductors. *Physical Review*, 83:879, 1951.

[24] J. Bourgoin and M. Lannoo. *Point Defects in Semiconductors II*. Springer, Berlin, 1983.

[25] S.S. Li and W.R. Thurber. The dopant density and temperature dependence of electron mobility and resistivity in n-type silicon. *Solid State Electronics*, 20:609–616, 1977.

[26] N.H. Fletcher. In *Proc. I.R.E.*, page 862, June 1957.

[27] S. Chapman and T.G. Cowling. *The Mathematical Theory of Non-uniform Gases*. Cambridge University Press, Cambridge, 1952.

[28] S.C. Choo. *IEEE Transactions on Electron Devices*, ED-19:954, 1972.

[29] P.P. Debye and E.M. Conwell. Electrical properties of n-type germanium. *Physical Review*, 93:693–706, 1954.

[30] M. Luong and A.W. Shaw. *Physical Review*, B4, 1971.

[31] J.M. Dorkel and Ph Leturcq. Carrier mobilities in silicon: Semi-empirically related to temperature, doping and injection level. *Solid State Electronics*, 24(9):821–825, 1981.

[32] W. Shockley. Hot electrons in germanium and ohm's law. *Bell System Technical Journal*, 30:990–1034, 1951.

[33] D.M. Caughey and R.E. Thomas. Carrier mobilities in silicon empirically related to doping and field. *Proceedings of the IEEE*, 55:2192–2193, 1967.

[34] S.M. Sze. *Physics of Semiconductor Devices*. Wiley, New York, 1981.

[35] J.J. Barnes, R.J. Lomax, and G.I. Haddad. Finite-element simulation of gaas mesfets with lateral doping profiles and sub-micron gates. *IEEE Transactions on Electron Devices*, ED-23:1042–1048, 1976.

[36] M.J. LittleJohn, J.R. Hauser, and T.H. Glisson. Velocity-field characteristics of gaas with $\Gamma_6^e - L_6^e - X_6^e$ conduction band ordering. *Journal of Applied Physics*, 48:4587–4590, 1977.

[37] N.F. Mott. Recombination: A survey. *Solid State Electronics*, 21:1275–1280, 1978.

[38] W. Shockley and W.T. Read. Statistics of the recombination of holes and electrons. *Physical Review*, 87(5):835–842, 1952.

[39] R.N. Hall. Electron-hole recombination in germanium. *Physical Review*, 87:387, 1952.

[40] S.R. Dhariwal, L.S. Kothari, and S.C. Jain. On the recombination of electrons and holes at traps with finite relaxation time. *Solid State Electronics*, 24(8):749–752, 1981.

[41] A.S. Grove. *Physics and Technology of Semiconductor Devices*. Wiley, New York, 1967.

[42] J.G. Fossum, R.P. Mertens, D.S. Lee, and J.F. Jijs. Carrier recombination lifetime in highly doped silicon. *Solid State Electronics*, 26(6):569–576, 1983.

[43] P.T. Landsberg and D.J. Robbins. The first 70 semiconductor auger processes. *Solid State Electronics*, 26:1289–1294, 1978.

[44] J. Dziewior and W. Schmid. Auger coefficients for highly doped and highly excited silicon. *Applied Physics Letters*, 31:346–348, 1977.

[45] M.S. Tyagi and R. Van Overstraeten. Minority carrier recombination in heavily doped silicon. *Solid State Electronics*, 26(6):577–597, 1983.

[46] W. Shockley. Problems related to p-n junctions in silicon. *Solid State Electronics*, 2:35–67, 1961.

[47] P.A. Wolff. Theory and multiplication in silicon and germanium. *Physical Review*, 95:1415–1420, 1954.

[48] G.A. Baraff. Distribution functions and ionization rates for hot electrons in semiconductors. *Physical Review*, 128:2507–2517, 1962.

[49] C.R. Crowell and S.M. Sze. Temperature dependence of avalanche multiplication in semiconductors. *Applied Physics Letters*, 9:242–244, 1966.

[50] A.D. Sutherland. An improved empirical fit to baraff's universal curves for the ionization coefficients of electron and hole multiplication in semiconductors. *IEEE Transactions on Electron Devices*, ED-27(7):1299–1300, 1980.

[51] R. Van Overstraeten and H. De Man. Measurement of the ionization rates in diffused silicon p-n junctions. *Solid State Electronics*, 13:583–608, 1970.

[52] R.F. Pierret. *Semiconductor Fundamentals*, volume 1 of *Modular Series on Solid State Devices*. Addison-Wesley, Reading, MA, second edition, 1988.

[53] D.J. Roulson, N.D. Arora, and S.G. Chamberlain. Modeling and measurement of minority-carrier lifetime versus doping in diffused layers of n+-p silicon diodes. *IEEE Transactions on Electron Devices*, ED-29:284–291, February 1982.

[54] J.G. Fossum. Computer-aided numerical analysis of silicon solar cells. *Solid State Electronics*, 19:269–277, 1976.

[55] H.T. Weaver and R.D. Nasby. Analysis of high efficiency solar cells. *IEEE Transactions on Electron Devices*, ED-28(5):465–472, 1981.

[56] M.S. Adler. Accurate calculations of the forward drop and power dissipation in thyristors. *IEEE Transactions on Electron Devices*, ED-25(1):16–22, 1978.

[57] W.L. Engl and H. Dirks. Models of physical parameters. In *Introduction to the Numerical Analysis of Semiconductor Devices and Integrated Circuits*, pages 42–46. Boole Press, Dublin, 1981.

[58] J.W. Slotboom. The pn-product in silicon. *Solid State Electronics*, 20:279–283, 1977.

[59] C.J. Glassenbrenner and G.A. Slack. Thermal conductivity of silicon and germanium from 3k to the melting point. *Physical Review*, 134:A1058–A1069, 1964.

[60] R.K. Cook. Numerical simulation of hot-carrier transport in silicon bipolar transistors. *IEEE Transactions on Electron Devices*, ED-30(9):1103–1110, September 1983.

[61] J. Lindhard, V. Nielsen, M. Scharff, and K. Dan Vidensk. *Selsk. Mat. Fys. Medd.*, 36(10), 1968.

[62] O. B. Firsov. *Zh. Eksp. Teor. Fiz.*, 32:1464, 1957.

[63] O.B. Firsov. *Zh. Eksp. Teor. Fiz.*, 34:447, 1958.

[64] O.B. Firsov. *Zh. Eksp. Teor. Fiz.*, 36:1517, 1959.

[65] L.D. Kovach. *Boundary-value Problems.* Addison-Wesley, Reading, MA, 1984.

[66] M.N.O. Sadiku. *Numerical Techniques in Electromagnetics.* CRC Press, Boca Raton, 1992.

[67] A. Thom and C.J. Apelt. *Field Computations in Engineering and Physics.* Van Nostrand, London, 1961.

[68] J.C. Strikwerda. *Finite Difference Schemes and Partial Differential Equations.* Wadsworth and Brooks/Cole, Pacific Grove, CA, 1989.

[69] R.D. Richtmyer and K.W. Morton. *Difference Methods for Initial-value Problems.* Interscience Publishers, New York, 1976.

[70] G.D. Smith. *Numerical Solution of Partial Differential Equations: Finite Difference Methods.* Clarendon Press, Oxford, 1978.

[71] J.D. Logan. *Applied Mathematics: A Contemporary Approach.* Wiley, New York, 1987.

[72] G. Strang and G.J. Fix. *An Analysis of the Finite Element Method.* Prentice Hall, Upper Saddle River, NJ, 1973.

[73] C.S. Desai and J.F. Abel. *Introduction to the Finite Element Method: A Numerical Approach for Engineering Analysis.* Van Nostrand Reinhold, New York, 1972.

[74] A.F. Franz, G.A. Franz, S. Selberherr, C. Ringhofer, and P. Markowich. Finite boxes—a generalization of the finite-difference method suitable for semiconductor device simulation. *IEEE Transactions on Electron Devices*, ED-30(9):1070–1082, September 1983.

[75] M.S. Adler. A method for achieving and choosing variable density grids in finite difference formulations and the importance of degeneracy and band gap narrowing in device modeling. In *Proceedings of the Nasecode 1 Conference*, pages 3–30, 1979.

[76] V. Axelrad. Fourier approach to semiconductor device modelling. *International Journal of Numerical Modelling: Electronic Networks, Devices and Fields*, 2:31–52, 1989.

[77] V. Axelrad. Fourier method modeling of semiconductor devices. *IEEE Transactions on Computer-Aided Design*, 9(11):1225–1237, November 1990.

[78] A. DeMari. An accurate numerical steady-state one-dimensional solution of the p-n junction. *Solid State Electronics*, 11:33–58, 1968.

[79] A. DeMari. An accurate numerical one-dimensional solution of the p-n junction under arbitrary transient conditions. *Solid State Electronics*, 11:1021–2053, 1968.

[80] A.B. Vasil'eva and V.G. Stel'makh. Singularly disturbed systems of the theory of semiconductor devices. *Math. Fiz.*, 17(2):339–348, 1977.

[81] A.B. Vasil'eva and V.F. Butuzov. Singularly perturbed equations in the critical case. Translated Report MRC 2039, University of Wisconsin, 1978.

[82] P.A. Markowich, C.A. Ringhofer, E. Langer, and S. Selberherr. A singularly perturbed boundary value problem modelling a semiconductor device. Report MRC 2388, University of Wisconsin, 1982.

[83] P.A. Markowich, C.A. Ringhofer, and S. Selberherr. An asymptotic analysis of single-junction semiconductor devices. Report MRC 2527, University of Wisconsin, 1983.

[84] S.J. Polak, C. Den Heijer, and W.H.A. Schilders. Semiconductor modelling from the numerical point of view. *International Journal for Numerical Methods in Engineering*, 24:763–838, 1987.

[85] P.A. Markowich. *The Stationary Semiconductor Equations*. Computational Microelectronics. Springer Verlag, Wien, 1986.

[86] E.P. Doolan, J.J.H Miller, and W.H.A. Schilder. *Uniform Numerical Methods for Problems with Initial and Boundary Layers*. Boole Press, Dublin, 1980.

[87] S.D. Conte and C. de Boor. *Elementary Numerical Analysis: An Algorithmic Approach*. McGraw-Hill Book Company, New York, 3d edition, 1980.

[88] J.M. Ortega and W.C. Rheinboldt. *Iterative Solution of Nonlinear Equations in Several Variables*. Academic Press, New York, 1970.

[89] S.P. Edwards and K. DeMeyer. A charge damping algorithm applied to a newton solver for solving soi devices and other ill-conditioned problems. In *Proceedings of the Nasecode 3 Conference*, pages 174–186, 1981.

[90] I.S. Duff, A.M. Erisman, and J.K. Reid. *Direct Method for Sparse Matrices*. Monographs on Numerical Analysis. Clarendon Press, Oxford, 1986.

[91] D.J. Hartfiel and A.M. Hobbs. *Elementary Linear Algebra*. PWS Publishing Company, Boston, 1987.

[92] W.F. Tinney and J.W. Walker. Direct solutions of sparse network equations by optimally ordered triangular factorization. *Proceedings of the IEEE*, 55(11):1801–1809, November 1967.

[93] J.W.H. Liu. Modification of the minimum degree algorithm by multiple elimination. *ACM Transactions on Mathematical Software*, 11:141–153, 1985.

[94] P. Amestoy, T.A. Davis, and I.S. Duff. An approximate minimum degree ordering algorithm. Technical Report TR-94-039, Computer and Information Sciences Dept., University of Florida, December 1994.

[95] W. Cheney and D. Kincaid. *Numerical Mathematics and Computing*. Brooks/Cole Publishing Company, Pacific Grove, CA, 2d edition, 1985.

[96] Y. Saad. The lanczos biorthogonalization algorithm and other oblique projection methods for solving large unsymmetric systems. *SIAM Journal of Numerical Analysis*, 19(3):470–484, 1982.

[97] Y. Saad and M.H. Schultz. Gmres: A generalized minimal residual algorithm for solving nonsymmetric linear systems. *SIAM J. Sci. Stat. Comp.*, 7:856–869, 1986.

[98] P. Sonneveld. Cgs, a fast lanczos-type solver for nonsymmetric linear systems. *SIAM J. Sci. Stat. Comp.*, 10(1):36–52, 1989.

[99] Z.Y. Zhao, Q.M. Zhang, G.L. Tan, and J.M Xu. A new preconditioner for cgs iteration in solving large sparse nonsymmetric linear equations in semiconductor device simulation. *IEEE Transactions on Computer-Aided Design*, 10(11), November 1991.

[100] Silvaco international web page. http://www.silvaco.com/atlas/atlas.html.

[101] Ansys web page. http://www.ansys.com:80/htdocs/profea2.html.

[102] B.L. Gates. *GENTRAN User's Manual REDUCE Version*. Information Sciences Department — The Rand Corporation, 1987.

[103] P.K. Moore, C. Ozturan, and J.E. Flaherty. Towards the automatic numerical solution of partial differential equations. *Mathematics and Computers in Simulation*, 31:325–332, 1989.

[104] Pdease web page. http://www.macsyma.com/pdease.html.

[105] J.B. Adams and W.N.G. Hitchon. *J. Comput. Phys.*, 76:159, 1988.

[106] K.M. Kramer. PhD thesis, University of Wisconsin at Madison, 1996.

[107] R. Sedgewick. *Algorithms in C++*. Addison Wesley, Reading, MA, 1992.

[108] SILVACO International, 4701 Patrick Henry Drive, Bldg. 1, Santa Clara, CA 94054. *ATLAS User's Manual: Device Simulation Software*, 4.0 edition, June 1995.

[109] F.L. Alvarado. Manipulation and visualization of sparse matrices. *ORSA Journal on Computing*, 2(2):186–207, 1990.

[110] F.L. Alvarado. *The Sparse Matrix Manipulation System User and Reference Manual*. The University of Wisconsin, Madison, WI 53706, March 1992.

[111] W.M. Coughran, M.R. Pinto, and R.K. Smith. Adaptive grid generation for vlsi simulation. *IEEE Transactions on Computer-Aided Design*, 10(10):1259–1275, October 1991.

[112] R.E. Bank. *PLTMG, A Software Package for Solving Elliptic Partial Differential Equations*. Society for Industrial and Applied Mathematics, Philadelphia, 1990.

[113] G. Dahlquist and A. Bjorck. *Numerical Methods*. Prentice-Hall, Upper Saddle River, NJ, 1974.

[114] C.S. Rafferty, M.R. Pinto, and R.W. Dutton. Iterative methods in semiconductor device simulation. *IEEE Transactions on Computer-Aided Design*, CAD-4(4):462–471, October 1985.

[115] A. Bricker, M.J. Litzkow, et al. Condor. UNIX MAN Page.

[116] L. Solymar and D. Walsh. *Lectures on the Electrical Properties of Materials*. Oxford University Press, Oxford, 1988.

[117] R.F. Pierret. *Semiconductor Device Fundamentals*. Addison-Wesley, Reading,MA, 1996.

[118] A. Bar-Lev. *Semiconductors and Electronic Devices*. Prentice Hall, Hemel Hempstead, 1993.

[119] D.A. Hodges and H.G. Jackson. *Analysis and Design of Digital Integrated Circuits*. McGraw-Hill, New York, 1988.

[120] K. Lee, M. Shur, T.A. Fjeldly, and T. Ytterdal. *Semiconductor Device Modeling for VLSI*. Prentice Hall, Upper Saddle River, NJ, 1993.

[121] B.J. Baliga. *Power Semiconductor Devices*. PWS Publishing Company, Boston, 1996.

[122] C.B. Goud. *Breakdown Voltage Analysis of Planar Junctions with Field Plate*. PhD thesis, Indian Institute of Technology, April 1991.

[123] C.B. Goud and K.N. Bhat. On the choice of appropriate ionization coefficients for breakdown voltage calculations. *Solid State Electronics*, 30(8):787–792, 1987.

[124] V.A.K. Temple and M.S. Adler. Calculation of the diffusion curvature related avalanche breakdown in high-voltage planar p-n junctions. *IEEE Transactions on Electron Devices*, ED-22(10):910–916, October 1975.

[125] M.S. Adler, V.A.K. Temple, A.P. Ferro, and R.C. Rustay. Theory and breakdown voltage for planar devices with a single field limiting ring. *IEEE Transactions on Electron Devices*, ED-24(2):107–113, February 1977.

[126] M.S. Adler, V.A.K. Temple, and R.C. Rustay. Theoretical basis for field calculations on multi-dimensional reverse biased semiconductor devices. *Solid State Electronics*, 25(12):1179–1186, 1982.

[127] N. Mohan, T.M. Undeland, and W.P. Robbins. *Power Electronics: Converters, Applications and Design*. Wiley, New York, 1989.

[128] K.M. Kramer, W.N.G. Hitchon, and D.M. Divan. Conductivity modulation in p-i-n diodes simulated using a highly flexible approach. In *Conference Record of the 1993 IEEE IAS*, volume 28, pages 1196–1201. IEEE Industry Applications Society, 1993.

[129] K.P. Brieger, W. Gerlach, and J. Pelka. Blocking capability of planar devices with field limiting rings. *Solid State Electronics*, 26(8):739–745, 1983.

[130] K.R. Whight and D.J. Coe. Numerical analysis of multiple field limiting ring systems. *Solid State Electronics*, 27(11):1021–1027, 1984.

[131] F. Conti and M. Conti. Surface breakdown in silicon planar diodes equipped with field plate. *Solid State Electronics*, 15:93–105, 1972.

[132] K.P. Brieger, E. Falck, and W. Gerlach. The contour of an optimal field plate–an analytical approach. *IEEE Transactions on Electron Devices*, ED-35(5):684–688, May 1988.

[133] D. Jaume, G. Charitat, J.M. Reynes, and P. Rossel. High-voltage planar devices using field plate and semi-resistive layers. *IEEE Transactions on Electron Devices*, ED-38(7):1681–1684, 1991.

[134] G. Charitat, D. Jaume, A. Peyre-Lavigne, and P. Rossel. 2d modeling of high voltage bipolar planar transistors using sipos layer and field plate as junction termination extensions. In *Proc. Symp. MADEP'91*, pages 158–164, Florence, September 1991.

[135] X.B. Chen, B. Zhang, and Z.J. Li. Theory of optimum design of reverse-biased p-n junctions using resistive field plates and variation lateral doping. *Solid State Electronics*, 35(9):1365–1370, 1992.

[136] W. Feiler, E. Falck, and W. Gerlach. Multistep field plates for high-voltage planar p-n junctions. *IEEE Transactions on Electron Devices*, ED-39(6):1514–1520, June 1992.

[137] C.B. Goud and K.N. Bhat. Analysis and optimal design of semi-insulator passivated high-voltage field plate structures and comparison with dielectric passivated structures. *IEEE Transactions on Electron Devices*, ED-41(10):1856–1865, October 1994.

[138] H. Yilmaz. Optimization and surface charge sensitivity of high-voltage blocking structures with shallow junctions. *IEEE Transactions on Electron Devices*, ED-38(7):1666–1675, July 1991.

[139] J.A. Appels, M.G. Collet, P.A.H. Hart, H.M.J. Vaes, and J.F. Verhoeven. Thin layer high-voltage devices (resurf devices). *Philips J. Res.*, 35, 1980.

[140] Y. Koshino and Y. Baba. Obic measurement of high voltage resurf devices having sipos passivation. In *Proceedings of the Symposium of High Voltage and Smart Power Devices*, volume 89-15, pages 200–205. Electrochemical Society, 1989.

[141] V.A.K. Temple and W. Tantraporn. Junction termination extension for near-ideal breakdown voltage in p-n junctions. *IEEE Transactions on Electron Devices*, ED-33(10):1601–1608, October 1986.

[142] V. Boisson, M.L. Helley, and J.P. Chante. Computer study of high voltage p-π-n-n$^+$ diode and comparison with a field limiting ring structure. *IEEE Transactions on Electron Devices*, ED-33(1):80–84, 1986.

[143] T. Stockmeier and P. Roggwiller. Novel planar junction termination technique for high voltage power devices. In *Proceedings of the 2nd International Symposium of Power Semiconductor Devices and Integrated Circuits*, pages 236–239, Tokyo, April 1990.

[144] K.P. Brieger, J. Korec, and B. Thomas. Comparison of different planar passivation techniques for semiconductor power devices. *Archiv. für Elektrotechnik*, 72:89, 1989. In German.

[145] C.B. Goud. Two-dimensional analysis and design considerations of high-voltage planar junctions equipped with field plate and guard ring. *IEEE Transactions on Electron Devices*, ED-38(6):1497–1504, June 1991.

[146] C.B. Goud and K.N. Bhat. Breakdown voltage and field plate and field-limiting ring techniques: Numerical comparison. *IEEE Transactions on Electron Devices*, ED-39(7):1768–1770, July 1990.

[147] J. Korec and R. Held. Comparison of dmos/igbt-compatible high voltage termination structures and passivation techniques. *IEEE Transactions on Electron Devices*, ED-40(10):1845–1854, October 1993.

[148] R.S. Ramshaw. *Power Electronics Semiconductor Switches*. Chapman and Hall, London, 1993.

[149] J. Albers. *IEEE Transactions on Electron Devices*, 32:1930, 1985.

[150] A.D. Boardman, W. Fawcett, and S. Swain. *J. Phys. Chem.*, 31:1963, 1970.

[151] W. Feller. *An Introduction to Probability Theory and its Applications*. Wiley, New York, 1968.

[152] D.K. Ferry. *Semiconductors*. Macmillan Publishing Company, New York, 1991.

[153] G.J. Parker, W.N.G. Hitchon, and E.R. Keiter. Transport of ions during ion implantation. *Physical Review E*, 54:938–945, 1996.

[154] H.D. Rees. *Solid State Commun.*, A26:416, 1968.

[155] H.D. Rees. *J. Phys*, C5:64, 1972.

[156] R.E.P. Harvey, W.N.G. Hitchon, and G.J. Parker. Plasma chemistry at long mean free path. *Journal of Applied Physics*, 75:1940, 1994.

[157] R.E.P. Harvey, W.N.G. Hitchon, and G.J. Parker. The role of the plasma in the chemistry of low pressure plasma etchers. *IEEE Transactions on Plasma Science*, 23:436, 1995.

[158] G.J. Parker, W.N.G. Hitchon, and D.J. Koch. Transport of sputtered neutral particles. *Physical Review E*, 51:3694, 1995.

[159] J.P. Biersack and L.G. Haggmark. *Nuclear Instruments and Methods*, 174:257, 1980.

[160] M.D. Giles and J.F. Gibbons. *IEEE Transactions on Electron Devices*, 32:1918, 1985.

[161] B.F. Beale. *Non-Local Electron Kinetics in an Inductively Coupled Plasma: 2D Model and Experimental Validation*. PhD thesis, University of Wisconsin at Madison, 1995.

[162] S.A. Freeman, J.H. Booske, and R.F. Cooper. Point defect interactions with microwave fields. Poster, July 1995.

[163] S.A. Freeman. *Investigations of Non-Thermal Interactions between Microwave Fields and Ionic Ceramic Materials*. PhD thesis, University of Wisconsin at Madison, 1996.

Index